Exploring Evolution

Exploring Evolution

Michael Alan Park

VIVAYS PUBLISHING

Published by Vivays Publishing Ltd
www.vivays-publishing.com

A catalogue record for this book is available from the British Library

ISBN 978-1-908126-25-2

Publishing Director: Lee Ripley
Designer: Nick Newton

Cover: Martin Bond/Science Photo Library
Frontispiece: Natural History Museum, London/Science Photo Library

Printed in China

Contents

Introduction   6

Chapter 1   What is the scientific method?   16

Chapter 2   How old is the earth and how has it changed?   28

Chapter 3   What do we need to know about genetics?   40

Chapter 4   How do species change through time?   52

Chapter 5   How do new species evolve?   72

Chapter 6   A look at the pageant of life   90

Chapter 7   What do we know about human evolution?   104

Chapter 8   What about the challenges to evolutionary theory?   124

Chapter 9   How has the theory of evolution influenced modern society?   140

Suggested readings   155
Picture Credits   157
Index   158

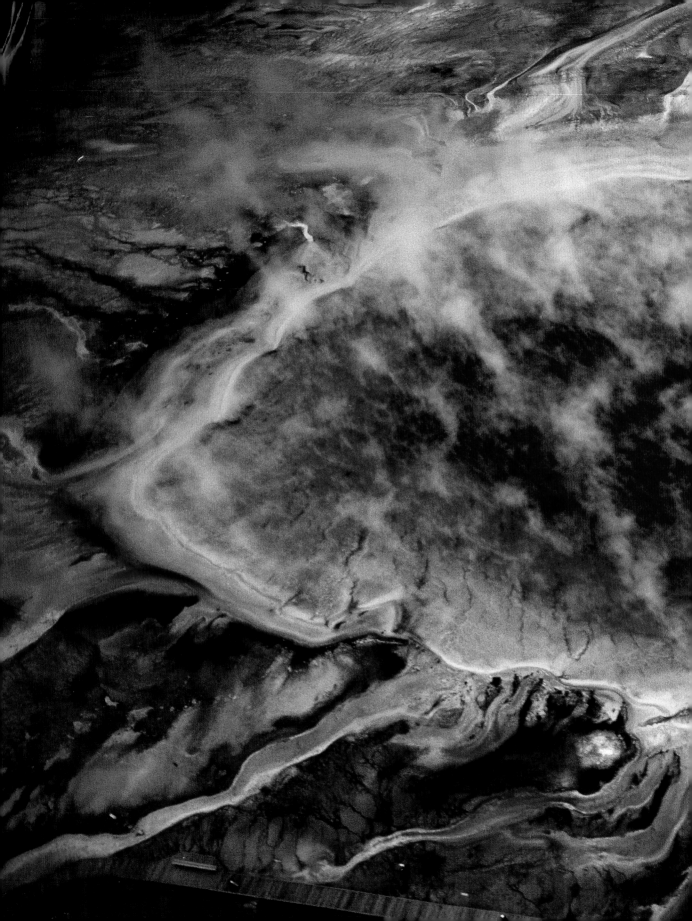

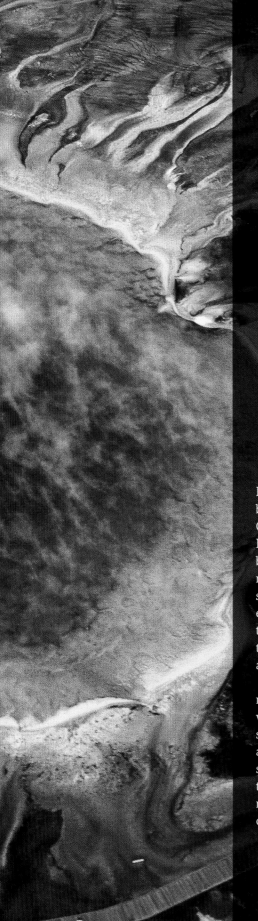

Introduction

It was the dinosaurs. It's always the dinosaurs.

Many fascinated children have blossomed into adult scientists because of them. For famed evolutionary biologist Stephen Jay Gould, it was the *Tyrannosaurus rex* skeleton in the American Museum of Natural History in New York. For me, it was a smaller but still impressive set of dinosaur bones in a museum in Cincinnati, Ohio. I had no idea—none of us kids did—where the dinosaurs came from, or how some got so big, or how they met their end. We simply accepted, with our youthful sense of wonder, that they existed and that something had killed them off. All very mysterious and romantic. Then and there, I was determined to become a paleontologist and study dinosaurs.

And yet, my early years of formal schooling taught me almost nothing about dinosaurs. Odd, since you'd think that schools would capitalize on teaching science using a subject that holds such fascination for so many youngsters. But there was a generally held and somewhat limited view of the nature of science. We studied chemistry, biology, and physics—fields that study things that can be seen in the present and are subject to direct experimentation. Dinosaurs, however, didn't fall into that category. In order to understand dinosaurs, we need to understand evolution,

PREVIOUS PAGE The Grand Prismatic Spring, Yellowstone National Park, is the largest hot spring in the United States; the bright colors result from the bacteria at the edges of the mineral-rich water. Life may have begun in just such an environment.

A computer image of a skeleton of *Tyrannosaurus rex*, the archetypal dinosaur. At 5 meters (16.5 feet) tall and weighing 7 tonnes, even the petrified bones are impressive.

a subject that deals with the past and thus with events that can't be manipulated in a science lab. Evolution was seldom taught because it wasn't a traditional laboratory science.

There may also have been another, more subtle, reason why evolution was not covered in my public school education. Evolutionary theory carries with it some baggage of misunderstandings and preconceptions. The word evolution itself often conjures up thoughts of *human* evolution, and it is the notion that *we humans* evolved that has stirred controversy. The processes of evolution apply equally to humans, jellyfish, and geraniums—as well as dinosaurs—but only when they are discussed in relation to humans does it raise visceral or philosophical concerns. Whether it was a conscious decision to avoid controversy or not, public schools usually gave short shrift to evolution.

Thus, I found myself in college, still with no better idea about the dinosaurs, when I stumbled into a biological anthropology course. Biological anthropology is a subfield of anthropology that focuses on humans as a biological species and studies such topics as human genetics, human evolution, the fossil record, and the biology of living human populations. There, for the first time, I was introduced to the details of evolutionary theory. In that course the theory was, naturally, applied to our species but I quickly saw that if the principles of evolutionary theory could explain the evolution of a species as complex as ours, it could explain the history of life in general, including my beloved dinosaurs. Indeed, nothing in biology makes sense *without* evolutionary theory. I had found a calling beyond paleontology, and I never gave a second thought to another career.

This brief book explains the essential principles of evolutionary theory and attempts to convey the beauty of life's evolutionary story. Evolutionary theory is not all that complex, relatively speaking. These pages lay out the basics—and with a welcome minimum of mathematics.

But before we get started, we need to address the controversy that seems to follow the topic of evolution. It is often perceived as a conflict between science and other forms of knowledge, with faith or religion being foremost among them. The genesis of the controversy is a simple lack of understanding about how science works, and, more

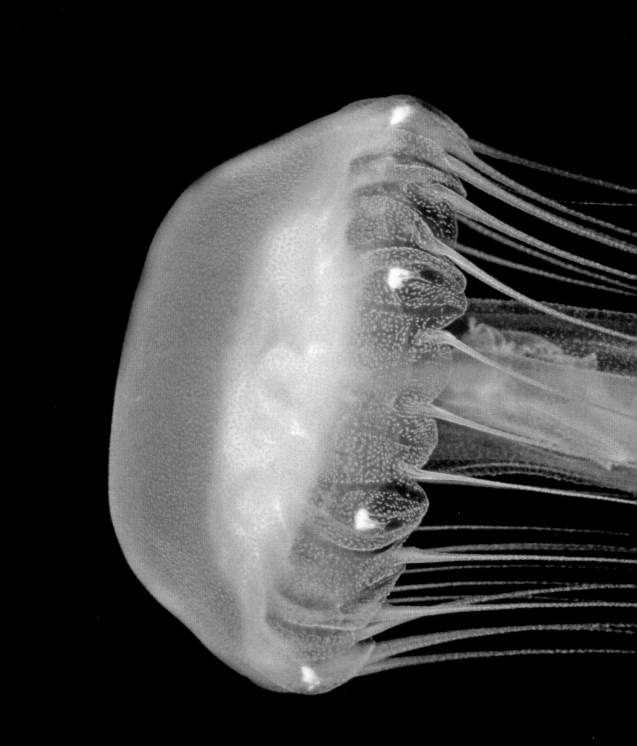

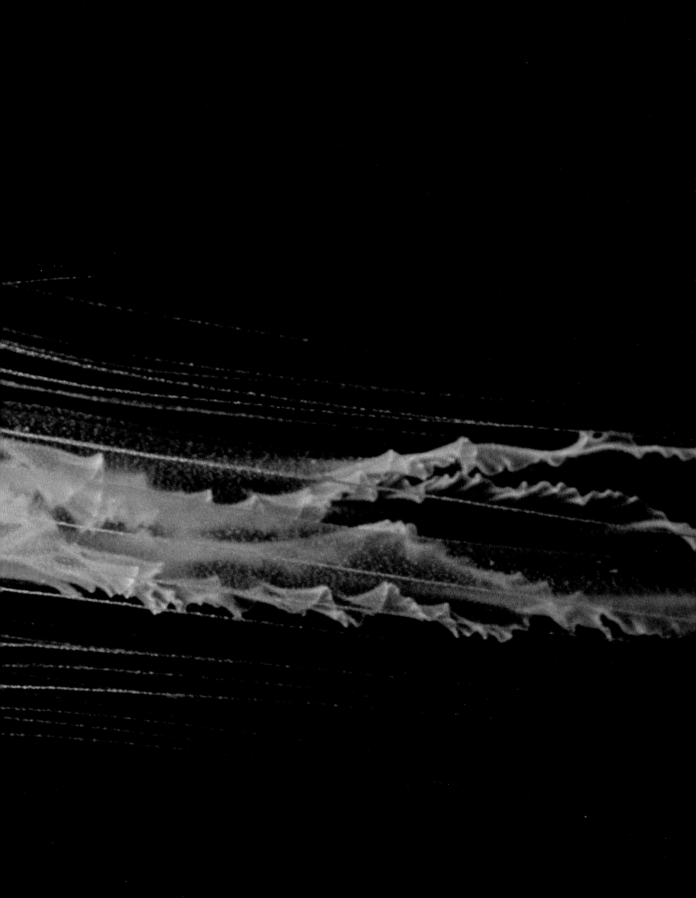

The Grand Gallery of Evolution at the *Muséum d'Histoire Naturelle* in Paris, where Lamarck, a predecessor of Darwin, taught (see Chapter 4). Although Lamarck was incorrect in some aspects of his evolution model, this modern museum—which he would not remotely recognize—is a part of Lamarck's legacy to evolutionary biology.

specifically, the science of evolution—the idea that the processes I'll describe *do* apply to us as well as to jellyfish and geraniums and dinosaurs. Once the principles are understood, any perceived contradiction between the different forms of knowledge vanishes. Although we still investigate and debate the details, evolution is as factual as any of the scientific concepts, such as gravity, that we take for granted and that are thus manifestly uncontroversial. I'll return to this issue in detail in a later chapter.

So we'll begin with a look at science itself. I once told one of my relatives that ever since I was a kid I'd always wanted to be a scientist of some sort and that I was pleased I ended up as one. "Oh," one of them responded, "I thought you were an anthropologist." That limited impression of science just won't go away. But science is more complex and broader than the stereotype. It is, simply, a specific process of inquiry that tells us both *what* we know about the real world and *how* we know it. And the scientific method, accurately defined, works for studying the evolutionary past as well as it does for mixing chemicals, observing stars, and smashing atoms.

After describing science and the scientific method in Chapter 1, we'll look briefly in Chapter 2 at how we've historically interpreted the earth's past, because that set the stage for our current understanding of life's history. Chapters 3 through 6 delve into evolutionary theory. Chapter 3 talks about genetics; it is genetic change through time that is the basis for physical change. Chapter 4 goes back to Darwin's century to show how evolutionary theory developed; we then explore our current knowledge of what Darwin explained so well—how species change through time as a result of natural selection. Chapter 5 talks about the origin of species—how new species evolve from existing ones, and how life on earth is like one huge, complex family tree. Chapter 6 is a brief narrative of the entire history of life on this planet, including a description of how the continents have drifted and the subsequent effects on evolution. I conclude this overarching examination with a few philosophical considerations. An overview of evolution as it pertains to humans is the subject of Chapter 7. The assumption that evolutionary theory and other areas of knowledge are eternally at odds is the topic of Chapter 8, where we'll look at scientific creationism and intelligent design. Finally, the influence of

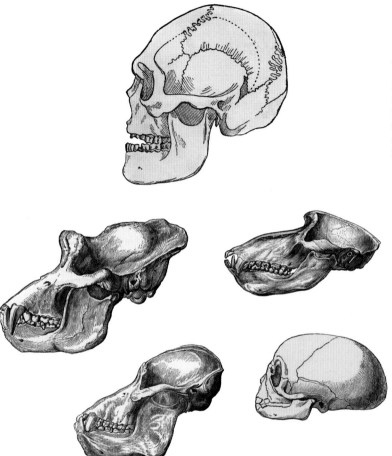

An image from *Primitive Man*, published in 1876 by the French scientist and writer Louis Figuier (1819–1894), showing a human skull (top) and those of some other primates. By this time, the fact that humans had evolved from a nonhuman primate was accepted among the scientific community and a formal study of our evolution had begun.

evolutionary theory on science itself and on other aspects of modern society and culture will be discussed in Chapter 9.

I have been involved now for more than 40 years in the field, doing research and introducing college students and the public to this fascinating area of science. There have been great advances in the tools we have available to study evolution and thus in our knowledge of the subject. This evolution of evolution itself has been one source of constant excitement. When I first taught genetics, for example, the idea that the human genome could be mapped through complex biochemical techniques was almost science fiction. It is now commonplace. Similarly, each year, it seems, a new fossil—of a dinosaur, a human ancestor, or some other creature—comes to light that makes

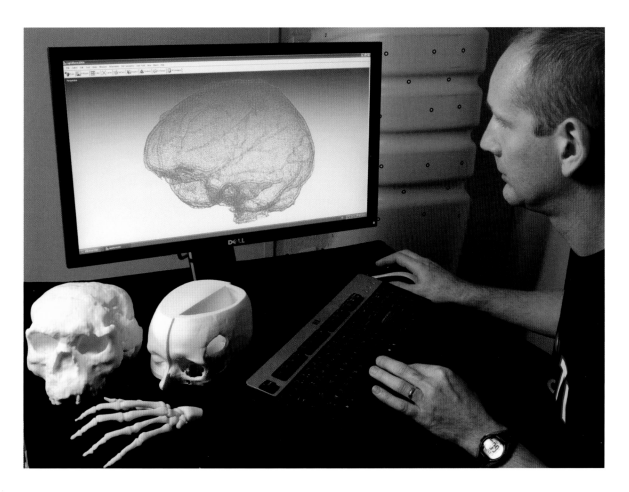

Researcher working with a 3D scan of the inside of a Cro-Magnon fossil skull. These data will be used to study the structure of its brain. Cro-Magnons were an early European culture of modern humans. On the left is a skull of *Homo erectus*, an earlier human form.

us question our existing models of life's history. And we date fossils with increasingly accurate technologies from physics. Just keeping up with all this is an enjoyable challenge. Most profoundly, however, are the constant reminders that follow from an understanding of the processes of evolution, of the unity of life on earth, and of the fragility and uniqueness of that life.

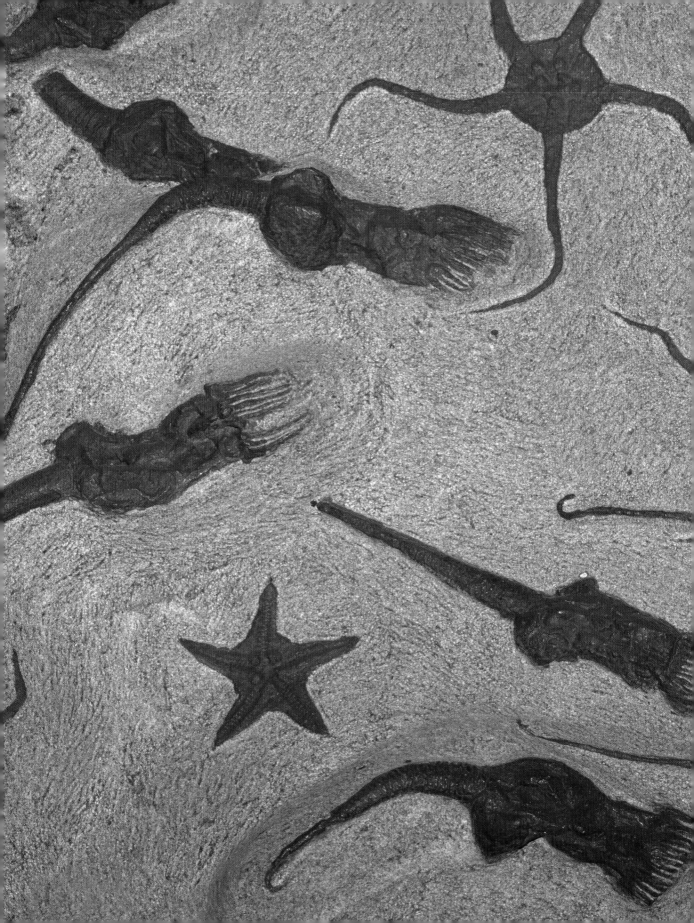

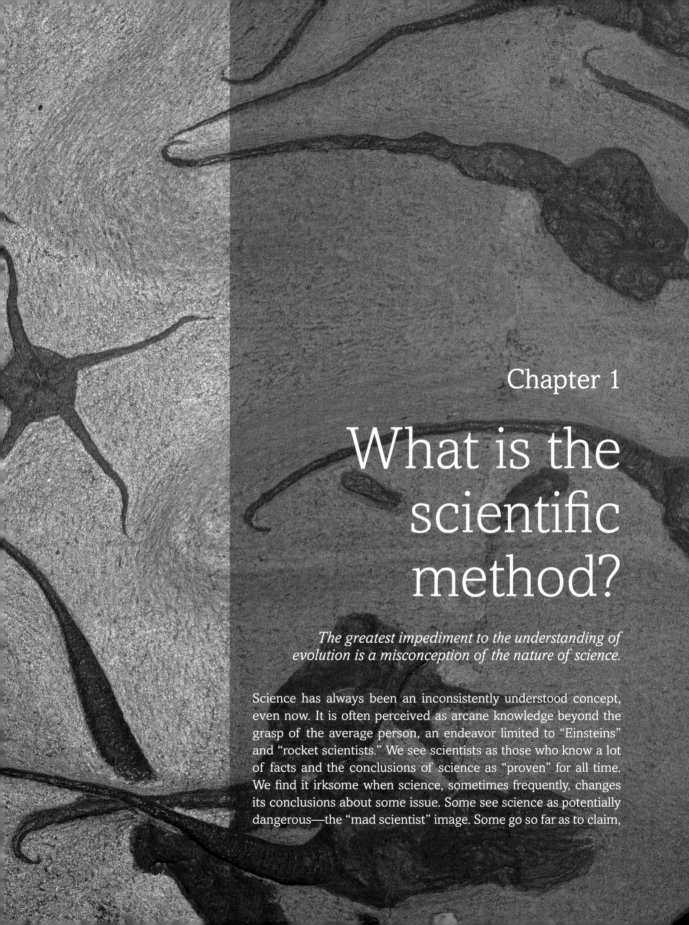

Chapter 1

What is the scientific method?

The greatest impediment to the understanding of evolution is a misconception of the nature of science.

Science has always been an inconsistently understood concept, even now. It is often perceived as arcane knowledge beyond the grasp of the average person, an endeavor limited to "Einsteins" and "rocket scientists." We see scientists as those who know a lot of facts and the conclusions of science as "proven" for all time. We find it irksome when science, sometimes frequently, changes its conclusions about some issue. Some see science as potentially dangerous—the "mad scientist" image. Some go so far as to claim, as a recent online article did, that all science and scientists are culpable because scientists invented nuclear weapons. Science is ignored in the name of political or religious ideology. Little wonder we are often confused.

I recently asked a small sample of people—as varied as I could find—to quickly name a real scientist, alive or dead, and then a fictional scientist. Almost 38% named Albert Einstein as the real scientist, and 65% came up with either Doctor Jekyll or Doctor Frankenstein as the fictional scientist. What sticks in our minds, it seems, is that scientists are quirky men who collect facts and turn them into arcane laws of nature that have little bearing on our day-to-day lives, or are demented men who turn themselves, or assorted body parts, into monsters.

Even when someone has a deeper understanding of science it's often based on the template of high school lab reports: Hypothesis, Materials, Methods, Data, Conclusions.

But science is not always quite that neat and clearcut. Nor does one have to be an "Einstein" to be a scientist.

What, then, *is* science and this mysterious scientific method?

Caricature of Albert Einstein, many people's archetype of the real-life scientist.

How science really works

Simply put, science is a method of inquiry, a way of asking and answering questions about our universe, whose goal is the most accurate picture of that universe possible at the moment. Science is conducted according to a set of rules, the most important of which is that a scientific idea must be empirically *testable* and potentially *refutable*. We must be able to evaluate concrete, material evidence to support, or deny, a scientific conjecture. Here, in general, is how that works.

We often begin with a set of observations we wish to explain and we generate a general explanation for those specific observations. This is known as induction, and the result of inductive reasoning

The largest air-insulated Van de Graaff generator in the world, in the Museum of Science in Boston. Constructed in 1931 it was used in early atom-smashing and high energy X-ray experiments, a classic example of the scientific method in action.

is a tentative general explanation, a hypothesis. Not all hypotheses, however, are scientific. If I claim, for example, that apples fall from trees toward the earth because there are spirits who dwell in the center of the earth that magically draw objects down toward them, that *is* a hypothesis, but it can't fulfill the defining criterion of science: It can't

be tested empirically. There's no way I can scientifically support that idea. There's also no sure way you could deny that my hypothesis is correct. Even if you counter that there's some natural force at work, I can still insist that those spirits created and are guiding that natural force. Do those spirits exist? That's not even an *answerable* question.

Suppose, however, you've proposed that a natural force accounts for the fall of apples. That can be materially tested and so it fulfills the definition of science. Note that I didn't say your idea need be "proved" to be scientific. So far all we're concerned about is the testability and the potential for the idea's empirical refutability. The question of the existence of a natural force is, at this point, simply an answerable question.

We test a scientific hypothesis using deduction—in a way, the opposite of induction. It goes from the general (your hypothesis) back to the specific. We ask: *If* your hypothesis is correct, *then* what else would we expect to find? If your hypothesis is incorrect, what should we find? Then, we check out these predictions in whatever way makes sense for the topic at hand. Clearly your natural force—obviously, gravity—passes every test of repetition, universality, and prediction we can apply to it.

Isaac Newton's rooms at Trinity College, Cambridge, with an apple tree planted later. The tree that reportedly gave rise to Newton's thoughts about gravity is in the front yard of his family home in Woolsthorpe-by-Colsterworth, Lincolnshire (where, home because Cambridge closed during a plague outbreak, he invented the calculus).

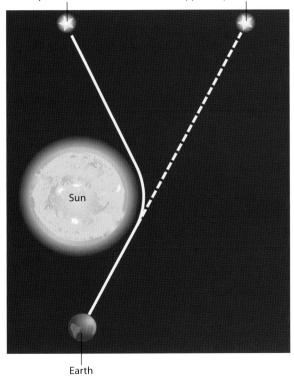

Actual position of star   **Apparent position of star**

Sun

Earth

A test for gravity: Being able to see a star during a solar eclipse in 1919 that was physically behind the sun verified Einstein's 1905 prediction that gravity generated by a large enough object deflects even light and thus bends its path. (The effect is greatly exaggerated here and the scale obviously condensed.)

There are no hard and fast rules for scientific testing. The specific tests vary with the particular topic. But the tests must be objective (that is, involving empirical data), and others must get the same results, under the same conditions, for the tests to be valid.

Notice that I haven't yet used the term *theory*. The term has a vernacular meaning as a synonym for hypothesis or guess: We all say things such as "that's just a theory" or "I have a theory about that." But the term *theory* has a formal meaning within science. A theory is a well-supported idea, but it is more than a hypothesis. A theory is a well-supported general idea that explains a large set of factual patterns.

Thus, in science, theory is a powerful term, and there are relatively few ideas that really deserve the title. The Theory of Gravity is one because it explains everything from the fall of objects on the earth to the motion of the planets around the sun. Einstein's Theory of Relativity (both the special and general versions) describes the geometry of four-dimensional space-time and how it affects and is affected by the unity of mass and energy (his famous formula $E=mc^2$). The Theory of Plate Tectonics explains the motion of continents through time and, thus, much of geological, environmental, and biological history. And the Theory of Evolution explains, well, all of biology.

That's basically what science *is*, but we can enhance our understanding by seeing what science is *not*.

Two misconceptions about science

Science proves ideas for all time. Some scientific theories are so well supported we may consider them to be facts. The earth really revolves around the sun and not the other way around. Gravity is a real force. The continents are really moving. Evolution is really happening.

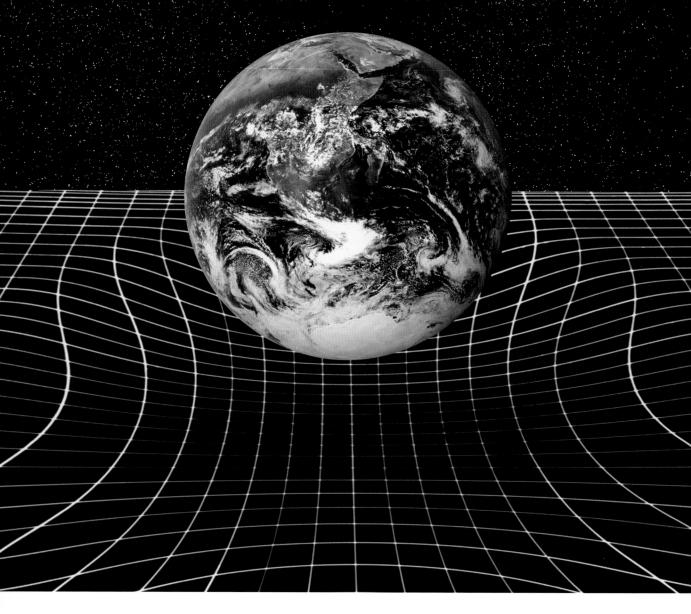

No theory, however, is ever complete. We know a lot about gravity, but debates continue about the origin of that force and its relationship to the other primary forces that drive the universe.

Evolution is a fact, but we still investigate—as you will see—how it works in general and with respect to particular organisms, as well as its overall history and pattern in the pageant of life. A theory is not the end product. There are always new data, new interpretations and new technologies. And this means that science can be, and often is, wrong about things in hindsight.

What science seeks are statements and explanations about the world that are as accurate as possible at the time. We always

An artist's depiction of the earth's gravity well, a two-dimensional analogy to the warping of three-dimensional space, Einstein's explanation for gravity. As much as we know about gravity, there are other models—and complications to this one—still being investigated.

understand and accept a degree of uncertainty and we evaluate ideas on the relative probability of their being accurate.

And this is one of the strengths of science: It is self-correcting. We don't hold tenaciously onto ideas just because we like them. Scientists, in fact, embrace change—even if it means giving up ideas we personally hold dear—if that change can give us a more accurate picture of the world we live in.

Scientists always deal in objective truth. Yes, of course, we try to be objective. But science is an endeavor performed within cultural and social contexts, and this places limits and constraints on our objectivity.

First, there is the matter of technology. The real nature of the solar system was out of reach until the telescope was invented. Continental drift, for which there was ample physical evidence, could not be supported until new technologies showed us what the crust of the earth really looked like. New techniques have recently shown that the human genome (the total genetic code for our species) is far more complex than we thought just a few decades ago, even though our previous view was generally considered accurate in its time.

Depiction by Galileo of his demonstration of the telescope to the Doge of Venice in 1609. The appeal was that Venetian merchants could see trade ships carrying goods coming from a long way off and manipulate prices. Often, practical matters give rise to inventions that are later used, as Galileo did the telescope, for scientific research.

Second, we can be constrained by existing preconceptions. Prior to Darwin, there was clear evidence for biological evolution but no mechanism had been proposed that reasonably explained how it worked. Moreover, the pervasive influence of a Judeo-Christian world view on our knowledge of the universe, the earth, and its life, meant that, for many, the very idea that life evolved was untenable. Most biology was conducted without this concept at its core. With Darwin's model of natural selection (which we will discuss in Chapter 4), there was a proposed mechanism which enabled scientists and the public to see the evidence in a new light, and it all made perfect sense.

Finally, cultural trends can have an influence. The Salem witch trials that took place in Massachusetts in 1692 were explained once as the result of the young women accusers having eaten products from wheat tainted with ergot, a fungus that contains a chemical akin to the hallucinogen LSD. In other words, the girls were on "acid trips." It doesn't take much imagination to guess when this idea was popular, even though it makes no sense, assuming as it does that only those handful of girls ate the tainted products.

In this scene from the Hartford (Connecticut) Stage production of Arthur Miller's play *The Crucible*, some of the young girls involved, played by (l to r) Erin Fitzpatrick, Lilli Jacobs, Eileen Conneelly, and Rachel Mewbron, are ostensibly overcome by the presence of evil in the courtroom.

Science is a human endeavor, and thus open to all the mistakes, and even abuses, of any area of human activity. Its uncertainty, its openness to correction, and its constant change in search of accuracy would seem, perhaps, to be weaknesses. Those features, however, are the very strengths of science. It is still the best method we have for understanding the real world.

But we have one more question to ask, the one most relevant to our overall topic.

How do we apply science to past events and complex phenomena?

Studying gravity is not like experimenting with chemicals, blasting atoms with fast moving subatomic particles, or using new high-tech methods to analyze genes. Gravity can't be seen or felt in the way that we can both see and feel the water that causes erosion. We study gravity through its results. We see its effect and the fact that it does the same thing under the same conditions over and over. For example, if we throw an object into the air it will always fall back down again – unless we happen to be in a space ship (which the theory of gravity also explains). Thus, our theory of a force called gravity is an induction that coordinates and explains a huge set of observable facts.

It is similar, although far more complex, with evolution. Most evolution took place in the past so we cannot go back and observe it directly. We cannot experiment on it by altering conditions to see what *would* have happened under different circumstances. Nor can we begin to replicate or even imagine the complex specific interactions of biology, climate, and geology that have affected the evolution of even a single species over perhaps millions of years.

So how did Darwin generate his model?

I'll let the late biologist Stephen Jay Gould have the last word on this. Darwin used "thousands of well-attested facts drawn from every subdiscipline of the biological sciences" to acknowledge that "only one conclusion about the causes and changes of life can possibly coordinate all these … various items under a common explanation." That explanation is natural selection. As that induced hypothesis is tested and modified in various ways which we'll talk about as

we proceed, we will see clearly that it is well-supported and that nothing refutes it. It is (Gould again) "granted the favor of probable truth." This is how we study evolution, by deciding what induced explanations account for what we see as the *results* of evolution—in the fossil record as

ABOVE The rainforests of South America, this one in Peru, are some of the most complex ecosystems on earth, with their countless species, each with innumerable and almost unimaginable adaptive relationships to climate, resources, other species, and, here, the vertical dimension of the layers of the rainforest canopy. The complexity is magnified when evolutionary time is factored in and, yet, such complex systems may still be studied scientifically.

well as in visible and genetic changes in living species. Studying the past and studying complex phenomena are no less scientific than, for example, understanding the effects of mixing two chemicals together.

With that background in science, and with that final point about the science of evolution, we can now plunge into our subject and see how it works in detail.

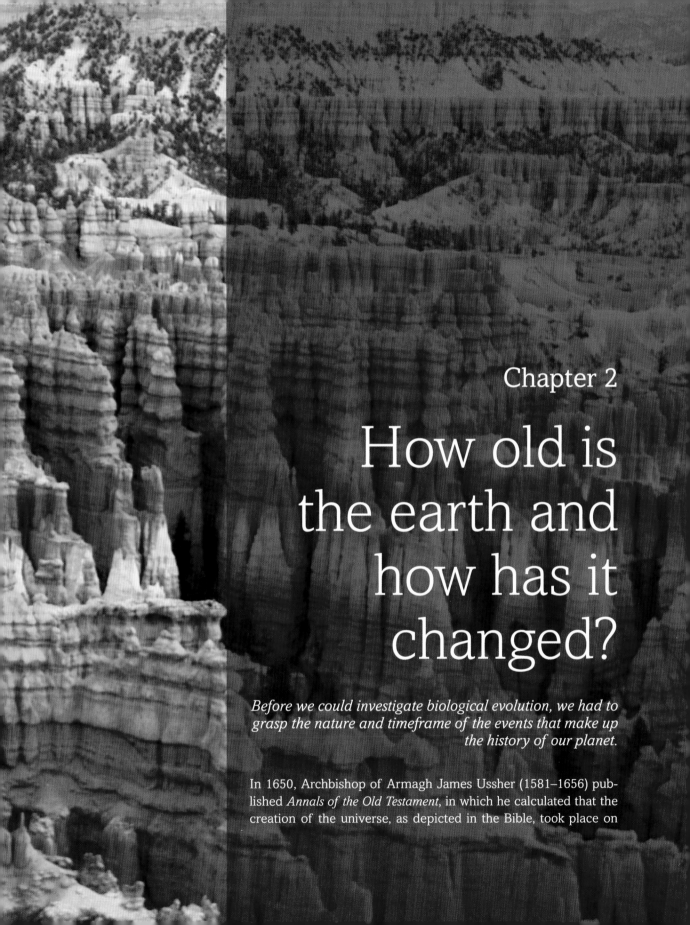

Chapter 2

How old is the earth and how has it changed?

Before we could investigate biological evolution, we had to grasp the nature and timeframe of the events that make up the history of our planet.

In 1650, Archbishop of Armagh James Ussher (1581–1656) published *Annals of the Old Testament*, in which he calculated that the creation of the universe, as depicted in the Bible, took place on

October 23 4004 B.C. He was off by 13 billion years or so, and we do have to wonder how on earth he came up with such a specific date.

We must remember as we smile at Ussher's conclusion, that he was a product of his time. He was influenced by existing assumptions that the earth was 6000 years old, based on earlier calculations (there's even a mention in Shakespeare's *As You Like It* of that age). Moreover, he wasn't just guessing or arguing toward a preconceived idea. He literally calculated the date—using information in the Bible and filling in gaps with historical records from other cultures. He chose the autumn because that's when the Jewish year begins and he derived the date using the old Julian (Roman) calendar.

The point is that while he used the Bible as his basis—what more authoritative source would he have chosen at that time and place?—he did perform an act of scholarship. He and his predecessors were simply without the observations, data, technology, and mindset that would allow them more accurate calculations. It was only in the 20th century, in fact, that an accurate age of the universe, the solar system, and the earth could be derived.

Archbishop Ussher's book dating the creation.

Evidence for an old and changeable earth

By the second half of the 18th century what we might consider "field science" was a common practice. Scientists or "natural philosophers," as they were called then, were explaining the natural world by *observing* the natural world, not just by thinking about it and using limited sources such as the Bible for reference. Two general observations started to shatter the idea of a young earth, unchanged since the biblical creation.

The first was the recognition of what the layers of rock and soil beneath the earth's surface actually meant. These layers are *strata* and their study *stratigraphy*. It became clear that the strata in any area were laid down sequentially, the oldest strata deeper and the younger ones closer to the surface. Furthermore, the strata were not all the same. The nature of the differing rocks in each layer indicated that there were very different conditions on the surface of the earth when each of those strata was laid down. Most startling of all, the sheer number of strata in some geographical areas provided more than a

OPPOSITE A rock in Sedona, Arizona, the strata accumulated over millions of years exposed by erosion.

hint that the earth must be far older than 6000 years. It soon became apparent that the earth was not young and had undergone major changes.

The second observation involved fossils found within the strata. Once explained as everything from deformed individuals to "tricks of nature," fossils came to be seen as the remains of creatures that once existed but existed no longer, at least not in that form. And if the strata were seen as the pages of the book of earth's geological history, the fossils had to be seen as the record of earth's biological history. Life, too, had changed and was very old.

The famous Burgess Shale fossil beds in British Columbia with well preserved marine fossils dated to approximately 530 million years ago. Because the fossil record is always incomplete (few organisms are actually preserved) the strata seem to show abrupt and sudden changes. This was used as evidence by the catastrophists.

Catastrophism offers an explanation

The stratigraphic and accompanying fossil records are imperfect and incomplete. Most creatures that have lived leave behind no trace. To

Localized catastrophes, such as volcanic eruptions, were not unknown to the ancient world or to modern Europeans. Mt Etna on Sicily, shown here, still erupts and Mt Vesuvius in Italy destroyed Pompeii in 79 A.D. Catastrophists imagined such cataclysms on a worldwide scale in the past.

be preserved as a fossil is a rare event, to find a fossil even rarer. There are, no doubt, whole groups of organisms about which we have literally no clue. Similarly, the record in the rock strata can't show every page of the story, every day-to-day event. Even some changes that took place over millennia are missing from our view. Thus, the strata seem to show abrupt alterations in types of rock and in the life forms therein, with few transitional fossil types or geological states to imply a gradual change.

A number of scientists in the 18th and early-19th centuries felt the obvious explanation was that, given the number of strata, the earth was ancient, but that it had undergone a series of world-wide cataclysms—floods, earthquakes, volcanoes, upheavals in the crust—that radically and quickly changed the surface of the earth and brought about mass extinctions of life forms. New life forms then appeared. This is the concept of *catastrophism* and perhaps its most famous proponent was the French naturalist George Cuvier (1769–1832).

Catastrophism was not, as it is often described, necessarily a religious idea. While Cuvier and others would have accepted some manner of intervention by a supreme being, their various models were naturalistic, logical, and literal descriptions of what the fossil and stratigraphic records appeared to show. In fact, Cuvier felt that new fossils in new strata represented species that had migrated into an area from elsewhere after a cataclysm occurred. The religious connection comes from the idea that the most recent cataclysm was Noah's Flood, then still taken as literally true. Thus, the Bible was thought to record only part of the earth's history, including the only catastrophe in human memory, but not nearly all of earth's history. And the catastrophists' acknowledgement of the sudden appearance of new organisms in new strata could be interpreted as supporting claims of divine creations.

A caricature of what we now know happened to the dinosaurs and many other forms of life. In Cuvier's time, however, the demise of the dinosaurs was explained in a biblical context—that they died during a great cataclysm, The Flood, and were not taken aboard the Ark.

Uniformitarianism

Catastrophism enjoyed a degree of popularity because, for many, it seemed to reconcile natural evidence with a biblical timeframe. But it did not hold up to close scrutiny. One of the first to take a different position on earth's history was another Frenchman, the Comte Georges-Louis Leclerc de Buffon (1707–1788).

Note that Buffon was born well before Cuvier. The history of scientific ideas is not always linear, not always a case of incomplete or incorrect ideas being older and then being corrected by subsequent

The Comte de Buffon by François-Hubert Drouais (1723–1775). His comment about uniform processes gave the name to Uniformitarianism and the concept that the earth was very old. He also founded the study of "biogeography," the relation of certain species to certain areas, and he espoused the idea that all human races had a common origin. At the same time, he felt species were immutable and so rejected the suggestion of a common ancestor for humans and apes.

NEXT PAGE The Mississippi Delta, formed over 7000 years by the slow, steady deposition of sediments at the mouth of the river, is a classic example of uniformitarianism.

ideas. Conflicting thoughts can overlap in time and this is a case in point.

Buffon concluded that although catastrophic events do occur, they are rare and localized and so "have no place in the ordinary course of nature." Rather, the earth's history is explained by "operations uniformly repeated, motions which succeed one another without interruption." Thus, uniformitarianism. The operations he refers to are normal, everyday processes such as wind and water erosion, and deposition of sediments in bodies of water; things we can all observe in the landscapes around us. Of course, such processes take place in small increments adding up to big overall changes, so Buffon was one of the first to propose an earth history longer than the commonly accepted 6000 years.

A Scotsman, James Hutton (1726–1797), elaborated. He saw the earth as a self-regulating system where, for example, new land was constantly formed under the sea from sediments produced by erosion, and this land would itself eventually provide new sources of soil. Changes were historically related, one leading to the next. And Hutton also knew that the earth was very old. Indeed, he thought the planet's history had "no vestige of a beginning—no prospect of an end."

The British surveyor and geologist William Smith (1769–1839) formalized the description of the evidence for uniformitarianism—the strata and their fossil contents. Smith, whose work earned him the nickname "Strata," documented patterns of strata and associated fossils across England, Wales, and Scotland, and produced the first geological map of any area of the world in 1815.

Another Scotsman, Charles Lyell (1797–1875), had an extreme version of uniformitarianism. He believed that the earth itself was uniform across vast spans of time. Changes occur, he said, and usually by small, slow, and steady processes, but they occur in great cycles, happening over and over. The dinosaurs, he thought, might someday return! Still, he provided a great deal of empirical evidence

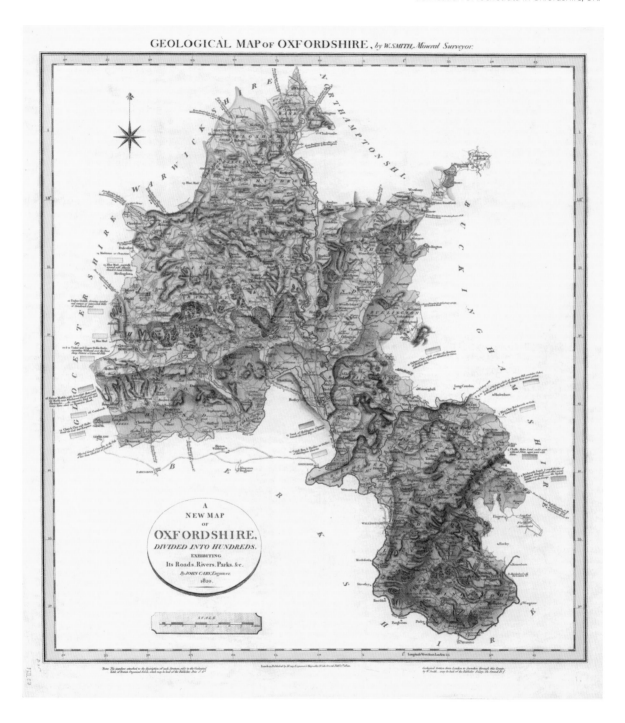

William Smith's 1815 map showing the delineation of rock strata in Oxfordshire, UK.

One facet of Lyell's uniformitarianism proposed that "the huge iguanodon [striped, on the left] might reappear in the woods … while the [pterodactyl] might flit again through [the] trees."

for the concept of a general uniformitarianism and laid out his model in his three-volume *Principles of Geology*, published between 1830 and 1833.

The stage was now set for the next big question: How had *life* on earth changed? It obviously had, and over millions of years, but that could not be explained as easily as geological change, which invoked observable and mechanical processes such as erosion and deposition. Change in living things was far more mysterious and problematic. Among those influenced by Lyell's work, and who, in fact, took the first of Lyell's volumes with him on his famous 'round-the-world' voyage on the HMS *Beagle*, was young Charles Darwin. On that voyage Darwin's observations and thoughts laid the groundwork for our answer to that big question.

Chapter 3

What do we need to know about genetics?

Modern genetics is an important tool for understanding the processes of evolution.

Every week the journals *Science* and *Nature* contain something new and interesting about genetics. Our knowledge of genetics goes back to the work of the Czech monk Gregor Mendel (1822–1884) who, in his Brno monastery, conducted experiments on pea plants; from these, he derived the basic laws of inheritance and of the interaction of "formative elements", what we now call genes. He crossed pea plants with different expressions of seven characteristics to see in what proportions the characteristics would be inherited. From this he discerned that these "elements" come in pairs, one from

LEFT Books from the library of the Abbey of St. Thomas at Brno, Czech Republic, where Mendel worked. Mendel's own book is on the left. On the right is the German edition of Darwin's *Origin of Species*. BELOW The seven traits of the pea plant Mendel used to derive the basic laws of genetics.

Seed		Flower	Pod		Stem	
Form	Cotyledons	Color	Form	Color	Place	Size
Grey & Round	Yellow	White	Full	Yellow	Axial pods, Flowers along	Long (approx 2m)
White & Wrinkled	Green	Violet	Constricted	Green	Terminal pods, Flowers top	Short (approx 0.3m)
1	2	3	4	5	6	7

each parent, and he described how they interact to produce traits (in modern terminology, phenotypes).

Mendel was dead-on with what he was able to discern about the mechanism of inheritance, but he couldn't possibly understand how his "elements" worked. What we learned over the next century elaborated on his discoveries, and in the 21st century our knowledge of genetics has taken off, with huge strides in technology allowing us to understand and test ideas in evolution in ways we never could before.

The *real* definition of a gene

There is a great deal of misinformation in the popular media about the nature of genes. We often read about the discovery of a gene "for" something, everything from diseases (breast cancer, for instance) to behavioral traits (sexual orientation or language). While there could well be a genetic *component* to these traits, there is no single gene for them. Genes don't work that way.

A gene is a pattern or template for building a protein. The cells of living organisms are made of proteins and proteins run the machinery

The structure of a typical animal cell showing the location of the genetic material in the nucleus and the ribosomes where the genetic code is turned into proteins.

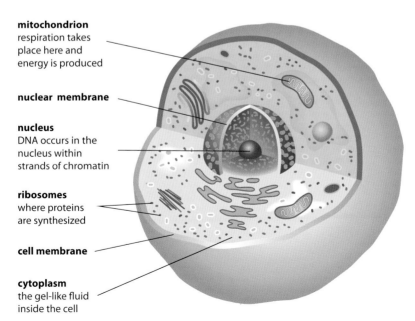

mitochondrion
respiration takes place here and energy is produced

nuclear membrane

nucleus
DNA occurs in the nucleus within strands of chromatin

ribosomes
where proteins are synthesized

cell membrane

cytoplasm
the gel-like fluid inside the cell

of life. Human genes make about 90,000 proteins. Even a non-protein component, such as the calcium in our bones, requires proteins to do its job.

The patterns for building proteins are in bonded pairs of bases (a family of chemicals) within the famous double helix molecule of deoxyribonucleic acid, DNA, described by Watson and Crick in 1953. To imagine a double helix, picture a flexible ladder with its ends twisted in opposite directions. These base pairs are the "rungs" of the DNA ladder. (Only one string of bases is the actual pattern. The other string helps hold the molecule together and facilitates copying of the pattern when cells divide and when it is "read" in the process of making a protein.)

These "rungs" are made up of combinations of four bases, which we can designate by their first letters: A, T, G, and C, and the system is beautifully analogous to a written language. English, for example, has 26 letters that are combined in myriad ways to make up words, the basic units of meaning. A word doesn't convey much information, but string words together into sentences and they impart whole concepts.

A portion of a DNA molecule showing the pairs of bases—always A-T, T-A, G-C, and C-G— that make up the "rungs" of the ladder. The bases on one side of the molecule constitute the genetic code.

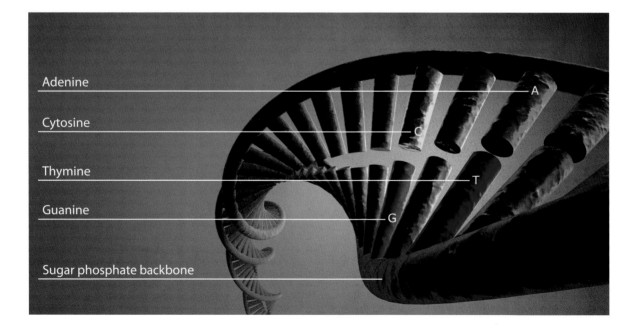

Adenine

Cytosine

Thymine

Guanine

Sugar phosphate backbone

The alphabet-gene analogy.

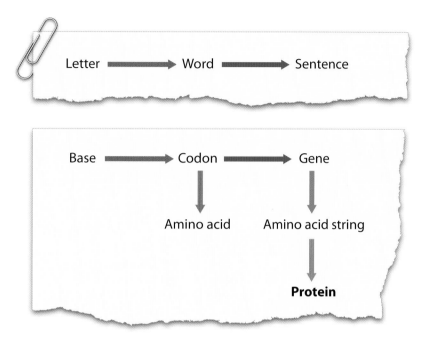

The beauty of the genetic pattern is that it is so much simpler. The genetic alphabet contains only those four letters: A, T, C, and G. The words, or *codons*, in the genetic language all contain only three letters each. Thus, there are a total of 64 words in the genetic dictionary ($4^3=64$). Each codon stands for an amino acid. The meaning of a string of codons is, thus, a string of amino acids. And a string of amino acids is a protein. So, a gene is a genetic "sentence," a pattern for synthesizing one of a species' required proteins.

Since a gene simply makes a protein, it should be clear that there are few single genes directly responsible for a finished trait. Most traits are the result of many proteins acting in complex ways and usually in interaction with factors outside the genetic pattern. Thus, most phenotypes (our observable characteristics) involve many genes. Even something as clearly inherited as human skin color is the result of several pigments as well as other biological and environmental factors; there is no single gene for skin color. How, then, could there be "a" gene for language or for sexual orientation? We need to keep this in mind as we discuss the processes of evolution and the physical changes through time that are the evidence and results of evolution.

The lessons from the Human Genome Project

Despite predictions that the completion of the Human Genome Project, announced in 2000, would result in everything from cures for diseases to "designer children," the technology involved and what we've learned from it have shown that genetics is more complicated than we imagined. And that's why those predictions are a long way from coming true. But what *have* we learned that is relevant to our discussion of evolution?

Most of the genome is not genes. The basic goal of the Human Genome Project was to figure out the sequence of those four bases (A, T, G, and C) for several sample humans and then to use the technology to (1) compare individuals, populations, and, eventually, species and (2) to figure out where the genes are and what they code for, with an emphasis on genes related to diseases. To our surprise, we learned that, at least in our species, 98% of the genome does not code for proteins. It is *non-coding*. Put even more startlingly, only about 2% of our genome makes the proteins that build our cells and run our bodies.

The non-coding DNA used to be called "junk DNA" because it was thought to literally do nothing. Much of it doesn't. Some of it may even be leftover DNA from the early stages of life billions of years ago. But some of this non-coding DNA does have a function: it might be marking the start and end of coding sequences; regulating coding sequences by turning them on or off and by determining their activity level; or jumping around within the genome to reshuffle its elements and even occasionally becoming part of a coding gene.

Thus, as we try to explain the physical changes of evolution in terms of the genetic changes underlying them, we need to remember that only a small part of our genome is directly involved. The same is true, although at different percentages, for all species.

The arrangement of coding genes is not sequential. If 2% of this book were made up of the sentences you've been reading, but 98% of it consisted of nonsense syllables between the sentences, you'd have a difficult time, but the sentences would still stand out. The genome is not arranged that way. Genetic sentences overlap, with the middle

OPPOSITE A computer image of a human DNA sequence. The colors represent the sequence of bases. The coding sections are not, however, sequential. Noncoding sections are interspersed throughout, making the actual genetic code difficult to read.

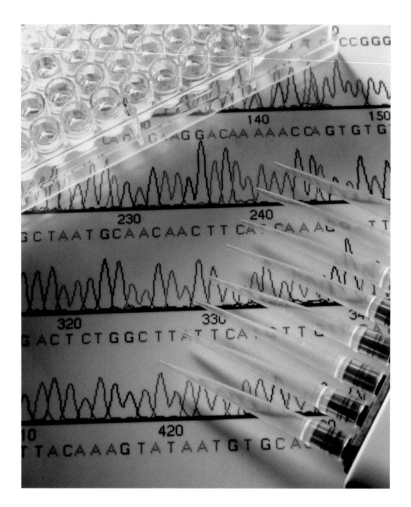

Computer printout of a sequence of human DNA with the bases indicated. Also in the photo is a sample tray and pipettes used in the study of DNA.

of one sentence also being the beginning of another. There are sentences within sentences. Non-coding sequences are found *within* coding sequences. The coding sequences can be *spliced* together in different ways to make different proteins. Non-coding sequences can sometimes code, and coding sequences can become non-coding. If you can imagine the difficulty of reading a book written that way, imagine how difficult it is to read the genetic book.

The result is that decoding the genome is difficult and we're still a long way from having all the details of most organisms' genomes. We're certainly a long way from "designer children."

BELOW Within this seemingly nonsense sentence are six complete and grammatical sentences of varying lengths. Note that "who" and "which" are only used in one sentence each. Otherwise they, too, are nonsense. This is analogous to the complex arrangement of coding and noncoding DNA sequences in the genome.

Thequickbrownfoxjumped**jgfuhfiohetgdij***overthen*fjrndyskcncu*lazydog*
kdheivcjfgyewioejicdheofjrtijtojwho**restedin***thesunwhichwasverybright*

Increasing knowledge of genomes allows us to better understand the relationships among species, current and past. The technology that allows us to sequence the genomes of species, allows us to compare species and populations within species at the most detailed genetic level, the very letters of the code. Sometimes phenotypes can

These two strikingly different looking butterflies are members of the same species, the Eastern Swallowtail. Some females have the dark form, which mimics another species of butterfly that is poisonous.

A simple evolutionary tree of the canine family, based on fossil evidence. We can now construct a more complex and accurate tree based on genetic data (see the figure on page 51).

be misleading. For example, two species can look almost identical but have significant genetic differences, while other species may have populations within them that are so phenotypically different we might mistake them for an entirely different species. Genetic differences, therefore, can give a more precise measure of relatedness.

We can even work backward in time, estimating how long it would have taken a common gene pool in the past to diverge into the separate species we observe today, thus helping us determine a time frame for evolution. This is one way, for example, that we have concluded

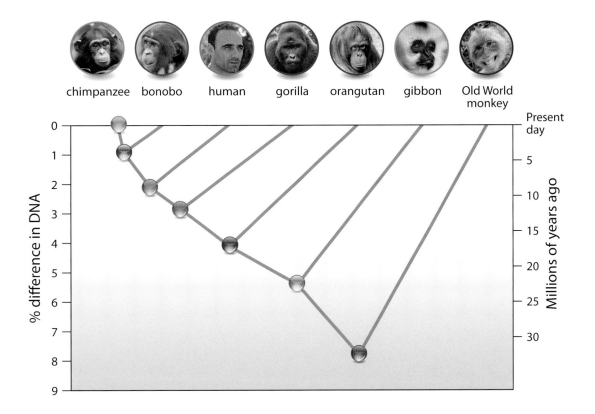

chimpanzee bonobo human gorilla orangutan gibbon Old World monkey

% difference in DNA

Present day

Millions of years ago

An evolutionary tree of some of the primates, based on genetic differences which can then be translated into divergence times.

that humans and our closest relatives, the chimps, had a common ancestor between five and seven million years ago.

Factors outside the genome itself influence the operation of our genes. A molecule related to DNA, called RNA (ribonucleic acid), was once thought to come in two forms, which together were responsible for reading the DNA code, carrying it out of the cell nucleus, and translating it into a protein. Now, it turns out there are many other forms of RNA (themselves produced by genes) that are non-coding, but that have functions such as knocking out disease-causing genes, determining which genes are active and what proteins are produced, and even turning genes on and off. They may also play roles in immune responses, brain function, and sex cell production. These regulatory processes are referred to as *epigenetic* and may be more important to the innovations in the evolution of species than are changes in the proteins themselves. Research on this continues.

All the above helps us fully understand and interpret the phenotypic changes that are the visible results of the processes of evolution. And now, with that background, let's rejoin Charles Darwin on board the HMS *Beagle*.

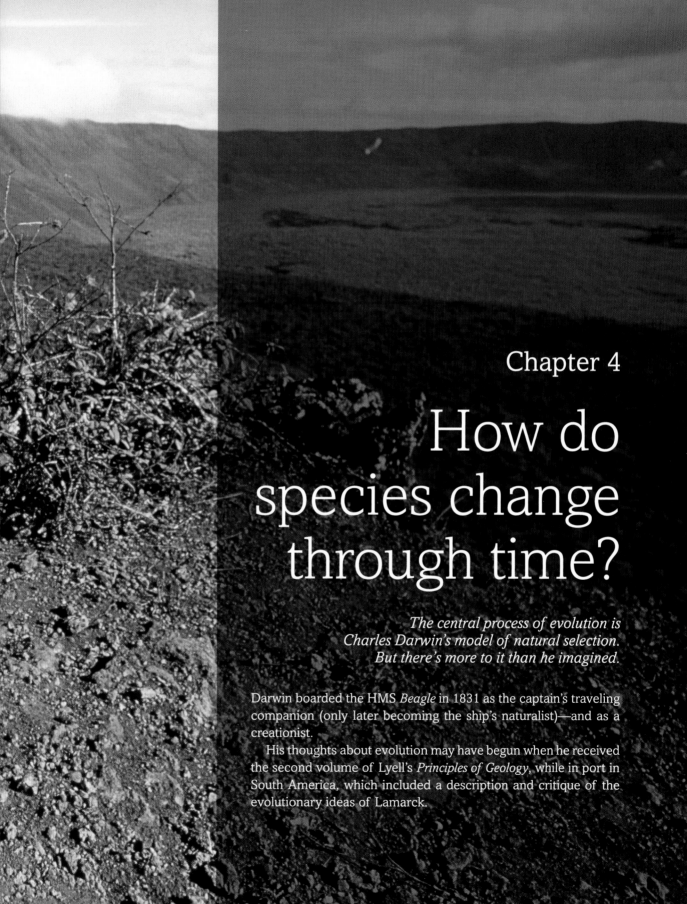

Chapter 4

How do species change through time?

The central process of evolution is Charles Darwin's model of natural selection. But there's more to it than he imagined.

Darwin boarded the HMS *Beagle* in 1831 as the captain's traveling companion (only later becoming the ship's naturalist)—and as a creationist.

His thoughts about evolution may have begun when he received the second volume of Lyell's *Principles of Geology*, while in port in South America, which included a description and critique of the evolutionary ideas of Lamarck.

The HMS *Beagle*

In 1809, the year Darwin was born, Jean-Baptiste Pierre Antoine de Monet de Lamarck (1744–1829) of the *Muséum d'Histoire Naturelle* in Paris published a text for his students, *Philosophie Zoologique*, in which he explained his model for how species change through time.

Although parts of his model were plainly incorrect and criticized during his lifetime, it was, in the words of evolutionary biologist Stephen Jay Gould, the first "consistent and comprehensive evolutionary theory ... in modern Western thought." Lamarck's work influenced, in one way or another, other thinkers in the field—especially Darwin.

Jean-Baptiste de Lamarck, who, among his other contributions, coined the word *biology*.

What did Lamarck say?

The centerpiece of Lamarck's theory of evolution is *adaptation*. The characteristics of any plant or animal are such that they aid the organism in surviving in a particular environment. So, Lamarck reasoned, if

A plate showing barnacles from *Encyclopedie Methodique – Histoire Naturelle les Vers* (1792), on which Lamarck worked. His knowledge of invertebrates led Lamarck to his 1809 publication in which he described his mechanism of evolution. Darwin, who came to very different conclusions about evolution, also worked with this family of barnacles.

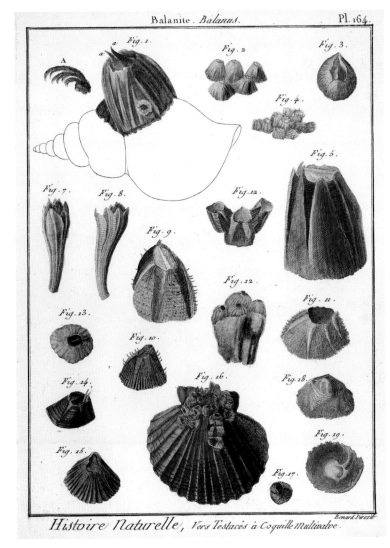

environments change then species must change to keep pace. As we have seen, the stratigraphic record shows that environments did indeed change. With important modifications, this is the centerpiece of evolutionary theory today. And it was this logic that may have initially stimulated Darwin's thinking.

But what Lamarck is known for, unfairly perhaps, are the parts he got wrong. First, he denied that extinction could take place (except, he offered presciently, for large animals wiped out by humans!). Second, he believed evolution in general was progressive, that organisms always evolved toward greater complexity. If this is the case, why are there still simple organisms? Lamarck argued for continuous spontaneous generation of new simple forms.

Finally, and what he is most remembered for, is *how* he thought adaptive change took place. He espoused a process called *the inheritance of acquired characteristics*. The idea is that individual organisms can adapt in response to environmental change—even forming new organs that help in their survival These changes are then passed on to their offspring who need not, then, go through the process themselves. Organisms do this by changing their habits in response to new "needs," using or not using certain responses or behaviors. Then, "these changes give rise to modifications or developments in their organs and the shape of their parts" (from Lamarck's 1809 document). Environmental change comes first, then a change in how individual organisms respond, then physical changes that are passed

In Lamarck's most famous example he suggested the giraffe evolved its tall stature by constantly stretching to reach higher and higher leaves. Individual giraffes thus got taller and passed this increased stature on to their offspring.

on to future generations. For example, offered Lamarck, the black-smith's work causes his muscles to grow and become strong and these acquired traits are passed on to his offspring who then don't have to go through their own adaptive changes. Clearly, this isn't how it works.

And Lamarck never said just *how* organisms change their habits, nor how these changed habits "give rise to modification" of their physical bodies. He simply said it was the case, which prompted Charles Lyell to accuse him of making claims with no concrete evidence.

While Darwin partially agreed with the idea of "use and disuse," it was Lamarck's idea of progression with no extinction and his undocumented "means of change" (as Darwin put it) that most disturbed him and that contrasts most strongly with Darwin's eventual model. Darwin even wrote, in an uncharacteristically churlish moment, that Lamarck's book was "veritable rubbish."

What did Darwin say?

Darwin returned to England in 1836, became involved in the scientific culture of London, and soon began an account of his voyage and a series of notebooks.

Map of the *Beagle* voyage.

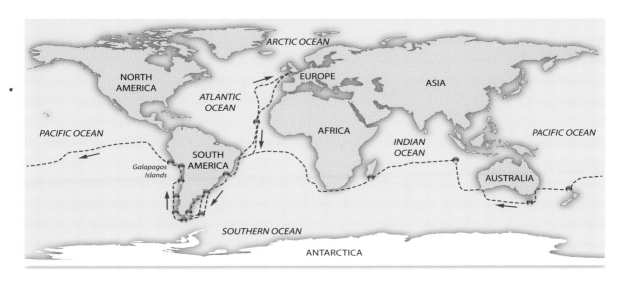

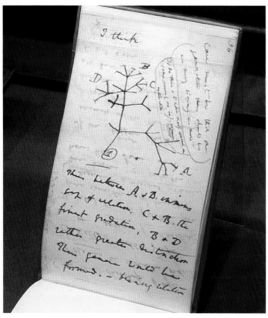

Darwin's notebook with his famous diagram. The words at the top could be read as "I **think**" but I believe they read "**I** think".

In a short time, it's clear he had a context for all that he had observed: an idea for how species change through time that addressed what he and others saw as the problems with Lamarck's model. In his *Notebook B* from 1837–38 he sketched his famous diagram, showing not only that species give rise to other species but also that extinction is "required."

And in 1842, in a short "*Sketch*" (on page 7 to be exact), he gave his model a name: *Natural Selection.*

Darwin's life in brief. Charles Darwin was born on February 12, 1809, the same day as Abraham Lincoln, into a well-to-do British family.

His grandfather, Erasmus Darwin (1731–1802), a physician, amateur scientist, and (bad) poet, had a special influence on Charles's life, even though Charles never knew him. Among the elder Darwin's interests was evolution, which he thought proceeded by the inheritance of acquired characteristics, as did Lamarck later on. Charles

Watercolor of Darwin from the late 1830s by George Richmond (1809–1896).

Darwin's desk in his room in Christ's College, Cambridge University. The room, still inhabited, was reconstructed, for the Darwin Festival in 2009, to look as it did when Darwin was there.

studied medicine at Edinburgh University and, when that failed to interest him, moved to Cambridge University to study for life as a country clergyman.

At Cambridge he began to formalize his interest in natural history, and in 1831 had the rare opportunity to observe nature as few others had by embarking on his famous voyage. As noted, within a few years of his return he had formulated his own theory of how species change through time and are related to one another. As confident as he was in this theory, however, he was unsure of its reception by colleagues and the public, and so he moved on to other scientific pursuits and left his evolutionary ideas unpublished. Until, that is, he received a paper in 1858 from young Alfred Russel Wallace in which Wallace described *his* new model for species change. It was almost exactly like Darwin's. Darwin's hand was called and the next year he published *The Origin*

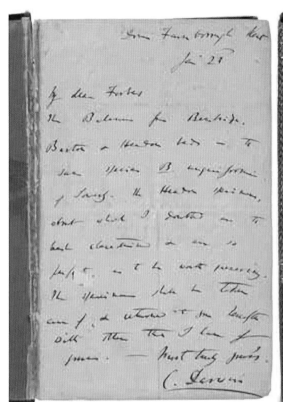

A rare copy of the first edition (1859). Only 1250 were printed and they sold out in a single day. The present location of about half of those is unknown. First editions now sell for nearly $200,000 (£120,000). This copy, at the Strozier Library at Florida State University, includes a letter signed by Darwin on the inside cover.

Darwin's tombstone in Westminster Abbey. The modest inscription (so unlike many of the others) reflects Darwin's lifelong personality.

of Species by Means of Natural Selection or the Preservation of Favoured Races in the Struggle for Life, which he said was an abstract of a forthcoming, larger work. (The *Origin*, as it is known, was nearly 600 pages long when published.)

Possibly to Darwin's surprise, the first edition sold out in a single day and was, for the most part, well received. Darwin spent the rest of his life researching and writing about numerous scientific topics, including the application of natural selection to humans in *The Descent of Man* (1871). He died in 1882 and was buried in Westminster Abbey.

Natural selection. A logical prediction from Lamarck's model was that all members of a particular species would look and behave the same, since they would all have responded to that species' environment in the same way. Darwin realized—to a large extent from observations of plant and animal breeding—that species show variation. Not every member of a species looks the same.

Moreover, a species' reproductive potential greatly exceeds the number of members an environment can support. Thus, there is competition for resources, and here's where the variation comes in: Some members of a species are simply better adapted to an environment; some are less well adapted. This is not a matter of differences in Lamarckian use-and-disuse responses but, rather, a matter of chance. Although Darwin didn't understand genetics and did not know where this constant source of variation came from, he realized it was vital to his model. Thus—putting it all together—natural selection is, in the words of Oxford biologist Richard Dawkins, "the nonrandom survival of randomly varying hereditary instructions …" As a result, the better adapted individuals will, on average, produce more offspring and the poorly adapted ones fewer offspring. The adaptive traits thus accumulate over time and the poorly adaptive ones decrease in frequency and may disappear altogether. And, should some environmental change be so fast or extensive that no member of the species can reproduce in sufficient numbers, the species may become extinct. This seems a simple and, in fact, rather obvious idea. But it was revolutionary in Darwin's time and *Origin of Species* is one of the most important scientific books ever written.

NEXT PAGE Variation in one species of European snail, Cepaea nemoralis. The differences may be adaptive responses to better camouflage in different parts of the species' range.

A secondary result of the process of natural selection, said Darwin, is the divergence of new species from old. But we'll take that up in the next chapter. Here, we'll stick with evolution *within* species, or microevolution in modern parlance.

Modern microevolutionary theory

Let's build a diagram that captures what we now know about microevolution.

Natural selection is the central process. As Lamarck noted, species are adapted to their environments, and a major driver of change through time is maintaining this adaptive relationship. The differences between Lamarck's and Darwin's models are in *how* species adapt, in the acknowledgment of extinction, and in the fact that evolution is not progressive. Adaptation can be toward *simpler* body plans and less complexity depending on the variation available and the nature of environmental change.

A classic example of natural selection comes from Darwin's finches, a group of related birds on the Galápagos, one of Darwin's stops on the *Beagle* voyage.

Natural selection: the central process of evolution.

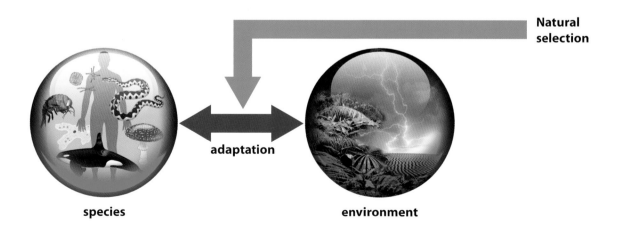

The Processes of Evolution

Natural selection

adaptation

species   **environment**

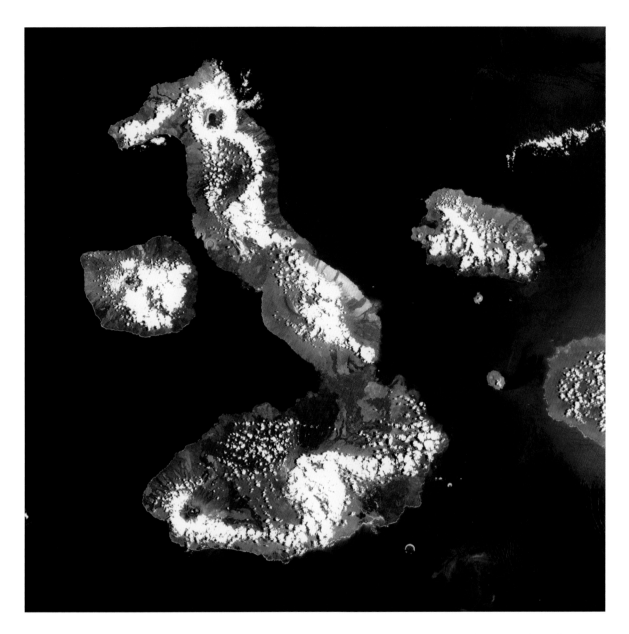

The Galápagos Islands, about 500 miles off the coast of Ecuador, one of the most influential stops on Darwin's voyage.

Since 1973, Rosemary and Peter Grant have conducted a long-term study of a species of finch on one small island that showed selection for what, to those birds, was a vital phenotypic trait—beak size.

The Grants saw the average beak size change back and forth from year to year and across generations (by no more than a millimeter) in response to rainfall amounts. The selective factor was the size and exterior toughness of the seeds the birds ate, as the seeds responded to changes in moisture. In drought years those birds with

1. Geospiza magnirostris.
2. Geospiza fortis.
3. Geospiza parvula.
4. Certhidea olivacea.

Some of the variation in beak shape and size of the Galápagos finches. An illustration used by Darwin in a book describing his famous voyage.

The medium ground finch studied by the Grants.

slightly bigger beaks had an advantage and so produced, on average, more offspring. The offspring would tend to inherit the larger beaks. In addition, the birds with the smallest beaks not only produced fewer offspring, but may themselves have died. As a result, over a couple of years the average beak size increased. In years of normal or above average rainfall the bigger beaks afforded no benefit and so the birds with smaller beaks reproduced as well as those with larger ones. The average size then returned to its previous norm.

The important point here is that the beaks did not get bigger *in response to* the environmental change. Variation in beak size was already present and so, under the stress of the change, larger beaks were "selected" by the natural situation to become more frequent.

An implication of this process is that species become extinct due to various factors. Take the famous extinction of the dinosaurs as an example. When a six-mile-wide asteroid smashed into the earth 65 million years ago, bringing about extensive and worldwide environmental change, no member of any dinosaur species was able to reproduce enough offspring to perpetuate its species. Many, of course, were killed outright. All the dinosaurs, as well as many other forms, died out. (We'll talk about this in more detail in the next two chapters). Had Lamarck been correct, the dinosaurs would have automatically been able to change, as much as needed.

Blind cave fish evolved, in a sense, a less complex body plan by losing their eyes, unnecessary in the complete darkness in which they live.

So, natural selection attempts to maintain the balance between species and environment in our diagram. It's almost impossible to talk about natural selection without sounding as if there is some agency in the process, but natural selection doesn't "attempt" anything: It just happens. Environments change, so natural selection has a constant job to do.

But species also change from within, and this greatly adds to the complexity of microevolutionary theory. Changes occur within a species' total genetic endowment, known as the gene pool. This in turn affects the species' phenotypic features. How does this happen?

Mutation. The initial source of all genetic variation is mutation—spontaneous and unpredictable changes in the genome. These changes can occur in whole chromosomes, one base pair, or anything in between. While most mutations affect only individuals, the ones we're interested in here are those that are passed on and thus become incorporated into a species' gene pool.

Mutations occur frequently, caused by chemical pollutants, radiation (even normal background radiation from the earth), and even a chance hit from the tiny particles known as cosmic rays (millions have passed through your body since you started reading this sentence). Most mutations, however, are the result of mistakes in the genetic mechanism itself. Molecules are moving around, splitting, and regrouping in various ways. And mistakes happen. While mutations occur all the time, *what* mutation takes place is random, a matter of pure chance.

Mutations are deviations in the genetic mechanism and therefore most are deleterious. And yet, mutations are the price we pay for evolution. If the first forms of life—simple, single-celled organisms such as bacteria—reproduced their genomes perfectly, with no errors, there would be no variation. No variation, no natural selection, no evolution.

The distribution of variation. Once genetic variation has entered a species' gene pool, how it is distributed makes a difference. Is variation evenly distributed throughout a species? Does the new variation become common or stay uncommon? Is it distributed in clusters? All

The Processes of Evolution

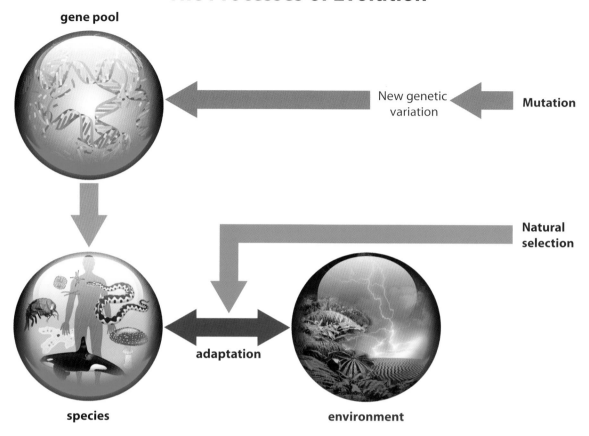

gene pool

New genetic variation

Mutation

Natural selection

adaptation

species

environment

Mutation: The primary source of diversity in a species' gene pool.

this will affect the phenotypic expressions within the species that the gene pool gives rise to.

Here's an analogy. If I have a gallon of white paint and I add an eyedropper full of red paint and do nothing, I'll have a can of white paint with a red dot. If I add the red paint and make a few stirs with a paint paddle, I'll have a can of white paint with red swirls. But if I add the red, close the can, and shake it, I'll end up with pink paint.

The paint can analogy of the distribution of genetic variation.

The Processes of Evolution

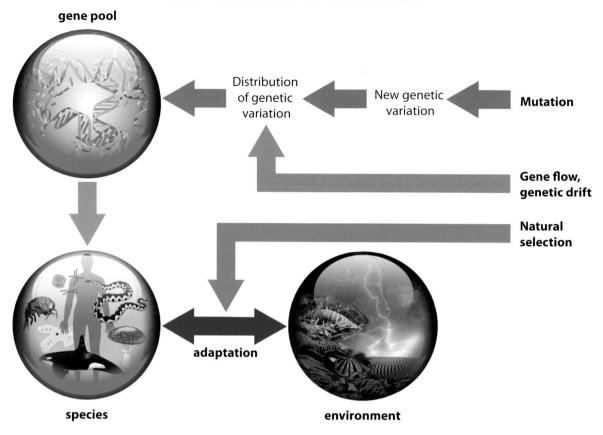

The processes of microevolution and their relationships.

The following three processes affect the distribution of genetic variation within a species as with the paint example.

The first process is called *gene flow*. This is defined as simply the movement of genes throughout a species as individuals move around or as adjacent populations within a species share genes. We humans are an example. We move around all the time and when we reproduce that results in the exchange of genes. Thus, gene flow is particularly influential in our species and is one of the reasons that—despite some superficial phenotypic differences we make much too big a deal about—genetically we are surprisingly homogeneous. We are really like that third can of paint, the pink one.

The next two processes fall under the term *random genetic drift*. In fact, they are two separate phenomena. The first is the two-part process of *fission and the founder effect*. Just as populations can merge, bringing about gene flow, populations can also split, called fission. This split results in two or more populations, each of which will have different genetic variations from the others as well as from the original population. Thus, the "founded" populations will be genetically unique, affecting the overall genetic nature of the species. This too is influential in our species, as human populations have always migrated—merging with other groups, splitting, merging again in different combinations—throughout our history

The final process of microevolution is *random gamete sampling*. Just as genes are not distributed representatively when a population splits, they are not distributed representatively when two individuals reproduce. In sexually reproducing species, each organism passes on only one of each pair of genes at a time, and only chance dictates which one will be involved in the fertilization that begins a new individual.

As a result, some genetic variants may actually disappear and some others may become more frequent, or even the *only* variant, through nothing more than chance.

These, then, are the processes that change a species through time. But Darwin's book was about the origin of species. How does *that* happen?

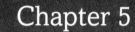

Chapter 5

How do new species evolve?

The descent of new species from existing species is, to some people, the controversial part of evolution, but it follows naturally from the processes that change individual species through time.

Natural selection is so obvious that few take issue with it. The problem for many people—philosophically or simply in understanding it scientifically—is the idea that species give rise to new species meaning that all life on earth is related in a huge, very complex family tree. Therein lies the implication that we humans are descended from some nonhuman species, and that we share

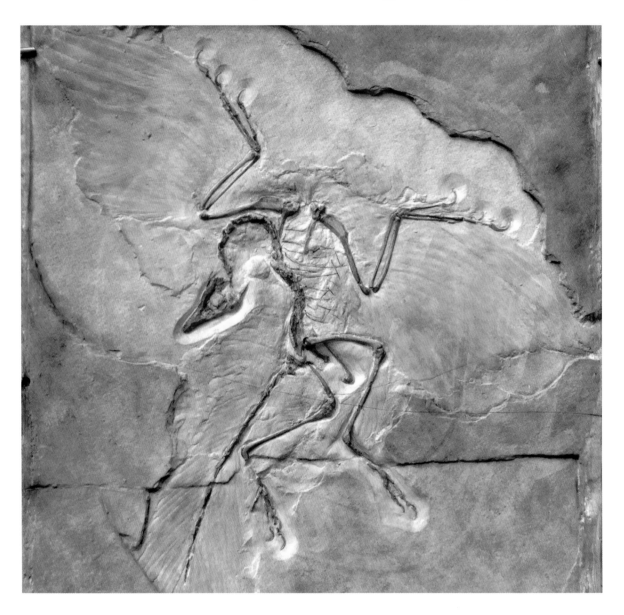

a common ancestor with existing nonhuman species. In other words, a primitive single-celled organism from the distant past gave rise to oak trees, fish, and humans. In this chapter we will explore how new species evolve and discuss the factual basis behind it.

Archeopteryx, the 135 million-year-old fossil that first clearly showed the common ancestry of birds and dinosaurs. It is, essentially, a dinosaur with feathers.

What is a species exactly?

As with any category that we impose on the chaos of nature, there is some debate over the most useful definition of species. While there are several points of view, we will use what's called the biological species definition. Under this model a species is a reproductively isolated population, meaning that members of one species can produce fertile offspring among themselves but can't reproduce with other species.

Let's look at some examples. As an obvious case, despite our rather amazing genetic similarity to chimpanzees (by one measure we are 96% genetically identical and our proteins, the products of our genes, are 29% identical), we are separate species. We can't interbreed.

Horses and donkeys can interbreed but their progeny, mules, are almost always sterile. Two mules can't make more mules. Thus, horses and donkeys are separate species, unable to combine their genetic endowments for more than one generation.

Despite a remarkable genetic similarity, we and our closest relatives are clearly separate species, with a last common ancestor five to seven million years ago.

A hybrid between a zebra and a horse. They are sterile because zebras have 32 or 46 chromosomes (depending upon the species) and horses have 64.

There are, of course, examples that challenge this neat definition. Caribou and reindeer are demonstrably the same species. Under artificial conditions they'll mate and produce fertile offspring. But in the wild they can't because they live in now-separate hemispheres So should we consider caribou and reindeer as two species?

Coyotes and wolves look distinctly different when you see them together, but they can interbreed and produce fertile offspring, although they tend not to. But, according to recent studies, the coyotes in New England were found to have about 15% wolf genes. Both species were exterminated in this area, but as coyotes moved back in they mated with some wolves along the way and so the New England coyotes are phenotypically different from the coyotes, say, in the American southwest. Are coyotes and wolves one species or two?

The reason the line blurs is simple: Speciation, the evolution of new species, is a process. It is not instantaneous and it takes time for two

A caribou (*top*) is the same species as the reindeer (*bottom*), although they never interbreed in nature. The caribou is North American and the reindeer Eurasian. The hemispheres have been completely separate since the last recession of the Ice Age glaciers, 10,000 years ago.

An eastern coyote or, according to some, a coywolf, bigger and more wolflike in appearance than its western cousins.

or more populations to diverge and achieve the reproductive isolation required by our definition. What we are seeing with the caribou and reindeer, and with the coyote and wolf, is speciation caught in the act (although, as I'll explain, that act needn't always go to completion).

How, then, does speciation occur?

Isolation: the key to speciation

For a new species to evolve from an existing species, one or more populations of that species must be isolated from the parent population. Gene flow, in other words, must be profoundly interrupted so that all the processes addressed in the previous chapter are happening separately in the isolated populations and in the parent population. What happens in one cannot be shared with the others by genetic exchange. As a result, the populations will accumulate genetic and, thus, phenotypic differences. If—and it's a big "if"—one of those differences prevents the populations from reproducing with each other, they are, by our definition, separate species.

But keep in mind that a reproductive isolating mechanism does not necessarily have to evolve. This explains the caribou and reindeer. Over the 10,000 years they have been geographically isolated (since the end of the Ice Ages when sea levels rose and separated

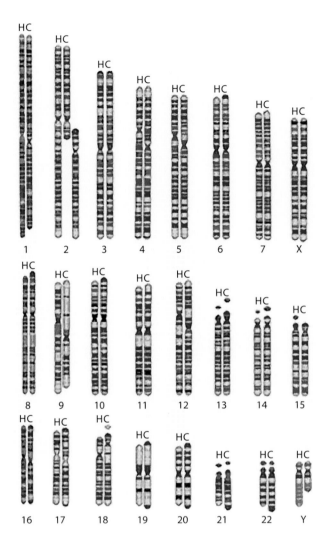

HC HC
1   2

HC
3

HC
4

HC
5

HC
6

HC
7

HC
X

HC
8

HC
9

HC
10

HC
11

HC
12

HC
13

HC
14

HC
15

HC
16

HC
17

HC
18

HC
19

HC
20

HC
21

HC
22

HC
Y

Human chromosomes on the left, compared with chimp chromosomes on the right, for "banding pattern," an indication of the relative similarities of the frequencies of the four bases in the genes. Note that human chromosome #2 is two chromosomes in the chimp. We infer that the greater number of chromosomes was the ancestral condition but we can't be certain. The difference in chromosome number is probably the main reason our two species can't interbreed.

once-connected Siberia and Alaska), the two populations have acquired phenotypic distinctions. Among those, however, is *not* one that prevents their reproducing (which, however, they can only do with our intervention).

What kinds of reproductive isolating mechanisms do we see in nature? It may be that two related populations have evolved so that they reproduce at different times of the year. In plants, two related species might have different pollinators (say, bats for one, insects for another). In some related species, the cells of reproduction are simply incompatible, so even if mating occurs, fertilization does not. (This is probably the case with humans and chimps. Chimps have one more pair of chromosomes than we do so the recombining of genes that defines fertilization won't happen.)

Or maybe fertilization can take place, but the fertilized egg is too genetically mixed and so cannot survive. Or, as in the case of horses and donkeys, the hybrid offspring are sterile. If just one of these differences evolves as two or more populations are going off in their own evolutionary directions, a new species is born.

What types of isolation provide the opportunity for this differentiation to happen?

Mechanisms of isolation

Isolation in different environments. This is by far the most common form. Simply, two or more populations of a species are physically separated from one another by some barrier that prevents gene flow between them.

The classic example is, again, Darwin's finches. There are 13 recognized species of finches on the Galápagos. Genetic studies have confirmed that they have all descended from a single species that

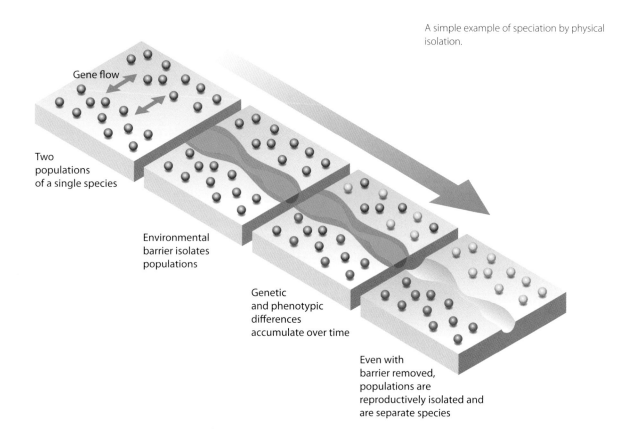

A simple example of speciation by physical isolation.

Gene flow

Two populations of a single species

Environmental barrier isolates populations

Genetic and phenotypic differences accumulate over time

Even with barrier removed, populations are reproductively isolated and are separate species

inhabits the west coast of South America. Over millennia, birds have been blown out to sea on strong air currents, or floated on mats of vegetation, and the lucky ones came to land on the islands. Largely isolated on different islands, they differentiated over the years into separate species.

Isolation within the same environment. This is when a species diverges into two or more separate species within one contiguous environment. For example, there is a species of fly in North America that is diverging because the flies mate only on the fruit in which they lived as larvae, so flies that develop on different fruits don't mix their genes, even when those fruits grow side by side. These flies are considered subspecies at the moment—a hawthorn population and an apple population and perhaps some others. But other differences are arising, such as the timing of mating, which is increasing their differentiation.

The classic example comes from Lake Victoria in East Africa. The lake was dry until about 13,000 years ago, and yet it is home to over 300 species of cichlid fish.

Recent genetic studies indicate that the cichlids are derived from a single common ancestor (making this one of the fastest examples of speciation in a large group of vertebrates). Although they all live in the same lake, populations responded to different food preferences, which led to different mate choices, which eventually isolated small groups and ultimately produced separate species.

A tank of African cichlids, showing the great variations in color, pattern, and size that characterize different but closely related species.

An axolotl, looking very much like the larval stage of the closely related tiger salamander. And yet the axolotl is an adult and capable of reproducing.

Isolation by genetics. This is the rarest form of isolation. It involves a mutation that radically changes those members of a species that possess it, but without affecting their viability and fertility. Mutations that affect development would seem the most likely. For example, the axolotl, a large (and critically endangered) Mexican salamander, is neotenous, meaning that it never metamorphoses beyond the strictly aquatic, larval stage of those amphibians. And yet it becomes sexually mature and reproduces. It is a separate species. The reason for the axolotl's neoteny is the lack of a single hormone, so the genetic change involved must have been relatively simple.

Clearly, such a mechanism of speciation would not occur often. Mutations that make such large changes without adversely affecting reproductive ability are exceedingly rare. But it does happen.

These, then, are the ways in which species change and new species evolve from existing species. Over the 3.6 billion year history of life on earth, what is the overall pattern of evolution these processes have brought about and what does this tell us about evolution in general?

The grand pattern of evolution

Microevolution is change over time within individual species. When new species branch off from existing species, the resulting pattern—the tree of life—is macroevolution. Let's begin by seeing what Darwin said about it, and then we'll see how it's been updated in modern times.

Darwinian macroevolution. Darwin was a strict uniformitarian, seeing evolution as slow and steady. He even denied that catastrophic mass extinctions could occur, because they were not gradual and were not related to the adaptations of species (but, rather, to chance).

Macroevolution, for Darwin (who did not use the term), was an extrapolation of microevolution by natural selection, just over long spans of time. It involved changes *within* species where local populations of a species would constantly become better adapted and increasingly different. The best adapted of those populations would spread and eventually "supplant and exterminate the older, less improved … varieties." This model is depicted in the only diagram in *Origin of Species*, with its "little fans of diverging dotted lines." The line that survived in each species would eventually become a new species.

We can interpret Darwin's overall view of evolution as one of linear change, with previous forms necessarily becoming extinct.

But here is where Darwin was incomplete. The overall pattern of evolution doesn't look like this.

The only diagram in *Origin of Species*. It shows some branching evolution—where, for example, *a6* and *f6* branch from the descendent of *a4*— but is mostly the evolution of single species through time. For example, the rest of the *a* line, where one population within a species is more successful and supplants the rest of the species.

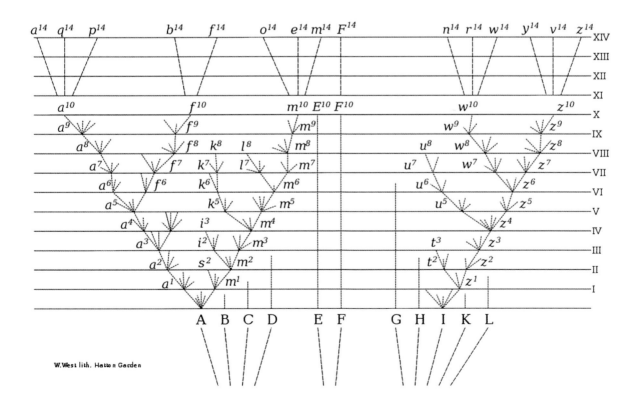

W.West lith. Hatton Garden

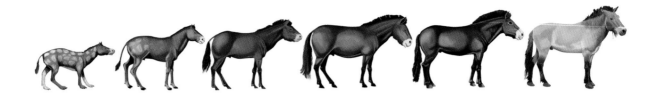

Stereotyped depictions of horse and human evolution as straight lines of gradual change through time.

Modern macroevolution A modern theory of macroevolution is described by evolutionary biologist Stephen Jay Gould. It involves three major concepts:

Speciation is branching, not linear. If a single species changes through time to evolve into a new species, as Darwin proposed, at what point does it become that new species? Where is the dividing line? In fact, when a new species evolves it requires the isolation of one or more portions of that species with a resulting interruption of gene flow. Extinction does not necessarily occur. The original species may well continue to exist after one or more new species have branched from it. Thus, evolution is a thick, luxuriant bush with many branches. What appears in the fossil record to be straight-line change within individual species is really a sampling of fossils from

Evolutionary biologist Stephen Jay Gould (1941–2002). A prolific writer and science popularizer, his work covered many areas of biology, anthropology, science history, and even baseball. But his major contribution to evolutionary theory is his model of macroevolution, the centerpiece of which is punctuated equilibrium.

Gradualism

Early division produces two lines and undergoes gradual transitions: accumulation of minute changes leads to modern representatives.

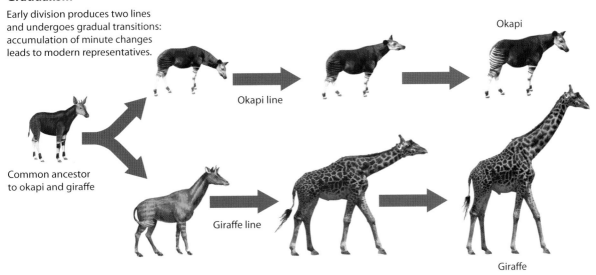

Common ancestor to okapi and giraffe

Okapi line

Okapi

Giraffe line

Giraffe

Punctuated equilibrium

Long periods of little change are interrupted by widely-separated episodes of speciation, each resulting in a sudden rise of new species.

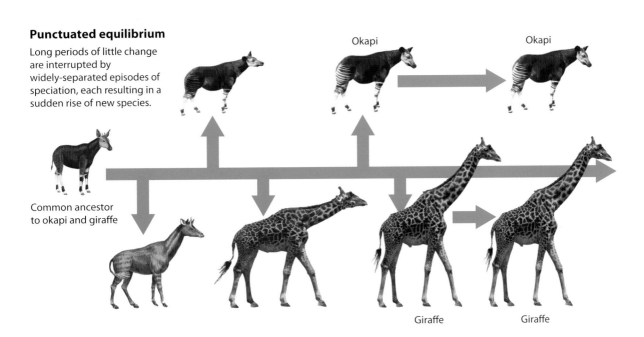

Common ancestor to okapi and giraffe

Okapi

Okapi

Giraffe

Giraffe

Punctuated equilibrium says that species tend to change little during their tenure on earth (equilibrium) and that new species are always the results of branching due to some form of isolation (punctuation). This contrasts with Darwin's gradualist model of constant, small changes pushing new species to arise when enough change has taken place in existing species.

different branches through time. Moreover, natural selection is conservative. It "tries" to *maintain* a species' adaptation to its environment, for the most part "editing out" what *doesn't* work. Thus, there is relatively little overall change within a species through time and, if there is, it is change around a basic theme or limited change within that theme. This is the model of macroevolution known as punctuated equilibrium.

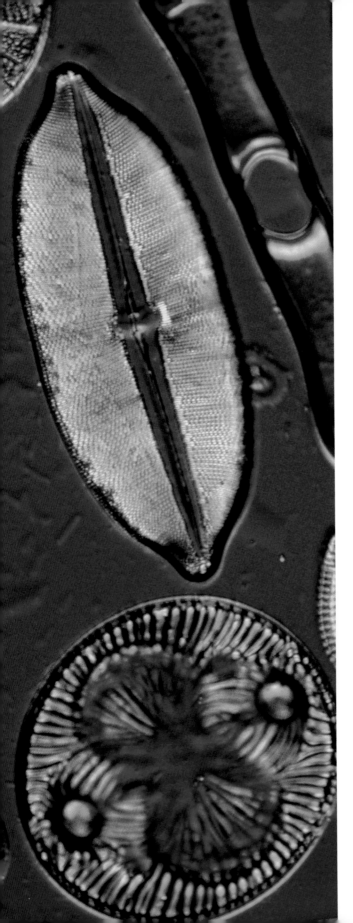

Diatoms, showing their complex siliceous cell walls that helped them survive the mass extinction 65 million years ago, a matter of simple good luck.

A form of selection may take place at the level of the species and larger groups. Darwin felt that selection took place at the level of individual organisms within a species where one organism competed with others within its species for reproductive success; in other words, by natural selection as it is normally defined. But species or related groups of species may also compete with other species. For example, some species may have traits that make them diversify into new species more rapidly than other species in the same environment. Their descendant species, then, will eventually outnumber those of other species and will be more prominent, and thus *appear* to be better adapted. The cichlids described above, for example, are not necessarily better adapted to the East African lakes than fish of other groups, but are simply more prolific in branching into new species. This phenomenon is known as species selection.

This factor can greatly influence larger patterns of evolutionary change through time. For example, Gould proposed that many groups of organisms experience an initial period of maximum diversity that gets restricted as some forms persist and others don't, for a variety of reasons. Thus, the overall shape of the evolutionary tree is an inverted pyramid, but look at any one section and it might look like a right-side-up pyramid. We will see this in action when we look at human evolution in Chapter 7.

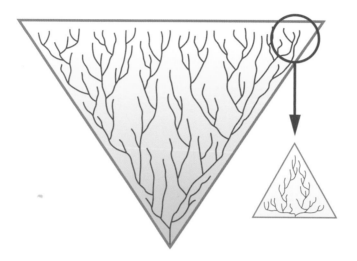

The overall tree of life is one of increasing diversity. Look closely, however, at one small section of that tree and note that diversity sometimes decreases over time as some lines within an adaptive theme persist and others die out. (After S. J. Gould, *Wonderful Life: The Burgess Shale and the Nature of History*, p. 46.)

Catastrophic mass extinctions have played a major role in the history of life. This is not the catastrophism of earlier centuries as described in Chapter 2—that is, a long series of cataclysms that accounts for the entire geological and paleontological history of the earth. Nor does it contradict the general idea of uniformitarianism. It simply acknowledges that about five times in the history of our planet, an unpredictable and sudden event brought about the extinction of large numbers of species, while others survived, and these evolutionary changes were "nonselective" in the Darwinian sense. That is, species that became extinct may have been perfectly well adapted before the catastrophe but just couldn't survive it. They had, in the words of Stephen Jay Gould, "bad luck" not "bad genes." Similarly, species that survived did so not because they suddenly adapted to the conditions of the catastrophe, but because they *already* had adaptations that turned out, by luck, to be beneficial when the catastrophe hit.

As an example, we can look to the mass extinction of 65 million years ago when all of the dinosaur species became extinct. They were presumably doing well before the event, but no members of any of their species had traits that allowed them to survive the major environmental changes caused by the asteroid strike that set off a chain of worldwide calamities. Diatoms, a form of photosynthesizing algae and a major source of earth's oxygen, did survive, while other forms of algae did not, because diatoms already had the ability to withstand adverse conditions through their hard cell walls made from silica.

On a geological time scale, then, evolution is uniformitarian and gradual but species do arise relatively suddenly as a result of

The five mass extinctions

A mass extinction is defined as when the earth loses more than 75% of its species in a relatively short period of geological time.

Extinction	End date (mya)	% of species lost	Proposed causes
Ordovician	~ 443	86	Alternating glacial and interglacial periods Mountain uplift affecting atmosphere and ocean chemistry Loss of CO_2
Devonian	~ 359	75	Global cooling Loss of CO_2 Deep water anoxia Bolide impact (debated)
Permian	~ 251	96	Volcanism Global warming Deep water anoxia Elevated hydrogen sulfide and CO_2 Acidic ocean water Bolide impact (debated)
Triassic	~ 200	80	Elevated CO_2 Global warming
Cretaceous	~ 65	76	Bolide impact (confirmed) Global cooling Volcanism then warming Tectonic uplift CO_2 drop

Mya = million years ago. Anoxia is oxygen loss.
A bolide is an extraterrestial object such as an asteroid that strikes the earth.

We are experiencing the beginning of a sixth mass extinction at present. Studies indicate that extinction rates for vertebrates over the last 500 years are as fast as, or faster than, extinction rates for the above five events. And projections for the future, although obviously conjectural, make a sixth major event seem likely.

(Data and analysis from Barnosky et al. 2011. *Nature* 471: 51–57.)

branching; species and larger groups within the same environment do undergo a form of selection; and there have been a few major "glitches" in the history of life that have reorganized the world's living things. All these have directed and redirected evolution as a whole.

Now, what have all these processes produced in the history of our planet, and what can we learn from all this?

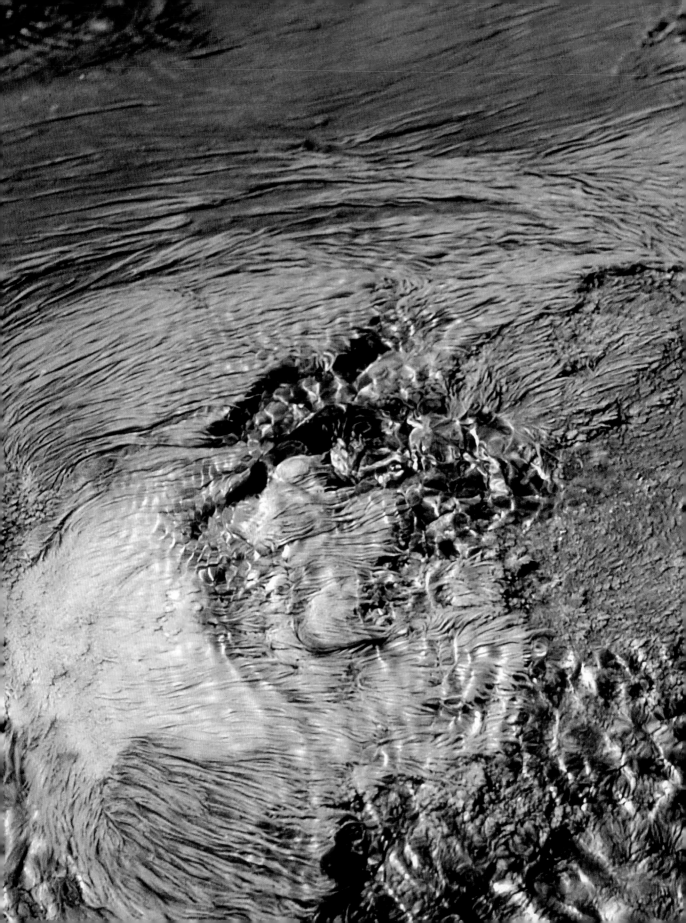

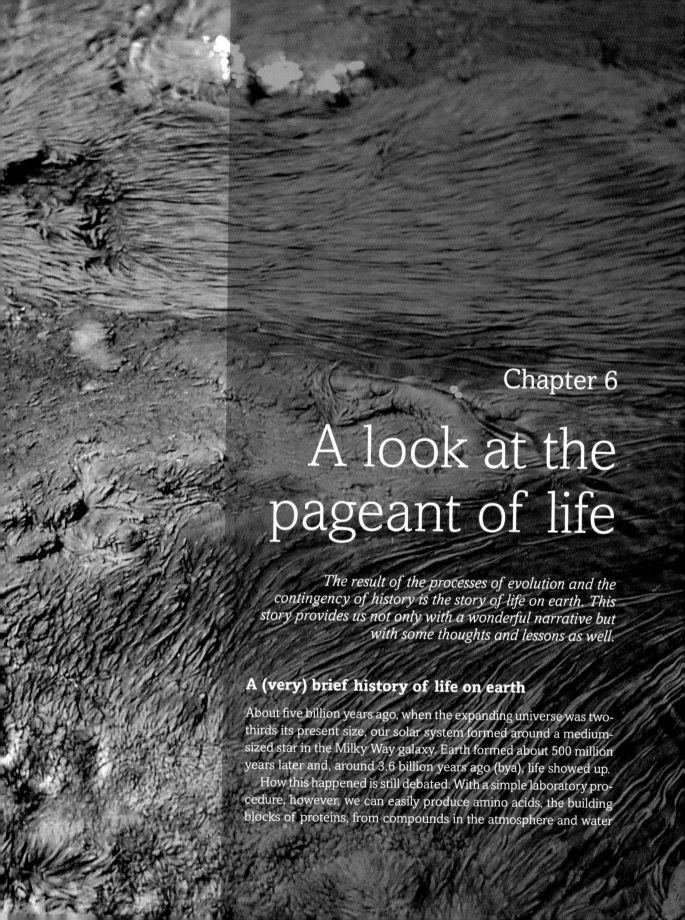

Chapter 6

A look at the pageant of life

The result of the processes of evolution and the contingency of history is the story of life on earth. This story provides us not only with a wonderful narrative but with some thoughts and lessons as well.

A (very) brief history of life on earth

About five billion years ago, when the expanding universe was two-thirds its present size, our solar system formed around a medium-sized star in the Milky Way galaxy. Earth formed about 500 million years later and, around 3.6 billion years ago (bya), life showed up.

How this happened is still debated. With a simple laboratory procedure, however, we can easily produce amino acids, the building blocks of proteins, from compounds in the atmosphere and water of the early earth. Add almost a billion years and truly countless molecules and their interactions, and the odds are that anything that could happen did. The evidence for the earliest life is indirect but compelling: fossils of accumulations of photosynthesizing single-celled organisms from Greenland, southern Africa, and Australia. There are even fossils of single-celled bacteria-like organisms themselves.

At two bya, we find evidence of complex single-celled organisms (those with nuclei and other components), and simple multicellular organisms appear about 1.7 bya. By 1.2 bya, enough free oxygen had

Living stromatolites in Australia, formed when mats of blue-green algae (photosynthesizing cyanobacteria) are covered with sand, silt, and mud which the algae cement down and then grow over.

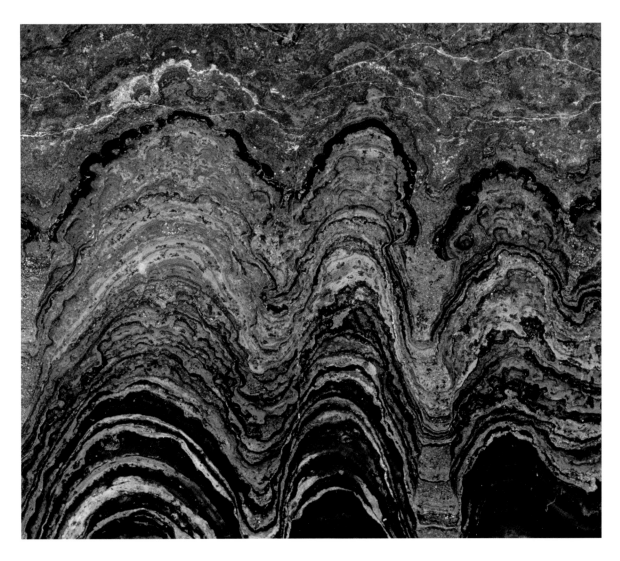

Fossil stromatolites from Australia, over three billion years old.

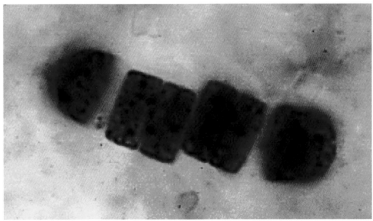

Fossil of a chain of cyanobacteria cells from one-billion-year-old rock in Australia, photographed using electron microscopy.

accumulated in the atmosphere in the form of ozone (O_3) to block ultraviolet radiation and allow the first creatures, still bacteria-like, to live on land.

All of these earliest organisms reproduced asexually, that is, by copying their DNA and splitting, making clones of themselves. As a result, variation was limited to mutations. But about one bya, some organisms began to reproduce sexually. Sexual reproduction—combining portions of two parental genomes in offspring—increases genetic variation, the raw material of natural selection, and evolution begins to speed up.

About 543 million years ago (mya), complex multicellular organisms seem to burst onto the scene. Many possessed shells and other hard exterior parts. We refer to this relatively sudden change as the Cambrian Explosion, but there is no agreement as to its cause. We do know, however, that in just a few million years all the major body plans

Artist's reconstruction of some of the bizarre creatures of the Cambrian. The light-colored fishlike organism at top-left is *pikaia*, the first fossil chordate, the group that gave rise to the vertebrates, represented today by fishes, amphibians, reptiles, birds, and mammals.

Artist's reconstruction of *Anteosaurus*, a mammal-like reptile from 266–260 mya. There were many mammal-like reptiles, some of which were the remote ancestors of today's mammals. Many, however, were wiped out in the greatest of all mass extinctions, 250 mya.

of multicellular animals had evolved, to include the ancestors of the vertebrates.

By 470 mya plants and fungi colonized the land. Fish and land animals appear in the fossil record around 425 mya, and insects by 400 mya; by 350 mya some of these insects had evolved wings. About the same time as flying insects appear, so do reptiles. The reptilian forms thought to have given rise to mammals are from around 260 mya.

Dinosaurs, a specialized form of reptile, make their appearance about 235 mya and true mammals show up a little later, about 220 mya. Sometime around 160 mya a group of dinosaurs evolved feathers, marking the ancestors of the birds.

Flowering plants are fairly late in the story; they appear about 100 mya. The possible precursors of the primates, the group to which we humans belong, may be as old as 80 million years. And with the extinction of the dinosaurs and many species of other groups 65 mya—when that asteroid slammed into the earth—the more immediate and recognizable ancestors of existing forms of life are found in the fossil record.

To put this all in perspective it's useful to compare the history of life on earth to some more conceivable frame of reference. I like astronomer Carl Sagan's "cosmic calendar" the best. It maps the history of

January	February	March
Big Bang		Milky Way forms
First stars		
First galaxies		April

May	June	July
		August
		Solar System forms

September	October	November	December
First lifeforms			

December 10	First animals	
December 25	First mammals	
December 29	Dinosaur extinction	
December 31 11.30:00 p.m.	First humans	
December 31 11.59:00 p.m.	First civilisations	
December 31 11.59:59 p.m.	Humans on Moon	

Artist's reconstruction of *Anchiornis*, a feathered dinosaur from China dated at 160 mya. Knowledge of the feather colors comes from the remains of pigment sacs in the preserved feathers. These could be compared to those of modern birds. The pigment is melanin, the same pigment that gives humans their skin color.

the universe onto a calendar year with major events placed in their relative positions.

The evolving earth

We have noted direct environmental changes that evolving species must adapt to or perish. Most, of course, eventually perish. And then there are those punctuations in uniformitarian evolution—many local events such as floods, volcanoes, and earthquakes, as well as those five global catastrophes—that profoundly reorganized the earth's life. But there is another source of environmental change, one that is both uniformitarian in its slow and imperceptible nature and global in its reach. This is continental drift, the movement of land masses over time, which operates through the process of plate tectonics.

Simply put, the crust of the earth is not a solid shell. It is divided into some 16 plates of various sizes, which fit together as in a giant, spherical jigsaw puzzle. The molten rock below the crust is constantly in motion and interacts with the plates, pushing them apart in some locations, pushing them together in others, sliding one plate under another in still other places. As a result, the continents (where the crust protrudes above sea level) change shape and position. The motion itself is slow—an average of two centimeters per year (about the speed at which your fingernails grow)—but over tens or hundreds of millions of years, imagine how these changes can affect global and local environments and, thus, the evolutionary histories of life on earth.

OPPOSITE The history of the universe reduced to a single year.

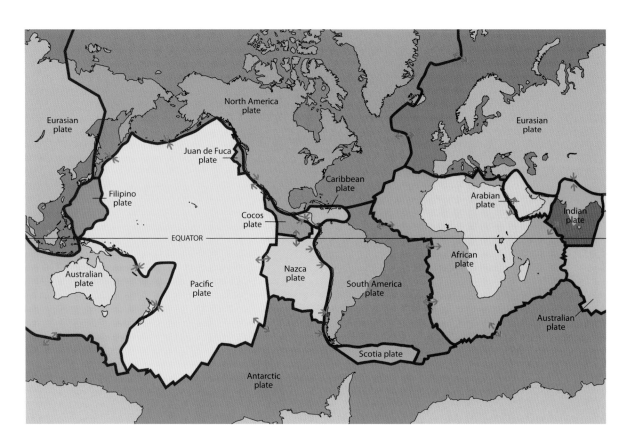

The major plates that make up the present crust of the earth.

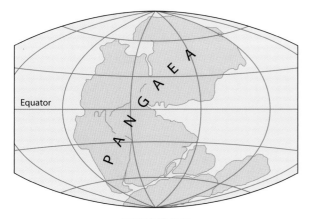

PERMIAN
225 MILLION YEARS AGO

TRIASSIC
200 MILLION YEARS AGO

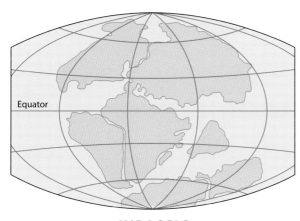

JURASSIC
150 MILLION YEARS AGO

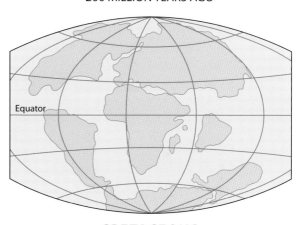

CRETACEOUS
65 MILLION YEARS AGO

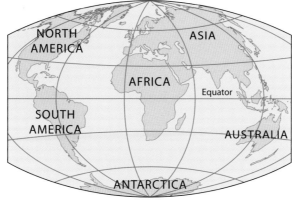

PRESENT DAY

The position and shape of the continents through time.

Not to mention how it affects our task of piecing together those histories. There are dinosaur fossils found in what is now the cold desert of Antarctica. Some of the rock strata under my feet as I write this in Connecticut end just to my east and pick up again in what is now Morocco.

So, we must see the pageant of life in a multidimensional way, with lines of influence running in every direction spatially and temporally; a complex and asymmetrical cobweb of relationships, with generalizations that define our theory of evolution but with details idiosyncratic to each case. Let me leave you with three more things to ponder.

Philosophical considerations

The pageant of life is not linear or progressive. Many visual depictions of evolution, and many accounts of it (including mine), are necessarily narrative in their format, telling a story of change through time as if there were a relatively simple and linear progression. We used to divide life's history into "ages": The Age of Reptiles, The Age of Mammals, The Age of Man.

But these ages, in fact, overlap. Mammals have been around almost as long as dinosaurs and didn't just suddenly appear when the dinosaurs met their demise. Furthermore, all forms of life are always evolving. The bacteria around today are not all just relics of the past. They have evolved too, as have geraniums and jellyfish and everything else. As we noted above, the overall tree of life is one of increasing diversity, since diversity gives rise to further diversity. Even the oldest forms of life are still represented in today's biota. Indeed, if there is any "age" of some form of life on earth, it is the "Age of Bacteria," and in a way it's the only "age," because all life relies on bacteria and bacteria are the most common life form on this planet.

"There is no new thing under the sun" (*Ecclesiastes* 1:9). We speak of origins in evolution: the origin of the earth, the origin of life, the origin of humans. But the only real origin is that of the universe (and *that* we don't fully understand). Since the universe began—with the Big Bang some 13.7 billion years ago—all subsequent events have been *rearrangements* of what already existed: matter condensed from

A narrative visual of the history of life. Such depictions note the first appearances of new forms of life but don't—almost necessarily because of the complexity—indicate that the history of life is an incredibly complex tree with many twigs and branches, not all becoming extinct when a new form evolves.

energy as it cooled; large atomic particles formed from smaller ones; stars came together from cosmic dust; heavy elements were created from lighter ones in the nuclear furnaces of stars; simple molecules were rearranged to form the organic molecules of life; and the genetic code was constantly shuffled to produce the extraordinary multitude of living things on this planet. When Carl Sagan said we are all "star stuff," he was being literal.

History is contingent. Nothing in history, whether it is your personal history, the cultural history of humans, or the evolutionary history of life, is inevitable. Each event is contingent upon all the events that came before. If one event had not happened as it had, the course of subsequent history would be changed. So, history makes sense when we work backward, but it can't be predicted looking forward. Imagine if that asteroid had missed the earth 65 million years ago; or if the continents had drifted in other directions; or if some other chemistry

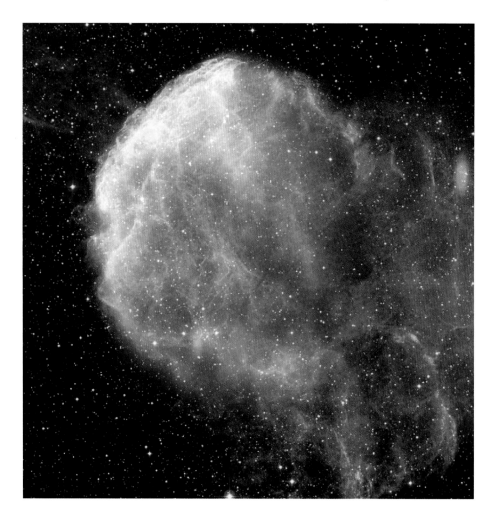

Supernova remnant IC 443 in a composite image from different data. It formed some 8000 years ago and is around 5000 light years (29 quadrillion miles) from earth in the constellation of Gemini. Supernovae are massive stars that have exploded as they collapse under their own gravity, spewing their heavy elements out into the rest of the galaxy. These supernovae winds eventually collide with primordial hydrogen clouds and trigger the formation of new second generation stars.

An imaginative depiction of a *Tyrannosaurus rex* watching the asteroid impact. Imagine if the asteroid had missed the earth.

had been behind the self-replicating system we call life. Things would have been different. But how, we can't know.

Or, what if some events millions of years ago in East Africa had been different? We humans might not be as we are, or might not be here at all. We'll take up the topic of our own evolution next, as an example of all these concepts we've been discussing.

Chapter 7

What do we know about human evolution?

To many people, this is the most controversial aspect of evolution, but the human story is, in fact, one of the best examples of evolutionary theory in action.

"Light will be thrown on the origin of man and his history," wrote Darwin almost at the end of *Origin of Species*. And that's all he published on the subject for a dozen years. Despite his confidence in his model of evolution by natural selection, he couldn't quite take that obvious next step and apply it to humans. Darwin always

PREVIOUS PAGE Olduvai Gorge, Tanzania, one of the most famous and productive of fossil human sites.

avoided controversy and he was sensitive to his wife Emma's concern about what she saw as the implication of his theory.

And yet, by 1871, the theory *and* its implications for us were generally accepted, so his next big book was titled *The Descent of Man and Selection in Relation to Sex*. The end of the 19th century saw the beginning of the search for the evidence of our evolutionary story in the fossil record. And now we know so much about our evolution that it does, indeed, provide some of the best concrete evidence for the reality of the theory.

Let's look at what we know, what we don't know, and what is still being debated.

Our place in nature

We humans are primates, one of about 17 major types of existing mammals. There are several hundred living species of primates (estimates vary)—monkeys, apes, and some creatures that look squirrel-like at first glance but are, in fact, our close relatives.

The primates are arboreal (tree-dwelling) mammals with visual, manual, and mental dexterity. The traits that facilitate these behaviors are stereoscopic (three-dimensional) vision (and various degrees of color vision); grasping hands (and in most primates, grasping feet)

Emma Wedgwood Darwin in 1840 by George Richmond. Emma wrote to Charles, with reference to his theory, "Every thing that concerns you concerns me & I should be most unhappy if I thought we did not belong to each other forever." (That is, would not meet in Heaven.) To which Charles replied, "When I am dead, know that many times, I have kissed & cryed over this. C. D."

The grasping hands of the primates, a human and an orangutan.

Dr Kenneth Oakley and Mr L. E. Parsons (right) prepare the Piltdown fossil for fluorine tests in 1949. It was such tests that showed, by 1953, that the fossil was a hoax. "Discovered" in 1912, the fossil was accepted for so long because it conformed to expectations of the day that it was the big brain that initiated human evolution. It appeared to have an ape-like jaw with a very human brain case. It was found to be literally the cranial bones of a modern human paired with the jaw of an orangutan, filed and painted to look the same age. To this day the identity of the perpetrator is uncertain.

and opposable thumbs; and large complex brains. Primates normally have single births with a long period of infant dependency and, thus, of direct and extensive infant care; and complex social organization based on differential relationships among individuals.

The human primate has its own unique expression of this combination of traits (as do, of course, all species of primates). We have the greatest degree of manual dexterity with our long thumbs, precision grip, and individual flexibility and movement of the fingers. We don't, of course, have prehensile feet because we are bipedal (walk on two feet), unlike the other primates who are quadrupeds. We are not arboreal. Our brains are three times the size one would expect of a primate of our body mass. We mature more slowly than even other primates of our size. And our social units are structured and maintained by cultural values—ideas, rules, and behavioral norms that we have created and that we share through a symbolic communication system.

These distinctions between the human primate and other primates are important to remember as we look into the story of our species' evolution.

The evolution of the human lineage

Based on our system of classifying living organisms, we and our direct ancestors are members of Family Hominidae, the hominids. How do we define the hominids in ways that would be displayed in the fossil record?

At the turn of the 20th century it was thought that our defining trait and, thus, our *first* trait, was our big brain.

Three endangered primates: LEFT a black lemur from Madagascar, a prosimian; BELOW a red shanked douc langur from Southeast Asia, a monkey; OPPOSITE a western lowland gorilla from Central Africa, a great ape.

But as more fossil evidence accumulated it became clear that the adaptation that kicked off our evolution was bipedalism. Our big brains were evolutionary afterthoughts. So when we've looked into the fossil record for our antecedents, we've searched for what could best be described as bipedal apes.

The very beginnings of our evolutionary line are still a mystery, with but a few tantalizing bits of evidence. We infer from the genetic comparison of ourselves with our very closest relatives, the chimps and bonobos, that our common ancestor lived somewhere between seven and five million years ago.

But the earliest fully agreed-upon hominids date to a little over four mya, and they come in two types, usually distinguished at the level of the genus. (We are *Homo sapiens*. *Homo* is our genus name. A genus may contain multiple species.)

Genus *Australopithecus* ("southern ape" even though we consider them hominids) is probably our most immediate early ancestor. Members of this genus lived in eastern and southern Africa from about 4.2 mya to a little under two mya.

They were small, maybe an average of five feet tall and 100 pounds. They had the brain size of chimps, averaging about 440 ml. (Brain sizes are measured in milliliters (ml) or cubic centimeters (cm^3). Modern human brains range from 1000 ml to 2000 ml.) Their skulls, faces, and bodies were quite apelike, with one exception: They walked bipedally.

We know this from two important body parts, both preserved in the fossil record often enough for us to make the assessment. First, the hole in the base

The fossil of "Lucy," the most famous member of genus *Australopithecus,* found in 1974 and dated at 3.2 mya. Her skeleton— nearly 40% complete and with almost every part well represented—showed definitively the antiquity of bipedalism.

Reconstruction of *Australopithecus* couple who made the famous Laetoli footprints, verifying that our early ancestors walked upright.

The famous Laetoli footprints found in 1978 and dated to about 3.7 mya. They were left in fresh volcanic ash that then hardened. They show a very human shaped foot and a human gait. They were probably left by two adult members of Lucy's species.

of the skull, the *foramen magnum* ("big hole" in Latin), from which the spinal cord extends from the brain and around which the vertebral column sits, is underneath the skull, not in the back as in a quadruped. Second, we have bones of the pelvises and they are far more like ours than they are like a chimp's. We even have 3.7-million-year-old footprints preserved in hardened volcanic ash showing a stride much like our own.

And yet, their upper bodies are more like those of an arboreal ape—heavy shoulder and arm muscles, relatively long arms, curved finger (and toe) bones. They had, it seems, a mixed adaptation to a mixed environment: arboreal ability to exploit the trees and bipedalism for moving efficiently on the ground while carrying food and infants at the same time. The fossils of these australopithecines, especially the early ones, are found in areas that were a mixture of forest and open space at the time.

About three mya there was a rapid climatic change in Africa, shrinking the forests and expanding the dry savannas. This alteration seems to have triggered three major changes in hominid evolution.

Open acacia woodland in Tanzania. It may have been in just such a mixed forest–open space environment that the hominid family first evolved.

Cast of a skull of *Paranthropus* from Tanzania on left, with a skull of almost contemporaneous early *Homo* on right, from the same general site. Both may be responses to a climatic change. Note, on *Paranthropus*, the huge teeth, heavy cheekbones (chewing muscle attachment areas), massive upper jaw, heavy brow ridges (possibly anatomical shock-absorbers for chewing), and ridge along the top of the skull for the attachment of the major muscle of chewing. These two individuals may have had similar sized bodies but the head of *Paranthropus* is much larger.

Australopithecus disappears from the fossil record in about a million years, but not before giving rise to two new hominid forms. One is us, genus *Homo*, which we'll consider in the next section. The second was *Paranthropus* ("near human"). In brain and body they were much like *Australopithecus*, but their heads had adaptations such as large teeth and muscles for more powerful chewing. *Paranthropus* seems to have lived in more open areas where, at least part of the year, they had to rely on the tougher sorts of foods, such as hard-shelled nuts and underground tubers, found on the open plains. This genus also is found in eastern and southern Africa and lived from about 2.8 mya to perhaps as recently as one mya.

Prior to *Australopithecus* and *Paranthropus* the hominid fossil record is more vague and contentious. *Ardipithecus* ("ground ape"), from more than five mya to 4.5 mya, would seem to be a good candidate for an early hominid. It is even more apelike than the australopithecines but is still pretty clearly a biped.

Candidates for even earlier hominids are still under debate. In time, they range from 5.6 mya to seven mya and were found in Kenya and Chad. They are claimed by their discoverers to be hominids because they show signs of bipedalism. Others disagree on the classification or even if they were bipeds.

So, if we must "connect the dots" and draw a tentative family tree, this one seems safe.

But to include the other forms is at present premature. It has been speculated that some could have been literally bipedal apes. If bipedalism was such a good adaptation that it started our evolution, perhaps similar lines also evolved in that mixed environment, but only our bipedal primate line persisted beyond about one million years ago. We do, after all, see some bipedal ability in our close relatives, the chimps and bonobos.

A simple and reasonably acceptable family tree of the hominid genera.

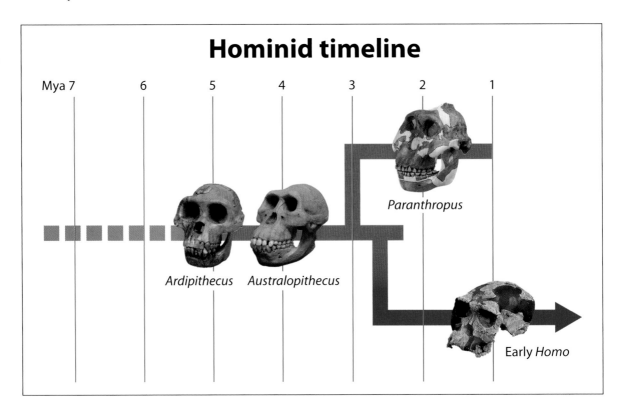

Hominid timeline

Mya 7   6   5   4   3   2   1

Paranthropus

Ardipithecus   *Australopithecus*

Early *Homo*

The evolution of genus *Homo*

Oldowan tools from Olduvai Gorge, Tanzania, some nearly 2 mya. These are core tools, made by taking off a few flakes to make a large, all-purpose device. The same tool makers also made flake tools, flakes taken off cores and themselves used as tools for more specialized purposes, such as taking flesh off bone or taking apart an animal carcass at the joints.

Among the early hominids there was certainly variation in brain size, and in one line this became adaptively significant. About 2.5 mya we find the first evidence of an increase in brain size as well as evidence for what those slightly bigger brains could do.

From eastern Africa come skulls with brains of 680 ml average along with the first stone tools. Certainly our earlier ancestors were making tools of bone and wood and using tools of stone.

Chimps also make tools of wood and use stones for tools. But *modifying* a stone requires a leap of mental imagination and of manual dexterity.

Why were those bigger brains and their behavioral and technological products adaptively advantageous? In the face of increasingly dry conditions, perhaps they helped our ancestors to exploit a food source on the plains more reliable than plants: the huge herds of grass-eating animals. But sitting down to feast on a dead animal (we were no doubt just scavengers early on and not big-game hunters) would almost certainly assure that *you* would be the main course in the feast of a lion or other predator. Better to quickly cut off what's left of the carcass and take the pieces back to a safe location. That's where those stone tools come in. The fossil record shows evidence of animal bones—exactly

The dangers and rigors of life on the savannas were certainly selective factors in the evolution of genus *Homo*.

Timeline of Genus *Homo*

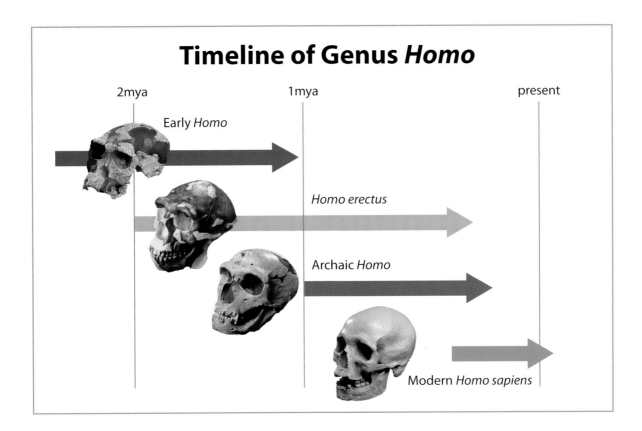

2mya   1mya   present

Early *Homo*

Homo erectus

Archaic *Homo*

Modern *Homo sapiens*

Major groups of the fossils of genus *Homo*. The lines are not connected into a family tree because there is still no agreement on how the groups are related. Some recognize a complex tree with as many as ten species of our genus. Others feel there are no distinctions at the species level and suggest a single species, *Homo sapiens*, all the way back to two million years ago. The debate continues.

those that would have been left after the predators and scavengers were done—with cut marks made from stone tools.

Moreover, there was an interesting feedback loop in operation. Bigger brains helped make the exploitation of animals possible. But bigger brains require more energy. Meat, however, puts energy into the body faster and more efficiently than plant material, so adding meat to the diet made possible even bigger brains. An evolutionary trend was in place.

The very earliest fossils included in our genus have an appearance that is transitional between *Australopithecus* and what would become us, so much so that there is debate as to how best to classify them. But by two million years ago we find fossils that are inarguably us—members of *Homo*.

While they still had jutting faces and receding chins, heavy brow ridges, and sloping foreheads, they had even bigger brains, modern body proportions, and made more sophisticated stone tools. They may even have had a refined form of bipedalism that allowed for "endurance running," that is, for a bouncing, jogging gait. This would be a useful adaptation on the open plains with which these ancestors were associated.

And here is where the biggest remaining debate in human evolution comes in. It seems we can't agree on just how many species of genus *Homo* there have been. Some anthropologists think that each group of fossils represents a different species, with *Homo sapiens* being a distinct newcomer, evolving about 200,000 years ago and replacing earlier species. Evidence for this is the often physically, geographically, and temporally distinct nature of each fossil group. Others feel that there is no firm evidence for species distinctions, and see all members of genus *Homo* for the last two million years as members of the same species, with change across space and through time around the *Homo* theme. (For the record, I belong to this latter group.)

The details of the debate, however, are beyond our scope here, and the ultimate test of ability to produce fertile offspring is impossible to conduct. So, as we did with the earlier hominids, we will discuss our genus at that level.

We may define genus *Homo* by the following characteristics:

- Enlargement of the brain
- Flattening of the face with a decrease in the size of jaws and teeth
- Refined bipedalism with the added ability for endurance running
- Modern body proportions (notably relatively shorter arms)
- Increased reliance on culture rather than biological change as a means of adaptation
- Increasing control of and influence over the environment
- Migration eventually to all habitable areas of the planet

The last two million years of our evolution have seen change, but it has been change involving these basic features.

All evidence points to the evolution of the first members of our genus as having taken place in Africa. The mix of adaptations proved very successful, so much so that in a few tens of thousands of years our genus had to spread out beyond Africa—no doubt in search of food, water, and shelter, and we find their fossils in the Republic of Georgia, China, and Java. Shortly after—"shortly" keeping in mind the time scale—*Homo* shows up in Western Asia and Europe as well.

A map of important sites in the hominid fossil record, showing general locations. Note that the earliest sites are concentrated in eastern and southern Africa. The earliest sites for *Homo erectus* are African; that group spread from there. Archaic *Homo* is also found all over the Eastern Hemisphere. The earliest sites for modern *Homo sapiens* are African. Debate continues, however, about whether the groups of *Homo* are separate species or one species changing through time and across space, with its species identity maintained by extensive gene flow.

This initial dispersal occurred during the phase commonly referred to as *Homo erectus* (a reference to the fact that when the first of these fossils were found, around the turn of the 20th century in Java, it was thought that they must represent the earliest "erect" or bipedal ancestors). At this point we still had smaller brains on average than today, a mean of about 980 ml (but with a maximum of about 1200 ml, within the modern human range). What this means for intelligence is difficult to determine. What to us are crude stone tools were, for their time,

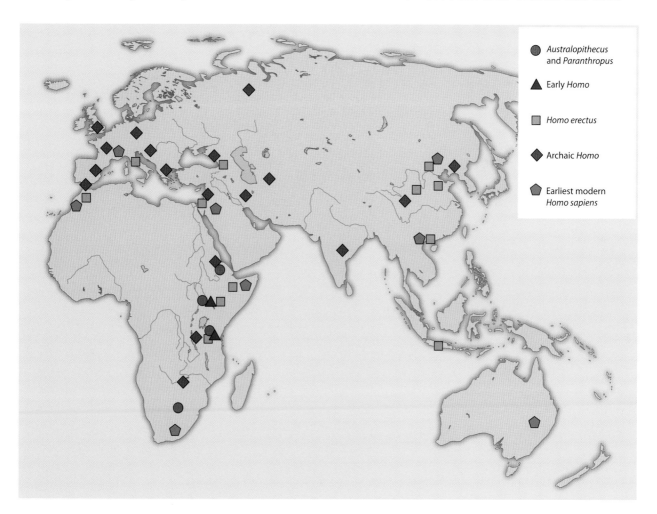

● Australopithecus and *Paranthropus*

▲ Early *Homo*

■ *Homo erectus*

◆ Archaic *Homo*

⬟ Earliest modern *Homo sapiens*

brilliant inventions. One can't expect *Homo erectus* to walk out of his cave, throw away his stone axe and invent a tablet computer. Culture evolves as well.

Our genus spread all over the Eastern Hemisphere, and about 780,000 years ago, we find the first evidence of brains of a modern human range from a site in Spain. By this time stone tools were taking on regional differences, fire had been tamed, and it appears big-game hunting (as opposed to small game and scavenging) provided a major food source.

And now, as the story gets more complex in its details, it becomes easier to summarize. Humans continued to spread out and to elaborate on and improve technology. Even the Neandertals of Ice Age Europe—so often the epitome of the "primitive cave man"—may have been the first to attach stone points to wooden shafts, and they buried their dead in some locations and took care of the ill and injured—clearly recognizable "human" traits. And art soon appears, in the form of the famous cave paintings from Europe, but also in stone tools so finely made they could have none but esthetic functions.

Perhaps 30,000 years ago, some wandering ancestors crossed a land bridge between Siberia and Alaska (exposed when sea levels dropped during glacial advances) and began to populate the Western Hemisphere. The islands of the Pacific followed. As we moved around, adapting culturally and mixing our genes, archaic physical traits began to disappear and humans began to look like our regional populations look today, with some pattern of differences but with an amazing degree of genetic homogeneity. The result is a species of seven billion, but

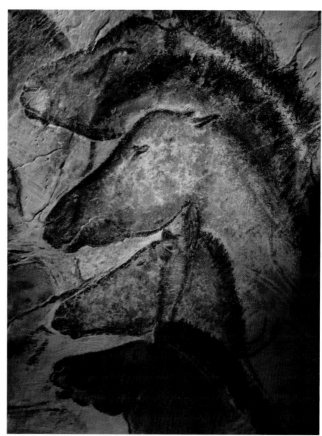

OPPOSITE A Neandertal burial from Kebara Cave, Israel, perhaps 60,000 years old.
ABOVE The Venus of Willendorf, possibly a fertility figurine from Austria, 24,000 years old. There were many such figurines from this time from Europe.
ABOVE RIGHT One of the earliest examples of cave art, some simple yet very realistic horses from Chauvet Cave in southern France, dated to a little over 30,000 years ago.
RIGHT Stone Age blade fragments found at the Blombos cave site in South Africa where beads and bone tools were also discovered and thought to date from around 75,000 and 80,000 years ago.

with no biologically distinct subspecies or racial groups. Race, and the other categories into which we divide ourselves, are sociocultural constructs.

Thus, despite some ongoing and intriguing debates, the story of human evolution is generally well understood, and provides us with perfect examples of natural selection and adaptation, speciation, gene flow, and aspects of the overall pattern of the family tree of life.

Despite a wide range of phenotypic variation in our species, as well as some pattern to its geographic distribution, meaningful biological categories of race (that is, subspecies) do not now exist. Phenotypic—as well as genetic—variation grades gradually across space such that no clear-cut dividing lines are possible. Race is a sociocultural construct.

Who are the "Hobbits" of Indonesia?

In 2004 an astonishing find was announced, from the Indonesian island of Flores—a partial human skeleton, dated to 18,000 years ago, of an adult female who stood only 106 cm tall (3'5") and had an estimated brain size of 380 ml, about the stature and brain size of *Australopithecus*. And yet, her physical features seem to clearly assign her to our genus, *Homo*, and she lived during a time when only modern humans populated the world. She was given the formal name *Homo floresiensis*.

Since her discovery, more specimens have been found, including arm, wrist, and foot bones of the original skeleton, a mandible from a second individual, and assorted bones of possibly seven other individuals. Dates range from possibly 90,000 years ago to 12,000 years ago. What *were* these strange looking people?

Hypotheses vary. Some claim they were modern *Homo sapiens* or a closely related species who responded to an evolutionary phenomenon called insular dwarfing, where animals on islands or even insulated environments inland with limited space and resources undergo selection for smaller size. We see this in a species of elephant on Flores and in animals and even humans in the dense rainforests of central Africa.

Others claim the Hobbits (named after J. R. R. Tolkien's characters) suffered from microcephaly, a genetic form of dwarfism. Still others even claim that *Australopithecus* might have left Africa and made it all the way to Southeast Asia, where they evolved in isolation on some of the islands.

Debate continues and many of us, pending further evidence, remain agnostic. For the record, though, I lean toward the insular dwarfing model. But I could turn out to be wrong—with the very next fossil find—and this is what makes science so exciting and such an important tool.

Homo floresiensis (left) with a skull of modern *Homo sapiens*.

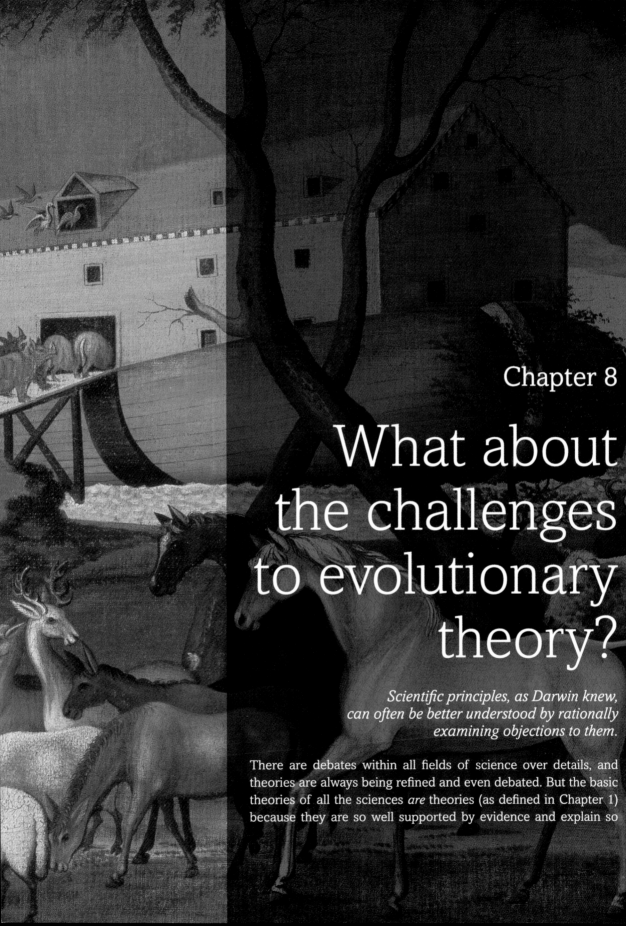

Chapter 8

What about the challenges to evolutionary theory?

Scientific principles, as Darwin knew, can often be better understood by rationally examining objections to them.

There are debates within all fields of science over details, and theories are always being refined and even debated. But the basic theories of all the sciences *are* theories (as defined in Chapter 1) because they are so well supported by evidence and explain so

PREVIOUS PAGE An 1846 painting by American artist Edward
Hicks (1780–1849) of the pairs of animals boarding Noah's Ark.
The reality of the Ark and the Flood is a crucial aspect of one
of the challenges to the theory of evolution.

many other observations that we consider them to be facts. No objections to the theory of gravity are raised, nor to the theory of plate tectonics, or (even if most of us don't understand them) to the two theories of relativity. What is it about evolutionary theory that stirs things up?

The theory of evolution is every bit as well supported and, thus, as factual as the other theories that inform our sciences, but it continues to be questioned by a doubting minority clever enough to make arguments that tap into some of the public's deeper concerns, questions, and misunderstandings.

What are these arguments and how does science address them? Why is evolution such a contentious topic? What is this minority view?

Pseudosciences

Galileo is famous for writing that science tells us "how Heaven goes" and religion tells us "how to go to Heaven" (although he was quoting someone else), the idea being the separate but necessary nature of both realms of knowledge. And yet, there is a popular conception that those two realms are eternally and inevitably at odds. In reality, religion and science can quite easily coexist; belief in one does not rule out belief in the other. So where does this misconception of an inevitable conflict come from? It comes largely from claims about the real world that fail all empirical tests, but that are accepted as matters of unquestioned faith. These are, thus, pseudosciences (literally, false sciences). A good example is the still extant Flat Earth Society, devoted to the idea that the earth is not a sphere but is, in fact, a flat disk.

So much irrefutable evidence exists for a spherical earth that it is as factual as a fact can get. But flat earthers adamantly insist on their point of view and incorporate all manner of arguments in support of it that *sound* scientific. By their own admission, however, the idea is based on literal interpretations of biblical passages. In Matthew 4:8, for example, it says "Again, the devil taketh him [Jesus] up into an exceeding high mountain, and sheweth him all the kingdoms of the world …" How, the flat earthers ask, could Jesus see all the kingdoms of the world unless the earth was flat?

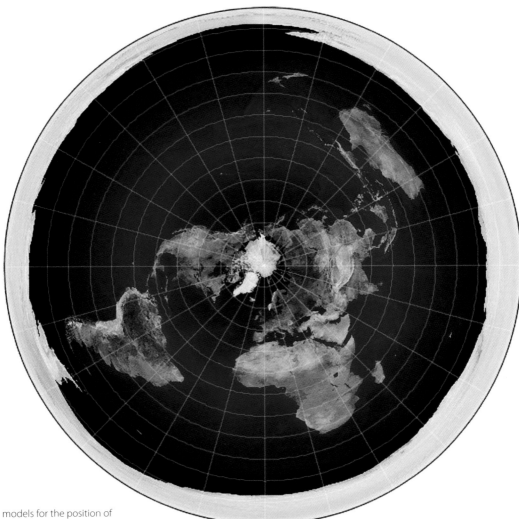

One of several models for the position of the continents on the earth within a Flat Earth theory. This one posits Antarctica as an "Ice Wall" around the edge of a disk-shaped planet. A serious problem with this model is that sailing the perimeter of the Ice Wall would take about three times as long as sailing around the real Antarctica.

The problem here is that some might well feel they have to make a choice—between clear scientific evidence and the chronicles of two major religions. While it is doubtful that many lose sleep over the matter of a flat earth, what about an issue that is more complex and for which the evidence is a bit harder to understand? What about evolution?

Scientific creationism

Creationism in general is simply the idea that a creator of some sort had something to do with the origin and history of the universe and everything in it. Roman Catholicism, for instance, officially feels that the deity it recognizes created everything, but that "everything"

Although there are scientific creationists worldwide it is a particularly United States-based phenomenon, perhaps initiated by the famous Scopes "Monkey" trial in Dayton, Tennessee, in 1925. Teacher John Scopes agreed to purposely violate a state law outlawing the teaching of evolution as a test case. The situation escalated into a national issue with the famous Clarence Darrow (left) for the defense and the equally famous William Jennings Bryan for the prosecution. The trial drew attention to a conflict that hadn't really existed before and evolution began to disappear from textbooks.

includes the processes of evolution we've been discussing. The Creator intervened again when humans came along, giving them a soul not possessed by other creatures. The two realms are otherwise kept separate, as Galileo would have had it.

Scientific creationism, however, pits the two realms of knowledge against one another. There are subtly different flavors of this idea but four claims are fairly common:

1 The universe and everything in it were created by "special powers."
2 The creation took place over a short period of time about 10,000 years ago.
3 Since the creation, no new "kinds" of organisms have appeared.
4 The geological and fossil records are the results of a "primeval watery cataclysm" or "a worldwide catastrophe of a hydraulic nature."

This "model" is derived from one literal interpretation of the first eight chapters of the book of Genesis and is in direct opposition to the well-supported scientific account of earth's history. Let's look at each point.

The invocation of "special powers" can be seen as beyond the realm of science. There can be no scientific evidence for or against

The site of Göbekli Tepe in southeastern Turkey consists of pillars decorated with carved reliefs linked by stone walls to make up perhaps as many as 16 buildings. It was built about 11,500 years ago, over a millennium before the creation of the whole universe as proposed by the scientific creationists.

the presence of our common conception of a deity with special powers. At the same time, there are well-established scientific models for how the universe and its components arose. So while science cannot deny the existence of a Creator, neither can the existence of rational explanations be ignored in the name of a Creator.

Sound scientific evidence—usually from multiple sources and technologies—has established the age of the universe at 13–14 billion years, the age of the earth at 4.5 billion, and life on earth somewhere around 3.6 billion. We also know that by 10,000 years ago sheep, dogs, wheat, rice, and goats had been domesticated, pottery had been invented, the glaciers of the Ice Ages were receding for the final time, and large farming villages were beginning to be built in Mesopotamia. And a lot had happened before all *that*.

How do we know how old things are?

There are two general methods that are used to arrive at the dates we sometimes so casually mention—absolute dating and relative dating. The basic techniques of absolute dating provide a specific date for an item, be it a fossil, a rock, or a human artifact. There are perhaps a dozen such techniques, based on physical properties of elements and atomic particles. The two best known are radiocarbon (14C) and potassium/argon (K/Ar). These work by measuring the decay of radioactive (unstable) isotopes of carbon or potassium into stable forms, nitrogen and argon respectively. The premise is that if you know the rate at which some natural process occurs, and if you know how much of that process has already occurred in a given situation, you can calculate the time at which the process started.

Radiocarbon is based on the fact that living things exchange carbon with the atmosphere at a fairly steady rate. Most of the carbon in the bodies of living things is the stable form, 12C. But a small percentage is unstable, 14C. When an organism dies, the 14C decays at a known rate into nitrogen. If we measure the amount of 14C in a bone, a piece of charcoal (formerly a tree), or a bit of cloth made from plant material, we know when the organism that gave rise to that object died, which is when it stopped exchanging carbon with the environment around it. Radiocarbon dating, then, only works on organic material, and only up to about 60,000 years ago; beyond that, there is not enough 14C remaining to detect accurately.

For materials older than 60,000 years, we need radioactive isotopes that decay more slowly. Radioactive potassium decays very slowly and its resultant product, argon gas, escapes into the atmosphere, unless it is trapped somehow. This trapping occurs in rock of volcanic origin, because when the molten rock in lava hardens, the decay process begins, but the argon can't escape from the lava rock. With K/Ar dating, we can date rocks (but only volcanic rocks) back to the very beginning of our planet.

Relative dating —so called because it provides dates based on comparisons— comes in two forms. The simplest is based on stratigraphy, the fact that deeper layers in the earth's surface are older. Thus, we can assume that a fossil or a rock in a lower layer is older than one in the layer just above it.

Relative dating is often used in conjunction with absolute dating. Suppose we have a fossil which itself cannot be dated. But suppose it is found in a stratigraphic layer between two layers of volcanic rock that *can* be dated. The fossil is then, logically, at an age between the dates of the volcanic layers. The hominid fossil Lucy was dated in this way. The petrified bones could not be dated directly, but they were found between volcanic layers dated at 3.4 and 3.0 million years old. The fossils were thus 3.2 million years old.

In many cases where we give dates for fossil or archaeological finds, more than one dating technique can be applied, so the veracity of the calculated date is supported.

The scientific creationists use "kind" instead of any formal category such as "species." What they mean by "kind" is somewhere between the broad categories from Genesis—"fish of the sea," "fowl of the air," "cattle"—and individual species. The idea is that there can be change *within* kinds, such as that brought about by human breeding or even natural selection, but that no *new* kinds have appeared. In other words, there are the same number of kinds now that the Creator created in the beginning. The issue, then, is not with microevolution but with macroevolution, the origin of species.

We now have more than ample evidence, as we've been discussing all along, that species *do* give rise to new species and that this process, over eons of time, *does* give rise to whole new "kinds" of living organisms.

The "cataclysm" or "catastrophe" the scientific creationists describe is a flood, and not just any flood. It's the flood for which

The numerous strata, with their fossils, in the sometimes mile-deep Grand Canyon of Arizona. The scientific creationists would have these strata laid down over a short period by a single worldwide flood.

Noah built his ark (Genesis: Chapters 7 and 8), a theme derived directly from the *Enûma Eliš*, the Babylonian creation story, as was much of the early part of Genesis. While some might say that the mention of a great flood in both the Bible and in the *Enûma Eliš* might indicate that such an event occurred, simply put, there is no evidence whatever for such a worldwide flood. Strata and their fossils represent eons of uniformitarian (and occasionally abrupt) environmental and biological change. They are not the results of a single worldwide catastrophe.

Scientific creationism is a pseudoscience, a set of testable ideas that even in the face of overwhelming contrary evidence are taken on faith, because they are seen as supporting tenets of a particular belief system.

Intelligent design

As with creationism, intelligent design (ID) has different variants, but basically those who believe in intelligent design accept that the earth is old, that natural selection is undeniable, and that it is likely that species do evolve from different species. However, they take the view that some living systems are too complex to have evolved through the processes put forward in evolutionary theory and, therefore, there must be an "intelligent designer." The human eye and even life itself are cited as examples of complex systems that they suggest must have been "designed."

ID proponents are careful not to identify the intelligent designer. One even claimed it *could* be extraterrestrials that came to earth, designed life, and left! But to a Western audience, the identity of the intelligent designer referred to is fairly obvious. While ID may be more sophisticated and seemingly more "rational," it is still actually a form of scientific creationism. There are three overlapping arguments that are put forward as support for intelligent design. They are:

1 The statistical odds against the evolution of something as complex as even a simple cell.
2 The concept of irreducible complexity, that a complex system will not function if even one part is removed; thus, complex

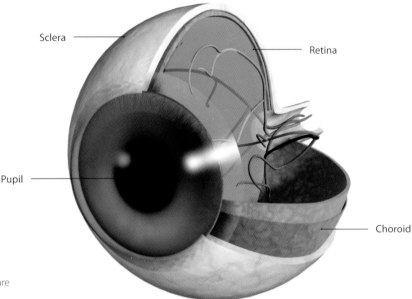

Basic anatomy of a complex eye (compare with the series of eyes on page 135).

Sclera

Retina

Pupil

Choroid

systems could not have evolved in a series of steps. As ID proponent Michael Behe puts it, "An irreducibly complex system cannot be produced directly … by slight, successive modifications of a precursor system, because any precursor to an irreducibly complex system that is missing a part is by definition non-functional."

3 That we simply recognize things that have been designed because we can see that they are too complex to have arisen by a series of chances.

Let's examine each:

The first argument proposes that the coming together of all of the chemicals that make up the genetic code, all the pieces of a single cell, or all the parts of complex eyes in precisely the way that they have is too statistically improbable for this to have happened without a designer. While this might seem to be a reasonable statement, in fact, it is a conscious twisting of statistics.

If one of my students were to come to me and say she would never come to class again because she had bought a lottery ticket and knew she was going to win, I would try to explain that the odds against her winning were too great to make that claim. But if another student told me he wouldn't be coming to class any more because

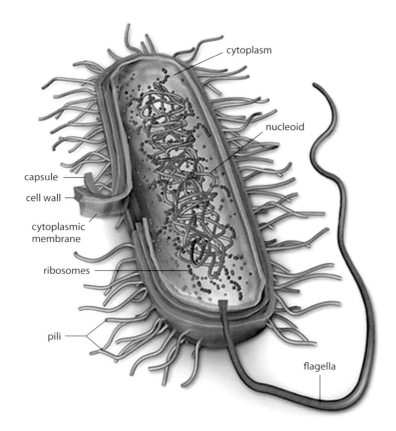

cytoplasm

nucleoid

capsule

cell wall

cytoplasmic
membrane

ribosomes

pili

flagella

Even a simple single-celled organism like a bacterium, with no nucleus or organelles (little organs), is, according to the intelligent design proponents, too complex to have arisen naturally.

he *had* won the lottery, I couldn't deny that he did on the grounds that the odds against him winning were great. The fact is, he did win. The initial improbability of something happening doesn't preclude it from happening. With genes, life, eyes, we have literally countless opportunities for various combinations of components over almost unimaginably long spans of time.

The idea that complex systems will not function if even one part is removed and that they are thus irreducible, doesn't work either, because complex systems *can be* reducible. A complex eye, for example, will still function as a light-gathering, image-producing organ even if some parts are simpler or not there. Complex eyes, in other words, *could have* evolved in a series of steps, enhancing or adding parts onto a biological system already in place and functioning. In fact, we can look at eyes of living mollusks and see, across that group of organisms, representatives of all the steps involved in going from very simple light receptor organs, such as those found in slugs, to eyes very much like our own, as found in the octopus.

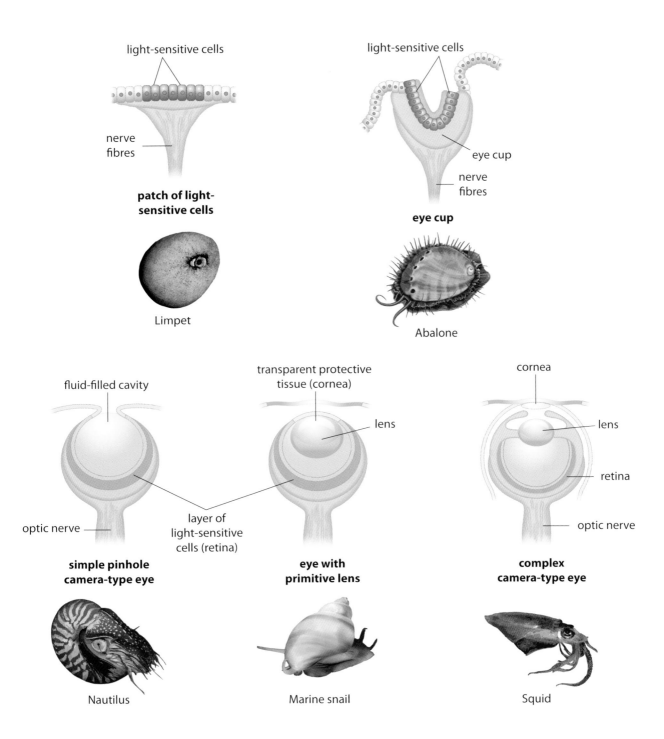

light-sensitive cells

nerve
fibres

**patch of light-
sensitive cells**

Limpet

light-sensitive cells

eye cup

nerve
fibres

eye cup

Abalone

fluid-filled cavity

optic nerve

layer of
light-sensitive
cells (retina)

**simple pinhole
camera-type eye**

Nautilus

transparent protective
tissue (cornea)

lens

**eye with
primitive lens**

Marine snail

cornea

lens

retina

optic nerve

**complex
camera-type eye**

Squid

Eye structures of living mollusks, showing possible evolutionary stages, each adaptively functional in its own way, of complex eyes.

The argument that we can simply 'recognize' that something has been designed goes back to William Paley (1743–1805) who used watches as an example.

The mechanism of a complex mechanical watch.

A timepiece is obviously designed and so requires a designer. Watches are a lot less complex than even simple living forms, thus, those too must require a designer. But the analogy is, in one important way, false, and, indeed, works to support just the opposite of what is claimed.

A particular watch is designed. That's obvious. But at no point did some designer wake up one morning, decide to invent a device that kept an account of time, and make up from scratch all the parts and their interactions that resulted in a watch. Rather, watchmakers appropriated parts and concepts already in place, many for other functions, and modified and cobbled them together in unique ways for the specific function they required. For example, the escapement mechanism of a timepiece resembles a waterwheel; perhaps the latter inspired the former.

And does this sound familiar—the appropriation of existing parts, slightly modified and put together in unique new ways resulting, after trial and error, in a new form? It's natural selection. So while *a* watch is designed, to be sure, the *idea* of various forms of watches *evolved*. The watch example, in fact, works against ID.

Finally, of course, the presence of an intelligent designer is both beyond scientific inquiry and not really necessary. We have models—testable and well supported—that account perfectly well for even complex biological systems.

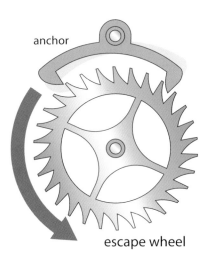

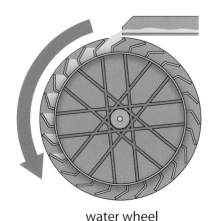

anchor

escape wheel

water wheel

The escapement in a timepiece transfers energy and controls speed. A waterwheel, similarly constructed, does the same.

For most people suspicious of the fact of evolutionary theory, the problem is simply a lack of understanding of science and, specifically, of science applied to this topic. But those who promote the ideas of scientific creationism and ID are attempting to proselytize a particular religious viewpoint, using the public's lack of knowledge of science and sense of fair play as devices.

The creationism movement

The self-proclaimed oldest creationist movement in the world, the Creation Science Movement, began in 1932 in Portsmouth, England, and exists alongside other organizations like it in the United Kingdom and elsewhere. But the movement is larger, more formally organized, better funded, and more political in the United States, where it has created several large creationist museums and has waged legal battles to get creationism or intelligent design into public school curricula.

Whatever the sociological reasons for this difference, it is also reflected in some recent polls. In the U.S. polls have shown about 45% of respondents support creationism, don't believe in evolution, or don't think humans have evolved from nonhuman species. (Questions vary but 45% as an average keeps showing up.) In contrast, one poll in the U.K. showed over 80% *not* believing in creationism or intelligent design. (Poll data, of course, are dependent upon the questions asked and the choice of respondents.)

In the U.S. there have been several court cases where creationist plaintiffs have asked for a ruling that would force the teaching of creationism or ID into public schools: all have failed. The courts have ruled that scientific creationism is religiously based. Where the creationists have had more success is in altering the treatment of evolution in science textbooks, or adding stickers in texts reminding readers that evolution is "just a theory" (a misuse of the term as we've defined it).

In cases such as this, both realms of knowledge suffer. Students being taught with such texts receive a poor science education and, at the same time, are presented with an interpretation that is held by a very limited sector of religious practice. Such conflicts have the result

A diorama in the Creation Museum in Petersburg, Kentucky, showing dinosaurs and humans living at the same time, a requirement if, as is claimed, the earth is only 10,000 years old and no evolution has taken place.

A sticker that the Cobb County Board of Education, Georgia, tried to have placed in public school biology texts. It clearly misuses the word "theory," and implies that there is no scientific consensus on the topic of evolution. After a complex court case in 2006, the board agreed not to use the stickers. The original court finding was that the sticker was in violation of the First Amendment of the U. S. Constitution that "Congress shall make no law respecting an establishment of religion."

> This textbook contains material on evolution. Evolution is a theory, not a fact, regarding the origin of living things. This material should be approached with an open mind, studied carefully, and critically considered.
>
> *Approved by*
> ***Cobb County Board of Education***
> ***Thursday, March 28, 2002***

of perpetuating the stereotype that science and belief are eternally at odds, when, in fact, they can—and should—be quite compatible. This is the danger of these pseudoscientific ideas.

What other interactions are there between the science of evolution and various aspects of society and culture? How has Darwin's great idea affected the other sciences and even the arts? We'll finish with a discussion of those questions.

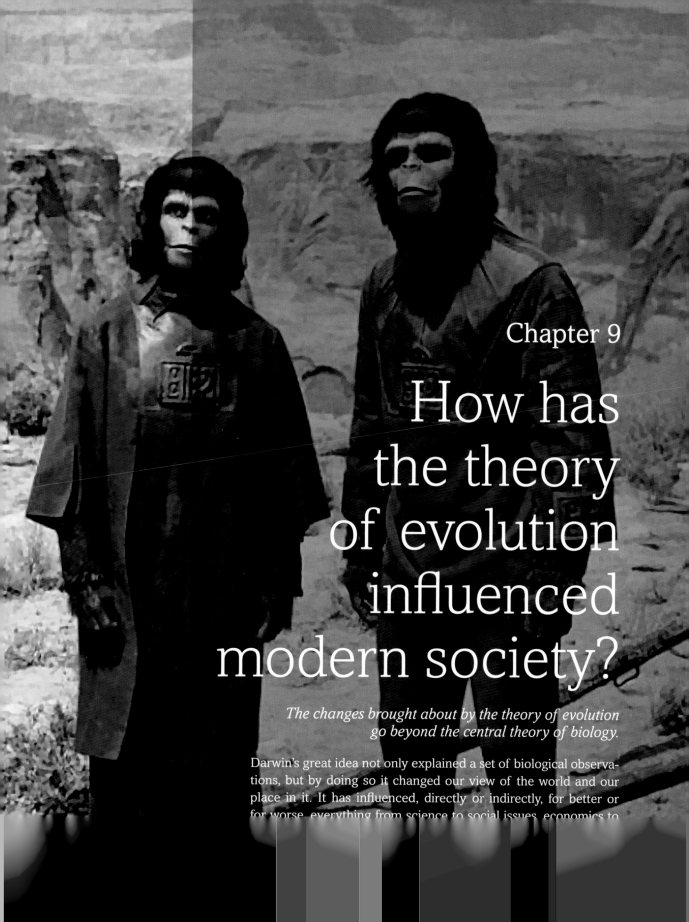

Chapter 9

How has the theory of evolution influenced modern society?

The changes brought about by the theory of evolution go beyond the central theory of biology.

Darwin's great idea not only explained a set of biological observations, but by doing so it changed our view of the world and our place in it. It has influenced, directly or indirectly, for better or for worse, everything from science to social issues, economics to philosophy, and even the arts. In addition, the theory of evolution has given us a vocabulary for discussing change and adaptation across a range of subjects.

Evolution and the sciences

It was Charles Darwin who, in his brilliant work, addressed the concerns about applying the scientific method to events of the past that could not be directly observed or experimented on. In doing so, in the words of historian Bert James Loewenberg, Darwin brought about "an epoch in the history of science and a landmark in the history of thought."

Using his expanded application of the scientific method, Darwin was even able to explain the cumbersome tails of peacocks. They are, he suggested (and this has recently been experimentally confirmed), a sign of health that is responded to by peahens in choosing mates.

The Darwin Telescope, launched by the European Space Agency to survey 1000 of the closest stars looking for small, rocky planets that could harbor life; research not possible without the concept that the universe and its contents have an evolutionary history.

Darwin showed that the method of inducing hypotheses and testing deductively could, indeed, be applied to a topic such as evolution. By amassing a monumental amount of data, he showed that natural selection and its results are, as Stephen Jay Gould noted, the "only ... conclusion about the causes and changes of life [that] can possibly coordinate all these ... various items under a common explanation." Darwin thus expanded the scientific method into all areas of the study of natural phenomena.

In addition, Darwin separated the consideration of a higher power from the study of the natural world. Just as Newton's laws of motion and gravity did not necessarily preclude a deity, neither did Darwin's evolution by natural selection. At the same time, said Darwin, his theory did not "require miraculous additions"; if it did, he said, he would consider it "rubbish." Considerations of ultimate cause and meaning were, he added, "beyond the human intellect."

So, by taking the "miraculous" out of the study of biology, and by opening up history to scientific investigation, Darwin changed the worldview that had informed and somewhat limited the sciences up to that point. Now, the universe and everything in it—once seen as a static system—could have a history of interconnecting facets at any given time and a causal sequence of events through time. This influenced both the theoretical and applied sciences.

For example, much of modern medicine would not be possible without an evolutionary perspective. Acknowledging and understanding that all living things are related, as in a huge family tree, allows us to understand diseases caused and transmitted by other organisms. Viruses are responsible for many diseases and they cause disease by commandeering the host's cell machinery, including the DNA, for their own purposes. They are able to do this because our genetic codes had a single origin in the distant past. Bacteria, that also share a

LEFT Colored scanning electron micrograph of a T-lymphocyte blood cell (green) infected with Human Immunodeficiency virus (HIV) (red), causative agent of AIDS. The HIV is another organism (although one not capable of its own reproduction) that itself can and does evolve. RIGHT A flea, carrier of many diseases that have had profound impact on humans and other species. One example is bubonic plague, which is directly caused by a bacterium and transmitted by fleas from rodents to humans.

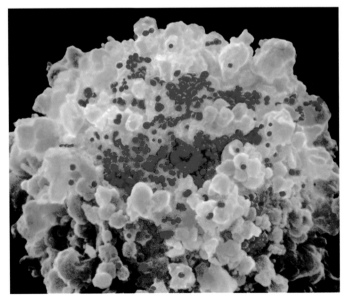

Normal red blood cells and a sickled cell (left). Sickle cell anemia is a potentially fatal genetic disease that is found in unexpectedly high frequencies in some areas of the world, particularly parts of Africa. The explanation for this is that persons with one mutant gene probably won't die from sickle cell but do have a resistance to malaria, a disease caused by a single-celled organism and transferred among humans by the bite of mosquitoes, an insect. Understanding the epidemiology of this disease would not be possible without an evolutionary perspective.

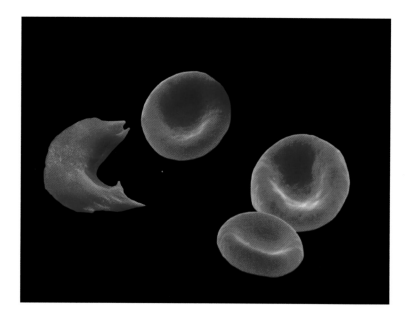

single origin with all life, are a major cause of disease as well, and are often transmitted by other organisms. Bubonic plague, a bacterial disease, is carried by fleas and transported by rats. Ticks and various insects are also disease carriers. And sometimes, as with sickle cell anemia, the interrelations among the organisms are quite complex and surprising.

Many of the drugs we use to fight disease are, of course, derived from other organisms and work because of the evolutionary relationships and similarities of the chemistry and physiology of living things. Penicillin, so important as an antibiotic, comes from a fungus. There are about 120 prescription drugs that are processed from plants. And some are from other animals.

Finally, the study of ecology is really the study of the complex and evolving adaptive relationships among organisms within ecosystems. Since the central process of evolution is adaptation, we understand ecological systems by studying how organisms have evolved to cope with the ecosystems in which they live, which include not only things such as temperature, altitude, and weather, but also other organisms, such as food sources, predators, and carriers of disease. We also understand that each organism is part of the environment to which it is adapted and so it may bring about changes to which it must in turn

re-adapt. This is known as "niche construction" and, of course, we humans are the ultimate niche constructors, consciously and radically altering the ecology of the entire planet. Without an evolutionary perspective we would see environments as static and so not understand how natural and human-induced changes come about and what their consequences might be.

The azure poison dart frog from Surinam, South America. Secretions from this frog have medical potential as muscle relaxants, heart stimulators, and appetite suppressants.

Evolution and social issues

For Darwin, an important influence on his theory of natural selection was *An Essay on the Principle of Population* by Thomas Malthus (1766–1834) who wrote that "Population, when unchecked, increases in a geometrical ratio. Subsistence increases only in an arithmetical ratio. … I can see no way by which man can escape from the weight of this law which pervades all animate nature." While Darwin did not apply natural selection to human biology in *Origin of Species*, he did apply it to humans in *The Descent of Man* (1871), and so it seems only natural that someone would try to apply the model to human social issues. But not always positively.

It was Darwin's half-cousin, Francis Galton (1822–1911), who coined the term eugenics and began the eugenics movement in Britain (which was later headed by Darwin's son, Leonard). Eugenics is the idea that certain human traits and behaviors have a genetic basis

| (a) | (b) | (c) |
| 3 COMPONENTS | 8 COMPONENTS | 7 COMPONENTS |

| (e) | COMPOSITE | (f) |
| 9 COMPONENTS | OF (e) & (f) | 5 COMPONENTS |

| (d) | (g) |
| 9 COMPONENTS | 9 COMPONENTS |

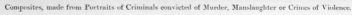

Composites, made from Portraits of Criminals convicted of Murder, Manslaughter or Crimes of Violence.

Composite portraits of criminals from a book on the life of Francis Galton. Galton developed the idea and method of combining images of the faces of a number of individuals sharing a 'characteristic' to gain an average type. Later the composite of the criminal type was taken by Galton as a failure and it was concluded that criminality was more a mental than physical trait, but Galton continued his support of eugenics.

and thus that selective breeding can improve our species in its social environment, just as natural selection acts to maintain and improve a species' adaptation to its natural environment.

Selective breeding in this context refers to preventing people with unwanted behaviors, or who are members of certain groups thought to possess unwanted behaviors, from reproducing. The motivation was to avoid the

The skull collection of Italian criminologist Cesare Lombroso in his Museum of Criminal Anthropology in Turin. Lombroso felt criminality was inborn and could be identified by physical features, especially of the head. This idea helped lay the groundwork for the more subtle but no less biologically deterministic assumptions of the eugenics movement. (Lombroso's own head is preserved in a jar in the museum!)

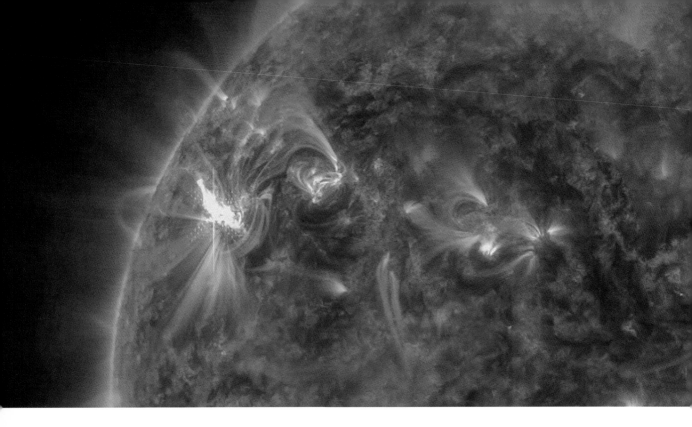

degradation of the species, the underlying assumption being that those who were demonstrably "fit"—that is, those with power, wealth, education—were genetically superior. The poor, on the other hand, were poor because they were less fit; it was suggested that they had genes that made them less intelligent and that's *why* they were poor.

In an interesting manipulation of logic, the perceived problem in some countries was that the poor (who were often equated with minority ethnic and racial groups, as well as those deemed to be mentally deficient) were outbreeding the rich and powerful. Thus, in a strictly Darwinian sense, the poor were *more* fit. Proponents of eugenics therefore thought that something had to be done about this perversion of nature, and forced sterilization programs were instituted. Some dozen countries had official sterilization programs, including Canada, Japan, Denmark, France, Iceland, Norway, and Switzerland. Between 1934 and 1975 more than 62,000 people were sterilized in Sweden. In the United States, after a Supreme Court decision that a sterilization program in the state of Virginia was constitutional, 7500 people were sterilized in that state between 1927 and 1972, and the last compulsory sterilization in the U.S. was performed in Oregon in 1981.

When the horrific behaviors of the Nazis in the name of social Darwinism were known, eugenics was rethought and, although it took time, practices in its name were gradually ended in the U.S. and

A solar eruption. In early September, 1859, just months before the publication of *Origin of Species*, a huge solar storm, observed first by British astronomers, sent charged particles toward the earth, disrupting telegraph communications with short circuits and fires, and causing auroras to appear as far south as Cuba and Hawaii. The high-tech system of the day was affected by an event millions of miles away—a lesson at the time in the interconnectedness of nature and the contingent nature of history.

other countries. But this did not end the thinking. There have been recent attempts to demonstrate profound biological differences in areas of "fitness" between racial groups, most notably in IQ (Intelligence Quotient). Although some of these arguments are filled with scientific sounding data, terminology, and quantitative formulas, they have been scientifically discredited and can be seen as clearly politically motivated, no less so than earlier social Darwinian proposals. The social legacy of Darwin's revolutionary idea will not die easily, it seems, and Charles would have been appalled by this application of his model. It is thus no reflection on Darwin's idea itself that it has been abused, sometimes tragically so.

The language of evolution and the evolution of language

Over time, Darwin's idea has been refined and enhanced as modern knowledge of the processes of evolution and of genetics has evolved, but the idea itself has had a lasting impact on our view of ourselves and our world. Just as we understand that the universe and life on earth have histories that can be examined scientifically and explained by empirical processes, we can also examine human cultures through the basic concepts of evolutionary theory—and indeed, using the very language of evolution.

Cultural systems are adaptive. They adapt groups of people to the natural environment (climate, food sources, and such) and to the cultural environment itself (for example, think of all the changes in cultures that have themselves been responses to the ever-changing technologies of digital communication: e-books, Facebook, smartphones). Moreover, new ideas gain acceptance or are rejected in terms of how well they work, and how adaptive they are at a given point in time. (As a classic example, the wheel, invented in the ancient Andes for use in children's clay toys, was never adopted for vehicles because wheeled vehicles would have been decidedly maladaptive on the treacherous roads of those mountains.)

Finally, cultural systems interact with one another and cultural ideas flow across geographic space and this process promotes cultural change as well. In short, cultures evolve.

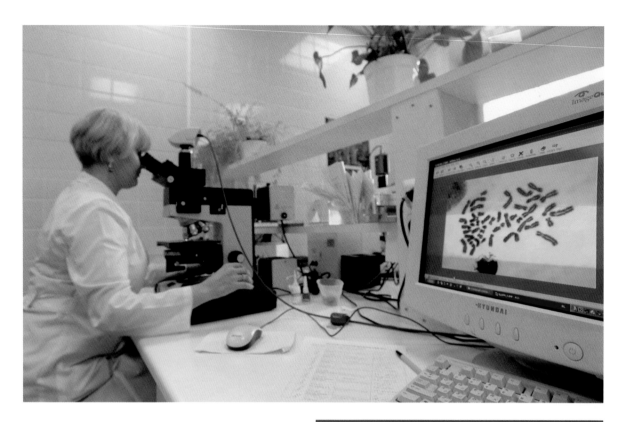

With qualifications—because we humans consciously bring about change—this is analogous to evolutionary change in nature, and our ability to study and analyze cultures and culture change would not be possible without an evolutionary world view. Let's look at a specific example.

Language is, so far as we know, an attribute only of the human species. Our ability to have language is, itself, a product of biological evolution. While there are thousands of different languages spoken now and in the past, the same basic structure is common to them all: an inventory of sounds (phonemes) that are combined to make basic units of meaning (morphemes) that are combined to make functional units of meaning (sentences, paragraphs, and so on). This structure allows us to produce a virtually infinite number of combinations; our means

Just two of many examples of how digital technology has changed our view of the world and of our lives.

of storing and sharing the complexities of our cultural systems. And we can locate our linguistic abilities in certain areas of the brain, an organization all humans share. At some point in our evolutionary past, these abilities, housed in a biological organ, proved adaptively important and those with a greater expression of these abilities were at a selective advantage. *What* languages you speak are learned, but your ability to learn those languages is biological.

Individual languages evolve. To the best of our knowledge, the sounds a language uses and the rules for combining them into meaningful units (grammar) are like genes that undergo mutation, drift, and flow—random changes not directly related to cultural adaptations. What *is* connected to cultural systems directly, however, are words themselves. The names people give thing tells us what sorts

The language centers of the brain.

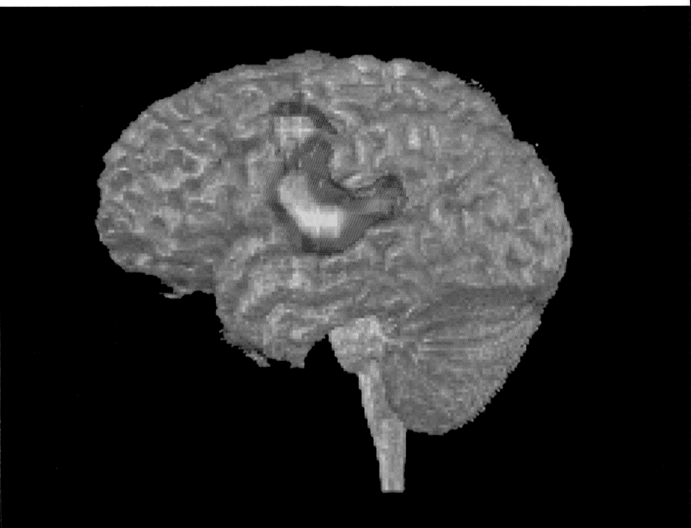

A family tree of Indo-European languages, showing that languages too have an evolutionary history.

of categories they recognize within their environments. In the classic example, Arctic peoples have complex categories and words for states of snow. A group of Native Americans in the Ecuadorian rainforest have one word, translated as something like "up there," to describe the Andes and all their features. This is an example of the adaptive function of language as it reflects and allows us to transmit our views of the world.

Languages, of course, change through time, adding new words and meanings in response to changing environments, including the cultural environment itself. For example, think of all the new words that have been added to our languages in response to the digital revolution of the last few decades. Alternatively, words that used to be common become extinct as they no longer serve a function or have been replaced by other words. For example while the objects still exist, the word 'charabanc' (a motor coach or bus) has vanished as has 'cyclogiro' (a type of aircraft propelled by rotating blades). And, as with species and cultures, languages are related to one another in a huge family tree. They can be arranged taxonomically according to similarities and differences. A common ancestor language gives rise to new languages. By comparing languages we can even attempt to date the common ancestor and possibly reconstruct it, even if it is now extinct.

Moreover, languages are shared. In evolutionary parlance, words and phrases flow among languages as cultures interact. American and British English, for example, contain words whose origins are French, Latin, Greek, Spanish, Celtic, Hebrew, Italian, even Sanskrit. The French use *le weekend*. My Japanese counterpart is now sitting at his *konpyutaa*.

The science of historical linguistics, in short, would not be possible without the world view that incorporates evolutionary change or without the basic theories of how evolutionary change comes about.

What does this all mean for us as human beings? Are we brought lower, as some argue, by accepting that we have evolved along with the rest of life, and that our cultures evolve not only by our conscious choice but also by the analogues of natural processes? Or is this, rather, an exulting thought?

We can do no better than to end with Darwin's final sentence from *Origin of Species*:

There is grandeur in this view of life, with its several powers, having been originally breathed into a few forms or into one; and that, whilst this planet has gone cycling on according to the fixed law of gravity, from so simple a beginning endless forms most beautiful and most wonderful have been, and are being, evolved.

Hand-colored photograph of Darwin in old age based on an 1879 photograph by the firm of Elliot and Fry. Darwin died three years after the photo was taken.

Suggested readings

There are hundreds of good books on the various facets of evolutionary theory. These are but a small sample that do not cover everything, but are ones I would particularly recommend to anyone interested to get started.

One must, of course, read Darwin. *The Origin of Species* as a whole is a commitment (although a worthwhile one) but Darwin kindly gave us a wonderful summary of his argument in the last chapter, XIV: Recapitulation and Conclusion. (I have quoted from the first edition, but this chapter is relatively the same in all six editions.)

There are several good biographies of Darwin. The most extensive is Janet Browne's two volumes, *Charles Darwin: Voyaging* (1995: Princeton) and *Charles Darwin: The Power of Place* (2002: Princeton).

For a history of evolutionary theory, try *Readings in the History of Evolutionary Theory: Selections from Primary Sources* (2012: Oxford). It has over 75 readings, from Plato to Stephen Jay Gould and includes a condensed version of Darwin's last chapter. Many of my quotes from scientists come from this book.

I mentioned Gould throughout and a good place to start is his first collection of essays, *Ever Since Darwin: Reflections in Natural History* (1977: Norton). If you like him, you can keep going through the subsequent nine collections. If you're intrigued by the idea of contingent history, try his *Wonderful Life: The Burgess Shale and the Nature of History* (1989: Norton). For another take on the same issue, though, which began a debate between the two, see *The Crucible of Creation: The Burgess Shale and the Rise of Animals* by Simon Conway Morris (1998: Oxford).

My best recommendation if you want to delve more deeply into genetics is to go to your nearest university bookstore and get the textbook for an introductory genetics course. But get a very recent one because the science of genetics is charging along at an unbelievable rate, which is probably why people are reluctant to write a popular book about it. For the place of genetics in evolutionary theory, however, I recommend *The Making of the Fittest: DNA and the Ultimate Forensic Record of Evolution* by Sean B. Carroll (2006: Norton).

For an introduction to the evolution of the human species, with more on genetics and evolutionary theory, I will shameless suggest my *Biological Anthropology*, 7th edition (2013: McGraw-Hill).

A fascinating book that includes the influence of Darwin on modern thought, and shows that Darwin and Lincoln shared more than a birthday, is *Angels and Ages: A Short Book About Darwin, Lincoln, and Modern Life* by Adam Gopnik (2009: Knopf).

The issues of scientific creationism and intelligent design are nicely covered in *Evolution vs. Creationism: An Introduction*, 2nd edition (2009: University of California) by Eugenie C. Scott. It includes a good discussion of the nature of science.

Picture credits

Every effort has been made to credit the appropriate copyright holder, however, if you find any omissions or errors please contact us and we will make any corrections in the next printing.

Courtesy of Janet M Beatty, **12**, **19**, **20**, **59**, **60(b)**; Courtesy the Great Geek Manual, **18**; Jerry Fowler/M.A. Park, **21**; T. Charles Erickson, **24**; Courtesy of Kenneth L. Feder, **28–29**; Jerry Fowler/after Mariana Ruiz, **42**; Jerry Fowler/after Infographics Diagrams (www.dipity.com) **43**; Jerry Fowler/M.A. Park, **45**, **51**, **64**, **69(t)**, **70**, **80**, **115**, **117**; Jerry Fowler, **57**, **69(b)**, **137**; © Natural History Museum, London, **58**; © G. Gerra and S. Sommazzi (www.justbirds.it), **66(b)**; Courtesy Strozier Library at Florida State University, **60(t)**; Courtesy of rooms.wikispaces.com, **60(b)**; © Jon Way/ECR, **78**; © The Tech Museum 2012, **79**; © Animals by Jamie, **81**; Charles Darwin, **83**; Courtesy Explow.com, **84(b)**; Jerry Fowler/after Tsjok45.multiply.com, **85**; Jerry Fowler/after S.J. Gould, **88**; Courtesy of J. William Schopf, **93(b)**; © John Sibbick, **94**, **95**; Courtesy USGS, **98**, **99**; Courtesy DataViz, **101**; © Noel Rowe, **108(t&b)**, **109**; M.A. Park, **119**; © 2012. Photo The Philadelphia Museum of Art/Art Resource/ Scala, Florence, **125–126**; Courtesy Wikimedia Commons/ Trekky0623, **127**; Getty Images/Hulton Archive/Stringer, **128**; Courtesy Wikimedia Commons/ Rolfcosar, **129**; Courtesy GeneralScienceInfo.blogspot.com/Science, **134**; Jerry Fowler/after Murphy, *Biology 102*, **135**; © Scoop Media, **138**; Film still *Planet of the Apes (1968)*/ © 20th Century Fox, **140–141**; Courtesy ESA 2002. Illustration by Medialab, **143**; Courtesy of NASA, **148**; Courtesy Museum of Criminal Anthropology, Turin, **147(b)**; © Facebook (opening page), **150(b)**; Jerry Fowler/after anthropology.net/2008/02/05/ the-indo-european-language-tree, **152**.

All Science Photo Library: Douglas Faulkner, **6–7**; Friedrich Saurer, **8**; David Wrobel, **10–11**; Sheila Terry, **14**; Pascal Goetgheluck, **15**, **114**, **117**; Dirk Wiersma, **16–17**, **93(t)**; Victor de Schwanberg, **22**; Science Photo Library, **23**,

30, **54**; Art Wolfe, **26–27**; Tony Craddock, **31**; L. Newman & A. Flowers, **32**; Dr. Juerg Alean, **33**; Lutz Lange/Detlev van Ravenswaay, **34**; Maria Platt-Evans, **35**; **NASA 36–37**; Natural History Museum, London, **38**, **39**, **55**, **107**, **113**; Dr. Tim Evans, **40–41**; Medical Ref.com, **44**, **133**; James King-Holmes, **42(t)**, **46**, **48**, **130**; Willard H. Sharp, **49(b)**; M. H. Sharp, **49(t)**; Laurie O'Keefe, **50**; John Beatty, **52–53**; Paul D. Stewart, **55**; William Ervin, **56**; Science Source, **58(r)**; George Bernard, **62–63**; PlanetObserver, **65**; Dr. Jeremy Burgess, **66(t)**; Dante Fenolio, **67**; Walter Myers, **73**; Chris Heller, **74**; Susan Kuklin, **75**; Mary Beth Angelo, **76**; Duncan Shaw, **77(l)**; Ron Sanford, **77(r)**; Claude Nuridsany & Marie Perennou, **82**; Paul Wootton, **84**; David Gifford, **84(c)**; Jan Hinsch, **86–87**; Georgette Douwma, **92**; Mark Garlick, **96**; Julius T. Csotonyi, **97**; Gensler et al/NASA/ CXC/ROSAT/ DRAO/ NRAO/DSS, **102**; Chris Butler, **103**; John Reader, **104–105**, **111**; Science Source, **106(t)**; John Reader, **121(b)**; Javier Trueba, **106(b)**, **110**, **115**, **120**, **121**; Martin Shields, **111**; Gregory Dimijian, **112**, **136**; Friedrich Saurer, **114**; John Devries, **116**; Jim Carter, **121**; Lawrence Migdale, **122**; Equinox Graphics, **123**; Stuart Wilson, **131**; Medical RF. Com, **133**; Paul D. Stewart, **142**, **147(t)**, **154**; NIBSC, **144(l)**; Dr. Tony Brain, **144(r)**; Dr. Stanley Flegler, **145**; Suzanne L. & Joseph T. Collins, **146**; Riva Novosti, **150**.

ACKNOWLEDGEMENTS

I would like to thank my friend and editor, Lee Ripley, for the idea for this book and for thinking of me to write it. All the people at Vivays Publishing did a great job of turning my ideas and words into a lovely final form. My wife, Jan Beatty, read the manuscript and made invaluable suggestions for content, wording, and illustrations. Bob Weinberger kept me accurate when I delved into the physical sciences. Finally, a far too belated thank you to my parents for encouraging my childlike sense of wonder and for the gift of reading so I could pursue it.

Index

Page numbers in **bold** include illustrations.

absolute dating techniques 130
adaptation 55–57, 64–68
alphabet-gene analogy **45**, **48**
amino acids 45, 91
amphibians 82, 94
An Essay on the Principle of Population (Malthus) 146
Anchiornis 97
animal cells **43**
Annals of the Old Testament (Ussher) 29–30
Anteosaurus **94**
arboreal mammals 106–107
archaic *Homo* **117**, 119
Archeopteryx, fossil **74**
Ardipithecus 113, **114**
Arizona **31**
art, appreciation of **120–121**
asexual reproduction 94
Australopithecus 110–113, **114**, **119**
axolotl fish **82**
azure poison dart frog **146**

Babylonian creation story 132
bacteria 68, **90–91**, 100
bacterial diseases 144–145
base pairs, in DNA 44, 45
Behe, Michael 133
Bertini, Giuseppi **23**
Bible 30, 100, 126, 131
Big Bang 100
biogeography 35
biological anthropology 9
biological species 75, 120, 122, 148, 149
bipedalism 107, 110–112, **111**, 114
birds 76, 80, 94
    and beak size 65–67, **66**
    earliest forms 95, 97
black lemurs **108**
brain size 107, 110, 115–117, 119–120
branching evolution **83**
branching speciation 84–85
Bryan, William Jennings 128
Bryce Canyon, Utah **28–29**
Bubonic plague 145
Burgess Shale fossil beds, British Columbia **32**
burial sites **120**
butterflies **49**

Cambrian explosion 94
Cambridge University 20, **59**
caribou 76, **77**, 78
cataclysms 88, 131
catastrophic mass extinctions 88–89, 97, **103**
catastrophism 34–35
cave art **121**
cave fish **67**
cell structure **43**
centrioles **43**
Cepaea nemoralis **62–63**
Chauvet Cave, Southern France **121**
chimpanzees 75, **79**
Christ's College, Cambridge University **59**
chromosones, and interbreeding 76
chronometers **136**
cichlids **81**, 87
coding sequences, genomes 48–49
codons, genetic pattern 45
Conneelly, Eileen 24
continental drift 97–100, **98**, **99**
cosmic calendar (Sagan) **96**
coyotes 76, **78**
Creation Science Movement 137
creationism 53, 137–139
Cretaceous period 89, **99**
criminality, and physicality 147
cultural evolution 149–150
    and language 150–153
Cuvier, George 34
cyanobacteria cells, fossils **93**
cytoplasms **43**

Darrow, Clarence **128**
Darwin, Charles
    biography 58–61, 106, 154
    influence on science 141–143
    influenced by 39, 53, 54–58
    Notebook B (1837–38) **58**
    political interpretations of 148–149
    portraits **58**, **154**
    suggested reading on 155
    see also Descent of Man (Darwin); finches, study of beak size; macro-evolution; natural selection; *Origin of Species* (Darwin)

Darwin, Emma *see* Wedgwood Darwin, Emma
Darwin, Leonard 146
dating techniques 130
Dawkins, Richard 61
deduction 20
Descent of Man (Darwin) 61, 106, 146
Devonian period 89
diatoms **86**, 88
dinosaurs
    Anchiornis **97**
    Archeopteryx **74**
    in education 7, 9
    extinction 67
    Iguanodon **39**
    Pterodactyl **39**
    time period 95
    Tyrannosaurus rex **103**
diseases 143–144
DNA **40–41**, 44, 47, 48
donkeys 75–76
double helix molecule **40–41**, 44
Drouais, François-Hubert 35
dwarfism 123

early *Homo* 113, **114**, **117**, **119**
earth
    age of 129
    history of life on 92–97, **101**
Eastern Swallowtail butterflies **49**
ecology 145
Einstein, Albert 18, **21**
elephants 123
Enûma Eliš 132
environment
    effects on isolation 79–81
    effects on species evolution 64–68, 92–97
epigenetic processes 51
esthetic appreciation **120–121**
eugenics 146–149
extinction 56, 57, 67, 88–89, 97, **103**
eyes **133**, 134, **135**

Family Hominidae 107, 110
feathered dinosaurs **97**
fertility figurines **121**
field science 30

finches, study of beak size 65–67, **66**
fish **67**, **81**, **82**, 87, 94
fission, of populations 70–71
Fitzpatrick, Erin 24
Flat Earth theory 126–127, **127**
floods, and scientific creationism 131
Flores, Indonesia, *Homo floresiensis* 123
flowering plants 95
fossils
    Archeopteryx **74**
    chordates **94**
    cyanobacteria cells **93**
    hominids 107, 110–114, 116–117, **119**
    importance of 32–34, 35
    Lucy (*Australopithecus*) **110**, 130
    in scientific creationism 128
    stromatolites **93**
founded populations 70–71
fungi 95, 145

Galapagos Islands **52–53**, **65**
Galileo **23**, 126
Galton, Francis 146
gene flow 70
genes 43–44, 45, 47, 147–149
genetic
    coding 47–48
    diseases 144–145
    laws of inheritance (Mendel) 42–43
    patterns 45
    random drift 70–71
    variation 68–71
genomes 47, 48–49, 49–51
genus *Homo* 115–123
geological history of earth 30–32
geological maps, earliest 35, **38**
geological periods **99**
giraffes **56**, 85
Göbekli Tepe, Turkey **129**
gorillas **109**
Gould, Stephen Jay **84**
    on Darwin 25–26, 143
    and dinosaurs 7
    on Lamarck 54
    model of macroevolution 84–89
    suggested reading on 155
gradualism 85
Grand Canyon, Arizona **131**
Grand Gallery of Evolution, Muséum
    d'Histoire Naturelle, Paris 12
Grant, Rosemary and Peter, study of beak
    size in finches 65–67
gravity **21**, **22**, 25

great apes **109**
ground apes 113

Heston, Charlton **140**
Hicks, Edward 126
history of linguistics 151–153
HMS *Beagle* 39, **54**, **57**, 136
hominids 107–114, **114**, **119**
Homo erectus **117**, **119**
Homo floresiensis **123**
Homo, genus *see* genus *Homo*
Homo sapiens 110, **117**, 118, **119**, **123**
homogenous populations, development of
    120, 122
horses 75–76
human
    chromosomes **79**
    evolution 9, **84**, 104, 105–123
    eyes **133**, 134
    genome 14, 23, 47–51
    traits 107, 120, 146–147
Human Genome Project 47–51
Hutton, James 35
hypotheses 19–20

Iguanodon **39**
induction 18
inheritance, laws of
    Father Gregor Mendel 42–43
    Lamarck 56–57
insects 95
insular dwarfing 123
intelligent design (ID) 132–137
interbreeding of species 75–79, **79**
isolation 77–82

Jacobs, Lilli 24
Java 119
Jennings Bryan, William *see* Bryan, William
    Jennings
Jurassic period **99**

Kebara Cave, Israel **120**

Laetoli footprints **111**
Lamarck, Jean-Baptiste de 53, 54–57, **55**
language evolution 150–153
Leclerc, Georges-Louis, Comte de Buffon
    34–35, **35**
life on earth, history of 92–97
Loewenberg, Bert James 142
Lombroso, Cesare 147
Lucy (*Australopithecus* fossil) **110**, 130

Lyell, Charles 35, 39, 57
lysosomes **43**

macroevolution 82–89
malaria 145
Malthus, Thomas 146
mammals 94, 95, 100, 106
marine snail 135
mass extinctions 88–89, 95, 97, **103**
medicine 143–144, 144–145
Mendel, Father Gregor 41–43
Mewbron, Rachel 24
Mexican salamander 82
microcephaly 123
microevolution 64–71, 82
Miller, Arthur 24
Mississippi Delta **36–37**
mitochondria **43**
mixed forest–open space environments **112**
monkeys **108**
morphemes 150
Mount Etna, Sicily **33**
Mount Vesuvius, Italy 33
mules 75
multicellular organisms 94–95
Museum of Criminal Anthropology, Turin **147**
Museum of Science, Boston 19
mutation 68–71, 82, 94

natural philosophers 30
natural selection
    Darwin 58, 61–64, 104, 143
    and new species 82–85
    within species 64–68
Nautilus **135**
Neandertals **120**
neoteny 82
Newton, Isaac 20
niche construction 146
Noah's Ark (Hicks) **124–125**
non-coding DNA 47, 48
nuclear membranes **43**
nuclei **43**
nucleoli **43**

objectivity 23
okapi 85
open space–mixed forest environments **112**
Ordovician period 89
organisms
    earliest 91–94
    multicellular organisms 94–95
    single-celled **134**

Origin of Species (Darwin) 59–61, **83**, 104, 154, 155

Paley, William 135
Paranthropus **113**, **114**, **119**
pea plants, traits of (Mendel) 42–43
Penicillin 145
Permian period 89, **99**
philosophy, and evolution 100–103
phonemes 150
pikaia (fossil chordate) **94**
Planet of the Apes (1968) **140–141**
plants
    earliest forms 95
    flowering 95
    as food source 116, 117
    and medicine 145
    reproduction in 79
    traits (Mendel) 42–43
plate tectonics 21, 97–100, **98**, **99**
potassium/argon dating 130
pre-conceptions, and progress of science 24
primates
    bipedalism in 114
    earliest forms 95
    endangered **108–109**
    traits 106–107
Principles of Geology (Lyell) 39, 53
progression in evolution, Lamarck 56
prosimians **108**
proteins 43–44, 45, 51
pseudosciences 126–127
Pterodactyl **39**
punctuated equilibrium 85

race
    and the eugenics movement 148, 149
    and evolution 122–123
radiocarbon dating 130
rainforests **26**, 123
random gamete sampling 71
random genetic drift 70–71
reading list 155–156
red shanked douc langurs **108**
reindeer 76, **77**, 78
relative dating techniques 130
relativity, theory of (Einstein) 21
religion, coexistence with science 126–127, 137–139

reptiles **94**, 95, 100
ribosomes **43**
Richmond, George 58, 106
RNA molecules 51
rock strata 9, **38**

Sagan, Carl 95, **96**, 97, 101
Salem witch trials (1692) 24
science
    coexistence with religion 126–127, 137–139
    definition 18
    objectivity in 23
    progress in 23–24
    scientific investigation 143–144
    scientific testing 18–21
scientific creationism 127–132
Scopes, John **128**
Scopes 'Monkey' trial, Dayton, Tennessee **128**
selective breeding 147–149
sexual reproduction 94
sickle cell anemia 145
single-celled organisms 92–94, **134**
skulls
    Australopithecus 110, 111
    Homo floresiensis **123**
    Homo sapiens **123**
    Museum of Criminal Anthropology, Turin **147**
    Paranthropus **113**
    size of 115
Smith, William 35
snails **62–63**
social behaviours, and eugenics 147–149
sociocultural development 122–123
solar eruptions **148**
Solutrean period **121**
southern apes 110
speciation
    and isolation 78–82
    macroevolution 82–85
    process of 76–78
species
    adaptation 61
    definition of 75
    genetic comparisons 49–51
    model for change (Wallace) 59
squid **135**
stereoscopic vision 106

sterilization programs 148
stone tools 115–117, 119, 120, **121**
stratigraphy 30–31, 35, 130
stromatolites **92**, **93**
Strozier Library, Florida State University 60
supernovae **102**

Tanzania, open acacia woodland **112**
technology, and progress of science 23–24
telescopes **23**
The Crucible (Miller) **24**
Theory of Gravity 21
Theory of Plate Tectonics 21
Theory of Relativity (Einstein) 21
thermophyllic bacteria **90–91**
timepieces, design of 136, **137**
tools 115–117, 119, 120, **121**
tortoises **52–53**
traits 45, 106, 107, 120
Triassic period 89, **99**
Trinity College, Cambridge University **20**
Tyrannosaurus rex **103**

uniformitarianism 34–35, 97
universe, age of 129
University of Cambridge *see* Cambridge University
Ussher, Archbishop of Armagh, James 29–30

Van de Graaff generators **19**
Venus of Willendorf **121**
vertebrates 81, 89, 94, 95
viruses 144
vision, in primates 106
visual history of life **101**
vocabulary, and language evolution 151, 153
volcanic eruptions **33**

Wallace, Alfred Russell 59, 152
watches, design of 136, **137**
waterwheels, design of **137**
Watson and Crick (1953) 44
Wedgwood Darwin, Emma **106**
western lowland gorillas **109**
wolves 76

Yellowstone National Park **90–91**

zoogeography 152

What Americans BELIEVE
and How They WORSHIP

REVISED EDITION

What Americans BELIEVE and How They WORSHIP

Revised Edition

by J. PAUL WILLIAMS

HARPER & ROW, PUBLISHERS
NEW YORK AND EVANSTON

C-R

LIBRARY OF CONGRESS CATALOG CARD NUMBER: 52-5477

TO
Clarence Milton Williams
My Father and Teacher

CONTENTS

ACKNOWLEDGMENTS   ix

MAPS   91

I. Preview   1

II. The Roman Catholic Church—Defender of a
Revelation   15

III. Protestantism (in general)—Which Reaffirms the
Faith   94

IV. Lutheran Churches—Guardians of Orthodoxy   152

V. The Protestant Episcopal Church—Which
Emphasizes Ritual   171

VI. Presbyterian Churches—At the Theological Center   196

VII. The United Church of Christ—Which Practices
Ecumenicity   216

VIII. Baptist Churches and Christian (Disciples) Churches
—Defenders of Religious Freedom   239

IX. The Quakers—Practicing Mystics   269

X. The Methodist Church—Evangelical Organization   286

XI. The Unitarian Universalists—Theologically Liberal   313

XII. Judaism—The Mother Institution   330

XIII. The Eastern Orthodox Churches—The Least
Americanized   362

XIV. The Mormons—Pioneers   378

XV. Christian Science—Which Emphasizes Healing   399

XVI. Other Religious Innovations—Seventh-day
        Adventists, The Pentecostal Holiness Church,
        The Salvation Army, Spiritualism, Jehovah's
        Witnesses, Unity School of Christianity, The
        Oxford Group (Moral Re-Armament), The
        Black Muslims   427

XVII. Some Nonecclesiastical Spiritual Movements—
        Astrology, Naturalistic Humanism, Hedonism,
        Alcoholics Anonymous, Beat Zen, Nationalism   453

XVIII. The Role of Religion in Shaping American Destiny 472

        NOTES   493

        INDEX   521

ACKNOWLEDGMENTS

IN PREPARING these pages, I have asked help from many people; it has always been given with great generosity. For a thoroughgoing review of the manuscript, I am especially indebted to Professor Winthrop S. Hudson, Professor Clarence P. Shedd, and Dr. Clarence M. Williams. Their many suggestions have been both cogent and detailed. My debt to them is large.

In addition two-score persons have given freely of their technical skill, or their knowledge, or their resources of books and magazines. Each of the major descriptive sections of the book has been read by two or more clergymen of the sect described. They have often read meticulously. I hope thus that errors of fact have been all but eliminated; what errors remain are of course wholly my own responsibility. (The interpretations are also my own; occasionally they have been retained in spite of questions raised by the denominational experts.) The following persons, in their several ways, have given assistance which has placed me greatly in their debt: Rev. Charles A. Anderson, Constance E. Bagg, Dr. C. Rankin Barnes, George Channing, Ruth Curby, Rev. D. Earl Daniel, Dean A. T. DeGroot, Dr. Frederick M. Eliot, Rev. George B. Ford, Rev. Donald H. Freeman, Captain Lucy V. Fusco, Rev. John J. Gearin, Rev. George C. Gutekunst, Professor S. Ralph Harlow, Rev. Duncan Howlett, Gordon B. Hinckley, Professor Mary I. Hussey, DeWitt John, Dr. Mordecai M. Kaplan, Bishop W. Appleton Lawrence, Rev. Timothy J. Leary, Rev. Eugene A. Luening, Bishop Francis J. McConnell, Professor Alan V. McGee, Professor Virginia P. Matthias, Rev.

Andrew A. Martin, Rev. James F. Madison, Professor James A. Martin, Dr. Daniel H. Miller, Rev. Francis D. Nichol, Gammar Paul, Rabbi Samuel Perlman, Dr. Herbert W. Prince, Rev. Emil C. Reichel, Dr. Elbert Russell, Bishop Joseph A. Synan, Rabbi Maurice H. Schatz, Helen H. Williams, L. Dwight Williams, Rev. Henry C. Wolk.

My thanks are also extended to the staff of the Williston Memorial Library, Mount Holyoke College, for much patient assistance.

Finally, I wish to express my appreciation to the several publishers who have granted permission for the inclusion of selections from their copyrighted works.

In the second edition, the material of the various chapters has been updated and three new chapters have been added—on Eastern Orthodoxy, Mormonism, and Christian Science. Short sections have also been added on Moral Re-Armament, Alcoholics Anonymous, Zen Buddhism, and the Black Muslims.

Each of the new chapters was read by two persons who know the group under discussion from personal experience. I am deeply indebted for the many suggestions they made—most of which I followed. Whatever errors of fact or interpretation remain are of course again wholly my own responsibility. Also, I have been loaned many books and magazines. To the following persons I wish to express my gratitude for the help they gave: Rev. Leslie C. Beale, Rev. Lee J. Beynon, Jr., Dean Walter H. Clark, Charles A. Davies, Prof. Ruth J. Dean, Nancy M. Devine, Arthur Dore, Dean Deane W. Ferm, Prof. Helen Griffith, Rev. John Paul Jones, Rev. Maurice A. Kidder, Robert Peel, Rev. Harold D. Smock, Virginia W. Sparrow, Rev. Neophytos Spyros, Prof. Ronald J. Tamblyn, Dr. Glen W. Trimble, A. Theodore Tuttle, Dr. Lauris B. Whitman.

In this edition as in the former one, I have written for the general reader as well as for the college student. My aim has been to present the material both accurately and readably.

What Americans BELIEVE
and How They WORSHIP

CHAPTER I

PREVIEW

AN ASTONISHING number of religious sects flourish in the United States, and they cover an extraordinary range of interests. To get a hint of this situation, it is but necessary to read the religious page of a city newspaper some Saturday evening. On this page will be displayed, in news item and advertisement, a range of religious belief and practice which runs all the way from the churches, long established and respectable, where the solid citizens have their membership, to the gospel missions which operate in vacant stores on the side streets, and the cults which meet in second-class hotels and which give their devotees not only promise of salvation in the next life but assurance of riches and prestige here on earth. In order to indicate the breadth of religious belief and practice in this country, let us look briefly at a few pictures of American sects in action.

In his newspaper the average American might find the announcement of a Quaker meeting. If he attended this service and if it followed the traditional Quaker pattern, he would enter a room that had no decorations whatever; at the front there would be no altar, no pulpit, no organ pipes to challenge his attention. In their place would be a simple bench set against the wall; on this bench would be seated two or three leaders of the congregation; the rest of the group would be seated facing the front in the manner of the usual church. After the beginning of the meeting, the worshipers would remain silent for a long time, anywhere from ten to thirty minutes. During this period each person would direct his own private devotions:

1

some would pray, others would meditate, some would think about passages of Scripture, others would ponder some social problem; and a few, no doubt, would just daydream or perhaps worry about the roast at home in the oven. Finally, one of the group would be "moved of God," as the Quakers say, to speak; without introduction he would rise to his feet and speak briefly. This "witness" would be short and it would be spontaneous; but it would speak of divine things: of the love of God and of His will that all men should live together as friends. Then the speaker would sit down and the silence would continue, a silence rich in fellowship and spiritual renewal. Probably others would be moved to speak. Worship after this extremely simple fashion would go on for an hour. Then one of the persons on the bench at the front would reach over and shake hands with the person nearest him. This action would be the signal for everyone, in friendly fashion, to shake hands with his neighbor. And the "meeting" would be over.

In their newspapers Americans can find from time to time announcements of the ordination of Roman Catholic priests. Here would be a religious service at the other extreme from the simplicity of the Quaker meeting; for the ceremony during which a young man enters into the priesthood is an occasion on which the Roman Catholic Church spares no pains. The ordination would usually take place in a cathedral, that is, in a church where a bishop has his headquarters. The building would be a vast structure, decorated with much care and expense. On the walls would hang paintings and plaques of scenes from the Bible; in various recesses and corners would stand statues of Jesus, of the Blessed Virgin, and of other saints.

As the stranger stood in the doorway and looked into the dim interior, past seemingly endless rows of pews, his eye would be fixed on a brilliantly lighted scene at the far end of the cathedral. Walking nearer, he would see that the central object is the high altar, placed at the back of a raised platform called the chancel. Along the sides of this platform are seats for the clergy, and among them, in a place where it can be seen by all, is the bishop's throne, his *cathedra*.

The candidates for the priesthood—all well-educated, of legitimate birth, over twenty-three years of age, and usually without physical disability—come to their ordination clad like Romans of the third century. First they kneel before the bishop, who charges them to speak boldly of any circumstance which might cause them to bring dishonor to the Church. Then they lie flat on the floor, face downward, for a full quarter of an hour, while the bishop leads them in prayers to God and to the saints. The ordinands—those being ordained—pray that God will have mercy upon them and that the saints will intercede before God's throne asking that temptations may not harass the priests of His Church. After these prayers are said, the ordinands rise and the bishop in silence lays his hands on the head of each in turn; thus they enter into the Apostolic Succession. (The meaning of this important term will be explained later.) Each of the many priests who are present then lays his hands on the head of each ordinand, and the central act of the ordination is over.

But the dramatic tempo of the service does not lessen. The hands of each candidate are anointed with holy oil in order that they may be hallowed and blessed, hands that will touch the Blessed Sacrament, baptize the newborn, anoint tenderly the dying, bless sadly the dead. Then the bishop gives what is perhaps the most important power of the priest: speaking for God, as Catholics believe, the bishop says in Latin, "Receive the power to offer Sacrifice to God, and to celebrate Mass." Immediately the young priests exercise their new power: they all celebrate their first Mass, together; kneeling behind the bishop as he faces the altar, they read and intone with him and bring to pass the astounding miracle of transubstantiation. (This term also will be defined later.)

After saying Mass, the power to forgive sins and to "retain" sins is given to the young priests. This is the power which, Catholics believe, God gives the priest to take away the guilt of sins that otherwise would send the soul to hell. The young priests then swear obedience to the authority of the bishop, present him a gift, and kiss his ring. The ceremony ends with the reading of the first verses of the Gospel according to St. John.

The whole of this elaborate service is carried forward with the vibrant intensity and solemn joy of those who feel they walk on the brink of eternity. The bishop is the wise father receiving the reverence of dutiful sons; both kindness and sternness sound forth in his voice, and in his eye is a gleam that matches the fire that flashes from the precious jewel in his episcopal ring. And the young priests walk forth from the chancel consecrated servants of Almighty God, with a glow on their faces like the glow on the faces of brides as they walk down the aisle on their wedding day.

Many American newspapers regularly carry announcements of the services of the Spiritualists, persons who believe it is possible for the living to communicate with the dead. One type of Spiritualist meeting is called a séance. A few earnest folk gather around a "medium," a person through whom, it is believed, messages can be sent to the next world and answers received. The place of meeting is usually a small room in which the chairs are arranged in a circle. After pleasantries have been exchanged and perhaps announcements made, the medium seats himself in a comfortable chair and the room is almost completely darkened. A solemn hush falls, the group begins quietly to sing hymns, and the medium seeks the mood which will bring the spirits: he may go into a trance, he may sleep, he may simply seek an attitude of composure. After a time, if the spirits are co-operative that night, strange things begin to happen. Perhaps raps will be heard, or tables will be moved, or voices will speak; if the medium is very powerful, images may appear. These mysterious "phenomena" indicate to believers that the spirits have come.

The members of the group begin then to ask questions eagerly; or the spirits may themselves take charge. The conversation usually goes forward just as a conversation would between two human beings; and the spirits are not above injecting an occasional wisecrack. Each spirit tells when and where he lived on earth, and gives some message to the group. Most of the messages are religious in character, urging the life of devotion, prayer and Bible reading. Sometimes the spirits tell about the next world, though usually their description has little detail in it: they say heaven is peaceful and Jesus beautiful. Occasionally the messages concern very mundane affairs: Martha, you will find that lost letter if you will look at the back of the middle drawer of your desk; John, you had better get rid of those oil stocks you bought last week. If, as is probable, a spirit comes who claims to be the soul of some relative who has but recently died, great tension is present in the meeting. It is a moving experience to hear a young wife talk lovingly to the voice she believes to be that of her departed husband, to feel how desperately she clings to the memory of his presence and affection, and to hear him calmly speak to her words of comfort and advice.

After perhaps an hour, the spirits seem to tire, the voices stop, the raps cease. The medium rouses himself, the lights are turned on, good nights are said, and the Spiritualists go out to confront the hard facts of this world fortified by the faith that they have been in touch with the sublime mysteries of the next world.

In the spring of the year the religious columns of the newspapers sometimes carry items which announce the summer conferences which the Student Christian Movement operates for college students. These conferences run for about a week, and they are held in some spot where nature herself seems part of a conspiracy to lift man's thoughts to God; perhaps the conference is housed in a mountain camp: near by is a lake, in the distance are rugged peaks, and in plain view are the beginnings of mountain trails which lead up and up until they reach summits from which the whole of God's creative power seems visible.

The young people who attend these conferences come from colleges where students are taught to exalt reason and distrust emotion. As a result most of them would no more make a display of their deepest religious feelings than they would of the skeletons in the family closet. However, perhaps as a sort of compensation for this strait-laced attitude, these students claim to have no use for the idea that religious people should be solemn in their deportment. Often they have a hilarious time: they swim, go canoeing, play tennis, ride horseback, dance. They sing around campfires, listen to the tall tales of the hills, and lampoon one another in dramatic skits.

But the center of their conference is not play. The center is worship. Quietly these young men and women bow together and pray. They seek to commune with God, to confess their sins, to rededicate themselves to living as they believe God would have them live. Often they feel themselves in the very presence of divinity. In some conferences an hour a day is set apart for silence; during this period everyone pursues his own devotion—he walks, or reads his Bible, or meditates, or kneels in the Chapel—seeking the strength which comes from searching for the highest good.

But these students also study. There is little room in their religion for miracles; but they have a tremendous respect for facts and for reason. Knowing this attitude, the Student Christian Movement brings to its conferences the best leaders it can find. Professors in the largest universities, pastors in the busiest churches give of their thought and energy to help these young minds find their way through the uncertainties and mysteries of America's many creeds, out onto what they believe is the solid ground where religious faith is both rational and free. In the long morning hours earnest groups sit under the trees or on the lake shore discussing Christianity, its dogmas, and its way of life. The students ask: How can science and religion be reconciled? Why do evil and suffering exist? What are the reasons for believing that Jesus was divine? What should be the Christian attitude toward disarmament? Long sessions are devoted to Bible reading and study.

Watching these discussions the superficial observer might conclude that the religion of college students is not very spiritual, that it tends to be chiefly a religion of the "higher thought processes." But if he looks more deeply, he will see that these academic questions are often but a superficial façade erected to hide deep-seated spiritual problems and emotions. These young people struggle intellectually; but down deep they are searching for the cause, the ideal, the view of life to which they can consecrate themselves. Many a student has come to the last conference session, his heart bursting with the realization that he has at last found his life's task, the vocation which is worthy of all his powers.

Occasionally the newspapers carry accounts of the services of some of the Holiness sects in the recesses of the southern mountains where worshipers handle snakes and fire. Professor Emeritus Virginia P. Matthias of Berea College attended one of these services; the following paragraphs are from her letter of description.

The meeting place is a small, square structure of wooden planks, set on four substantial stakes, high off the road. Inside are rough wooden benches, and at the front of the room is a large platform surrounded by benches for the more active members. On the front wall are a crucifix and several nails for hats and coats.

As we came in and sat down on the rear seat, the audience of about sixty people was singing and shouting to the accompaniment of a very loud guitar (accented on the first syllable) and extremely noisy cymbals. The tunes were in jig time, and the audience beat out the rhythm with hands and feet. . . .

After a particularly wild song the preacher reached under the pulpit and drew forth a milk bottle stuffed with paper, which he lighted. The paper had evidently been soaked with kerosene, for it burned a long time. This branch of the Holiness Church believes that no harm can come to a person who has been saved. Their cornerstone is Mark 16:18 which says: "They shall take up serpents; and if they drink any deadly thing, it shall not hurt them." I cannot find that Mark has a thing to say about handling fire; but I suppose it follows that if "they" handle fire it shall not burn them. At any rate, here came the fire. The bottle was passed from hand to hand among the people at the front of the room, and eager hands reached for the flame; people jumped about and shouted, their bodies jerked, and the fire seemed dangerously near being hurled to the floor. I noticed that most people did not hold their hands in the fire for more than an instant. Some did, however—for some time; and as nearly as I could see, they were not burned. I offer no explanation, but this is what appeared to be true: they handled fire and were not burned.

Suddenly everything quieted down. The fire was put out, the singing stopped, the people sat down. The preacher began to talk, and he kept on for about an hour. He kept the attention of the audience very cleverly —he kept asking them questions. He took us quickly from Daniel in the lions' den to John the Baptist. And what did John baptize with? With water, said the audience without hesitation. An' after Jesus come, what good was that there water? Twarn't no good, said the audience; twarn't no good at all—they had to be baptized with the spirit. . . . The preacher pointed out that today one says he's a Baptist and another that he's a Presbyterian. But was they ary Baptist or Presbyterian in the Bible? The audience said thar warn't nary one. The preacher said thar's jest one way —the spirit and the truth. He did not define the spirit, nor yet the truth, but added quickly that it don't do no good to preach nothing about the Bible less'n you preach about the spirit.

The preacher accomplished a sudden transition from the spirit to the verse in Mark, "They shall take up serpents." He said that serpents were rattlesnakes. We craned our necks uneasily to see whether any rattlesnakes were secreted behind the pulpit with the milk bottle of fire. The state law now prohibits snake handling in religious meetings, but the Holiness people say this obstructs religious freedom, and they do not intend to obey it. So we looked nervously about for the preacher to bring forth a snake. But he didn't bring out ary one. He was sorry he didn't have one. Several times he was sorry. In lieu of handling an actual rattlesnake he recounted how several of the saved had held snakes as big as your arm, wrapped 'em right around their necks. No harm came to them.

All at once everyone was singing again, and the preacher was exhorting the sinners. Several sinners flung themselves on their knees at the platform. The guitar beat out its heavy chords, the cymbals crashed, the people shouted and sang. The preacher knelt with the sinners—but whether or not they were saved that night we never knew; it was half past nine, and we went home.

These five pictures—the Quaker meeting, the Roman Catholic ordination, the Spiritualist séance, the student conference, the Holiness service—are but the briefest introduction to the vivid detail and wide divergence of religious life in America. If the reader of the average newspaper were to explore all the possibilities suggested by the religious news, he would witness baptisms, weddings, dedications, exorcisms, parades, last rites, faith healings, Children's Day exercises, revival meetings, prayer meetings, every-member canvasses, military funerals, national conventions—and much more. It would be an intensely interesting study. It would be an intensely vital one also. For these many ceremonies are but the symbols, the rough outer garments, which clothe the deepest convictions of the American people. In the last analysis, religious belief is the most important fact about a person or a nation; the way people live is determined by whatever religion they hold, by their basic values.

Yet a good many Americans today consider that religion in any form is no concern of theirs. They shrug their shoulders and declare that they have no objection to other people being religious. "Every person to his taste," they say. Religion to them is one of a thousand interests. Some people like golf, others go in for poetry. The church is a kind of club for people who like to sit and think about their sins of a Sunday morning when they might still be abed.

It is the fad in certain sections of the American intelligentsia to think of persons who are concerned about religion as being antiquarians, hangers-on to the vestiges of forgotten enthusiasms, to be viewed with the same tolerance that a commander of a tank corps would view a regiment of cavalry. H. L. Mencken put the point neatly: "The cosmos is a gigantic fly-wheel. . . . Man is a sick fly taking a dizzy ride on it. Religion is the theory that the wheel was designed and set spinning to give him the ride."

Widespread though this point of view is, it is surely in error. No careful thinker ought to permit his disapproval of specific religions to lead him to conclude that religion as such is either unimportant or unessential. A man's religion is whatever he does to relate himself to what he believes is the supreme reality in the universe. Religion in this sense is basic to all high living. The person whose religious awareness is strong has in his life a sense of direction and purpose while the person whose religious awareness is weak lives a life which is atomistic and essentially undirected. The person whose religion is strong is like the writer of a drama who strives to turn all the action in his play toward a definite goal; the person who has little or no religion is like a puppet which is ruled by the jerks of outside forces. Religion probably supplies life's strongest motivations; the conviction of being in the right, of conducting one's life in line with the will (or the laws) of supreme reality probably gives a person more staying power than anything else. The student, for example, whose integrity in examinations is predicated merely on the thesis "I owe it to myself to do good work" will cheat if he thinks it will pay him in the end to cheat. Ethical conviction, the power to see it through, is not the possession of men who base their code on a proposition like "A gentleman would not do that." Rather is it the possession of men who believe that their moral code is fixed in the eternal scheme of things and is not to be cast aside merely because freedom from punishment is fairly sure.

Religion is essential also to the welfare of societies. Every society is at bottom a spiritual entity. Every society is founded on some kind of religion, on some conception of the nature of the world we live in and of its demands on us. The religion of a people is what they think makes life worth living, what they really care for, and will if necessary die for. Thus the thing most worth knowing about any people is the actual status of their religious thinking. Are their corporate actions controlled in the last analysis by regard for themselves, or for their families, or for their nation, or for people as such? Do they place property above personality or personality above property? Is their form of government, in their belief, the will of God (or of Mother Nature) for their nation, or is it merely a set of temporarily convenient arrangements? Is their society integrated around shared ideals, or is the stability of their community life threatened by basic disagreements over what constitutes life's highest values? Questions of this order are the most important questions that can be asked about American society—or any other society. They explore the spiritual foundations on which national welfare rests.

When the people who live in a given locality begin to lose faith in the things their society stands for, that society is in grave peril—unless a compensating spiritual movement arises. Hope and common purpose are essential ingredients of the spiritual cement which holds a society together. Arnold J. Toynbee in his study of the earth's civilizations writes:

> The breakdowns of civilizations are not catastrophes of the same order as famines and floods and tornadoes and fires and shipwrecks and railway-accidents; and they are not the equivalent, in the experience of bodies social, of mortal injuries inflicted in homicidal assaults. . . . The broken-down civilizations have not met their death from an assassin's hand, . . . In almost every instance we have been led . . . to return a verdict of suicide.[1] . . . The breakdowns of civilizations are due to an inward loss of self-determination.[2]

As is written in the book of Proverbs, "Where there is no vision, the people perish." John Stuart Mill said, "In politics as in mechanics the power which is to keep the engine going must be sought outside the machinery." The absence in a society of religious integration—of ultimate values shared in common—first enervates and then destroys.

Americans live in the midst of a revolution; the discoveries of science and the theories of Karl Marx have combined to challenge spiritual values the world over. Traditional religions have been destroyed in many nations. In Europe, Asia, Africa, and Latin America, revolutionary forces have overturned ancient values and created new patterns of living. Only a person who is blind to the true nature of our times could think that revolutionary forces are not at work in this country also. Unfortunately revolutions often destroy the good along with the bad. Thus the survival in the United States of democracy and freedom is by no means certain. The future depends on many factors. Some of these are material in nature: America's ability to produce, her provision for research in the physical sciences, her skill in military defense, her concern (or lack of it) for the preservation of her natural resources. But more basic than any of these is the quality of her elementary spiritual convictions: such things as the genuineness of her dedication to human welfare, the persistence with which she fights for the eradication of special privilege, the spirit of cooperation she extends to other nations, the depth of her belief in freedom for the expression of strange opinions, the amount of her willingness to struggle against infringements on liberty—the liberty of others as well as her own. If democracy and freedom endure in America, it will be because the majority of her citizens believe that the nation which practices democracy and freedom works with the forces of the universe and will receive rewards denied to nations living under other systems. Persons who have such convictions possess at the center of their lives the spirit of devotion and sacrifice which gives vitality to a society.

Thus the continued growth of the idea that religion is unimportant forebodes nothing but ill since this idea leads to an inadequate understanding of life's dynamics, of the forces which work both in individuals and in societies. A knowledge of the religious beliefs and practices of our fellows is clearly one essential to an adequate view of the world in which we live. Every person who strives for a reasonably full education must be informed concerning the spiritual convictions from which spring the basic choices of his neighbors and his society.

This book will attempt to describe the spiritual forces now playing an important role in the United States. The book will indicate the intricacy of the spiritual problem which faces the American people and also furnish data and points of view from which to attack the problem. The traditional religions will receive the lion's share of the space. Chapters will be devoted to each of the following groups: Catholics, Lutherans, Episcopalians, Presbyterians, United Church of Christ, Baptists and Christians (Disciples), Quakers, Methodists, Unitarian-Universalists, Jews, Eastern Orthodox, Mormons, and Christian Scientists. Then will come a chapter on some recent religious innovations, groups like the Pentecostals, Adventists, and Moral Re-Armament.

A study of all these churches and societies would provide a view of the range of organized religion over the country. But organized religion by no means covers all of America's religious life; for, more than

one-third of the American people are members of no church or synagogue. In order to study the beliefs of this group, it will be necessary
to abandon the ecclesiastical approach and to study, all too briefly,
some dominant schools of thought, such movements as nationalism,
atheism, and (the most widespread of all devotions in America)
hedonism (the worship of pleasure).

The story of the various groups will be outlined broadly, except
that in most of the chapters a part of the narrative (usually the life of
a representative leader) will be told in some detail. This brief focusing
of attention on a small section of the story runs the risk of bringing
the hasty reader to think that the importance of these especially
illuminated episodes in religious history corresponds to the amount of
space they occupy in this text. This is not a correct conclusion; many
similar episodes might have been detailed for all the denominations.
These special sections are included in order to help the reader understand that religion is not something found chiefly in books; it is the
glowing center of vital experience. Theological principles are not mere
abstractions; they are the blueprints by means of which vigorous personalities order their lives.

In describing these various American faiths, my purpose will be to
deal fairly with each sect and school of thought. In order to accomplish
this purpose, I will try to set down the facts and let them speak for
themselves. Whatever our prejudice in religion—and most of us
harbor prejudice—our need for more information is clear. To all who
would understand, the facts, insofar as they are available, speak an
eloquent language. At points where there is disagreement as to the
facts, I will try to indicate clearly the opposing points of view. The
reader may upon opening the book turn first to those pages where his
own group is described and may conclude that the treatment is insufficient. *He is asked to note that, in general, material common to
two or more sects is not elaborated in the later sections; moreover, he
is asked to remember that material torn out of context is difficult to
appraise.*

As a part of the effort to keep always before the reader's mind a
clear indication of the sense in which statements are true, a large
number of phrases like "Catholics assert," "Methodists believe,"
"Mormons teach," have been inserted in the text. It is hoped that

these guideposts will reduce to a minimum misunderstandings on what is fact concerning which there is general agreement, and what is faith or opinion—and whose faith or opinion. I shall not hesitate upon occasion to express my own opinion. Since the reader will wish to know from what point of view the writing proceeds, it had better be made clear at the beginning that I am a Protestant: reared a Methodist, ordained a Congregationalist, and now holding membership in a Friends Meeting as well. I hope there are in the writing no hidden assumptions prejudicial to theologies or religious forms not my own, except two: that the freedom to follow conscience in private worship is among the most precious of our liberties, and that the duty to strengthen public morals is among the most essential of our responsibilities.

The effort to be unprejudiced in dealing with as controversial a subject as religion may fail; each reader will have to judge the matter for himself. However, it is not my intention to write for the person who wants to read only pleasant things about his church. One orthodox clergyman, after reading the chapter I had written about his denomination, said, "I could say very strong things about the people in the Church, but I would never dare say anything about the Church of Jesus Christ." Although this clergyman continues to be my friend, he is strongly displeased with the chapter as it stands. My opinion is that whatever one's theory of the origin and nature of the Church, the institutions which now exist are run by men and exhibit human frailty as well as human strength and wisdom. A rosy picture is probably an inaccurate one. No institution or way of thinking is all good and none is all bad. The Reverend P. Antonio Astrain, a Spanish Roman Catholic priest, states this proposition:

Catholics and Protestants have agreed in writing the history of the sixteenth century as it is said Apelles painted the portrait of his one-eyed friend—in profile. But with this difference that we Catholics present it from the side of the good eye, and the Protestants show it on the side of the blind eye. So long as history is written in that partial way, it will be impossible for us to understand each other. . . . It is necessary to examine the beautiful and the ugly, the good and the evil.[3]

Horace Greeley was joking when he said, "All Democrats may not be horse thieves; but all horse thieves are Democrats." Unfortunately most religious people are not joking when they say, "They're all out of step but us."

The mature way to study a religion is to try to find out what it is that makes other people act as they do. Our religious beliefs are often miles apart; but our religious needs are much the same. The more we know about a group of people the less extreme their ideas seem; the more we know about the struggles other people are making the more they seem like ourselves. We all, confronted by life's mystery, construct symbols to help us deal with it. But our symbols differ. While we are going through the process of getting acquainted with the faith of other people, we must struggle against the temptation to distrust their sincerity and also against the temptation to think them unintelligent for believing in what are obviously—to us—absurd notions.

"The things taught in colleges and schools," said Emerson, "are not an education, but the means of education." Similarly a study of the facts of the religions will not supply us with the warm garments of religious wisdom; but such a study ought to supply us with the materials out of which such garments can be fashioned.

CHAPTER II

THE ROMAN CATHOLIC CHURCH

IN MEDIEVAL times nearly everyone believed in miracles. Four or five centuries ago, however, the idea began to gain ground that every event is governed by natural laws and that God never breaks these laws. This idea has had a steady growth, until today it is believed by many people—just how many, no one knows.

One of the most revealing classifications of American religions is based on the degree to which churches teach belief in miracles. Some of the churches hold that when human beings get into trouble God will help them, just as an earthly father helps his children; God, they say, is not bound to follow "natural laws"; He made those laws. Other religious groups believe that a God who broke the laws He Himself had made would be fickle and unworthy of our highest devotion; members of these churches say that all the evidence of science points to the existence of a God whose love for man is expressed in His dependability, that what man must do is to find out what the laws of God are and obey them. A few religious groups hold it to be irrational to believe in any kind of personal God and so, of course, do not believe in the love of God as expressed in miracles; such groups contend that the universe is governed by natural laws which work mechanically.

THE FOUNDING OF THE ROMAN CATHOLIC CHURCH

No American group maintains the traditional emphasis on supernaturalism with more vigor than does the Roman Catholic Church. This emphasis can be seen at many points in Catholic teaching,* but

* Many Protestants hold that the single word *Catholic* should never be used to refer to the Roman Catholic Church. They believe that this Church is

is nowhere more strikingly in evidence than in the belief concerning the divine origin of the Roman Catholic Church—a miracle of awesome proportions. The basis for this belief is a passage found in Matthew 16:15–19—one of the most disputed stories in the Bible. In this passage Jesus asked his disciples:

> . . . But whom do you say that I am? Simon Peter answered and said: Thou art Christ, the Son of the Living God. And Jesus answering, said to him: Blessed art thou, Simon Bar-Jona: because flesh and blood hath not revealed it to thee, but my Father who is in heaven. And I say to thee: That thou art Peter, and upon this rock I will build my church. And the gates of hell shall not prevail against it. And I will give to thee the keys of the kingdom of heaven. And whatsoever thou shalt bind upon earth, it shall be bound also in heaven: and whatsoever thou shalt loose on earth, it shall be loosed also in heaven.

Catholics believe that this passage describes the origin of the Roman Catholic Church; its creation was a direct act of God. They assert that it was Jesus' intention to found just one Church, that He appointed Peter—and no one else—to be the ruler of the Church, that He gave Peter supernatural powers over the Church, and that He decreed that after His ascension all true religion was to be under the supervision of Peter and his successors.

To faithful Catholics this doctrine is the foundation on which rests all their hope of overcoming the temptations of the Devil and of being saved from the torments of hell. They are thrilled and awed by the fact that God could so love His children that once and for all, without uncertainty or equivocation, He should give to men the richest possible gift: the means of gaining heaven. They believe their Church has certain knowledge of all things necessary for salvation, and was given by God complete authority over all men in religious matters.

In the last years of his life Peter, according to Catholic belief, went to Rome where he became Bishop and where he was finally executed by crucifixion. According to tradition, he was crucified head downward, at his own request, since he did not feel worthy to die in the

---

catholic only in the sense that Protestant churches also are catholic. Nevertheless, in this chapter, the customary American usage will be followed and the proper name *Catholic* will be used interchangeably with the proper name *Roman Catholic,* except where the Roman Catholic Church might be confused with other groups to which the term *Catholic* as a proper name is also applied.

For definitions of the terms *catholic* and *protestant* see pages 95 f.

same manner as did his Lord. Catholics hold that the apostolic powers which Jesus had given Peter were passed on to the successive Roman bishops; they too receive from God authority over all Christians. In the nineteen hundred years since the death of St. Peter, 261 men have sat on his throne, an unbroken line, say the faithful, which runs straight back to the fount of all Catholic authority—the words of Jesus that day long ago in Palestine. Thus the Popes, bishops, and priests of Catholicism are believed to be in the Apostolic Succession— each one of them has received his supernatural power and authority from the hands of men who go back in a continuous line to God Himself.

The popular title of the Roman Bishop today is Pope, a word which means father; it comes from the Latin word *papa*. Catholics believe that the Pope is the spiritual father of all mankind, the representative of Christ on earth, and the sovereign ruler of all baptized Christians; to him obedience is due in matters of faith and morals. Today Catholics are debating whether the Pope shares authority with the college of bishops. Until recently practically all Catholics have held that other churches do not originate with Jesus Christ.

PAPAL INFALLIBILITY

The Roman Catholic Church is governed from the top; all formal authority in religion rests with its priests and none with its laymen. At the head the Pope has power in churchly matters. His decisions are law; Catholics believe that he is the agent through whom Christ works on earth.

The Pope's authority can best be illustrated by an examination of the doctrine of papal infallibility. Catholics believe that the Pope cannot make a mistake when he speaks intentionally and officially to the entire Church on matters of religious faith or morals. He is not infallible (note it carefully) in all his acts. He is not infallible, for example, in his decisions concerning business affairs, or in his judgments of the men he appoints to office. The opinions of the Pope concerning the merits of, say, Gilbert and Sullivan as opposed to the merits of Hammerstein and Rodgers would be of only incidental interest to Catholics. Emphatically, they do not think the Pope is God. But when he (1) exercises his supreme office, (2) addresses the entire

Church, (3) speaks on faith or morals, (4) interprets Scripture or tradition, and (5) intentionally formulates the doctrines of the Church —the means of man's eternal salvation—he is believed to be incapable of error. Catholics are certain this is true because they have, they believe, Christ's promise never to desert the Church; and on simple humanitarian grounds, if on no other, it is inconceivable to them that Christ should not use His supernatural powers to prevent error when the Pope is teaching mankind the truths which will determine whether man will spend eternity in heaven or in hell.

Non-Catholics have rather generally misunderstood the dogma of infallibility. This misunderstanding is illustrated by a story which once circulated widely among non-Catholics in the United States. It concerned James Cardinal Gibbons, for many years the famous Archbishop of Baltimore. After it had been announced in 1870 that the dogma of papal infallibility had been explicitly formulated, Gibbons is reported to have said, "Well, the last time I saw the Holy Father, he called me Jibbons."

This apocryphal story exhibits a serious misunderstanding. The conditions under which the Pope is believed to be incapable of error are carefully restricted, as has been noted. Moreover, it should be especially remembered that whenever the Pope uses his infallible powers Catholics believe he does not create a new dogma; rather he simply makes explicit doctrines which have been implicit in the revelation since apostolic times.

The Catholic Church insists that the Pope alone can determine what human activities are in the realm of "morals." Many non-Catholics feel that the Pope claims as his legitimate jurisdiction any political or economic issue about which he happens to be concerned.

Catholics do not believe that the Pope leads a sinless life. According to their teaching, Jesus and the Blessed Virgin were sinless, but no one else. Catholics admit that a few of the Popes have been unsavory characters. A contemporary American Catholic writes as follows:

In the tenth century, that period of profound humiliation for the Papacy, the papal chair fell into the hands of rival factions of the noble Roman families who contended in filling it with unworthy creatures. . . . The Papacy again fell into evil hands in the eleventh century. Benedict IX, a nephew of Benedict VIII and of Pope John XIX, was elected through machinations of the Counts of Tusculum of which family he was a member. A child of twelve years of age and already perverted in morals, his election was obtained by bribery and intimidation. His private life was scandalous and his public life ruled by family greed.[1]

Non-Catholics should not conclude from such frank statements as these that Catholics think that morals are unimportant in religion or that many of the Popes have been wicked men. Just the reverse is true. Catholics put a tremendous emphasis on right conduct and insist that no other line of rulers, civil or ecclesiastic, could begin to compare with the Popes for personal integrity or devotion to the welfare of mankind. *Our Sunday Visitor,* an official Catholic paper, contained the following statement:

> The Church's worst enemies can discover at most four or five who were unworthy of the Papacy, which means that the lives of 98% of all the Popes were not open to any severe criticism. The first 29 died martyrs, and everyone will agree that the Popes of the past century have been saints.[2]

Few non-Catholics today attack the characters of the modern Popes. Concerning Pius XII the Methodist Bishop G. Bromley Oxnam, in an address in which he spoke unfavorably of some aspects of Catholic action, claimed "no lack of respect for the distinguished, devoted, brilliant and brotherly Christian, who is the present Pope."

The point of all this discussion of the personal lives of the Popes is to make clear that their morals, whether good or bad, have no bearing on the belief in their infallibility. The worst Popes, though they no doubt are suffering in hell, are still considered to have been infallible; God miraculously protects the Church from theological and moral error when the Pope exercises his power of infallibility.

By no means all the teachings of the Church are thought to be infallible; there are various levels of certainty in Catholic theology. Infallible teachings, at the highest level, are called "Of Faith"; they must be believed by all Catholics. Teachings at the next level are called "certain"; their denial is "rash." Then come "common opinions"; their denial is "offensive to pious ears." Lower still are "more common opinions" and "less common opinions."

The infallibility of the Pope means, of course, that the fundamental dogmas of the Roman Catholic Church are believed to be perfect.

Traditionally faithful Catholics could speak as follows: "There is but one true Church. God did not found the churches you Protestants attend; your churches are the result of human invention. Some of you even insist that every man must make up his own creed—a horrible thought. How can you take such appalling chances with your eternal salvation?" Catholics frequently deny that their Church is one of the religious denominations as it is usually designated in American speech and law; they have held rather that Catholicism is The Church. Cardinal Cushing of Boston said in an address in Holyoke, Massachusetts, that the Catholic Church is God Himself on this earth; a member of the staff of St. Patrick's Cathedral, New York City, said, "The Catholic Church claims to be Christ."[3]

Although such attitudes have dominated Catholic thought for many years, a change may be in the making. In 1964, Pope Paul VI on a trip to the Holy Land implied that churches other than the Roman Catholic are true churches.

The belief in the perfection of the Catholic faith means that there is no place in the practice of ordinary Catholic priests and laymen for the analysis and appraisal of *basic religious* ideas. The position is that either the fundamentals of religion come straight from God or they do not. If they do not, then religion is false. If they do, then religion is something objective, a gift from God, and not subject to judgment. Said Cardinal Gibbons, "Would God have left the soul of men to chance? Must He not have provided absolute assurance on so grave a matter as the salvation of the soul?"

Clearly, in the light of these teachings, it is the duty of the individual Catholic to accept the basic teachings of the Church without questioning; it would be a grievous sin for a mere human to question the teachings of God. One of the American Catholic bishops said:

> To cause discussions on matters of our Holy Catholic Church, even if they are on "policies or methods" . . . is about as wise and appropriate as having such debates in an old-fashioned saloon, the debaters fortified with jugs of beer or something stronger.[4]

Yet in spite of all this restriction on the discussion of the religious fundamentals, Catholics feel sure that they are intellectually free. Hilaire Belloc, an English Catholic author, contended that only Catholics have true intellectual liberty; he wrote:

Nowhere outside the household of the Faith is the speculative reason fully active and completely free, save possibly as a rare exception, in a few of the more intelligent sceptics. . . . He [the Catholic] accepts very few non-Catholics as his intellectual equals.[5]

This position can be asserted because most Catholics in their thinking use an unusual definition of religious freedom. They hold it is not freedom to believe error. If something is untrue, they say, it is not freedom to have liberty to have faith in it. Would it be true freedom for a child to be permitted to play with the bottles in the medicine cabinet and to take a drink of any one he wishes, even though some are full of poison? As an editorial in a diocesan paper put it: "There is no place in freedom for that liberty which would destroy liberty itself."[6] Now in religion the Church has The Truth. Giving a person religious freedom means for most Catholics preventing him from making the wrong initial assumptions. Recently this commonly held position has been the subject of debate; some Catholic leaders now contend that true religious freedom consists in having immunity from all forms of coercion in religious matters.

But beyond these initial assumptions there is vigorous, and sometimes aggressive, disagreement among Roman Catholics. The bishop who, speaking about a conference of prelates, said, "We disagree on everything but the Apostles' Creed," spoke with obvious exaggeration, but with much truth. In fact the success of the Catholic Church has often been attributed to her ability to satisfy the needs of many types of personality. Two Protestant students of church life, speaking of the Roman Catholic Church, declared that "perhaps the superlative genius of the organization is its extraordinary capacity to deal with human idiosyncrasies."[7]

Catholics emphasize their belief in tolerance and contend that they are intolerant of religious and moral error only; they cannot, they say, see men speeding toward the chasm of hell and still, like good fellows, say "You go your way and I'll go mine." But if men are headed away from damnation and toward salvation and are following the commands of Almighty God as revealed to the Roman Catholic Church, wide tolerance in belief and action is granted. In all life's relations—in education, in politics, in business, in recreation—Catholics assert their belief in a live-and-let live policy, *except where the destiny of the*

human soul is at stake. But here too the Church claims a true tolerance; for, she protects her children against what she is certain is error.
Our Sunday Visitor states the Catholic opinion emphatically: "Whatever intolerance is manifest in the world today is not Catholic
intolerance towards others, but Communist, atheist, infidel and
Protestant intolerance towards Catholics."

THE HIERARCHY

The Catholic Church is ruled from Rome. The character and
extent of this rule is an important aspect of our study since many
non-Catholics in the United States show deep concern over the fact
that so many Americans give spiritual allegiance to a foreign prelate.

For many centuries the Pope was the monarch of a considerable
territory in Italy; in addition to being the spiritual ruler of world
Catholicism, he was the temporal ruler of an Italian principality.
However, the unification, in 1870, of the small Italian states into the
modern nation of Italy cost the Pope his "temporal power"; his kingdom was taken away. In protest, he shut himself up in his Roman
palace—the Vatican—and refused to come out. For over a half
century the successive Popes were voluntary prisoners. In 1929, however, Mussolini and the Pope came to an agreement by which the
Pope again became a temporal sovereign. An area of about one fourth
of a square mile, in the heart of Rome, was given to him. This section
of the city ceased to be Italian territory and became an independent
state; it is called the Vatican City, and the Pope is its ruler.

The affairs of the Vatican are run with great pomp and ceremony.
"Today, it is substantially an Italian court of the Renaissance type,
with titles and costumes used in the sixteenth century."[8] The Vatican
contains a throne room, the Pope wears a triple crown, and one of his
important assistants is a cardinal in charge of ceremonies. Protocol is
so strict at the Vatican that some of the faithful were shocked by the
fact that President Woodrow Wilson during a private audience with
the Pope kept on his overcoat. Etiquette requires that the Holy Father
eat alone.

The Pope carries tremendous responsibilities. As the spiritual leader
of half a billion people and the Bishop of the cathedral church of
Rome, he must officiate at a large number of churchly functions: the

canonization of saints, the elevation of cardinals, the celebration of Holy Days, the reception of pilgrims, the consecration of devotional objects. In addition he has the final authority in administering the vast affairs of a world-wide enterprise. His days are full of audiences with bishops and archbishops, of reading reports, of presiding at conclaves, of preparing sermons, of writing encyclical letters, of studying the complex problems which confront all religions today. In his personal life, the Pope has the same duties which are assumed by every priest: he says Mass each morning, he reads his Breviary (mostly selections from the Bible) for about three quarters of an hour each day, and he confesses his sins regularly to an especially appointed confessor.

The second highest dignity in the Roman Catholic Church is that of the cardinals; they are the "princes of the church," holding in the ecclesiastical court of the Vatican the rank which the "princes of the blood" hold in secular courts. Cardinals are appointed by the Pope and compose the College of Cardinals. Their number traditionally was limited to seventy but this limitation is no longer observed. The chief power of this group is to elect the Pope; aside from this function, the duties of the College are advisory. As individuals, however, the cardinals hold many important administrative posts. For example, each of the Roman Congregations—the major departments of the Church's government—has a cardinal at its head, and some of the archdioceses are also headed by cardinals.

From the fourteenth century until 1946 the College of Cardinals was dominated by Italians; for all these centuries, Italians composed a majority of the group. This situation disturbed American Catholics, who saw no reason to suppose that spiritual sensitivity was a special possession of persons who happened to be born in Italy. Americans have long felt that the growing strength of the Church in the United States was not sufficiently recognized in the Sacred College. They know that the Vatican leans heavily on the United States for its income. The extent of American support is not known to the public; the exact status of Vatican finances is a closely guarded secret. However, it is sometimes estimated that gifts from the United States to the Pope exceed in value the gifts from all other sources combined.

Pope Pius XII strove to give Catholics everywhere representation among the cardinals. When he assumed office in 1939, twenty-four out of thirty-eight cardinals were Italian. But at one stroke he elevated twenty-eight "foreigners" to the Sacred College, and only four Italians. Thus for the first time in the modern era the Italians formed less than a majority of the College. The representation of the Western Hemisphere was increased from only three to fourteen. The number of cardinals in the United States was increased to five. Pope John XXIII continued this policy. In 1961, of the eighty-six cardinals, thirty-two were Italian and fifty-four were non-Italian. Six were from the United States. In 1962, the number of Cardinals was increased to eighty-seven.

The Pope has, in addition to the counsel of the cardinals, the assistance of a multitude of departments, bureaus, and courts. Such interests as the following are under the supervision of especially qualified leaders: diplomatic relations, missions, education, finance, doctrinal problems, direction of bishops, direction of orders of monks and nuns, administration of the sacraments, conduct of worship, annulment of marriages, disputes among the faithful. Although final administrative power belongs to the Pope, in actual practice the affairs of the Church are so extensive that most of the decisions which come from the Vatican are made by lesser officials. This elaborate organization is usually referred to as the *Curia*.

The Roman Catholic Church is divided into territorial units called dioceses; a diocese is ruled by a bishop. Dioceses in the most important cities are given prestige by being called *arch*dioceses and by being ruled by an archbishop. In the United States there are 111 dioceses and 27 archdioceses. These territorial units are combined into larger units called provinces. Each province is presided over by an archbishop. The archbishop at the head of a province has great dignity; but he has relatively little power over the other bishops in the province; they are governed in most matters directly from Rome. The archbishop's function in the province is one of spiritual and intellectual leadership. In his own archdiocese, of course, the archbishop exercises the authority of any bishop.

The bishops are believed to be the successors of Christ's apostles and thus to have apostolic powers. A writer in *The Catholic Encyclopedia* says, "The episcopate is monarchial. By the will of Christ, the supreme authority in a diocese does not belong to a college of priests

or of bishops, but it resides in the single person of the chief."⁹ Thus
the bishop has the power to care for and direct the *spiritual* life of all
Catholics who dwell within the boundaries of his diocese. He does
this by organizing it into parishes, by visiting each of these parishes
periodically, by personally holding title to all Catholic property in
his diocese (except property belonging to the religious orders), by
establishing and maintaining schools, by ordaining priests, by providing agencies for the conversion of non-Catholics, by setting personally an example for the spiritual life of the faithful.

The bishop has the power to enforce his decisions; he can censure,
demote, excommunicate, and in the case of priests, he can "confine
for a time in a monastery." Since there are over two thousand bishops
and archbishops throughout the world, it would be foolish to claim that
these powers are never abused. But the bishop seldom finds it necessary to use the full weight of his power; priests and laymen alike are
so accustomed to expecting direction from him on all matters pertaining to faith and morals, that they seldom challenge his authority. Yet
Catholics do not usually consider him to be a stern ruler. Almost universally Catholic folk think of their bishop as a beloved leader who
has given his life to the Church and who needs their prayers.

The hierarchical system continues right to the bottom of the
Catholic governmental structure. Each parish is under the control of
a pastor. The pastor may have as his assistants ordained priests
(curates), and he can seek freely the advice of able laymen in difficult
matters of administration; but in the last analysis, he has final authority within the boundaries of his parish, unless he is overruled by
his bishop or by the Vatican. He has charge of the funds of the church
and he may disperse them without being under obligation to report
back to his parishioners. He has control of all services of worship and
he is the head of the parish school. Ordinary members of the hierarchy
may not come into his parish to perform religious functions or even to
speak on religious matters without his permission. Many a person
bent on decreasing religious prejudice has made the mistake of asking
a well-known priest to speak at a goodwill meeting, without first con-
sulting the man who is in control of the parish where the meeting will
take place. His permission is mandatory in this as in all other matters
pertaining to the spiritual welfare of the parish. The pastor has au-

thority only in ecclesiastical matters, and he is not thought to be infallible. In his conduct of parish affairs, he acts in accordance with the long-established principles of the Church and is under the direct and constant supervision of his superiors. Catholics put their trust in his spiritual wisdom and in his supernatural powers; for the "hands have been laid upon him" and he is a priest of God, and Catholics believe he acts for Christ.

The title *monsignor* is given to priests who have made exceptional contributions to the life of the Church. The monsignor ranks in dignity just below the bishop; but the title carries no specific duties.

The bishops and archbishops in the United States have banded together to form the National Catholic Welfare Conference, with headquarters in Washington. The Conference has great prestige and it works effectively in education, social service, and publicity; but it does not have jurisdiction over the American bishops. In Washington also is the office of the Apostolic Delegate. He is the direct representative of the Vatican, the leading member of the hierarchy in this country, and the wielder of a great deal of authority.

The cardinals, archbishops, and bishops of the whole Church are sometimes called into council. The Second Vatican Council came into existence in 1962; the last prior council was convened in 1869.

CATHOLIC THEOLOGY

Let us turn now to Catholic beliefs about God, Jesus Christ, angels, demons, saints, and the future life.

Catholics, along with most of the groups we will study, are *monotheists*—they believe in one God. The following statement is a simple exposition of Protestant and Catholic monotheistic belief: God is personal—He has more in common with a human being than with anything else we can liken Him to. God is pure spirit—He has no body and is not subject to the limitations of matter. God is eternal—past, present and future are one to Him. God is ubiquitous—He is present everywhere. God is omniscient—He knows everything. God is holy—His love is perfect. God is omnipotent—His power is boundless.

Many of these propositions are puzzling; thus they receive a great deal of exposition in Catholic (and other religious) writing. Limitations of space prevent the elaboration here of these discussions; they can be but illustrated by quoting a passage from a standard college textbook on the Catholic faith. Dr. Paul J. Glenn expounds the omnipotence of God in the following manner:

"God is omnipotent. He can do all things. Can God, then, make a square circle? Can God make an object that shall be entirely black and also entirely white? Can God utter a truth that is false or a lie that is true?" Certainly not. God can do all things, but what you suggest are not things, but *denials of things.* You suggest contradictions, that is two things, one of which negatives or cancels the other: the result is simply zero. A "square circle" is "a circle that is *not* a circle"; in other words, it is nothingness. Your suggestion is like this: you draw a circle on the blackboard. Then you erase it carefully leaving not a trace of the drawing. Then you stand back, and, pointing to empty space, you say, "Can God make that?" Make *what?*[10]

Catholics are *trinitarian,* as are most Protestants. Trinitarians believe that in the Godhead there are three persons: the Father, the Son, and the Holy Ghost. God is *one* substance but contains *three* persons. An inquiring mind is moved to ask, "How is it possible for three persons to be one God? Such a proposition is contrary to the human reason." But the dogma of the Trinity, say the faithful, was not conceived by the human reason; the Trinity is one of the truths about supernatural things which God Himself revealed to mankind. One Catholic writer explains the Trinity thus:

While we grant that the dogma of the Blessed Trinity is an absolute mystery, which unassisted reason could never discover, nor even recognize as possible, we deny that it involves any contradiction. . . . All things in God are common to the Three Persons, and are one and the same, except where there is the opposition of relation. The divine activity is common to the Three Persons, who are the One Principle of all things. As the words *one* and *three* refer to essentially different things, NATURE and PERSON, there can be no question of any contradiction of terms. . . . The Three Persons who have the Divine Nature are not three gods, because the Divine Nature is numerically the same in each one of them. How this can be we can never comprehend. We accept this doctrine only because it has been revealed to us by God himself.[11]

The doctrine of the Trinity means that Jesus Christ is God. Jesus is the Son, the second person of the Trinity. Trinitarians believe that God in His infinite love for mankind wished to make clear all of the supernatural truths essential to man's salvation. Therefore, He took upon Himself the form of a human being. The Son lived on earth, walked over the dusty roads of Palestine, instructed His disciples, was executed, rose from the dead, and now reigns in heaven, where He is enthroned at the right hand of the Father. Only the *second* person of the Trinity did these things. This teaching about Jesus is. called the Doctrine of the Incarnation.

The orthodox, Catholic and Protestant alike, believe, however, that Jesus Christ was man as well as God. They believe that he had the attributes of man—he suffered, became tired, was tempted, hungered. Thus Jesus Christ is believed to have had two natures—one human and the other divine. "Yet," says one writer, "these two Natures remain strictly distinct; the lower does not in any way influence the higher, while the higher only influences the lower as it would do even if it were separated."[12] This doctrine is called the dogma of the dual nature of Jesus Christ. How the doctrine can be true is a transcendent mystery.

The nature of the Holy Ghost is also a mystery. It is believed, however, that this third person of the Trinity works in the world: He appeared in the form of a dove at the baptism of Jesus; He descended in tongues of fire at Pentecost. (See Acts 2.) He is present in the soul of every true Christian, where He accomplishes the cleansing and strengthening work of salvation.

THE ANGELS

Catholics believe that many beings other than the Blessed Trinity inhabit the celestial regions. Next to God are the angels which, according to Catholic faith, exist in prodigious numbers. Angels are beings intermediate between God and man, and they are purely spiritual, having no bodies. There are three hierarchies of angels and each hierarchy has three choirs. The function of the first hierarchy is the contemplation of the Godhead; of the second, the arrangement of things in the universe; of the third, the execution of God's orders. The names of these choirs, from top to bottom, are said to be as follows: seraphim, cherubim, thrones, dominations, virtues, powers, principalities, archangels, angels. The names of only four individual angels are known: Raphael, Michael, Gabriel, and Lucifer (Satan). Pius XII near the close of his reign urged some pilgrims from New York to seek "a certain familiar acquaintance with the angels," and added, "No one is so humble but he has angels to attend him. So glorious, so pure, so wonderful they are, and yet they are given to be your fellow wayfarers, charged to watch carefully over you lest you fall away from Christ, their Lord."[13]

The Catholic Church teaches that sometimes demons (fallen angels) take possession of a human being and control his body, talking through it and committing crimes. "Authentic cases of [demon] possession sometimes occur and every priest, especially if he be a parish priest, or pastor, is liable to be called upon to perform his duty as exorcist."[14] An exorcist is a person who casts out demons. A priest is considered to have received the power of exorcism at one of the rites which are preliminary to his final ordination. He finds frequent use for this power. For example, a child about to be baptized is exorcised as a preparation for entry into the Church. The priest puts salt into the child's mouth, rubs spittle from his own mouth on the child's nose and ears, blows his breath in the child's face, and says, "Depart, thou evil spirit, and give place to the Holy Ghost." The press noted a non-baptismal exorcism when a fourteen-year-old St. Louis boy was freed of an evil spirit.[15] Another example of the use of the power of exorcism is the preparation of holy water. Found near the entrance to Catholic churches, holy water is a mixture of exorcised water and exorcised salt. Physical contact with this water is believed to have a purifying and strengthening effect. A leaflet published by the Benedictine Fathers says that Catholics "sprinkle the interior of their homes with holy water in times of severe storms and similar dangers, and in many places it is customary for the Priest to bless the whole house with the newly-blessed Easter water on Holy Saturday. . . . At the time of death, holy water is frequently sprinkled over the bed of the sick person and around the room to ward off all the wiles and temptations of the Evil Spirit."

Catholics believe that at the end of time—the Judgment Day—Satan and all his demons will be confined in hell and never again will be allowed to pass beyond its boundaries.

THE SAINTS

In addition to the angels, God has with Him in heaven, according to Catholic teaching, the souls of men and women who are saved;

these are the saints. Since God is holy, no soul which has not been made spotlessly clean can enter heaven. Very few human beings live lives of such purity that they can be purged in this life of their sins and thus go to heaven immediately after death; most souls must go first to purgatory to be cleansed of the stains of earthly sins. When, after purgation, the soul finally reaches heaven, it has the most glorious experience possible: it sees God, it has the beatific vision.

Catholics say prayers through souls the Church has declared to be saints; Catholics believe it is entirely rational to present their petitions to God through the souls who stand before His throne. Since the prayers of a saint are said directly to God, they are thought to be much more effective than the prayers of one who is still confined to the earth.

The greatest of all the saints is the Virgin Mary, the Mother of God.

Mary is God's choicest handiwork, as the most perfect and most holy person God ever made.[16]

We Catholics love her because we love Jesus. Jesus was God, so that His mother must be called the Mother of God. We love her because God loved her first. It was He, after all, who chose her out of all the girls in Palestine; it was He who set her apart from everlasting to be the mother of His Son in this world; it was He who granted her the unique privilege of begetting a child without the cooperation of any natural father. He it was who spared her soul the withering blight of Adam's sin; who ornamented her with every virtue; who lifted her body up from the tomb, lest it become the food of worms and decay, and brought it direct to heaven.[17]

Catholics believe in the *Virgin Birth,* that is, in the dogma that Jesus had no earthly father. Furthermore, they teach that Mary and Joseph remained virgins all their lives. Non-Catholics point out that the brothers of Jesus are mentioned several times in the Bible.[18] Catholics reply that the word *brother* proves nothing; it had wide meaning among the Jews. The men whom our English translations call brother—James, Joseph, Simon, and Jude—were probably cousins of Jesus, say Catholics, though we cannot be certain about it. But we can be certain that they were not sons of Mary; for, it is a part of divine revelation that Mary was perpetually a virgin.

Catholics believe in the *Assumption of the Blessed Virgin Mary,* that is, in the preservation of her body after her death and its transportation to heaven where it was reunited with her soul. Two centuries ago denial of the Assumption was declared to be *impious* and *blasphemous*. In 1946 Pius XII asked all bishops for their opinion on making the Assumption a part of the *infallible* teaching of the Church. In 1950 the dogma was declared to be definitely part of the revelation, and belief in it became mandatory.

Mary is honored in many ways: her statue is found in all Catholic churches (or "should be," says my neighbor, the Catholic curate); Catholics consider that she is the patron saint of the United States; she is honored every time a Catholic says the Rosary. The Rosary is a prayer composed in part of a portion of the Annunciation to Mary of the coming birth of Jesus. The full prayer follows:

Hail Mary, full of grace! The Lord is with thee: blessed art thou amongst women, and blessed is the fruit of thy womb, Jesus. Holy Mary, Mother of God, pray for us sinners, now and at the hour of our death. Amen.

In praying the Rosary, a worshiper repeats the Lord's Prayer, and then repeats the prayer printed above ten times. This sequence is repeated fifteen times. During these prayers, the worshiper meditates on fifteen different events in the life of Jesus.

Rosary beads are used to count the prayers. Strictly speaking, the Rosary is the spiritual exercise and not the string of beads.

It would be well at this point to note that the Catholic version of the Lord's Prayer differs from the Protestant version. The difference is that Catholics leave out the words which Protestants use to conclude the prayer, "For thine is the kingdom, and the power, and the glory, for ever." These words are found in the Protestant *King James Version* of the Bible, but not in the Catholic *Douay Version* (Matthew 6:13). Biblical scholars generally agree that these words were not spoken by Jesus, but were added by a later editor.

Next to the Virgin Mary the most beloved saint is Joseph, her husband. A priest writes:

The dignity of St. Joseph in the last analysis stems from the fact that he was the virginal husband of the Mother of God. It was because of his marriage to our Lady that he possessed the rights of a father over Jesus her Son. . . . Joseph obtains graces and favors for his clients by going to Jesus through Mary. . . . Since the Holy Family was one on earth in love and mutual confidence, they have the same close intimacy in Heaven.[19]

Catholic veneration of the saints is evidenced by the fact that faithful Catholic parents name each of their children after a saint and by the fact that in Catholic churches many works of art—paintings, windows, statues—do the saints honor. It should be carefully noted, however, that Catholics *adore* God alone; only *honor* and *reverence* are given the saints. Special emphasis should be put on the fact that Catholics are supposed never to worship the *statue* of a saint; any statue is but an *aid* to worship. It is my impression, however, that lay Catholics sometimes violate this rule.

THE NEXT LIFE

Catholics believe that there are four possible places (or states) for the soul after death: heaven, hell, purgatory, and limbo.

Heaven is generally considered to be a definite place—just where is not known. It is the place where the souls go who have been made pure of sin; there they experience a supernatural happiness. This happiness consists chiefly of seeing God face to face, that is, of experiencing the beatific vision. At the "last day," it is believed, the souls in heaven will receive back their bodies. These bodies will be glorified, will be immortal, will be beyond the reach of pain, will shine like the sun, will be able to move with instantaneous speed, will be incapable of sin.

Hell is conceived as a place or state where the damned—demons and men—are punished for their sins. From certain passages in the Bible it is generally thought that hell is inside the earth; no direct revelation on the point has been made.

Purgatory is believed to be a place (or perhaps a state) of cleansing where souls go immediately after death to atone for the sins they have committed on earth. The soul suffers torment in purgatory: the beatific vision is denied and the purgatorial fire causes suffering more severe than anything which can be experienced in this life. Some Catholics believe that the sufferings of purgatory are as painful as those of hell; others, however, assert that the severity of purgatorial punishment corresponds to the wickedness of the sins which were committed on earth. Catholic teaching is that the punishment of purgatory will have an end; every soul will be released as soon as its debt for sin has been paid. Furthermore, purgatory itself will one day cease to exist.

Limbo is of two kinds. The limbo of the fathers is an abode of natural happiness where souls went if they were justified before Christ opened the gates of heaven at the time of His resurrection; the saints of the Old Testament, for example, went at first to limbo and then entered heaven after the resurrection. The limbo of the infants is the place of eternal abode for unbaptized infants. Jesus said, "Unless a man be born again of water and the Holy Ghost, he cannot enter into the kingdom of God." However, that saying does not mean that all unbaptized persons go to hell; hell is thought to be reserved for those who have willfully disobeyed God's law. A person cannot be willfully disobedient before he has reached the age of reason. Therefore, God provides in the limbo of the infants a place where unbaptized persons, who have not achieved the knowledge of right and wrong, are sent. It is a place of much greater natural happiness than the earth, and those who dwell there are free from any positive spiritual anguish for the loss of the beatific vision.

Immediately after death, God decides the soul's fate, decides to which of the abodes for the dead the soul will go. This decision is the *particular judgment.*

Theologians suppose that the particular judgment will be instantaneous, that in the moment of death the separated soul is internally illuminated as to its own guilt or innocence, and of its own initiation takes its course either to hell, or to purgatory, or to heaven.[20]

The Catholic Church teaches that at the end of time will occur the *last judgment,* sometimes called the *general judgment;* then Jesus Christ will return to earth and will judge all men. The time when this event will happen no one knows; but the place where it will occur is thought to be in the air just above the earth. Evidence for this faith is found at many points in the Bible. Perhaps the best-known passage is I Thessalonians 4:15–16 where Paul wrote:

For the Lord himself shall come down from heaven with commandment, and with the voice of an archangel, and with the trumpet of God: and the dead who are in Christ, shall rise first. Then we who are alive, who are left, shall be taken up together with them in the clouds to meet Christ, into the air, and so shall we be always with the Lord.

Catholic dogma holds that the last judgment will be preceded by many signs and wonders; among these forebodings are the following: the Jews will be converted to Christianity; Enoch and Elijah (the two men who are said in the Bible never to have experienced death but to have gone directly to heaven) will return to earth; there will be a great reduction in the number of Christian people through the abandonment of the faith by many nations; the world will be ruled for a time by the Antichrist (the archenemy of Christ); extraordinary natural disturbances will occur—pestilences, famines, earthquakes, fires; the trumpet of the resurrection will sound, a trumpet which will awaken the dead; the "sign" of the Son of Man (perhaps the cross) will appear in the heavens.

At death the soul and the body are separated. On the last day the bodies of men, both the saved and the damned, will be resurrected and will be reunited with the souls to which they were joined before death.

THE MASS

No part of Catholic belief and practice when fully understood is more astonishing to Protestants and Jews or more satisfying to Catholics than the Mass—the central act of Catholic worship. The first saying of the Mass, according to the Catholic faith, was on the night before Jesus was betrayed, at the Last Supper.

And whilst they were at supper, Jesus took bread, and blessed, and broke: and gave to his disciples and said: Take ye and eat. This is my body. And taking the chalice, he gave thanks and gave to them, saying: Drink ye all of this. For this is my blood of the new testament, which shall be shed for many unto remission of sins.

The Mass is the re-enactment of the Last Supper; included are the saying of many prayers, the giving of many blessings, the reading of Scripture lessons, and frequently the preaching of a sermon; but all these are of minor significance when compared to the Consecration of the elements, the repetition by the priest of the words which Jesus said at the Last Supper. The priest takes unleavened bread and grape wine to the altar; then, as he says in Latin, "This is my body," the bread ceases to be bread, according to the Catholic faith, and becomes the Body of Jesus Christ. As he says, "This is my blood," the wine ceases to be wine and becomes the Blood of Jesus Christ.

Non-Catholics, unless they read carefully, are apt to miss the full significance of Catholic faith at this point. The bread does not become the *symbol* of Jesus' Body; nor is His Body present merely after a *spiritual* manner; nor is it present *along* with the bread. The substance *bread* ceases to be bread and becomes flesh, the flesh which walked the earth as perfect God and perfect man. Likewise, the wine becomes the actual Blood of Jesus. A priest writes, "Jesus Christ is present in this Sacrament *truly,* not merely in a figurative manner, *really,* not merely by faith, *substantially,* not merely by His power."[21] This stupendous miracle is called transubstantiation—the changing over of the substance.

The reasonableness of belief in transubstantiation is defended in the following ways. First, its reality is a part of the revelation, part of the infallible teachings about which there can be no mistake. Second, at the Last Supper Jesus did not say, "This bread is a *symbol* of my body"; He said, "This *is* my body." Third, Jesus performed another miracle of transubstantiation when He changed water into wine at the wedding feast. Fourth, it is not considered that the miracle is wrought by the priest; rather the miracle is wrought by God who uses the words of the priest as the occasion for using His supernatural power.

Catholic writers sometimes make even greater claims for the powers of the priest. A passage from a theological handbook reads:

The priest has power over the lifeless created thing and over the Creator Himself, and that just when he pleases. One word out of his mouth compels the Creator of the Universe and of Heaven to come down to earth, strips Him of His greatness and hides Him under the form of the Bread.[22]

One of the Cardinals, in a pastoral letter written in 1906, said:

Where even in heaven is there such power as that of the Catholic priest? Once did Mary bring the divine Child into the world, and behold, the priest does it not once but a hundred, a thousand times, as often as he celebrates. To the priests has Christ handed over the right over His holy humanity, to them he has similarly given control over His body. The Catholic Priest can . . . make it present upon the altar, shut it up in the tabernacle, take it out again, and give it to the faithful to enjoy. . . . Christ the only begotten Son of God the Father is thus at his disposal.[23]

After the Mass has been said, the Body of Our Lord looks like bread and tastes like bread. But this fact, acknowledged freely by Catholics, does not destroy their faith. They say that the miracle deals with the *substance* and not with the *accidents* of the bread and wine. The real substance of any object, they believe, can never be directly experienced by the senses; the senses experience only the form, the accidents. For example, a student has on his desk a paperweight. It is made of iron. But he cannot really experience iron. He can see that the iron is perhaps round and brown; he can feel that it is heavy, and cold. He can strike it with a pencil and discover that it has a dull sound. But with all these sensations he has not experienced iron; he has experienced only the *accidents* which surround the iron. The iron in the paperweight might have had other accidents: it might have been square, red, hot, and have a hollow sound. Similarly, say Catholics, with the bread and wine on the altar. When God performs the miracle of the Mass He changes the physical substances *bread* and *wine* into the physical substances *flesh* and *blood;* but He leaves the accidents unchanged. This is the reason the Blessed Body and the Precious Blood continue to look and taste like bread and wine.

Non-Catholics can begin to get some idea of the attitude which faithful Catholics have toward their church building if it is realized that they believe the building is the house of God in a literal sense; God is physically present there. Thus ecclesiastical law requires that churches which contain the Blessed Sacrament be kept open if possible for at least a few hours every day; most Catholic churches are in fact open every day in the year. Of course, Catholics believe that God is everywhere, just as other Christian people do. But in addition they believe that after the elements have been consecrated, God is really present on the altar. The actual words of the Consecration are said by the priest with his back turned toward the audience; but immediately after the miracle has been performed, he elevates the wafer of bread, now called the Host, high above his head for all to see. The worshipers then look with awe upon the Body of God and devoutly say, "My Lord and My God."

In receiving Communion, Catholic laymen are permitted to partake only of the Body of Our Lord; since the twelfth century the drinking of the Precious Blood has been reserved for the priest. Communion in but "one kind" means no spiritual loss to the layman; for, the infi-

nite Christ is believed to be in the Blessed Body, whole and entire—
the divine being cannot be divided into parts.

The non-Catholic reader should ponder the spiritual and physical
effects on devout Catholics of receiving into their own bodies what
they believe to be the Body of Almighty God Himself. Their own weak
human selves thus become the very tabernacle of the infinite Christ
and continue so until the digestive processes significantly change the
accidents of bread. How this wonder can be, Catholics do not understand. It is a mystery. It is beyond reason, though they consider it is
not contradictory to reason. The infallible Church is the guarantor of
their faith.

Another significant teaching about the Mass is that each time it is
said, the death on Calvary is presented anew. Traditional Christians—
both Catholic and Protestant—hold that sin so estranged man from
God that it was necessary for some act of sacrifice to be made to win
back for man the favor of God; the Crucifixion was this sacrifice.
Catholics believe that the Mass repeats the sacrifice of the Crucifixion.
There on the altar Jesus Christ takes again upon Himself the sins of
men and again is offered up to appease the wrath of a God angry with
the wickedness of His children. There is but one difference, in Catholic belief, between the sacrifice on the cross and the sacrifice on the
altar: the one was bloody and the other is not.

Having considered briefly the central doctrines of the Mass, a description of it and of its celebration will give some understanding of
Catholic moods and piety, and of the kind of impressions made on
worshipers by the great rite. Any description for non-Catholics must
include much detail because they are usually greatly confused by the
elaborate ceremonies which surround this sacrament.

The Mass takes place on an altar. This sacred structure must have
a stone that has been consecrated by a bishop and it must contain a
sealed sepulcher in which are fragments of the bodies of at least one
and preferably two or more martyred saints. Above or on the altar
must be either a crucifix or a painting of the Crucifixion. Some altars
are movable; but those seen in most Catholic churches are fixed. On
fixed altars there is a little safe called the tabernacle in which the
Sacred Hosts are reserved, if any remain after Communion.

As worshipers gather in the church for the Mass, they sometimes make an offering before they enter the aisles; then, just before they enter the pews, they pause to show their reverence for the reserved Host in the tabernacle by genuflecting. One genuflects by a momentary bending of the knee. Once in the pew the Catholic worshiper is supposed to begin his own private devotions: he may meditate, he may say his own prayers, he may pray the Rosary. Many Catholics continue these private devotions right through the Mass, pausing only to observe its most solemn moments. Other Catholics follow the reading of the priest in a Missal, a book which contains the Latin worship forms which the priest is using and also English translations. Under usual conditions, Catholic worship is much less "congregational" than Protestant worship.

Before Mass begins the priest puts on the vestments—the garments —in which it is customary for Mass to be said; these vestments were adopted for ecclesiastical use from the secular dress of the Romans in the third and fourth centuries. As the priest puts on each garment he prays that he may be made more worthy of his divine responsibilities. The vestments are in six colors; each color is symbolic of a different aspect of the spiritual life and is used at different seasons of the Church year. White is a symbol of joy; red, of blood; green, of hope; violet, of penance; rose, of joy during penitential seasons; black, of mourning.

At the beginning of the Mass the priest and his server enter, the priest carrying the chalice. At the foot of the altar the priest, in Latin, prays forgiveness for his own sins and for the sins of the people.

I have sinned exceedingly in thought, word and deed: through my fault, through my fault, through my most grievous fault. Therefore I beseech blessed Mary, ever virgin, blessed Michael the Archangel, blessed John the Baptist, the holy Apostles Peter and Paul, all the saints, and you, brethren, to pray to the Lord our God for me. . . . May the almighty and merciful Lord grant us pardon, absolution, and remission of our sins.

The priest then ascends the altar steps, and bending over the altar kisses it; this kiss is symbolic of a greeting on the part of the Church, the bride (represented by the priest) to Christ, the bridegroom (represented by the altar). Then the priest asks that God will have mercy and recites the Gloria.

Glory be to God on high, and on earth peace to men of good will. We praise Thee; we bless Thee; we adore Thee; we glorify Thee. We give Thee thanks for Thy great glory. . . . Thou only art the Lord: Thou only, O Jesus Christ, art most high, together with the Holy Ghost, in the glory of God the Father.

Then follows the service of instruction: the priest reads from one of the Epistles, from some of the Psalms, from one of the Gospels, and then preaches a sermon. The sermon is in the vernacular, in English in most American churches. On solemn occasions the priest recites the Nicene Creed. At the end of the service of instruction, an offering is sometimes taken.

After the service of instruction is over, the priest prepares the elements for the Consecration. He places a wafer on the paten—a small plate—and then pours wine and a few drops of water into the chalice —the sacred cup. Wine and water are mixed in the chalice to signify the mystical union of both the divine and human elements in Jesus Christ. The priest then washes carefully his thumbs and forefingers; for, they will touch the Blessed Body. During the cleansing of his fingers he recites part of the 25th Psalm. Then by solemn prayers he prepares himself for the Consecration.

At the beginning of the Canon—the most sacred part of the Mass— three bells are rung to warn the faithful that the Consecration is about to take place. The worshipers kneel and the priest proceeds in a low voice; for it is fitting that things most sacred and holy be celebrated in near silence. However, he speaks loudly enough so that he can hear his own voice. He prays to God to remember especially certain living persons; for example, those who are ill in the parish, those who may have given a stipend for the Mass, those who are present. He then prays for protection through the merit of the saints and asks that the sacrifice about to be made be acceptable unto God. The priest then takes the wafer in both his hands and says in Latin the words of Jesus at the Last Supper, "HOC EST ENIM CORPUS MEUM" ("This is my body"). This sentence is believed to perform the miracle of transubstantiation.

Immediately after these words of Consecration the priest elevates the Sacred Host high above his head for all to see. The faithful look at it, confess it as their God, and bow in humble adoration. This is a supreme moment. For those who truly believe, how could it help but set everything right, remove temptation, quicken resolve, make all the sordidness of living endurable?

Then the wine is consecrated and the chalice is elevated. Periodically throughout this service of Consecration, bells have been rung in order that the faithful may be on the alert at the time of the divine advent. The thoughtful worshiper is deeply aware of the presence of God. He is also taught to be aware of the presence of thousands of celestial spirits.

It was revealed to St. Mechtilde that three thousand angels from the seventh choir, the Thrones, are ever in devout attendance around every tabernacle where the Blessed Sacrament is reserved. Doubtless a much greater number are present at Holy Mass, which is not merely a sacrament but also a sacrifice.[24]

Several prayers follow the Consecration and then the priest himself takes Communion: he consumes the Host and drinks the Blood. He is very careful that every particle of the Host shall be eaten and that no drop shall remain in the chalice. He scrupulously cleans the paten and, before Mass is over, he drinks wine which his server pours into the chalice; this wine mingles with such small portions of the precious Blood as may have remained after his Communion.

The sacrificial aspect of the Mass is completed by the Communion of the priest. Every sacrifice demands the annihilation of a victim. In the Mass, Christ's Body and Blood are destroyed by human eating and drinking. Thus, since Christ has again taken our sins upon himself, God's anger at the sins of the worshipers has been appeased anew.

After the priest's Communion, those among the worshipers who also wish to receive Communion come to the altar rail and kneel. The priest then picks up the ciborium, a vessel containing Hosts which were consecrated when he consecrated the Host he himself consumed. He carries the ciborium to the altar rail; then, grasping one of the Hosts between his thumb and forefinger, he places the Blessed Body on the tongue of the communicant. Meanwhile, the server holds a large plate under the chin of the communicant in order to guard against the danger that the Sacred Host might fall to the floor and be contaminated.

When the last person wishing to receive Communion has been

served, the priest returns to the altar, places any unconsumed Hosts in the tabernacle, locks it, reads further prayers, blesses the people, reads a passage from the Gospel, takes the sacred vessels in his hands, withdraws from the sanctuary, and Mass is over.

Many interesting regulations surround the saying of the Mass. Unless a special dispensation is made, Mass is said in the interval between two hours before dawn and noon—except at Christmas when a special Mass is said at midnight. Anyone who receives Communion is required to abstain from eating solid food and from drinking alcoholic beverages for the three hours previous, and from drinking nonalcoholic beverages for one hour previous. Water does not break this "eucharistic fast."

Catholics are urged to take Communion frequently—several times a week—and violate Catholic teachings unless they take it at least once every year, during the Easter season. Before Communion can be received the communicant must be in a state of grace; this regulation will be explained when the Sacrament of Penance is discussed. Children are not permitted to receive Communion until they have reached the age of reason and can understand the significance of what they are doing; a child usually receives first Communion at about the age of seven. Only Catholics are permitted at the Lord's table. Whenever the Host is reserved in the tabernacle, a lamp near the altar must be lit to make known its presence. The door of the tabernacle is always open if the Host is not present. By the payment of a stipend for the support of the priests, the layman can have masses directed toward any worthy cause. Some of the religious orders offer to place the names of deceased loved ones on a permanent roll; Mass is then said for these persons many times each year.

Prior to the Second Vatican Council, Mass was said in Latin, a practice which puzzled many Protestants. In explaining it, Catholics pointed out that their services were uniform throughout the world, except for a few Oriental groups to whom the Pope had granted the use of another tongue. An excellent defense of the use of Latin was made by a Catholic chemistry professor, a colleague of mine, who lectured to a class in religion. He said, "What do I care what language the service is in so long as I know that God is there!"

This vigorous defense of the Latin Rite on the part of most Catholics did not prevent a spirited campaign on the part of some other Catholics for the increased use of English in services of worship. One priest wrote as follows:

People . . . want to pray in the vernacular. That is the only language they understand. . . . They need to participate in all prayers and ceremonies in their own tongue.[25]

Another priest wrote:

What could help us Catholics in this effort [to make religion penetrate the whole of life] more than taking our sacrifices, our most loved and efficacious prayers, out from behind the language barrier; having us pray together in the familiar language[?][26]

This latter opinion won the approval of an overwhelming majority of the bishops of the Second Vatican Council. In the *Constitution on the Liturgy,* extensive use of the vernacular in all the sacraments was authorized as was also a limited reception by the laity of communion under both species. The *Constitution* also provided for greater use of Bible reading and sermons. The actual implementation in Catholic worship of these decrees may take a considerable period of time.

Catholics are taught that two or more Masses can be heard at the same time with as much profit to the worshiper as though they were heard separately, provided Christ is adored on each altar. All but the smallest Catholic churches have more than one altar and on special occasions two or more Masses are said simultaneously. *The Register,* a Catholic newspaper, reported that at a memorial service held at the grave of a Catholic chaplain killed on Okinawa, forty-five jeeps held forty-five altar stones and Mass was said on each one. Occasionally, Mass is said with great pageantry. Some years ago thousands of worshipers gathered in the Harvard Stadium for a military memorial Mass at which cannons were used instead of bells to warn of the approach of Christ to the altar.

Catholics estimate that on the average the Host is elevated four times every second of the day and night the year around, except on Good Friday, when Mass is never said. The Church teaches that attendance at Mass is not optional but obligatory.

BAPTISM

Catholics are taken into their Church soon after they are born, by means of the Sacrament of Baptism. Baptism is usually performed by the priest. For it to be valid, he must cause baptismal water to flow over some portion of the body, if possible the head, and say at the same time, "I baptize thee in the name of the Father, and of the Son, and of the Holy Ghost."

The effect of baptism, according to Catholic belief, is to purify the soul of sin and to make it fit to enter into the presence of God. If the soul died immediately after baptism, it can have committed no postbaptismal sin, and thus it goes directly to heaven. Catholics teach that every child from the moment he is conceived is tainted by original sin. This sin was the first sin of Adam, the sin which lost for human beings the supernatural gifts with which they had been originally endowed. Adam and Eve are said to have committed an infinite sin when they ate of the fruit of the tree of knowledge; they disobeyed an infinite command of God. As a result they became hateful to God, were driven out of the Garden of Eden, and were required to live lives of toil and suffering. "The human race," says a Catholic writer, "is a unit, summed up in its head Adam, and, therefore, the Church has ever taught that Adam's sin with all its effects was transmitted to all mankind"[27]—except to the Virgin Mary and to Jesus. They are the only human beings not stained with original sin. The Church teaches that when Mary was conceived in the womb of her mother, St. Anne, God exempted her soul from original sin from the moment of her conception, since she would be the vehicle for the divine Incarnation. This teaching is called the dogma of the Immaculate Conception. It should not be confused with the Virgin Birth.

No human being, weak and ignorant as we all are, could rid himself of the burden of an infinite guilt. Therefore, Catholics teach that it is necessary for God to use His power to take from men the weight of Adam's sin. God does this through giving supernatural aid to men; this aid is called God's *Grace*. Baptism is the method which Jesus Christ directed the Church to use in order to repair, through the instrumentality of God's Grace, the damage of original sin.

Catholics believe that unless a person has been baptized he cannot go to heaven. But that does not mean he will go to hell; he may go to limbo. Hell is reserved for those who sin willfully. Infants, before they achieve the age of accountability, cannot commit sins worthy of hell. Nevertheless, they cannot achieve heaven unless they have been baptized.

Catholics believe that baptism is not limited to the baptism of water. Two other kinds are possible: the baptism of blood and that of desire. The baptism of blood is martyrdom for the Church. Catholics hold that the baptism of desire comes automatically to those who sincerely wish to be received into the Church (or who would wish to be received if they knew the supernatural truths of religion) but are prevented from receiving the baptism of water. Catholics believe that the baptism of desire comes also to every person who sincerely and honestly lives up to his best lights. Thus, persons who have never heard of Christ or His Church receive God's Grace and can attain salvation, provided they have *perfect* contrition. A Boston priest who insisted there is no salvation outside the Roman Catholic Church was excommunicated.

Since a person must have reached the age of reason before he can experience the baptism of desire, Catholics put great emphasis on the baptism of infants by water. It is considered to be so essential that in cases of extreme necessity anyone can baptize—a layman, a Protestant, even an atheist. Priests carefully instruct physicians who deliver Catholic mothers how to baptize newborn infants who are in danger of imminent death. If a pregnant mother dies, a Caesarean operation must be performed provided there is hope that the child still lives and can be baptized. The Church teaches that in the event of premature birth through miscarriage, the fetus, no matter at how early a state of development, must be baptized. A fetus may never be killed by direct abortion, even though such an operation would save the life of the mother, because not only is abortion murder but an infant life could then never receive baptism and thus could never gain heaven.

Baptism is one of the sacraments. Since the idea of God's Grace has now been defined, the term *sacrament,* which involves the term *grace,* can also be defined. A sacrament, according to Catholic theology, is an external sign, instituted by Jesus Christ, for mediating the

Grace of God. Catholics believe that Jesus Christ designated seven signs through which men can receive the Grace of God. These signs are all observable; that is, they are external. The seven Catholic sacraments are: the eucharist (Communion), baptism, confirmation, penance, extreme unction, marriage, and ordination.

CONFIRMATION

Ordinarily after a Roman Catholic has received his first Communion, he is eligible for the Sacrament of Confirmation. This sacrament increases the Grace of God first received in baptism. Confirmation is sometimes called the Sacrament of the Holy Ghost. During it the Holy Ghost is said to come "in power," as He did on the Feast of Pentecost described in the book of Acts. On that day the disciples "were filled with the Holy Ghost."

In former times confirmation was administered immediately after baptism; today, however, only those who have the use of reason are eligible to receive it. Generally it is not now administered earlier than age seven, and usually not until adolescence. But it is not put off too long; for, as one writer says, "children carry within themselves the elements of the passions, which if not promptly eradicated, will gradually grow."[28]

As the date set for the administration of the Sacrament of Confirmation draws near, the children of a parish are in a state of considerable excitement. They have studied earnestly for many months in order to be worthy of the Sacrament; and their mothers have gone to much care and expense in order to make adequate material preparations for the event—the girls usually wear simple bridal veils, as they do at their first Communion, and sometimes they wear a large red ribbon to symbolize the tongues of fire at Pentecost.

Confirmation usually takes place before the high altar in the parish church. There the bishop of the diocese confirms by using three external signs: he imposes his hands, that is, he raises his hands up and out over the waiting children and prays; he anoints each child on the forehead with the sign of the cross; and he slaps each child gently on the cheek, thus symbolizing the persecution and suffering which a true Christian must bear.

In emergencies confirmation can be administered by pastors.

PENANCE

Priests are believed to have the power to forgive the eternal punishment due for sin; this power is exercised through the Sacrament of Penance. Sins are *willful* thoughts, words, deeds and omissions which are contrary to the Law of God. The sins which men commit are of two kinds: venial and mortal. Mortal sin is considered to be so serious that one would have to spend eternity in hell, unless his punishment is taken away by God's Grace. Venial sin is less serious and can be atoned for in this life or in purgatory.

Three things are necessary before a person sins mortally: a grievous offense, sufficient reflection, and full consent of the will. Most people commit many sins which are considered mortal: they break the Ten Commandments, yield to anger and hate, stay away from Mass and the other sacraments, fail to keep the Friday abstinence, engage in indecent amusements, etc. For these mortal sins two kinds of punishment must be suffered: one is eternal (in hell) and the other is temporal (it will end). A "grievous offense" may be but a venial sin if either sufficient reflection or full consent of the will is lacking. For venial sins only temporal punishment (in this life or in purgatory) is imposed by God. The primary function of the Sacrament of Penance is to forgive mortal sin and to take away *eternal* punishment—the Sacrament keeps one from going to hell. It may also affect favorably the amount of one's punishment in this life or it may shorten the length of one's stay in purgatory.

There are four parts to penance: contrition, confession, satisfaction, absolution. Contrition is the part of penance in which a Catholic is genuinely sorry for his sins. The Catholic Church teaches that perfect contrition is essential for the full effect of the Sacrament. For a sorrow to be perfect, it must be supernatural—it must refer to God—and it must not be sullied by personal motives.

[If a penitent] be sorry for his sins because by them he has offended an all-good God to whom he owes everything, the sorrow is perfect; if he be sorry for some more selfish reason, as because he must undergo a supernatural punishment for his sins, either on earth or in the next life, the sorrow is imperfect but supernatural, and, with the Sacrament of Penance, is sufficient to obtain God's forgiveness for any sin, no matter how grave nor how often committed.[29]

Contrition must be sincere. God knows the heart and cannot be deceived. Any Catholic who thinks that he can escape from the punishment of hell simply by going through the form of penance, is very far from the teachings of his Church. A priest has the power to forgive sins only if the penitent is sincere. A penitent must intend, at the time of confession, to mend his ways. A hypocrite in the confessional may deceive the priest and hear pronounced the words of forgiveness, but in reality he still faces the awful prospect of eternal damnation. But a penitent whose confession is sincere and supernatural can be forgiven, even though the sin he confesses is habitual and will probably be repeated; for, it is the disposition of the heart *at the time of confession* that determines the will of God toward the sinner.

At confession the penitent orally and within the hearing of a priest accuses himself of all the mortal sins he has not previously confessed; he should also confess grievous offenses among his venial sins, but confession of lesser offenses is not required.

Confession usually takes place in a confessional, a small booth located near the back of most Catholic churches. This booth is so arranged that the penitent can keep his identity secret from his confessor; the booth has at least two compartments: one for the priest and one for the penitent. Between these compartments is a small opening covered by a screen; words can pass through this opening but nothing can be seen through it. Men may confess anywhere, but women are required to use the confessional. A penitent is not required to confess to his own pastor; he may go to confession in any church. Some Catholics make frequent use of this privilege of confessing to a stranger. However, nothing prevents the worshiper in the confessional from divulging his name to the priest, though he seldom does.

During confession Catholics can repeat their inmost thoughts without fear of exposure. Stern regulations prevent the priest from divulging any information learned during confession. He may speak or write nothing of what his penitents have told him; he must never act in such a way as to bring suspicion upon penitents. He must never take advantage of information he has gained through the confessional; in writing recommendations, in making financial decisions, even in saving his own life, he must never use such information.

After a worshiper has confessed his sins, he is counseled by his confessor; and then he is instructed what "satisfaction" he must make, that is, what penance he must perform. If a grievous wrong has been done another person, the penitent is required to make restitution, if that is possible and expedient. But for most sins the satisfaction required is of a religious nature; some spiritual devotion is imposed. When the penance required has been stated by the priest, he usually gives absolution at once, that is, he pronounces the words of the forgiveness of sins. Then the worshiper leaves the confessional and performs the satisfaction.

At the end of the Sacrament of Penance, after the guilt of mortal sins has been taken away, the worshiper is said to be in a "state of grace." If one were to die in a state of grace he would be certain of escaping hell. However, he would have to face the punishments of purgatory, since the Sacrament of Penance deals primarily with eternal punishment.

Catholics are required to go to confession at least once a year. Many Catholics go much more frequently; some of the religious orders require their members to go once a week in order that they may be almost continuously in a state of grace. One does not fall from grace unless he commits a mortal sin; but many Catholics make a practice of always going to confession just before receiving Communion.

INDULGENCES

The subject of indulgences is especially interesting because it played a prominent part in the events which began the Reformation. Catholics believe that an indulgence shortens the time which a sinner must suffer, in this life or in purgatory, for his misdeeds. Since all sins— both mortal and venial—which are committed after baptism are believed to have a *temporal* punishment attached to them, most people must suffer a very long time in the appalling conditions of purgatory. Just how long this time will be the Roman Catholic Church does not pretend to know; God has not revealed all things to His Church; He has revealed simply all things necessary to salvation.

The purgatorial punishment (and the punishments suffered on earth for sin) can, according to the Catholic religion, be lessened by the application of the merit which Jesus Christ and the saints built up during their lives on earth. Just as sins have temporal punishment attached to them, so good deeds performed in a state of grace have merit attached to them. In heaven there is an immense treasury of merit; it was put there by the sinless life of Jesus, by His sacrifice on the cross, and by the excess of good deeds over sins in the lives of the saints. "An indulgence is not a permission to commit sin, nor absolution in advance of sin, but it is a commutation of the temporal punishment due to sin"[30]—by drawing on the treasury of merit.

The granting of an indulgence might be likened to a king remitting all or part of the taxes of a citizen who had performed meritorious service for the government or who had given significant indications of loyalty. For example, a worshiper can gain "seven years indulgence" when the priest elevates the Host at Mass, by looking at the Blessed Body and saying devoutly, "My Lord and My God." Other good deeds to which indulgences might be attached by the Pope are saying certain prayers, giving money for certain causes, going on certain pilgrimages.

Indulgences vary in value; their value is stated in days or years. The Church is uncertain concerning the exact meaning of these temporal terms. Some indulgences are plenary—they remit the whole of the temporal punishment which has been incurred by a sinner up to the time the indulgence is received. However, indulgences are effective only to the extent that a person has purity in his dispositions—a condition seldom perfectly achieved. Careful Catholic writers make clear that when an indulgence is procured for the benefit of a soul in purgatory, "the degree of its acceptance depends on the will of God, so that there is no certainty that the penalty of these souls is fully remitted."[31]

Non-Catholic authors frequently assert that indulgences are "sold" and point to the fact that the Reformation was begun by the belief of Luther that the monk Tetzel was selling indulgences in order to get money for the building of St. Peter's Church in Rome. Catholic historians admit that in some times and places the doctrine concerning indulgences has not been properly understood and practiced; but Catholics deny that indulgences are abused in the Catholic Church today. Father Conway makes the following statements on this point:

Pardons are not sold today in the Catholic Church. . . . A man does not purchase a wife, if at marriage he signs over to her a portion of his property. A Protestant does not purchase a wife for ten dollars, because he gives a fee of that amount to the minister who performs the ceremony. . . . Catholic historians—Gasquet, Pastor, Janssen, Michaels, Paulus—have frequently mentioned the abuses connected with the preaching of Indulgences in the Middle Ages. The medieval pardoner, depicted by Chaucer in the *Pardoner's Tale,* was often an unscrupulous rascal. . . . We must carefully distinguish between Tetzel's teaching with regard to Indulgences for the living, and Indulgences applicable to the dead. With regard to Indulgences for the living, his teaching . . . was perfectly Catholic. . . . "As regards Indulgences for the dead," Pastor writes, "there is no doubt that Tetzel did, according to what he considered his authoritative instructions, proclaim as Christian doctrine that nothing but an offering of money was required to gain the Indulgence for the dead, without there being any question of contrition or confession. He also taught, in accordance with an opinion then held, that an Indulgence could be applied to any given soul with unfailing effect."[32]

Some non-Catholics hold the opinion that the theory of indulgences is so difficult to understand that even today average Catholics frequently misunderstand it and that some members of the hierarchy occasionally take advantage of this misunderstanding. *The Christian Century,* a Protestant magazine, wrote editorially as follows:

The theory of "indulgences" is somewhat intricate and lends itself both to misrepresentation by critics and to abuse by agents of the church. . . . The eagerness of ecclesiastics to collect money for the church, even after they have ceased to be eager to collect it for themselves, has led them into exploiting the relation between payment and salvation in terms which seem clearly designed to create an impression which is at variance with strict Roman Catholic theory on this point.[33]

ANOINTING OF THE SICK

When Catholics are ill they may receive the Sacrament of Anointing of the Sick. The priest goes to the sick room and anoints the patient with Holy Oil, olive oil which has been blessed. With the tip of his finger he places a small amount of oil on the lids of the eyes, the lobes of the ears, the nostrils, the closed lips, the hands, and the feet. While he does this he says:

Through this holy unction [anointing] and His most tender mercy may the Lord pardon thee whatever sins or faults thou hast committed by sight, hearing, smell, taste, touch, and walking.

It is held that through this rite a person who is in danger

of death from illness receives spiritual aid for his final struggle with the Devil. Catholics believe that a return to physical health also is often a result of this Sacrament. A priest once said in my presence that he believed there were men walking the streets of our town who would be dead if he had not given them the Sacrament of Anointing of the Sick.

This Sacrament is never administered to soldiers going into battle, or to criminals about to be executed. Penance is the Sacrament used in such cases. The Anointing of the Sick is reserved for persons who are ill.

After accidents involving death, priests are hurriedly summoned. On these occasions they sometimes anoint persons whom bystanders think are already dead. The Catholic Church, however, instructs its priests not to assume too quickly that the soul has departed the body. Only rigor mortis and the beginnings of putrefaction constitute certain signs of death in the eyes of the Church.

CATHOLIC MARRIAGE

Celibacy is a more blessed state than marriage, according to Catholic teaching. Catholics point out that Jesus never married and assert that St. Paul also remained single. Catholics remind us of their belief that Mary and Joseph, the parents of Jesus, were perpetual virgins. In his letter to the Corinthians, Paul urges the single life: "I would that all men were even as myself. . . . I say to the unmarried, and to the widows: It is good for them if they so continue even as I." Thus a celibacy *dedicated to God* is thought to be definitely superior to marriage. Late in his reign, Pope Pius XII expressed the hope that widows would not remarry, saying: "Though the Church does not condemn a second marriage, she expresses her predilection for the souls who wish to remain faithful to their spouses."[34] Nevertheless, Catholics hold that marriage is an exalted state, instituted of God for the perpetuation of the human race, and for His own glory. Marriage has the status of one of the seven sacraments, and its effect is to bestow upon the bride and groom the Grace of God.

Catholics believe that a truly Christian marriage can take place only with the permission of and under the guidance of the Church. A mar-

riage between two Catholics takes place in the church building before
an altar, and the Mass and Communion are important parts of the
ceremony.

The Marriage ceremony, apart from the Mass, is in the vernacular.
Non-Catholics attending Catholic weddings are often struck by the
similarity between the most frequently used Protestant form of the
marriage ceremony and the Catholic form. This similarity is a good
indication of the similar origin of many Christian churches and of the
many things they have in common.

The Catholic Church does not approve of divorce under any circumstances if by divorce is meant the dissolution (with the right of
remarriage) of what the Church considers to be a valid Christian
marriage which has been consummated by sexual intercourse. The
Church may infrequently consent to a *civil* divorce, in order to ensure
protection from molestation, or to assure alimony, or the like.

However, the hierarchy does grant annulments. A person sometimes thinks he has received valid Christian marriage when his union,
according to the Church, lacks some element essential to its genuineness. Genuine marriage must be voluntary: neither party may be
coerced in any way. It must also involve persons capable of sexual
intercourse, persons not already married, persons not too young, persons not too closely related, etc. If any condition essential to valid
Christian marriage is absent, an annulment can be procured, usually
from the local bishop; and it may be procured with relative ease if
definite proof can be produced that the marriage was indeed invalid
at any point—who would hesitate to tear up a counterfeit bill? Tyrone
Power, the movie actor, for example, was granted an annulment of
his marriage to his Catholic wife because the ceremony had been
performed by a civil official; subsequently Power was married to another Catholic woman with the full blessing of the Church. In addition
to marriage before anyone but a Catholic priest, Catholics can obtain
annulments for such reasons as marrying a Protestant without a special dispensation from the bishop, marrying a Jew without a special
dispensation from the Apostolic Delegate, withholding "interior consent" at the time the ceremony was performed, and agreeing before
the marriage to practice birth control, or abortion, or to get a divorce
if the marriage proved unsatisfactory.

If a voluntary, Christian marriage has been consummated by physical union, it becomes spiritually indissoluble. Adultery, criminality, incurable insanity may be adequate grounds for *separation,* but not for real divorce. In Italy, where the state follows the Catholic Church at this point, there is no governmental provision for divorce, although legal separation is possible.

Non-Catholics sometimes assert that the condemnation of divorce does not prevent Catholic couples from seeking divorce in the civil courts. Reliable statistics on this matter are almost nonexistent. Unfortunately, marriage and divorce records in the United States do not give an accurate picture of the influence of religious affiliation on civil divorce. The results of some samplings are available, however. One careful study, reported in 1938, surveyed the family backgrounds of over thirteen thousand young people in Maryland; this research discovered that the rate of civil divorce among parents of the three major religious groups was: Jewish 4.6 per cent, Catholic 6.4 per cent, Protestant 6.8 per cent; but that the rate of divorce when one parent was Protestant and the other Catholic was 15.2 per cent, and when neither parent had a religious affiliation, 16.7 per cent.[35] Another study, reported in 1949, surveyed the family backgrounds of about four thousand Michigan State College students; in this study the percentages of divorce were: Jewish, 5.2; Catholic, 4.4; Protestant, 6.0; mixed Catholic and Protestant, 14.1; no religious affiliation, 17.9.[36]

The Catholic Church persists in its refusal to grant divorces even though it deprives itself of much apparent advantage. Catholic apologists claim, no doubt correctly, that their Church loses thousands of members annually because of its teaching on this subject, and would gain one hundred thousand members overnight if divorce were sanctioned. Sanctioning divorce, however, is never considered by the hierarchy; for, according to Catholic teachings, Almighty God Himself, during His Incarnation, taught that remarriage after divorce is nothing but legalized adultery. (Read Matthew 5:32; 19:9.) That settles the matter. It is unthinkable, says the hierarchy, to seek material advantage in the face of such a command.

Mixed marriages—marriages between persons of differing creeds —are opposed by the Roman Catholic Church, even as they are opposed by most churches and synagogues. Such marriages frequently result in indifference to religion and in actual loss of faith. However, in spite of vigorous propaganda urging marriage within one's own communion, young people of different faiths often fall in love and insist on getting married. A Protestant writer estimated from official Catholic statistics that 36 per cent of the marriages performed by priests in the United States are mixed marriages, and this estimate "does not include the large number of marriages of Catholics with non-Catholics before justices of the peace and non-Catholic clergymen."[37] A sociologist puts the estimate of Catholics who marry outside their Church at "approximately one-fourth."[38]

The Catholic Church has found effective methods of protecting its interests in mixed marriages, methods which have stirred up much discussion. The Church rules that the non-Catholic party must undergo a series of instructions in the Catholic faith and must sign a statement in which a promise is made to bring up all children as Catholics. Different forms of this antenuptial agreement are used in various dioceses in the United States. One form is as follows:

I, the undersigned, not a member of the Catholic Church, wishing to contract marriage with _____, a member of the Catholic Church, propose to do so with the understanding that the marriage bond thus contracted is indissoluble, except by death. I promise on my word and honor that I will not in any way hinder or obstruct the said _____ in the exercise of _____ religion and that all children of either sex born of our marriage shall be baptized and educated in the Catholic faith and according to the teachings of the Catholic Church, even though the said _____ should be taken away by death.

I further promise that I will marry _____ only according to the marriage rite of the Catholic Church; that I will not, either before or after the Catholic ceremony, present myself with _____ for marriage before a civil magistrate or minister of the gospel.

In some sections the following additional promise is required:

I will not interfere in the least with the free exercise of the Catholic party's religion and I will lead a married life in conformity with the attitude of the Roman Catholic Church regarding artificial birth control, contraception or so-called "planned parenthood," realizing fully that these practices are against the natural and divine law.

The requiring of such promises has stirred up a heated argument. In order to give the reader an indication of the extent of this debate and of the reasoning which prompts it, the following four statements are quoted: Two were written by Catholics and two were written by Protestants. In defense of requiring the signing of this statement, one Catholic apologist wrote as follows:

> Protestant opposition to mixed marriages isn't logical. Catholic opposition is, and for these reasons: Because she is the only church which holds that marriage is a Sacrament in the true sense, one of seven Sacraments instituted by Christ, and committed to her keeping. . . . Because, honestly believing that hers is the very religion which Christ founded, she must require the observance of "all things whatsoever she has been commanded." . . . The average Protestant sacrifices no principle when he yields to the Catholic in this matter because he usually regards membership in a particular religious organization as a matter of preference rather than of obligation.[39]

The Protestant position on the antenuptial agreements was stated by Leland Foster Wood, former head of the Commission on Marriage and the Home of the Federal Council of the Churches of Christ in America (now the National Council).

> Such demands mean that a Christian person who believes his own church to be a true church of Christ is asked, when he marries a Roman Catholic, to act as if his church were no church at all but a dangerous organization. He is required to proceed as if he had no faith in the adequacy of Jesus Christ as Savior and Guide but rather must assume that only in the Roman Catholic Church could his children have assurance of salvation. . . .
> There should be in any case an understanding that when children have reached a suitable age they shall be free to determine their own faith in order that their religious allegiance may be a matter of inner conviction and self-dedication and not the outcome of any kind of constraint. . . .
> No more can we tolerate the idea that it is the duty of the Roman Catholic member to do everything possible through his home to proselyte while a Christian of another church must avoid even expressing his deepest religious convictions as if they were some kind of poison that would destroy his children.[40]

The English Archbishop of York commented on the antenuptial agreements, saying:

> I feel it necessary to warn Anglicans against signing this document, and to ask them to do their utmost to dissuade members of our Church from doing so. It means that Anglican fathers or mothers married to Roman Catholics are deprived of the right to influence the spiritual and religious upbringing of their children. It means disloyalty to the Church of their baptism and of their fathers. It is a humiliating condition.[41]

Father Bertrand L. Conway writing in a book designed to answer the questions of inquiring non-Catholics wrote that the chief reason for the resistance to mixed marriage

. . . is the danger of loss of faith on the part of the Catholic party and the children born of the marriage. Frequently a bitter unbeliever or a bigoted Protestant manifests his hatred of the Church after marriage, and by ridicule, bad example, and moral pressure of various kinds, occasions the apostasy of a weak-minded, ill-instructed or careless consort. The Catholic Church in the United States loses thousands annually by mixed marriages.[42]

Recently Cardinal Cushing of Boston and Father Hans Küng of Tübingen University have questioned certain aspects of the Church's position on intermarriage; their statements have had only the status of personal opinions.

BIRTH PREVENTION

Few Catholic pronouncements have gained more public attention than the invectives directed against chemical, mechanical, or other artificial methods of birth control. Such methods are believed to be a frustration of a natural human act, to be contrary to God's divinely ordered laws of nature, and to be sinful and degrading vices committed by people who are "eaten up with selfishness and the love of pleasure."

There is unquestionably a widespread, and almost universal sense of sin and shame . . . in men and women, who practice birth control.[43]
The husband who practices birth control regards his wife not as his wife bound to him by a love both sacred and holy, but as his mistress. . . . The wife who practices birth control regards her husband not as a husband united to her in the noblest of human loves, but as a paramour.[44]

The uncompromising view of the Catholic Church toward birth control is nowhere more clearly seen than in the statement of Father Francis J. Connell in the *American Ecclesiastical Review:*

In the third decision rendered by the Sacred Penitentiary [an agency of the Vatican] . . . in response to the third question, which asked whether a man [a husband] who used contraceptive devices should be likened to an assailant to whom the wife must oppose the same resistance as a virgin to one attacking her, the Sacred Penitentiary replied in the affirmative.[45]

The "population explosion" does not affect most Catholic teaching on birth control. Many Catholics have denied that there is any real chance of overpopulation and insist that food technology can keep well ahead of increases. In a 1960 Palm Sunday sermon, the Pope urged Catholics to have large families;[46] and in the same year the Archbishop of Delhi on tour in the United States asserted that India may come to have too few people, saying, "We are convinced that within about four years India will be producing more food than its people require."[47]

In recent years some priests have advocated the rhythm system, the utilization of the "sterile period," the days during the menstrual cycle when conception is least likely. The late Pope Pius XII ruled that this method and sexual abstinence are acceptable if they are used for "serious motives," but said that it is a sin to limit intercourse in these ways if the motive is simply to avoid procreation or to satisfy "sensuality." The primary purpose of marriage is "the procreation and education of children."[48] Intensive research on the use of the "sterile period" in order to make it a more reliable method of control is being conducted under Catholic auspices.

The opposition to Catholic teachings about birth control centers in the Planned Parenthood Federation of America, an organization which has the support of many non-Catholic religious leaders. The Federation's position is that bringing children into the world should not be a matter of chance; every child should be planned for and should be born into a family where the mother's strength has not been depleted by too much childbearing and where the father's capacity to provide financially is not endangered by too many mouths to feed. One of the Federation's leaflets asserts that birth control is no more contrary to the law of nature than are "anesthesia, immunization against disease, control of infection or any other great advance in medical science." The Protestant Episcopal Bishop James A. Pike in a statement distributed by the Federation writes:

Sexual intercourse has *two* primary functions in marriage: procreational and sacramental. Neither is secondary. A sacrament, of course, is "an outward and visible sign of an inward and spiritual grace." The inward and spiritual requisite is the total and permanent pooling, in love under God, of hopes and fears, of strength and weakness. The outward and visible sign is, as in other sacramental relationships, both expression of spirit and means of grace. The sexual act expresses the love and commitment the couple already possesses; it also strengthens and inspires that commitment.

The Planned Parenthood Federation maintains that the overwhelming majority of America's physicians approve birth control as a physical and psychological benefit. The Federation also asserts that the experience of birth-control clinics is that planned parenthood is apparently practiced almost as much by Catholic as by non-Catholic parents of comparable economic and educational status.[49]

The hierarchy insists that the Church never teaches that parents should bring into the world children who cannot be properly cared for. The Church has no desire to threaten the health of families or to endanger their financial status. In those circumstances where fewer children seem mandatory, let parents practice abstinence. "By mutual consent married people are always allowed to live as brother and sister, and their conduct will be very pleasing to God."[50] Self-restraint is a source of much ethical and spiritual blessing.

> The immediate purpose and primary end of Marriage is the begetting of children. When the marital relation is so used as to render the fulfillment of its purpose impossible, it is used unethically and unnaturally. The pleasures of marriage are innocent in view of legitimate childbearing; they become sinful and degrading, only when separated from the sacrifices and responsibilities of parenthood.[51]

A number of studies have sampled Catholic lay opinion concerning family limitation. Studies made between 1936 and 1948 show "a progressive increase of favorable Catholic response concerning birth control."[52] For example, a survey of public opinion by *Fortune* indicated 69 per cent of Catholic women, twenty to thirty-five years of age, believed that knowledge of birth control should be made available to all married women.[53] The term *birth control* in these studies may have included in the minds of some of the respondents the rhythm and abstinence methods. A careful survey in which definitions were more explicit gave somewhat different results. Over twenty-seven hundred white married women, eighteen to thirty-nine years of age, "selected in such a way as to constitute a scientific probability sample of the approximately 17 million wives in our national population," were questioned concerning their birth-control practices.[54] Only 13 per cent of the Catholic wives (1 per cent of the Protestant wives) gave unqualified disapproval of the general idea of family limitation.[55] Moreover, 50 per cent of the

Catholic couples who had been married ten or more years and who had had no evidence of impaired fecundity used a method of limitation other than rhythm (1 per cent used abstinence).[56] The study showed "that many Catholics—even those who attend Church regularly—do use types of contraception which are unacceptable to the Church."[57] The Catholic journal, the *Commonweal,* commenting on the fact that the twenty-five leading American cities were not producing enough children to keep the population stationary, said, "Unhappily the proportion of Catholics in the cities under study has no observable effect . . . upon the birthrate."[58]

Catholic leaders admit that some Catholic parents use contraceptives, but declare that that fact is no argument in favor of their use; such couples go directly against the commands of God and sin grievously. Persons who use chemical or mechanical methods of preventing conception endanger their soul's salvation.

In the last three decades, approval by Protestant groups of birth control has greatly increased. Official statements have been made by the Episcopalians, Methodists, Presbyterians, Congregationalists, Unitarians, and others. This position is a relatively recent one. The Jesuit weekly, *America,* noted that "in 1908 the Anglican Bishops were unequivocally opposed to contraception; opposed it again, though less vigorously, in 1920; gave permission for its 'conscientious use' in 1930; and appeared as its advocates in 1958."[59]

HOLY ORDERS

Multitudes of Christians—both Protestants and Catholics—believe that the most exalted station for which a person can be chosen in this life is the ministry or the priesthood of Jesus Christ. This choice is made, they believe, by God Himself, perhaps "in the Councils of the Trinity." As traditional Protestants would say, a person is "called of God to preach Christ"; as Catholics would say, he "receives a divine vocation." Catholics are taught to look on the priest as the director of their spiritual and moral lives, as a constant source of comfort and security, and as a living evidence of Christ's love for sinners. One Catholic writer, going considerably beyond the official dogmas of his Church, even asserted that with the laying on of the bishop's hands in ordination the priest "becomes another Christ."[60]

The priest's life is arduous and full of cares. Like a physician he is constantly on call to give succor to the sick and the dying. Yet early each morning he is up preparing to celebrate the Mass, in order that the members of his parish on their way to work may visit their Lord and partake of His Blessed Body. This ministry accomplished, the priest's day is full of activities: he comforts the sorrowful, hears confessions, visits the sick, buries the dead, solicits funds, distributes alms, administers schools, directs clubs, consults his superiors. And every day he must spend about three quarters of an hour in personal prayer, devotion and reading of the Breviary; and regularly he must himself go to confession.

Every priest takes the vow of absolute obedience; he must conduct his life exactly as his superiors require in all matters over which they have legitimate jurisdiction. Every priest also takes a vow of chastity. The taking of this vow is not believed to be a divine requirement. Chastity is simply a rule of discipline; it has been imposed only since the fourth century. The rule may be abrogated in special cases. For example, a former Lutheran pastor, a resident of Denmark, was ordained a Roman Catholic priest and was permitted to remain in the married state; similarly "a half dozen" married German Lutheran ministers are reported to have been ordained by the Roman Catholic Church.[61] These rare instances point up the fact that clerical celibacy is a matter of discipline. The Church asks her leaders to remain single in order that she may be strong, in order that she may have from her priests service that knows no stint. An unmarried priest has fewer obligations than a minister with a wife and family. An unmarried clergy also costs less to maintain.

But clerical celibacy is not simply a negative thing; priests make of it a positive dedication to the glory of God. Theirs is a celibacy "for the kingdom of heaven's sake." Religion is always strongest when it is put into positive rather than into negative terms; the non-Catholic will fail to understand clerical celibacy if he thinks in terms of prohibition. The priest through his ordination becomes the "Spouse of the Church."

Non-Catholics sometimes contend that lifelong continence is unnatural and that its enforcement has caused widespread immorality. Catholics deny both assertions.

Laxity of observance at certain periods will, of course, be admitted by any candid historian, but no one who knows the facts can deny that the law of celibacy has been faithfully observed from the fourth century by the vast majority of the clergy of the West. . . . Celibacy is not impossible, for the grace of God is given abundantly to all His priests to keep them chaste. Daily Mass, the recitation of the divine Office, the frequent meditation on divine truths, the consolations of the confessional, the intimate contact with the sick and dying—all these are aids to keep every priest faithful to his vow.[62]

MONKS AND NUNS

Catholic priests are of two general types: the "secular" or "diocesan" priests; and the "regular" priests, the members of the "religious" orders. The spiritual powers of both of these groups are the same but their habits of life and their vows differ. A secular priest is one who lives "in the world," among the people. A "religious" (used as a noun) is one who has become a member of a religious community, who lives usually with his fellows in a monastery, and who in addition to the vows of chastity and obedience has also taken the vow of poverty. There are over one hundred fifty orders of regular priests in the United States. The monastic life is available to women also. In this country there are over seven hundred religious orders for women. In 1959 there were fifty-three thousand Roman Catholic priests (secular and religious) in the United States but in that year there were one hundred sixty-five thousand sisters (all religious).

Some of the most famous of the Catholic orders are the Franciscans, the Dominicans, the Benedictines, the Jesuits. Such groups as these have houses scattered throughout the world and have an elaborate system of internal government. Some of the orders are *contemplative;* that is, the members give themselves over completely to the cultivation of the spiritual life, seldom leaving their communal homes. In solitude and retirement they seek union with God. Most of the orders in the modern day, however, are *active;* they engage in a wide variety of social services: caring for the sick, the orphaned, the aged, the blind, the destitute. Some of the orders specialize in preaching, others in foreign missions, others in education. Among the sisters the most common activities are teaching and nursing. Abraham Lincoln said of the nursing during the Civil War, "More lovely than anything I have ever seen . . . are those modest sisters going on their errands of mercy."

Persons who have never come into close contact with individual members of the religious orders are apt to view this type of life with a mixture of awe and astonishment, and to wonder how normal human beings can renounce the freedom of everyday living and "imprison" themselves in the convent.

The answer to such views must begin with the assertion that of course there is no such thing today as imprisonment. Members of the religious orders are free to leave their communities at any time. Leaving would not be easy, however. Once perpetual vows have been taken, the breaking of them is a serious matter; much moral and spiritual pressure is used in order to persuade a person not to break his vows and thus commit mortal sin. But physical force is not used.

Nor is moral suasion often necessary. The religious conduct "a war against nature" with "virginity and continence as means, and charity as the end."[63] Some joys of ordinary living are doubtless forgone, but in their place the monks and sisters feel that they have the most rewarding experiences which humans can know: giving over oneself wholly to a divine vocation. The very ceremony of entry into the religious life is made to symbolize a joyous and not a sorrowful experience; it is considered to be the beginning of an exciting and romantic adventure. For many a sister, the most thrilling event of her life was her act of "renunciation" when she took the veil, thus becoming "the bride of Christ," joined mystically in marriage with Him. During that ceremony she wore a wedding gown, was attended by a bridesmaid and a trainbearer, and, to symbolize her spiritual and perpetual union with Christ, she was given a plain gold wedding ring to wear on the third finger of her left hand.

The joys and opportunities offered by a religious vocation are well illustrated by the experiences of Cardinal Gibbons, perhaps the outstanding Catholic leader who has lived in this country.

CARDINAL GIBBONS

"His words . . . had more weight in the country at large than any other man's, except the President's,"[64] claimed the biographer of

Cardinal Gibbons (1834–1921). Certainly his active influence lasted far longer than that of any political leader of his time: he was Archbishop of Baltimore for forty-three years, and during thirty-five years of that time he held the rank of Cardinal.

James Gibbons was born but a half mile from the cathedral over which he was to preside for so many years, and in it he was baptized. His father's ill-health, when James was three, caused the family to return to Ireland, from whence the parents had migrated a few years before. Some fifteen years later, after the father's death, the family returned to America. A year later James Gibbons began studying for the priesthood.

He finished a "full course" of "six years" in two, despite such frail health that "some made the prediction" that he could "not long survive." He was ordained in 1861 by the Archbishop of Baltimore and was assigned as pastor of St. Bridget's, a parish located in Canton, an isolated section "where the hand of the law seemed not to reach." The Know-Nothing, anti-Catholic frenzy—strong in Maryland—had but recently passed; so bigoted was the neighborhood that on an Election Day one group "carried half-hogsheads of beef blood to the polls and bespattered with the contents citizens who would not vote the anti-foreign ticket."[65] Consequently the members of the parish thought it unwise for the young priest to sleep in the rectory alone and unprotected. But no injury came to him during his pastorate.

His work soon became arduous. St. Lawrence's Church, a mile across the Patapsco River, was added to his parish.

Every Sunday morning, in midwinter snows no less than in the zephyrs of summer, he was accustomed to leave Canton at six o'clock [to begin] . . . his double task of the day. . . . As no Catholic clergyman may celebrate Mass except while fasting, it was generally about one o'clock in the afternoon when, after a morning's arduous labor, he could eat. His digestion was permanently wrecked by this ordeal, which compelled him to observe great care in diet throughout his life.[66]

After four years' work in Canton, Gibbons was made secretary to the Archbishop of Baltimore, "traditionally a steppingstone to promotion in the Church." Three years later, at age thirty-four, he was consecrated a bishop—the youngest in the world-wide hierarchy—and assigned to the missionary diocese of North Carolina. The Catholic population of that state numbered only eight hundred, and in all of it there were but three priests. Since the diocese had no episcopal residence, the young bishop moved in with the pastor of the largest church; they lived in a "lean-to" of four rooms, built against the rear wall of the church building. The floors were bare, the furniture rough, and often the priests had to prepare their own meals.

Almost immediately after his installation, Gibbons began missionary journeys among the overwhelmingly Protestant population. He preached in all sorts of places: public halls, homes, court houses, Masonic lodge rooms, fire engine houses, Protestant churches. Many were the Catholics who heard him who had not had contact with the Church in years; many more were the Protestants who came to hear Catholic dogmas expounded with forthrightness, winsomeness and good will. In one town the trustees of the Methodist church were so moved by Gibbons' irenic spirit that they offered him their house of worship, rang the church bell, and called together a congregation almost wholly Protestant. For his part Gibbons stood in a Methodist pulpit, accepted the assistance of a Methodist choir, and read from a Protestant Bible. Henceforth in his ministry, he was to preach to large crowds of Protestants, usually stressing dogmas which were accepted by Christians generally. On one occasion, he asked a priest to preach to a certain congregation on a very hot Sunday afternoon; the priest did so, ardently. But immediately after he was finished, Gibbons ascended the pulpit and preached another and very different sermon. "Did you not see," he explained afterwards, "that more than half of the congregation were Protestants." His biographer says that he "actually made Protestant denominations more tolerant of each other."

Yet he never soft-pedaled the Catholic claim to be the only True Church. Whenever he was asked about the movement for unity among the churches, he insisted that Protestants must return to Rome. Gibbons wrote a book, *The Faith of Our Fathers,* which was directed primarily to Protestants. It sold over two million copies and had a vast influence. In it he said:

I heartily join in this prayer for Christian unity, and gladly would surrender my life for such a consummation. But I tell you that Jesus Christ has pointed out the only means by which this unity can be maintained, viz: the recognition of Peter and his successors as the Head of the Church.

In coming to the Church, you are not entering a strange place, but you are returning to your Father's home. . . . You come back like the Prodigal Son to the home of your father and mother.[67]

During his second year in North Carolina, a call came from Rome to attend along with his fellow bishops the Vatican Council, the first ecumenical assembly of the Roman Catholic Church since the Council of Trent, three hundred years before. He attended as the youngest of all the bishops and voted for the decree declaring the Pope's infallibility. Returning from the Council he was appointed, at age thirty-seven, Bishop of Richmond, Virginia. Six years later he became Archbishop of Baltimore, and nine years after that was elevated to the cardinalate.

Gibbons was as vigorously American as he was Catholic; speaking in Rome, he said:

As a citizen of the United States, . . . I say, with a deep sense of pride and gratitude, that I belong to a country where the civil government holds over us the aegis of its protection, without interfering with us in the legitimate exercise of our sublime mission as ministers of the Gospel of Christ. Our country has liberty without license, and authority without despotism.[68]

Gibbons once said he would not alter one word of the Constitution, and over and over asserted that there is "no antagonism" between the "laws, institutions and spirit" of the Catholic Church and those of the United States.

American Catholics rejoice in our separation of Church and State, and I can conceive no combination of circumstances likely to arise which would make a union desirable to either Church or State.[69]

One incident illustrates his strict observance of the line between church and state. In 1911 the City Council of Baltimore declared a civic holiday to honor in a municipal celebration the golden jubilee of his ordination to the priesthood and the silver jubilee of his elevation to the Sacred College. The celebration was attended by the President and Vice-President of the United States, the only living ex-President, the Chief Justice of the Supreme Court, the Speaker of the House, many members of the House and Senate, and a large number of other dignitaries including distinguished members of the Protestant clergy. A short time after this occasion, the Council was on the point of declaring another holiday as part of the ecclesiastical celebration which the priests of the Church arranged to observe the jubilee. Gibbons promptly sent a message to the Council asking that the day not be set aside. Thus he in effect agreed with the Baltimore Ministerial Union which had protested saying, "In this [ecclesiastical] celebration we cannot be expected to take part. . . . We regard such proposed action as a direct violation of . . . complete separation of Church and State."[70]

Yet Gibbons did not hesitate to make his voice heard on matters which he felt involved public morals. During his tenure, the Archdiocese of Baltimore included Washington, D. C. in its boundaries. Gibbons came to know personally large numbers of our national leaders, including most of the men who served in the White House. He spoke out on many national issues. He supported temperance and opposed prohibition. He said, "I regard 'woman's rights' women as the worst enemies of the female sex" (but when woman's suffrage finally came, he urged the sisters in the orders of the Church to vote). He warned against "race suicide," saying, "Marriage . . . is not intended for self-indulgence, but for the rearing of children." He condemned lynching, communism, socialism, the persecution of the Jews, the secular public school, the Louisiana Lottery, the revolutionary movement in Mexico, the independence movement in the Philippines, the election of United States Senators by popular vote. He defended the right of labor to organize and persuaded the Vatican in an unprecedented action to rescind its ban on the Knights of Labor.

Gibbons rejoiced greatly when Pope Leo XIII issued his great *Encyclical Letter, Rerum Novarum,* on the condition of labor. "Some remedy must be found quickly," said Leo, "for the misery and wretchedness present so heavily and unjustly at this moment on the vast majority of the working classes." Gibbons said, Christ "has thrown a halo around the workshop, and has lightened the workman's tools by assuming the trade of an artisan. . . . A conflict of labor and capital is as unreasonable as would be a contention between the head and the hands."[71]

Catholics remember Gibbons as their outstanding leader during a period of great Catholic expansion; the Church in America trebled during his episcopate. Non-Catholics remember Gibbons for the creative leadership he gave the movement to develop interreligious amity; today they hope another Catholic leader of like temper and ability will arise. A humble man, one never insisting on his own dignity nor trying to impress others, Gibbons always conducted himself with simplicity and accessibility. He loved after Mass to romp with the altar boys, and he took long walks about the streets of Baltimore during which he engaged in many friendly conversations. One day a family applied at his residence for the privilege of making confession to a priest. They were informed that all the priests were resting and that confessions could be heard later. The father of the family persisted, saying they had a long journey to make before nightfall and they must soon be on their way. The doorkeeper went to the private apartments in the house and soon returned with the Cardinal himself. On another occasion the planners of a civic meeting were hesitating over how to place Gibbons and the Episcopal Bishop of Maryland in a procession. The Cardinal solved the difficulty by taking the Episcopal bishop's arm and saying, "My dear brother, we will walk together."

The breadth of his sympathies became proverbial. One of his close friends was Joseph Friedenwald, a leading Baltimore businessman of Jewish faith. Gibbons once asked a businessman to give a friend of "education, refinement, and character" a job; the friend turned out to be a retired Protestant minister. Walking one day with a man from another city, they passed near a church from which a large congregation was just emerging; Gibbons was saluted by so many people that his companion said, "You seem to be well acquainted in this parish." "Ah," said Gibbons, "these are our Episcopal friends." He took the lead among the hierarchy in securing Catholic participation in the Parliament of Religions which was part of the Columbian Exposition and "could see no merit in the suggestion that the part which Catholics would take in the convention would involve any recognition or approval of the numerous sects within and without the circle of Christianity that were to be represented there."[72]

But for all his tolerance and understanding, his public contacts and responsibilities, Gibbons was first of all a priest. It was his regular practice to spend three to four hours daily in private devotions. "He went to confession once a week at St. Mary's Seminary, and annually joined in the retreat there."[73] And he never lost sight of the purpose to win converts to the Catholic Church. In the *Faith of Our Fathers* he wrote:

Remember that nothing is so essential as the salvation of your immortal soul. . . . Let not, therefore, the fear of offending friends and relatives, the persecution of men, the loss of earthly possessions, nor any other temporal calamity, deter you from investigating and embracing the true religion.[74]

MIRACLES

The possibility of miracles is emphatically taught by the Roman Catholic Church. The Church has, of course, the testimony of the Scriptures to defend this belief. But she does not hold, like the Protestant Fundamentalists, that miracles were confined to Biblical times. She teaches that throughout the Church's history God has performed providential acts for the edification and comfort of saintly people. Miracles happen all the time. "In devout minds," says one authority, "there is even a presumption for and an expectation of miracles."[75] The person who denies the occurrence of miracles, say Catholics, either denies the existence of God altogether or denies that God cares enough for His children to help them in their physical and spiritual needs.

But Catholics insist that they are not superstitious; the Roman Catholic Church fights superstition and magic with all its might. A miracle is defined by Catholics as an effect wrought in nature directly by God Himself without the use of ordinary natural means. Magic, on the other hand, is defined as the attempt of men to effect changes in nature through the co-operation of such supernatural agencies as demons or lost souls. Witchcraft is possible according to Catholic teaching, though concerning its extent the Church "observes the utmost reserve." Catholics hold that the present widespread skepticism concerning witchcraft is the reaction against the witch mania which broke out in the sixteenth and seventeenth centuries. But, Catholics contend, the teachings of the Bible and the experiences of Christian people must lead realistic minds to assert the possibility of witchcraft and at the same time to condemn severely its practice. Forbidden is the engaging in any type of superstition: palmistry, astrology, divination, spiritism, idolatry, the interpretation of dreams, the wearing of charms.

Non-Catholics frequently assert that the medals which some Catholics wear are charms which are thought to have the power to ward off disease, accident, and other unfortunate occurrences. No doubt many Catholics do have a superstitious faith in the medals they wear. But they sin in holding such a faith. One clergyman preaching in St. Patrick's Cathedral, New York City, attacked the abuse of "plastic piety." "It is about time," he said, "we shake off some of the nonsense surrounding medals, chain prayers, statues and the rest. . . . Wear medals, but understand them."[76] In strict Catholic teaching, a medal is but the badge of a saint; it is a reminder to the wearer that he should live a virtuous life, and a spiritual one. Living such a life, he may confidently hope that he will receive supernatural help in time of crisis. But he has *no guarantee* that the trouble he fears will be prevented.

Nevertheless, the practice of engaging in spiritual exercises in the hope of guarding against trouble is widespread and appears to receive clerical encouragement. In my community, Catholic churches are crowded every year in midwinter on the Feast of St. Blaise. He is the patron saint of throat diseases and on the day of his feast communicants, kneeling at the altar rail, receive from the priests a blessing in his name. The Carmelite Fathers of New York issue a leaflet which says that a scapular is available and that "whosoever dies clothed in this scapular shall not suffer eternal fire." Moreover, the fulfillment of two conditions (chastity according to one's state in life and the daily recitation of the Little Office) assures one of being freed from purgatory on the first Saturday after death. Each kissing of the scapular grants one five hundred days indulgence. However, says the leaflet, the scapular "is not a talisman. . . . It is the sign of devotedness to the Blessed Virgin. . . . *an habitual sinner will not persevere in wearing the Scapular.*"[77]

The Catholic Church officially is very skeptical concerning any claim that a miracle has occurred. A good example of this skepticism occurred at Gloucester, New Jersey, where more than a thousand people came to see what they took to be a vision of the Virgin in a light that appeared on the door of St. Mary's Church. The pastor put some black cloth over a window in the rectory and demonstrated that the vision was simply the light of a street lamp reflected by the window glass onto the freshly varnished church door. But if it can be demonstrated that an observed effect was not produced by natural means, nor by fraud, nor by magic, nor was the result of hallucination, the Catholic Church gladly assents to the faith that God has given another evidence of His miraculous power. These miraculous effects are said to happen every day—God lives! and he gives the faithful constant reminders of His love and care.

Miracles of healing are the best-known of the modern manifestations of God's intervention in the affairs of mankind. The most famous of all shrines where healings take place is at Lourdes in Southern France. There in 1858 a fourteen-year-old peasant girl, Bernadette Soubiroux, saw in a hollow of the rock a vision of the Blessed Virgin. The Virgin appeared to Bernadette nineteen times, a fountain miraculously gushed forth, and the Virgin told Bernadette to instruct the clergy to build a church at the spot. The clergy, incredulous at first because no one but Bernadette had seen the vision, were finally convinced because they felt compelled to accept the fact that miracles were taking place. The church was built and the Pilgrimage of Lourdes was recognized. Since that time thousands of healings have occurred, healings that could have no other possible cause, say the faithful, except the direct intervention of God. A writer in the *Catholic Encyclopedia* writes, "There exists no natural cause capable of producing the cures witnessed at Lourdes."[78]

A North American shrine which has attained great prestige is St. Anne de Beaupré, thirty miles north of the city of Quebec. As many as twenty-five thousand persons have visited this shrine on a single day and a large number of cures have been certified as genuine by the Catholic Church.

Catholics identify living persons as being particularly close to God, the recipients of His special favors. The following paragraphs are taken from an article on such a person. This article was written by a layman; as far as I know the Vatican has never indicated that the events narrated are truly miraculous. But the account was published in an official Catholic newspaper. The account concerns a stigmatic, that is, a person on whose body appear wounds like the wounds which Jesus Christ received at the crucifixion.

Theresa Neumann was born in 1898, the eldest of 10 children. Their father was both a tailor and farmer. She was a perfectly normal child. There was little about the youngster to distinguish her from others in the strongly Catholic village [in Bavaria].

In the spring of 1918 she injured her spine in a fire. The result was a complete paralysis of her limbs with blindness coming soon after. Then on May 17, 1923, the date of the beatification of St. Theresa, in Rome, the sick girl regained her eyesight. She had prayed to the young saint every day. . . .

It was on Thursday, March 5, 1926, that the first manifestation occurred. A wound suddenly opened [over] Theresa's heart and began to bleed profusely. From that moment on she suffered every agony that Christ experienced at His Crucifixion. The marks of the nails appeared on her hands and feet. There was an imprint of the crown of thorns on her head. She bled at the eyes. The stigmata lessened on Sunday but the following Wednesday the signs again appeared and she went through the Crucifixion for the second time. That was the beginning.

Since that time, on certain Fridays which fall on Church Days, other than joyous ones, Theresa Neumann goes through her ecstasies. . . . Theresa has not slept for eighteen years. Nor has she touched any food excepting the small wafer which she receives at Communion daily.[79] [Other accounts assert that she also has had no drink during this period.][80]

These illustrations indicate how strongly the Catholic Church holds to a belief in the supernatural. No one knows, of course, the extent to which individual Catholics believe in miracles; some Catholics no doubt reject the teaching of their Church at this point. But if they make this rejection known, they feel the full weight of ecclesiastical censure—for the Church herself takes her position squarely on the traditional belief. And from it the faithful receive much comfort. Miracles are an evidence to them that God is in heaven and has constant and watchful concern for His children here on earth.

THE BIBLE

Non-Catholics frequently assert that the Catholic hierarchy is opposed to the Bible. Such a statement is rather like saying that judges who preside in divorce courts are opposed to the institution of marriage. The hierarchy guards carefully against what it considers to be improper use of the Scriptures. But it vigorously asserts faith in the Bible, and considers it to be the major source of revelation.

Catholics believe that the Bible contains no "formal error"; this perfection is possible, they say, because God inspired it. One Catholic writer says:

> We cannot restrict inspiration to certain parts only. . . . We cannot restrict inspiration to faith and morals alone. . . . We do not look for precise scientific formulas in the Bible, for it does not teach science *ex professo*. Nothing in its pages contradicts the teachings of natural science, because the same God is the author of natural and supernatural truth. But the sacred writers generally speak of scientific matters in more or less figurative language, or in terms which were commonly used at the time they wrote.[81]

The idea that Catholics are opposed to the Bible arises from the fact that for many centuries the Catholic Church discouraged its laymen from reading the Bible. During much of this period the Bible was the chief instrument of the Protestant Reformation, and Catholics consequently opposed Biblical translation and dissemination. American Catholics are still forbidden to read Protestant translations of the Bible. But reading *a Catholic translation* is encouraged and is considered to be a pious act. Laymen who read the Bible are expected to familiarize themselves with the official interpretations. A considerable movement is afoot for Protestants and Catholics to publish joint translations of the Bible; in fact in England and Germany official authorization has been given to Roman Catholic scholars to make serious efforts in this direction; English Catholics now publish, with a few changes, the American Protestant *Revised Standard Version*.

Catholic piety for laymen, unlike Protestant and Jewish piety, does not ordinarily express itself in extensive devotional reading of the Scriptures. Catholic piety ordinarily finds expression in such devotions as the Rosary, meditation on the various incidents in the life of Jesus, contemplation of the Sacred Heart of Jesus, and preparation for confession and Communion.

CATHOLIC EDUCATION

Catholics pay their full share of taxes for the public schools in the United States; yet they are so convinced of the importance of religious education that they maintain through voluntary contributions a separate educational system. They contend that the Church cannot develop real religion in a community where the children get most of their education under secular auspices. The Church even contends that it rightfully should have control of all education, "public" as well as Catholic. One authoritative book declares, "We deny, of course, as Catholics, the right of the civil government to educate, for education is a function of the spiritual society, as much as preaching and the administration of the sacraments."[82]

Catholics aim to enroll every Catholic child in a full-time Catholic school and to develop in every section of the country a complete educational system—all the way from the nursery through the university. They are succeeding to a remarkable degree. More than 60 per cent of the Catholic students of elementary school age are now attending Catholic elementary schools; more than half of the Catholic students of high school age are attending Catholic high schools. The proportions enrolled in these schools have been steadily rising. The Church now has over two hundred and thirty colleges and universities in the United States; and new ones are planned every year. Catholics employ nearly four times as many teachers as they do diocesan priests.

Currently there is much controversy between Catholics and non-Catholics over the public support of parochial schools. The hierarchy in several states has succeeded in getting textbooks and bus transportation for parochial pupils paid for out of public funds. This development has been vigorously opposed by many Protestants and Jews. They believe in the public schools. They contend that sectarian education is the function of churches and synagogues and that any system of private education brings dissimilar training to the young, drains off interest in public education, and threatens the unity of the nation. The welfare of the nation, they say, requires that private education should be discouraged rather than encouraged. The right of any group of citizens to establish private schools, assert Protestants and Jews, must be maintained if we are to keep essential liberties; but such citizens ought to bear all the burdens of the venture and ought not to expect financial help from the public, any more than a businessman should expect financial help from the public if he hires his own detective agency instead of relying on the skills of the police.

Catholics retort to this line of reasoning that God has entrusted the Church with the salvation of the race, that the Church in an ideal society would have control of all education, that "the atmosphere of the public schools, is, in effect, atheistic, [since] not only is God ignored, but His laws are not even taught the child"[83] and that requiring Catholic parents to send their children to a secular school or else to endure "double taxation" is a violation of conscience rights. Professor John A. O'Brien of the University of Notre Dame said in 1961 that the Catholic school system in the United States is saving the nation's non-Catholic taxpayers at least $2,735,162,500 each year and that "because Catholic families bear a double burden, the educational taxes of each non-Catholic family in the U.S. are reduced $76.66 each year."[84] Archbishop William O. Brady of St. Paul charged that giving Federal aid to public education only would be "one more confirmation that we Catholics are second class citizens."[85] However, a Jesuit, the Dean of the Boston College Law School, said that although many Catholic parents "do resent deeply the denial of their claim to aid for schools of their choice, Catholic parents and Catholic educators should not be encouraged to think that any state or federal aid will be forthcoming for Catholic schools in this or even in the next generation."[86]

The hierarchy is widely credited with having defeated efforts in the Congress to pass bills providing for federal aid to education. These defeats were due to the opposition of conservative Republicans and Southern Democrats as well as of Roman Catholics. But no doubt a bill would have passed if it had received Catholic support.[87] Catholics contend that the Constitution does not forbid governmental aid to the religious school. They are struggling to find a formula which will provide such aid and also be generally regarded as constitutionally acceptable. After the defeat of the 1961 bill, a defeat in which a Catholic member of the House Rules Committee cast "the decisive vote," the Legal Department of the National Catholic Welfare Conference proposed federal aid to parochial schools in teaching secular subjects; the Department contended that there

exists no constitutional bar to aid to education in church-related schools in a degree proportionate to the value of the public function it performs. Such aid to the secular function may take the form of matching grants or long-term loans to institutions, or of scholarships, tuition payments, or tax benefits.

Leo Pfeffer, Director of the Commission on Law and Social Action of the American Jewish Congress, replied:

> The legal memorandum issued by the National Catholic Welfare Conference asserts that . . . all that need be done is to apply "the art of cost accounting" to draw a dividing line between costs attributable to secular aspects of education and those attributable to religious aspects.
>
> This assertion rests on a premise that has been uniformly and consistently denied by Catholic educators, theologians and philosophers, i.e., that the secular can be divided from the sacred and that the Catholic parochial school is nothing but a public school with religion added as a supplementary subject. Were this to be so, there would be no reason for parochial schools, since the religious instruction could easily be provided after regular public school hours. . . .
>
> The standard Catholic text on education, Redden and Ryan's, "A Catholic Philosophy of Education," states that "the only school approved by the Church is one . . . where the Catholic religion permeates the entire atmosphere, comprising, in truth and fact, the 'core curriculum' around which revolve all secular subjects."
>
> The "art of cost accounting" is indeed advanced. But it cannot make secular that which is sacred or constitutional that which violates the First Amendment.[88]

Strenuous efforts are made to secure the attendance of Catholic children at parochial schools, even to the extent of warning parents that failure to send children will result in a withdrawal of the sacraments.[89] Attendance at non-Catholic colleges and universities is sometimes viewed with alarm; for example, the Archbishop of St. Louis declared in a pastoral letter that Catholics in the Archdiocese may not in conscience attend non-Catholic institutions of higher education unless written permission is obtained from the Church and that such permission will be granted only for "just and serious reasons."[90]

Catholic schools are often charged with accepting lower academic standards than obtain in the public schools; one author wrote of the "obvious inferiority"[91] of the parochial schools. Catholic colleges are often said to produce fewer intellectual leaders than do non-Catholic colleges. One study of sixty-four of the "most eminent" scientists in this country as judged by other scientists found that "none of them came from Catholic homes."[92] A study of college graduates whose names appeared in *Who's Who in America, 1938,* revealed that the highest ranking Catholic college in terms of the number of graduates cited was only 137th on the list. A priest wrote an article which considered the problem of "the impoverishment of Catholic scholarship in this country, as well as the low state of Catholic leadership in most walks of national life."[93] It is surely true that many Catholic educational institutions when judged by ordinary secular standards do not measure up academically; it would be surprising if they did in view of their cramped budgets. But when Catholic schools are measured by Catholic standards, which make Catholic faith the end of living, Catholic schools have no peer in this country. Catholic schools make religion the center of the curriculum and annually present the Church with scores of thousands of devoted followers.

A paragraph about the Catholic press is in order at this point. The hierarchy has had marked success in persuading Catholic people to buy and to read Catholic newspapers and magazines. Over five hundred Catholic publications are issued in this country. They have a total circulation of over twenty-seven millions. Local diocesan papers have the help of the press department of the National Catholic Welfare Conference. The Conference sends out thousands of words each week to its clients.

Protestants think the Catholic Church gets a break in the secular press; Catholics think just the reverse. The Catholic Press Association estimated in 1945 that less than one-tenth of 1 per cent of the news in secular newspapers is Catholic news.

CATHOLIC STRENGTH

The Roman Catholic Church is the largest single religious body in the United States. It reported for the year 1960 forty-two million members. All of Protestantism in the United States had in that year perhaps eighty to eighty-five million members (using a Catholic definition of "membership"). However, Protestants *reported* only sixty-three and a half million members.[94] The discrepancy is due to the fact that Protestants unlike Catholics often do not count persons who have been baptized only, and more frequently drop an inactive person from the rolls. Catholics count as members any person who has been baptized by the Church, even though he may never have been confirmed and may never attend Mass. The churches and synagogues reported in 1960 a total membership of 63.6 per cent of the population.

As is indicated, the above figures are based on reports by the churches. Somewhat different results are obtained by polling individual Americans. In 1957, the Bureau of the Census polled a random sample of "about 35,000 households" asking about the religion of persons fourteen years old and over.[95] The Bureau's estimates (extrapolated from the poll results) of the number of persons (with their children under fourteen) who considered themselves to belong to various religious groups in 1957 were: Roman Catholic, forty-two million; Protestant, one hundred and ten million; Jewish, five million; some other religion, two million; no religion, three and a half million; religion not reported, one and two-tenths million. On the basis of these estimates the percentages of Americans who regard themselves as belonging to the various groups are as follows: Roman Catholic, 26 per cent; Protestant, 66 per cent; Jewish, 3.2 per cent; other religions, 1.3 per cent; no religion, 2.7 per cent; religion not reported, 0.9 per cent. However, the answers of a considerable percentage of Americans to such questions as "What is your religion?" mean little more than that they were born into a family of a certain religious background, that they have made no open break with their family tradition, and that they are not Jews, or not Catholics, or not Protestants. For a sizable proportion in all the sects, "membership" does not involve active support either by attendance or by contributions. Accurate comparisons of the number of persons actively following the various religious groupings have not been obtained.

The world membership of the major groups in the Judeo-Christian tradition is sometimes given as follows: Roman Catholic, five hundred and ten million; Protestant, two hundred and ten million; Eastern Orthodox, one hundred and thirty million; Jewish, twelve million.[96] (Before Hitler began his systematic extermination of the Jews, there were about seventeen million Jews.) These figures seriously underestimate the strength of Protestantism as compared to the other groups since the estimates for Protestantism often count only those persons who are on membership rolls, while the estimates for the other groups often count whole populations. In the Roman Catholic estimate, for example, only 8 per cent of the population of South America is counted as non-Catholic even though a much greater percentage thinks of itself as having no interest in any kind of religion whatever. The misrepresentation of Protestant strength is so marked that one highly placed Protestant could write, "According to the most reliable statistics available, the total Catholic constituency and Protestant constituency throughout the world are nearly equal."[97]

Catholics sometimes claim that the number of *practicing* Catholics exceeds the number of practicing Protestants; an editorial writer in a leading Catholic paper even asserted, "On any given Sunday there are more than twice as many Catholics attending divine services as there are members of all other religious organizations taken together."[98] These claims are certainly in error. (See pages 472 f.)

Unfortunately religious statistics in this country are often unreliable; there are no common definitions, no uniform methods of gathering data, and no official gathering agency. In this book statistics will be presented for most of the sects described in order to give the reader some indication of comparative sizes and tendencies. He will need, however, to allow for a margin of error.

Only in recent decades has the Catholic Church manifested real influence in this country. All through colonial times and during the first decades of our national life, the Catholic portion of our population was very small. The first period of rapid Catholic growth began only about 1830; it continued until the beginning of the Civil War. Another period of rapid growth was from about 1890 until the beginning of World War I. However, since the cessation of large-scale immigration, Catholic growth has slowed down to about the pace of Protestant growth. There is little evidence that Catholics are making spectacular gains as contrasted with Protestants. *The Yearbook of American Churches, Edition for 1962,* using reports from the churches, states that from 1926 to 1960 the Catholic percentage of the total United States population increased from 16.0 per cent to 23.6 per cent while the corresponding Protestant increase was from 27.0 per cent to 35.4 per cent. However, in the single year 1960, Catholics reported an increase of 3.2 per cent while Protestants reported an increase of only 1.8 per cent. The United States population in that year also increased 1.8 per cent.[99]

A number of well-advertised conversions to Catholicism have

seemed to indicate that Catholics are making serious inroads on Protestant membership lists. Such inroads are improbable. The publicity given to the conversion of prominent persons is primarily an indication of the influence of the Catholic Church with the secular press. Conversion is a two-way process. Surveys conducted by Protestant organizations indicate that Catholics are being converted to Protestantism faster than Protestants to Catholicism. The Omaha Council of Churches found in fifty-one Protestant churches a "ratio of 4.57 Roman Catholics received to one Protestant entering the Catholic Church."[100] A national survey of one-tenth of the Methodist churches for the year 1958 found that

more persons who were once Roman Catholics joined The Methodist Church than were dismissed by it to the Roman Catholic Church. . . . Of the 1,963 Catholics who joined The Methodist Church, 829 gave as their reason the fact that the tenets of Romanism no longer held nor attracted them, while a lesser number, 737 joined our church because of marriage. Conversely, an almost insignificant number, 39, left The Methodist Church to become Catholics because they were dissatisfied with the tenets of Methodism, while 463 left because of marriage, accounting in part for the larger number, 407, females who left our church as compared to the much lesser number 150 males.[101]

The number of conversions from one group to another probably does not represent much altering of the proportional strengths of the larger religious bodies.

A good many students of American religions assert that the Roman Catholic Church is the most vital religious organization in the United States today. I am inclined to agree with this opinion: as an institution the Catholic Church appears to receive more loyalty from its members, to come nearer to giving its members what they demand spiritually, and to show more promise of influencing national trends in the immediate future than does any other church.

Two facts, in my judgment, are primarily responsible for this situation. One is the system of Catholic schools; through these schools the *whole* educational experience of a large percentage of Catholics is dominated by a church point of view. Thus millions of Americans have been trained from infancy to revere the Catholic Church and to follow the decisions of the hierarchy.

The other major factor in Catholic strength in the United States is the excesses associated with radical anti-Catholicism. On at least two occasions opposition to Catholicism has assumed major national importance—in the 1850's as the Know-Nothing movement and in the 1920's in the second phase of the Ku-Klux Klan. Persecution, provided it is not too severe, tends to strengthen rather than to weaken a religious minority. Opposition has meant that American Catholics have developed a strong group consciousness and loyalty such as has been absent from the religious experience of many modern groups. Perhaps nowhere else in the world do Catholics have as intense loyalty to their Church as they have in America. In countries where the Catholic Church is theoretically dominant—in Italy, in France, in Latin America—the average citizen is often indifferent to religion. The Catholic Church is most vital in those countries where it has strong Protestant competition: Germany, England, Canada, the United States.

The average American Catholic's attitude toward his Church contrasts rather sharply with the thinking and actions of most American Protestants. H. Paul Douglass commented on this fact, saying, "The church as an institution is a sort of Protestant whipping-boy. It is beaten as disappointed heathens beat their idols." Over against this common Protestant attitude is "the invariable pity awakened in Catholics in behalf of the church 'sore oppressed.' "[102] Anyone who doubts Douglass' judgment needs but to attempt to discuss religions objectively before an audience composed of American Catholics, Jews and Protestants. Most Protestants and Jews will usually listen quietly to the expression of opinions adverse to Protestantism and Judaism; but many Catholics will enter vigorous objections to even mild criticisms of their religion.

Only since World War I have Catholics begun to rid themselves of this minority psychology. Prior to that time they tended to think of themselves as a foreign colony in the midst of "the Americans"; many of them were foreign born and half of them attended foreign language churches. Not until 1908 did the Vatican cease to class the United States as a mission field. But today Catholic leaders are vigorously combating this psychology; they assert that Catholics are in fact the religious majority. Nevertheless, the consciousness of being a minority still characterizes American Catholicism and doubtless will continue to characterize it for a considerable time to come. This consciousness, while it lasts, will increase the loyalty of Catholics to their Church.

The institutional consequences of a minority psychology are not all beneficial. A persecuted minority is usually conservative. The persecution which it endures, or which it fears, makes it wish to conserve its energies and not to embark on experiments. As a result of this attitude, the Catholic Church has a kind of institutional toughness which distrusts innovation. This conservatism tends also to give power to the clergy; in America the Catholic bishop has more authority than in European nations. Many Catholics would like to see a liberalization of policy. Such liberalization may be in process as the result of the labors of Popes John XXIII and Paul VI and of the Second Vatican Council.

The strength of the Catholic Church seems to be an internal strength. The Church has not and does not seem to be winning favor with the rest of the nation. It is true that Catholics are having more and more influence on national affairs. But this influence in most cases appears to be the result not of persuasion, but of power.

THE ANTI-CATHOLIC MOVEMENT

"A kind of Protestant underworld" exists, writes a Congregational clergyman, "an opposition that expresses itself in unsigned manifestoes and stirs up undisguised hatred of Catholics."[103] "The situation is bad and we might as well admit it," writes a well-known Catholic layman.[104] The anti-Catholic movement unfortunately is still a prominent part of American life. Two decades ago under the leadership of the National Conference of Christians and Jews a campaign to rid the country of religious prejudice seemed well on the way to success; today a campaign to arouse lethargic non-Catholics to the "danger of Catholic power" finds heavy financial support. Moreover, the persons who are aroused today are not limited to a lunatic fringe; many sober and fair-minded non-Catholic religious leaders express concern. Why this change?

One reason has been the refusal by recent Catholic leaders to support the movement to develop comity among the religious groups. Because they sincerely believe their religion to be The Truth, Catholics have thought they should be honored in their refusal to deal with other churches as equals. Some members of the hierarchy have shown a tendency to label as "bigotry" any opposition to Catholic policies.

Many non-Catholics are alarmed by the increasing power in politics of the Catholic Church. It succeeded, according to a widespread opinion, in preventing American support of the Republic in the Spanish Civil War. It persuaded the State Department to discriminate in granting passports against Protestants who sought to go as missionaries to South American countries.[105] It secured at the Vatican a diplomatic representative from the United States. Its priests occupied a heavily disproportionate number of executive positions in the Chaplains Corps.[106] It brought Massachusetts (and other states) to enforce laws which make religion a paramount concern in the adoption of children; a Massachusetts court issued orders requiring that two children be taken out of the custody of a home in which they had lived for years, even when the best interests of the children (other than the change of religion), the desires of the adoptive parents, and the desires of the natural mother all supported the plea that the children be allowed to remain in the home.[107] Paul Blanshard, a vigorous critic of the Catholic Church, writes that in many cities "there is a kind of unwritten political law that . . . no person in public life must ever say anything directly hostile to the Catholic hierarchy."[108] In many sections membership in the Catholic Church is essential to political success. On the other hand, in the national government Catholics hold less than their share of the offices. (See pages 286 f.)

Catholic influence on the press and on the circulation of newspapers and books disturbs and frightens many non-Catholics. Editors know that the hierarchy, like any American business, or college, or professional group, will do what it can to prevent unfavorable publicity. And the hierarchy is in a position to do a great deal. A Columbus, Ohio, newspaper published a story with a picture of a priest who had renounced his vows and married; the local bishop called for a boycott, and the paper lost thousands of subscribers.[109] A columnist writing in the European edition of the *New York Herald Tribune* criticized Cardinal Spellman's dealings with striking grave diggers; the Cardinal complained, and the columnist was promptly dropped.[110] In San Francisco a newspaper was boycotted and lost heavily because it printed a news item concerning a priest who was arrested for drunken driving with a female companion.[111] The New York Public Schools banned *The Nation* from its libraries because Catholics objected to a series of articles by Paul Blanshard. And Macy's department store in New York finally yielded to pressure and ceased to handle the best-selling *American Freedom and Catholic Power,* the book into which Blanshard's *Nation* articles were expanded.[112]

Many newspaper editors seem to be unwilling to review fairly or even to notice books which are critical of the Catholic Church. Mr. Blanshard's book was reviewed in *The New York Times* by "a devout Catholic"; and the best-known Catholic book in reply was reviewed not by a Protestant but again by a devout Catholic.[113] Emmett McLoughlin's book, *People's Padre,* which tells of his experiences as a priest and as an ex-priest, was reviewed by only two newspapers in the whole New England and Middle Atlantic area even when it was selling a thousand copies a week, and even though Reinhold Niebuhr could write of the book, "*People's Padre* deserves attention."[114] Thomas Sugrue's book, *A Catholic Speaks His Mind,* was reviewed in *The New York Times* by an editor of a leading Catholic journal. Sugrue wrote

Some of my non-Catholic friends said to me, "Did you have to do it?" They didn't like the trouble it made for them; they are editors and they were frightened at the idea of dealing with the book. If they printed one word for it they would be deluged with Catholic pressure, and they knew it. . . .

I have been a book reviewer for twenty years; I have written and talked about nearly two thousand volumes. . . . The act only proves what I have said about Catholic pressure; all editors and publishers of secular periodicals are afraid of it. Is that something of which the Catholics should be proud—that editors who are my friends and for whom I have worked for years are afraid to give a fair review to my book?[115]

Catholic influence on the movies is also very irritating to many non-Catholics. Through its Legion of Decency the Church instructs its members which motion pictures are objectionable and which proper for Catholic eyes. If the matter stopped there, few but the movie producers, who want their market to extend to the entire population, would complain; certainly the right of a church to set standards for its own members should be inviolable. But the Legion of Decency, like the Protestants in prohibition days, succeeds in setting standards for the general population. The Legion has upon occasion prevented the exhibition of pictures which clergymen of other sects found unobjectionable. It has also a large influence in Hollywood;[116] any casual observer can see that in picture after picture great effort is made to please and not to offend the Catholic Church, though no corresponding effort is extended in Protestant directions. The film *Elmer Gantry,* for example, was considered as among the best produced in its year by the New York Film Critics; but its treatment of the Protestant evangelist was most derogatory. One sarcastic reviewer wrote:

> It was cool and comfortable in the theater, and I was short of sleep. The leading character of the movie, posing as a priest, seemed to be deceiving a nun in the dark before the high altar of his church. Two reels later he was framed by a prostitute-with-photographer, and as the picture ended he threw off his robe with the announcement that he was putting away childish things. . . .
> By this time I knew I was dreaming; after all, Hollywood knows how to avoid sacrilege! So I awoke and found on the screen, thanks to United Artists, a Protestant evangelist betraying a deacon's daughter and turning her into a prostitute.[117]

Catholic pressures extend to television. The world TV premiere of the outstanding film *Martin Luther,* which had been produced by Lutheran Church Productions, was canceled by Chicago station WGN-TV because of pressure brought by Catholics.

Many writers contend that the Catholic Church is antidemocratic. This contention is certainly untrue for the Church in the United States, if by *democratic* is meant loyalty to the nation and to its form of government. American Catholics need take a back seat to no other group in their willingness to defend their country and its institutions. On the other hand, if by *democratic* is meant "popular determination of major policies," then the Catholic Church is not democratic; let it be noted, however, that the Church shares this characteristic with practically all of America's economic institutions and most of her educational institutions.

No doubt the factor in Catholic life which most disturbs informed non-Catholics is the statements which come from Catholic priests concerning the intention of the Catholic Church to take from non-Catholics full liberty of worship, if and when Catholics become the dominant group. Pope Pius IX insisted that man is not "free to embrace and to profess that religion which, guided by the light of reason, he judges true." Pope Leo XIII denied that "every one may, as he chooses, worship God." Two highly placed American priests, Dr. John A. Ryan and Dr. Francis J. Boland wrote:

Does State recognition of the Catholic religion necessarily imply that no other religion[s] should be tolerated? . . . If these are carried on within the family, or in such an inconspicuous manner as to be an occasion neither of scandal nor of perversion to the faithful, they may properly be tolerated by the State.[118]

A Jesuit writing in a magazine published in Rome declared:

The Roman Catholic Church . . . must demand the right of freedom for herself alone, because such a right can only be possessed by truth, never by error. . . . In a state where the majority of people are Catholic, the Church will require that legal existence be denied to error, and that if religious minorities actually exist, they shall have only a *de facto* existence without opportunity to spread their beliefs. . . . The Church cannot blush for her own want of tolerance, as she asserts it in principle and applies it in practice.[119]

The alarm of non-Catholics is increased by their observation of Spain where according to numerous Protestant reports no signs announcing services of worship are permitted on Protestant buildings, where all publicity announcing Protestant worship must be by word of mouth, where Protestants are not permitted to reply in the press to attacks made on them, where Protestants may not bury their dead with the rites of their church, where the printing of Protestant hymnbooks is forbidden,[120] where the printing or importing of Protestant Bibles is forbidden, where petitions for the opening of Protestant chapels are ignored, where thirty (perhaps more) chapels have been closed in recent years.[121] Nor is the status of non-Catholic liberties in Italy reassuring, where the public schools indoctrinate Catholicism, where parish priests are paid by the government as though they were civil servants, where priests who have been converted to Protestantism are denied employment which brings them into contact with the public ("a priest who renounces Catholicism, and chooses to become a teacher or a Protestant minister, is liable to arrest"),[122] where a Catholic prelate used his position as head of the Olympic Committee for Religious Assistance to bar all Protestant clergymen from Olympic Village until the last week of the games and to withhold from the athletes notices of services in Rome's Protestant churches.[123]

How shall all these non-Catholic fears be dealt with? "The first thing that must be said," writes Father George H. Dunne, S. J., "is that the question . . . needs to be honestly faced. It is no good merely to say that no American non-Catholic has reasonable ground for being concerned."[124] And Catholics have faced it; over and over American priests have declared their satisfaction with the American system. The basic thing which non-Catholics need to understand is that not all Catholics, nor all priests, agree on such matters. The Church is not the monolithic structure which outsiders so often assume. An excellent illustration is the attitude toward the public schools. One of the bishops wrote a pamphlet describing the public schools as "Our National Enemy No. 1." A priest answered saying that the best place for such pamphlets is "the ash-can."[125] Similarly with politics; one observer wrote that "a widespread discussion [is] now raging in the Roman Catholic Church on the proper relation of church with state."[126] In the midst of the 1960 preconvention campaigns one Catholic view was stated by an editorial in the Vatican journal *L'Osservatore Romano* to the considerable embarrassment of Catholic candidates for the presidential nomination.

An absurd distinction is made between a man's conscience as a Catholic and his conscience as a citizen. . . .
A Catholic can never depart from the teachings and directives of the Church. In every sector of his activity, his conduct, both private and public, must be motivated by the laws, orientation and instructions of the hierarchy. . . .
The problem of collaboration with those who do not recognize religious principles might arise in the political field. It is then up to the ecclesiastical authorities, and not to the arbitrary decisions of individual Catholics, to judge the moral licitness of such collaboration. . . .
It is highly deplorable . . . that some persons, though professing to be Catholics, not only dare to conduct their political and social activities in a way which is at variance with the teachings of the Church, but also take upon themselves the right to submit its norms and precepts to their own judgment, interpretation and evaluation. . . .[127]

"To a man, the American Catholic commentators on the editorial denied its relevancy or application to Sen. Kennedy or to the American situation."[128] *America,* a journal published by Jesuits, called the editorial "A Bewildering Article,"[129] and the *Commonweal,* published by lay Catholics, said, "It is obvious that . . . the views of American

Catholics do not always receive the consideration they deserve in some Catholic circles in Europe, and it would be foolish to ignore this fact— or to pretend that it poses no problems."[130] American Catholics quote a former head of the Society of Jesus:

American Catholics . . . have not the slightest desire to substitute for these advantages [of religious freedom] that "protection" by the State which in Europe has so often meant the oppression of the Church.[131]

Archbishop John T. McNicholas, as chairman of the Administrative Board of the National Catholic Welfare Conference, issued in 1948 a statement for the American bishops, in which he said:

We deny absolutely and without any qualification that the Catholic Bishops of the United States are seeking a union of Church and State by any endeavors whatsoever, either proximate or remote. If tomorrow Catholics constituted a majority in our country, they would not seek a union of Church and State.[132]

Doctors Ryan and Boland, cited above, wrote:

While all this [limitation of non-Catholic freedom of religion] is very true in logic and in theory, the event of its practical realization in any State or country is so remote in time and in probability that no practical man will let it disturb his equanimity or affect his attitude toward those who differ from him in religious faith.[133]

A former editor of *The Commonweal* said in a public address that while American Catholics are "stuck" with Vatican pronouncements concerning the proper relationships between church and state, he believes that none of the pronouncements would be put into effect even if America were to become overwhelmingly Catholic.[134] The alert observer will note the fact that some overwhelmingly Catholic countries—Ireland, for example—are very far from following Spain in its attitudes toward Protestants. The author of a study on *Roman Catholicism and Religious Liberty* published by the World Council of Churches, which is Protestant and Eastern Orthodox, wrote:

Roman Catholic literature representing this modern tendency [to defend religious liberty] has lately been so voluminous and of such quality that it would be an understatement to say that, for *one* book or article in favour of the traditional doctrine, *ten* have been published defending universal religious freedom.[135]

On the other hand, in Holyoke, Massachusetts, sometimes said to be "the most Catholic city in America," Catholics used economic

force in the effort to prevent a lecture on birth control by Margaret
Sanger; concerning this incident one priest said:

> In a Catholic community . . . where a Catholic moral code is accepted
> by the majority of the community, such people as Margaret Sanger
> should not be brought in to disturb the public. . . . We are against the
> Communists for the same reason. They have no right to speak wherever
> and whenever they please.

Another Holyoke priest said:

> In Italy, Catholic religion and morals are the basis of the country and
> of the people's way of life, for the church is in the majority. In a country
> where this is so, the church should be favored in the laws and opportuni-
> ties of the state. The Protestants in Italy have gone out of their way to be
> nasty. They say there is nothing to the Catholic religion. They put a
> Methodist school up right across the hill from the Vatican just as if
> flaunting us. This kind of action by a minority naturally makes the Vati-
> can doubt that they should have as much liberty as the Catholics in the
> country.[136]

The struggle between conservatives and liberals within the Church
has been brought sharply into world view by the sessions of the Second
Vatican Council. Although Pope John XXIII was widely acclaimed
as an "interim" pope, he proved to be a charismatic leader who
brought new perspectives to the Church. At this writing, two sessions
of the Council have been held, one in the fall of 1962 and one in the
fall of 1963. The formal accomplishments thus far have not been
great; but reports of the votes have made very clear that an overwhelming majority of the cardinals and bishops are eager for changes
in certain areas of the Church's life. For example, the Council indicated by an 84 per cent majority its approval of the thesis that the
college of bishops in union with the Pope enjoys full and supreme
authority over the Church. It approved by a 75 per cent vote the
thesis that in certain dioceses the ordination of married deacons should
be sanctioned. These actions were only straw votes and did not bind
the Church; nevertheless they indicated the presence of a strong liberal
majority in the Council. And yet the second session of the Council
adjourned with little accomplished, the liberal majority having been

blocked by a conservative minority which succeeded in preventing decisive action on many issues. The accusation was widely expressed that the conservative Curia was using obstructionist tactics. One priest wrote:

The delay in the work of the second session is mainly due to the obstruction on the part of a tiny minority, whose influence is out of proportion to their number. Curial cardinals have chaired the various conciliar commissions, and until now they have been able to keep their position.[137]

What the final result will be only time can tell. But the expectation is strong today that the liberals will eventually find a way to reshape the Church in some important particulars. Certainly many non-Catholic clergymen accustomed to having contact with Catholic priests can already sense a new respect for non-Catholic denominations, a new willingness to co-operate, a new freedom of movement. Yet many Protestants and Jews doubt that the new spirit, even if it becomes officially triumphant, can with any rapidity permeate the Church's life. Henry P. Van Dusen, former president of Union Theological Seminary, wrote:

For years to come many of us will continue to confront the obscurantist and exclusivistic Church to which we are accustomed.[138]

In the past the Catholic Church has refused any kind of co-operation with other religious agencies. To co-operate might imply a recognition that Protestant churches and Jewish synagogues do in fact represent true religions. In an authoritative journal for priests appeared the following statement:

It is precisely because a considerable proportion of our prominent and educated lay Catholics are inclined to "soft-pedal" the unqualified exclusiveness of the Catholic religion that it is dangerous for them to participate in "intercreedal" meetings, even when the purpose of these meetings is limited to the fostering of better understanding among citizens, the promotion of social welfare, or other like objectives of a purely natural character. Not a few of our Catholics could take occasion in such surroundings to state that everyone has the God-given right to practice any religion he chooses, that the most ideal type of religion between church and state is realized when a government accords equal rights to all forms of religion, that we all have the duty of promoting the religious activities of the various churches, etc.—statements which are being incessantly repeated in our land today, but which no Catholic can approve if he wishes to be consistent with the principles of his faith.[139]

The struggle against exclusiveness is doubtless just beginning in the Roman Catholic Church. Exclusiveness limits the role which the Church can play in American society. Exclusiveness creates tension; and tension creates fear; and fear creates a climate in which condemnations are made wholesale. The Catholic Church has had little influence on other religious groups; rather because of frequently expressed enmity to non-Catholic movements, she has been resisted on principle. Exclusiveness creates a mood in which proposals are appraised, not on merit, but on a judgment of the source from which they came. The resultant failure of communication makes almost inevitable a weakening of the spiritual fiber of the nation. And thus the reinforcement of America's spiritual core is made less likely.

If America can maintain such religious freedom as she now has, a type of freedom which permits parents and churches to indoctrinate children with sectarian beliefs, then the future of Catholicism—and of all churches which emphasize exclusiveness—is pretty much assured. But the question is: Can America preserve freedom without a widespread revitalization of her central spiritual dynamic and without a new and common commitment to the religious values at the heart of her culture? And what role can the many churches which emphasize exclusiveness play in the achievement of this commitment?

The following maps are from *Churches and Church Membership in the United States,* a series of bulletins prepared by the Bureau of Research and Survey, National Council of Churches. This study is based on reports by 114 denominations for the year 1952 and it relates the data compiled from these reports to certain aspects of the 1950 United States Census of Population. The 114 denominations reported a total membership which was 49.2 per cent of the United States population in 1950; this total was 80 per cent of the number of church members reported by *all* of the denominations which reported to the National Council in 1952.

Map 1 is from Series A, No. 3, 1956.
Map 2 is from Series A, No. 4, 1956.
Map 3 is from Series C, No. 1, 1957.

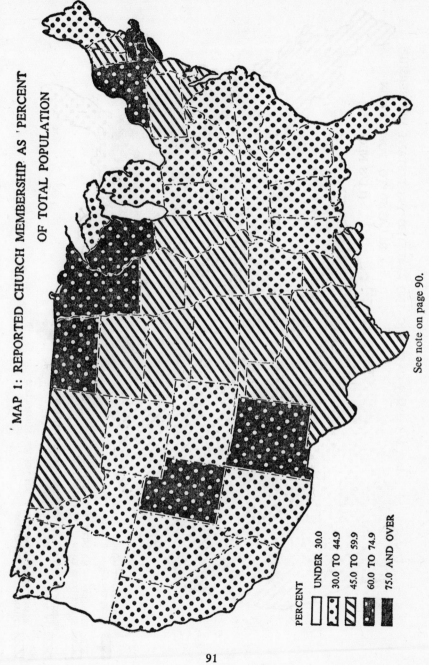

MAP 1: REPORTED CHURCH MEMBERSHIP AS 'PERCENT OF TOTAL POPULATION

PERCENT

UNDER 30.0
30.0 TO 44.9
45.0 TO 59.9
60.0 TO 74.9
75.0 AND OVER

See note on page 90.

91

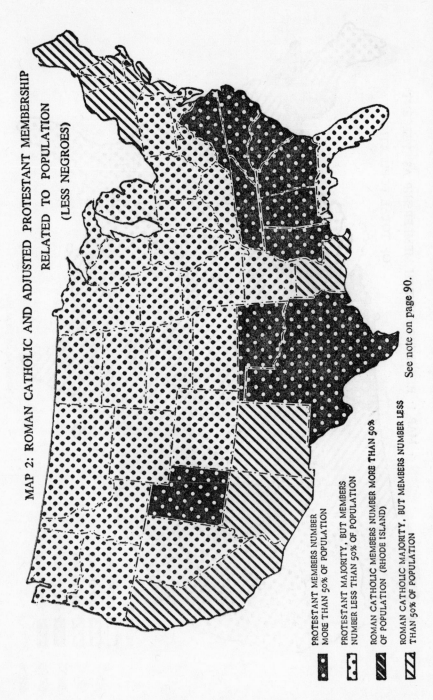

MAP 2: ROMAN CATHOLIC AND ADJUSTED PROTESTANT MEMBERSHIP
RELATED TO POPULATION
(LESS NEGROES)

See note on page 90.

• PROTESTANT MEMBERS NUMBER
  MORE THAN 50% OF POPULATION

• PROTESTANT MAJORITY, BUT MEMBERS
  NUMBER LESS THAN 50% OF POPULATION

ROMAN CATHOLIC MEMBERS NUMBER MORE THAN 50%
OF POPULATION (RHODE ISLAND)

ROMAN CATHOLIC MAJORITY, BUT MEMBERS NUMBER LESS
THAN 50% OF POPULATION

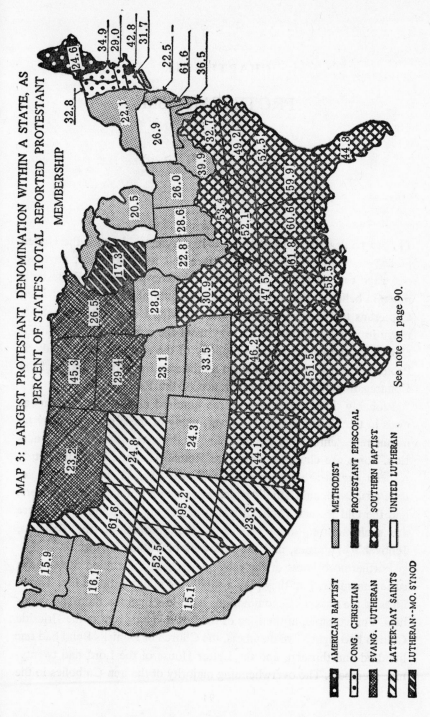

MAP 3: LARGEST PROTESTANT DENOMINATION WITHIN A STATE, AS PERCENT OF STATE'S TOTAL REPORTED PROTESTANT MEMBERSHIP

METHODIST

PROTESTANT EPISCOPAL

SOUTHERN BAPTIST

UNITED LUTHERAN

AMERICAN BAPTIST

CONG. CHRISTIAN

EVANG. LUTHERAN

LATTER-DAY SAINTS

LUTHERAN—MO. SYNOD

See note on page 90.

93

CHAPTER III

PROTESTANTISM

TRYING TO describe Protestantism is like trying to describe the United States; one can say almost anything about it—and almost anything one says can be shown to be false in some particular. Protestants range in belief all the way from the supernaturalism of the right-wing Lutherans to the agnosticism of the left-wing Unitarians. Protestants range in worship forms all the way from the complexity of the high-church Episcopalians to the simplicity of the silent-meeting Quakers. Protestants range in emotionalism all the way from the restraint of the Congregationalists to the exuberance of the Pentecostals.

Probably there are close to three hundred non-Roman denominations in the United States. However, many of these groups are by no means representative of the main line of Protestant thought and tradition. Some are churches which are Roman Catholic in all important respects except acceptance of the authority of the Pope; for example, the Polish National Catholic which has a quarter of a million members, and the Old Roman Catholic which has eighty thousand. Then there are such nontraditional groups as the Spiritualists, the Bahaists, Jehovah's Witnesses, and the followers of Father Divine.

Furthermore, most of the religious sects in the United States have so few members that they play a small role in our national life. Three-fifths of the nation's denominations have less than ten thousand members. For example, at the time of the last religious census the Erieside Church had eighty-five members, the Church of Daniel's Band had one hundred and thirteen, and the Latter House of the Lord had twenty-nine members. The overwhelming majority of the non-Catholics in the

United States have no connection with these small sects. Four-fifths of the members of Protestant churches belong to six great Protestant families: Lutheran, Presbyterian, Episcopal, Congregational, Baptist-Christian (Disciple), Methodist.

Protestants frequently deplore, and anti-Protestants frequently deride, the diversity within Protestantism. No doubt there have been instances where the formation of new Protestant denominations represented a thirst for power rather than a new spiritual insight—in religion as in politics there are people who follow a rule-or-divide policy. No doubt also the continuance of denominational divisions represents in many instances merely the perpetuation of tradition and is an unwise duplication of effort.

However, lamenting with too much emphasis the religious divisions in America tends to obscure the fact that these divisions prove the reality of religious freedom. It was but a brief period ago—in terms of total human history—that but one church was recognized in each nation. In Catholic countries everyone was forced to be Catholic. In Protestant countries everyone was forced to be Protestant. To oppose the established church was often equivalent to treason. Rebellion against ecclesiastical authority has sometimes been the only way sincere Christians could preserve their moral integrity and bring an established church to its senses. The call for unity in religion is sometimes simply an uncritical acceptance of the former standard which demanded religious uniformity. Often "unity" is the demand of leaders who themselves refuse to compromise but who condemn everyone else for declining to "come over and subject yourselves to us." Americans do not deplore the fact that there are over eighteen hundred colleges and universities in the United States, nor the fact that there are more than three hundred thousand manufacturing concerns. From some points of view, the number of our religious denominations is quite small. In recent years a strong movement for Protestant church union has set in, a movement which already has brought together a number of denominations. But in all efforts at union great care has been taken not to endanger religious liberty; Protestants universally prefer the continuance of division within the Church to any threat to freedom.

The term *Protestant* was first used in connection with a protest some of the German princes made against a decree, promulgated in 1529, which erected serious obstacles against the Protestant advance. However, the term came into common usage in the England of Queen Elizabeth's time. A person who "protested" in that day was one who bore witness, declared a belief. Today the verb *protest* has come to have a negative connotation, but the noun *Protestant* still refers to a person who affirms religion positively.

The term *catholic* means universal. It goes back to the belief that the true Church is one, is undivided, and is destined to cover the whole earth. Of course, the Roman Catholic Church today is not "catholic" in this sense; and the Protestant churches, likewise, have no monopoly on "protesting" their Christianity. Unfortunately, in common American usage a person who says he is a Roman Catholic sometimes means little more than non-Protestant and non-Jew; and Protestant sometimes means little more than non-Catholic and non-Jew.

HOW OLD ARE THE PROTESTANT CHURCHES?

"Martin Luther founded your Church; but Jesus Christ founded the Catholic Church," said my Catholic neighbor's son to my small daughter. All Protestant groups would disagree with this frequently heard statement; however, they would not all agree among themselves concerning the founding of the Christian churches.

A few Protestants, fired with anti-Catholic zeal, assert that the Roman Catholic Church has so far departed from the teachings of Jesus Christ that the term *Christian* should no longer be applied to it.

Also a few Protestants assert that Jesus never intended to found a church and did in fact found none. Thus, in their opinion, no church goes back to him.

Most Protestant churches hold that Jesus Christ was the founder of the Church, and that all Christian churches are descended from that beginning but that the Protestant churches are closer to His spirit and teachings than are any of the others.

Any candid Protestant must admit that there is much truth in the Catholic contention that the Protestant *governmental organizations* were developed within the past four centuries. But arguments about the origin of ecclesiastical governments are of little consequence to most Protestant minds. Protestants believe that the Church is primarily a matter of faith and action. The true Church is found wherever two or three are gathered together and seek earnestly to study the Scriptures and to follow after Jesus. The Christian Church is a fellowship of the spirit and not an organization. It is found, say Protestants, as truly in the outposts of civilization as it is in the most elaborate cathedral. It is found as truly in the hearts of untutored laymen as among the most learned clerics.

Answering those who wish to argue about the primacy of ecclesiastical governments, Protestants agree that for hundreds of years Christians in the West were governed by the Pope. They also agree that the present Roman Catholic organization is a direct descendant of the organization which was in control during this period. They still further agree that in the sixteenth century large numbers of Christians defied the authority of the Pope and formed different governmental authorities.

However, Protestants contend that the Pope's rule over all the Western churches obtained only during the latter part of the Middle Ages. During the first five hundred years of Christian history, the Bishop of Rome was but one of the Christian leaders, not *the* leader. During the next five hundred years he frequently *claimed* universal authority, but this authority was not acknowledged throughout Christendom. During the eleventh century his authority became dominant in the West; but it has never been dominant in the East. And beginning with the Reformation in the sixteenth century, his authority ceased to be dominant in the West.

To the mind of Protestants, the history of the Church might be likened to the route traveled by many pioneers in the old frontier days. A small wagon train formed, let us say, in Massachusetts. As it traveled west, more and more people decided to join it. It got larger and larger. After the group had traveled about a thousand miles, one family gained sole leadership. Finally the train reached eastern Kansas where the Sante Fe Trail went one direction and the Oregon Trail another. The leading family ordered the entire group to take the Sante Fe Trail; but half of the group defied this command, chose other leaders, and took the Oregon Trail. Saying that the Protestant churches had no history prior to the Reformation is like saying that the group on the Oregon Trail had no existence before the separation of the two groups.

At the time of the Reformation, Protestants, under the leadership of Martin Luther in Germany and John Calvin in Switzerland, claimed that the Pope was a usurper and that they had returned to the authority of Jesus Christ. They believed that the Pope had so far departed from the teachings of Christ as revealed in the Bible that he and all his followers were apostate. They contended that no human being should stand between Christ's Word, the Bible, and the individual Christian. They said that the Bible sanctioned but two sacraments: baptism and communion. They believed in the marriage of the clergy. They said that no one is saved by receiving the sacraments or by doing good, but only by receiving the unmerited Grace of God. (See pages 162 ff.) They affirmed the "priesthood of all believers"; Luther wrote:

If a little group of pious Christian laymen were taken captive and set down in a wilderness, and had among them no priest consecrated by a bishop, and if there in the wilderness they were to agree in choosing one of themselves, married or unmarried, and were to charge him with the office of baptizing, saying mass, absolving and preaching, such a man would be as truly a priest as though all bishops and popes had consecrated him.

Contemporary Protestant scholars hold that a large percentage of the distinctive Roman Catholic practices date only from the reforms of the Council of Trent, a sixteenth-century reaction within the Catholic Church to the Reformation. Most Protestants honor and revere Martin Luther, but they do not think he founded the Protestant churches, any more than Catholics think the Council of Trent founded the Catholic Church. Protestants hold that they have revived many ancient Christian practices which the Roman Catholic Church had discarded.

Protestants reject, of course, the Catholic dogma that the formal authority of the Bishop of Rome goes back to Jesus Christ. The passage in Matthew where Jesus speaks to Peter saying, "Thou art Peter, and upon this rock I will build my church," is dealt with variously. Some scholars assert that the passage is a gloss, an addition to the original text. They point out that the Gospel according to Mark, which most scholars believe to be the oldest biography of

Jesus and which is probably the written form of Peter's own reminiscences, does not contain the heart of the story in Matthew 16:16–19. In fact, nowhere else in the New Testament is that part of the story told, though both Mark and Luke tell the first part of it. This omission is very strange, if Jesus did in reality impress on all his disciples his intention to found one Church and to make Peter the head. One writer declares that it is "inconceivable that a saying of Christ so central . . . should have been left unrecorded by three out of four" of his biographers.[1] Other Protestant scholars believe that all of the passage in Matthew is genuine, but contend that its meaning has been grossly misinterpreted. They assert that Matthew says nothing about there being but one true Church, or about the successors of Peter, or about all Christians being ruled from Rome; and that there is no evidence that Peter was ever in Rome or even that there were bishops in Rome before the second century. The writer of Acts indicates that James the brother of Jesus, not Peter, was the leading apostle (Acts 15:6–21). Paul made it clear that he thought Peter had no special prerogatives, and criticized him severely.

The Duty of Private Judgment

Most Protestants contend that the best avenue of approach to God is the individual's own prayer and devotion. They believe that every person who earnestly strives to learn God's will—and then lives up to his understanding—has done everything possible to ensure his salvation.

The truths of religion are public, according to Protestant teaching. In the Bible and in nature God has presented mankind with divine revelation. This revelation is available to all men. No group or class of men has more direct access to this revelation than have other men, provided they are equally learned and saintly. God will protect one man from error in the same way that He will another.

At the time of the Reformation, most Protestants considered that the teachings of the Bible form one consistent whole and are crystal clear to all honest readers. Accordingly, after these teachings had been determined by the Church leaders, the individual Christian was expected to accept and believe. Some of the churches still hold to this idea (e.g., Lutheran and Fundamentalist). However, the majority of American Protestants today consider that each individual is responsible for working out his own faith. He must follow conscientiously such light as is available to him and decide for himself. Yet, he is not spiritually isolated; for, he shares the search for divine truth with all his fellows. Protestants discuss religion a great deal; they set one opinion over against another. And they listen to many a sermon —some of them rather longer than the pew bargains for. But most laymen expect in the end to make up their own minds; and most ministers present the gospel as to their spiritual peers.

Protestants seldom accept the thesis of a sermon uncritically; instead they ponder over it and think through its implications. Around the Sunday dinner table, the sermon is often the topic of conversation. Many a mature churchman remembers how in his youth he received his introduction to rigorous thinking by sitting in on these Sabbath conversations. But in most of Protestantism, after the sermons and the discussions are over, the individual himself decides what he will believe about religion, just as he decides what he will believe about politics or economics.

A logical development of this belief in private judgment is religious tolerance. Many Protestants today have real understanding of, and sympathy for, religious beliefs contrary to their own. It was not always so. In earlier centuries, Protestant sects viewed with suspicion not only the beliefs of Roman Catholicism but also the beliefs of other Protestant groups. Professor Arthur C. McGiffert wrote, "Intolerance was even more general and more bitter [in early Protestantism] than in Roman Catholicism."[2] A modern historian would hardly expect to find the situation otherwise. Protestantism in those days was an insecure movement; tolerance is a virtue usually limited to people who are fairly sure of themselves. Furthermore, four centuries ago everyone felt that bigotry was a religious duty. Killing for religious reasons was common. During the religious wars in France, "Protestants wore strings of priest's ears and a Catholic commander asked his men: 'Why do you crowd the prisons with Protestant captives? Is the river full?' "[3] Even as late as the eve of the American Revolution, "Except for Pennsylvania . . . there was no colony in which Catholics could live in comfort," writes a Catholic historian.[4]

But today most, though not all, of the Protestant churches recognize the implications for interfaith relations of the belief in the right of private judgment. Protestants usually show a much less sectarian spirit than formerly and are more tolerant of religious differences than are Roman Catholics. As a consequence, Catholics often say that Protestants believe that "one religion is as good as another." The misinterpretation here could hardly be worse. A proper phrasing might be, "Every man's right to hold his religion is as good as another's."

THE BIBLE

Protestantism stresses biblical study as does no other branch of Christendom. At the time of the Reformation, Protestants put the authority of the Bible in place of the authority of the Pope. Luther said, "The common man, the boy of nine, the miller's maid, with the Bible know more about divine truth than the Pope without the Bible."

This stress on the importance of biblical study does not mean that all Protestants hold the same view of the Scripture. A few carry devotion to it to the point of bibliolatry, that is, they worship the Bible. One of the seventeenth-century English theologians even went to the extent of asserting that the possession of a copy of the Bible is an indispensable means of getting to heaven. None of our contemporaries would take such an extreme position; but many of them hold that everything written in the Bible is literally true. Many believe literally, for example, the account in Genesis which says that God created the world in six days. And many believe literally the teaching in the same account that woman was created out of Adam's rib. William Jennings Bryan, leading politician and religious conservative, declared that he would consider no proposition whatever to be preposterous provided it were found in the Bible. In one state a bill was presented to the legislature proposing to alter the value of the mathematical symbol π to agree with the implication of the biblical statements that the circumference of a round object is three times its diameter. (See I Kings 7:23.)

Of course, most of the people who hold a literal view of the Bible recognize just as clearly as anyone that they believe things which are contrary to what goes for common sense. Their defense is the same as the Catholic defense under similar circumstances: man knows only what God has chosen to reveal; if some aspects of the revelation seem contradictory to our puny minds, that is no reason to doubt God; we may wonder why God has not revealed more truth to us, but we must accept what truth He has made known, even though it may appear to be unreasonable.

Most present-day Protestants are not biblical literalists. They believe that the Bible contains the divine word, that it is the revelation of God, that it should be studied by every Christian. But they do not look to the Bible for scientific truth, nor necessarily for historical accuracy. They do not believe in the "typewriter" kind of inspiration, the kind of inspiration which would use men simply as mechanical agents. The majority of Protestants believe that God inspired the biblical prophets and evangelists with noble spiritual ideals, and then let these authors use their own human skills in setting down their convictions and in recording history.

Study of the Bible by liberal Protestant (and Jewish) scholars has led to many startling conclusions. These conclusions are the product of the rigorous use of the scientific method, and they produced a revolution in religious thinking as fundamental for Protestant theology as were the discoveries of Newton for physics or of Pasteur for medicine. Liberal biblical research has gone forward on the assumption that the same kind of human skills and frailties were at work in the production of the Bible as were at work in the production of other great books, and that the rules of ordinary logic apply in the biblical field just as in other fields. Research based on such assumptions has brought many liberal scholars to conclusions sharply opposed to traditional ideas. The following are examples:

The books of Moses (the first five books of the Bible) were not in fact written by Moses; no man could tell the story of his own death. (See Deuteronomy 34.) It is not even certain that Moses could read and write.

Many of the books of the Bible were not originally produced in their present form. They are compilations of earlier and sometimes contradictory documents. The first two chapters of Genesis are an example. In these two chapters are contradictory accounts of the creation of the earth. The explanation of these differing accounts is that the editor of Genesis used two different stories of the creation which he found in the written records of the Hebrew people and used them side by side without ironing out the discrepancies.

The Hebrews, just like modern peoples, exaggerated their national greatness. Throughout most of their history they were subject to foreign powers. They also, like us, developed mythologies about their national heroes; for example, there is evidence which indicates that Goliath was killed by another man than David.

The Book of Daniel is not a true picture of the experiences of the Jews in Babylon. This book, appearing many centuries after the Babylonian period, was written in the midst of a fierce struggle for independence and was intended to give courage to Jewish revolutionaries.

Jonah and Ruth are not narratives of historical happenings, but tracts in which the authors strove to influence the direction of Jewish thought by telling a story.

The narratives telling of Jesus' birth are considered to be legends. No birth narratives are contained in the oldest Gospel, Mark. The later Gospels, Matthew and Luke, give accounts of Jesus' birth which are contradictory at important points.

Paul is frequently given credit for the writing of books of the New Testament which were written by others; for example, the books of Timothy and Hebrews are erroneously attributed to Paul.

Such conclusions as these are taught in a large majority of the Protestant colleges and universities of the United States; such conclusions are also taught by many American clergymen and are accepted by an increasing number of Protestant laymen.

Protestant emphasis on the necessity for each individual Christian to read the Bible for himself gave the public school its first great impetus. In medieval times peasants and laborers had no particular reason for getting an education. But the Reformation gave men a tremendous desire to read the Bible. In addition, the discoveries just prior to the Reformation of printing and of paper-making made possible the manufacture of cheap books. Thus four hundred years ago the common man, for the first time in history, was in a position where he wanted a Bible and could buy one. As a consequence the early Protestants set up schools in order to make it possible for him to study religion.

THREE PROTESTANT THEOLOGIES

Theological differences in Protestantism exist not so much between denominations as within denominations; each of the larger churches contains a wide range of theological points of view. In the Episcopal Church, for example, there is a group which is vigorously Catholic in belief, though no allegiance is given the Pope; in the same Church are some members who are militantly evangelical. In the Baptist Church, the weight of opinion is decidedly on the supernatural side; but that denomination gave birth to what was for three decades probably the most radical theological school in the country. In Massachusetts most of the Congregational churches are theologically liberal, but the Park Street Church in Boston, the best-known Fundamentalist church in New England, is Congregational.

From the point of view of theological teachings, the large Protestant churches today are much alike; most of them have their conservative and liberal wings, but the weight of theological opinion rests in the middle. As a consequence theological extremists commonly discover that they have more in common with extremists in other denominations than with the majority in their own communion. The similar theological structure of the churches also means that services of worship tend to be similar. A stranger might worship in many a Protestant church for weeks without knowing whether he attended a Methodist, a Baptist, a Congregational, a Christian (Disciples), or a Presbyterian church. Consequently, denominations exchange members freely. Many a minister in analyzing his church roll has discovered that the members of his flock were born into a score of denominations. It is about as rare to find a birthright member in some city churches as it once was to find a native son in California.

The theological diversity of Protestantism can be shown by describing three outstanding points of view: Fundamentalism, Liberalism, and Neo-orthodoxy. A description of these three will give the reader a grasp of the range of Protestant thought. However, most Protestants would feel that in no one of them is their faith accurately described. The fact is that *Protestant thinking cannot be adequately summed up in three or four categories.* Freedom of religious thought produces many individual formulations of belief. As a result, Protestant thought is more like a continuum extending from extreme right to extreme left, than it is like a series of discrete positions. We will study here three points on the continuum, but the variations go on almost endlessly.

FUNDAMENTALISM

A large group of present-day Protestants prides itself on maintaining the traditional theology intact; chief among this group in both numbers and influence are the Fundamentalists. Many of them do not like the name *Fundamentalism*. They preach, they insist, the Christian religion and dislike to have it called an "ism" by people who have rejected the "fundamental" Christian tenets. But the name sticks, as has many another disliked name in religion.

Socially speaking, Fundamentalism arose as a "defense of the agrarian culture of the nineteenth century against the developing urban culture."[5] Theologically speaking, Fundamentalism is an effort to assert traditional dogma, reaffirming the ancient and medieval view of the relation of God to nature. The Catholic view of God and of nature is accepted, except that most Fundamentalists deny that miracles have happened since Biblical times. God's purpose in performing the miracles, they say, was simply to make clear to mankind that the Bible is His special revelation and is to be taken literally. This purpose having been accomplished, there is no reason for God to interfere further with the course of nature. But Fundamentalists have no doubt about the Biblical miracles. The sun stood still, the ax floated, the water turned to wine, actually and literally, just as narrated in the Bible. Fundamentalists are just as sure these things happened as they are of the things that occur in their own living rooms, and surer than they are of what they read in the newspapers.

The five "minimum basic doctrines" of Fundamentalism are:

1. The inerrancy and infallibility of the Bible.
2. The virgin birth and the complete deity of Christ Jesus.
3. The resurrection of the same body of Jesus which was three days buried.
4. The substitutionary atonement of Jesus for the sins of the world.
5. The second coming of Jesus in bodily form.

Points four and five of this platform need some explanation. The substitutionary atonement is the dogma that Jesus appeased the displeasure of God for the sins of mankind by dying on the cross. By that act, Jesus took on himself the guilt of original sin and the guilt of all the sins men have committed. He shed his blood for us and through his Grace made possible our salvation. Fundamentalist clergymen speak much of God's Grace and of "the blood shed for our sins."

They also speak much of the sinfulness of man and of his utter inability to free himself of sin except through the power of Jesus Christ. J. Gresham Machen, a learned and able defender of Fundamentalism, wrote that sin "is a mighty power, which is dragging us resistlessly down into an abyss of evil that has no bottom."[6] He argued that men must be brought under a conviction of sin. "A man never accepts Christ as Savior unless he knows himself to be in the grip of the demon of sin and desires to be set free."[7] "Without the sense of dire need the stupendous, miraculous events of Jesus' coming and Jesus' resurrection are unbelievable. . . . The man who is under the conviction of sin can accept the supernatural; for he knows that there is an adequate occasion for its entrance into the course of this world."[8]

The Second Coming is a familiar part of the traditional supernaturalism. On the Day of Judgment, Jesus will come to earth again in bodily form, just as the New Testament writers said he would. Then he will separate the sheep from the goats; some will go to everlasting punishment and some to everlasting glory. That will be the end of the world. Some Fundamentalists think that by a careful reading of the Bible it is possible to predict just when the Second Coming will take place. Over and over men have predicted that the "time is at hand." But when the date predicted arrives and nothing happens, the calculators usually announce that they have made a mistake and calculate the date anew. Some of the New Testament writers prophesied that the Second Coming would be preceded by many disasters: famine, pestilence, earthquake, war and rumors of war. Today many people who believe in the Second Coming are doubly sure that it is near at hand because of the invention of the atomic bomb with its terrible powers of destruction.

Fundamentalists also hold to the traditional Protestant doctrine of salvation by faith; this teaching is discussed on pages 162 f.

Fundamentalists believe in a literal heaven and a literal hell, and in the existence of the Devil as a personal agency. Much of their missionary drive comes from the belief that persons who are not saved will be damned. This belief was common throughout Protestantism until the present century. In the eighteen-eighties a Congregational missionary of ten years' experience in India was not permitted to return to his post because he doubted the unconditional damnation of the "heathen." Most Fundamentalists still maintain this position.

The use of revival meetings is common among Fundamentalists. This method was once popular with most of the Protestant churches. Some of the most distinguished Protestant leaders—the Wesleys, Whitefield, Edwards, Finney, Moody—were ardent users of the method and it accounted for much of Protestantism's growth. But today many churches, with the waning of the Fundamentalist mood and mind, have reacted against the use of mass methods. Such methods were badly abused by many evangelists, notably Billy Sunday. Advertising with calliopes, taking the collection in small dishpans, smashing pulpit furniture, preaching a sort of combination life and fire insurance, Billy Sunday "soaked it into Satan" and persuaded thousands to hit the sawdust trail. In one community Sunday saw so much sin that in a prayer he said, "O Lord, the next time you come here bring along plenty of antiseptic and rubber gloves." This brazen clowning brought ridicule on revival meetings and as a result they have been abandoned by most Protestant groups. However, the Fundamentalists contend that a method should be judged by its best and not by its worst manifestations. They continue to hold revival meetings, usually conducting them with restraint and dignity, but nevertheless making an emotional appeal to sinners to give their hearts publicly to Jesus.

The theory of evolution, contradicting as it does the first chapter of Genesis, has drawn much Fundamentalist fire. In the twenties they succeeded in four states—Tennessee, Mississippi, Arkansas and Texas—in securing the passage of laws which forbade the teaching of evolution in the public schools, judging correctly that no religion can long endure without systematic support from schools and colleges. This law was a challenge both to those who believe in the principle of freedom of religion and to the growing number of theological Liberals, agnostics and atheists.

In Dayton, Tennessee, a young high school teacher, John Scopes, broke the antievolution law and was brought to trial. Clarence Darrow, a famous freethinker and the nation's best known criminal lawyer, came to Scopes' defense. William Jennings Bryan headed the prosecution. The resulting struggle is one of the famous episodes in American religious history. Twelve untutored jurymen were asked to decide between the "Bible and Darwin," between "God and the gorilla." A professional showman, scenting new business opportunity, brought a monkey to Dayton and charged the Tennessee farmers admission to see their "ancestor." The nation's newspapers, scenting larger circulations, made a circus of the event. Famous writers from northern cities went to Dayton as reporters and sent back mocking articles.

The sympathies of Tennessee were Fundamentalist and Scopes was found guilty and fined. The nation laughed. But the laughter and the newspaper reports failed to plumb the depths of the profound emotions aroused in the hearts of conservative churchmen all over America. They viewed with fear the future of their religion. At the trial the Tennessee lawyer in charge of the prosecution asked, almost tearfully,

Why have we not the right to bar science if it comes from the four corners of the earth to tear the vitals of our religion? . . . If we bar that upon which man's eternal hope is founded, then our civilization is about to crumble. Tell me that I was once a common worm that writhed in the dust? No! Tell me that I came from the cell of the ass and the monkey? No! I want to go beyond this world where there is eternal happiness for me and others. . . . Who says we can't bar science that deprives us of all hope of the future life to come?[9]

The issue raised here is not a simple one. The persistent and dogmatic teaching of evolution and kindred theories in the public schools will in the end destroy traditional supernaturalism. Does the religious liberty of Fundamentalists require that their children be protected from such teaching? If so, what about the religious liberty of scientists? And how can we gain spiritual unity in a nation with such divergent points of view? Do Americans yet understand the full implications of religious liberty?

Opposition on religious grounds to the teaching of evolution persists. In 1960, the state supervisor of public-school curriculum guides for the state of Washington wrote, "If Darwinian evolution is true then the Bible is untrue, and I prefer to hold by the Old Book rather than to accept a worthless theory." The supervisor's removal from office on account of this statement promised to have political repercussions because of Fundamentalist reactions.[10]

Some Fundamentalist leaders have set themselves belligerently against every other religious group, refusing all intercourse, claiming infallibility for their dogmas, misrepresenting the beliefs of others. This "tyrannical legalism" has brought forth a reaction. One theological conservative wrote:

Let me say a word about that anxious breed of younger men who are conservative in theology but are less than happy when they are called "fundamentalists." These men are both the cause and the effect of a radical atmospheric change within American orthodoxy. . . .
I call myself orthodox because I cordially assent to the great doctrines of the faith. But I do not for one moment suppose that assent to doctrine is either the instrumental cause of justification or the touchstone of Christian fellowship. . . .
Once we are done with the business of semantics, we can turn to the really exciting item on the agenda of faith: sharing fellowship with all who love Jesus Christ and who are willing to test and correct their partial insights by the full insight of God's Word.[11]

The more tolerant type of Protestant conservatism was strengthened in the nineteen-fifties by the appearance of the fortnightly journal *Christianity Today*. The orientation of this journal is strongly conservative, both in theology and in social philosophy. But it does not have the "fundamentalist" temper; it makes a studied effort to deal fairly with events and movements; its slant is no more in evidence than is that of most of the nonacademic, religious journals.

Christianity Today employed the Opinion Research Corporation to poll the Protestant clergy of the United States concerning their beliefs. This poll showed that 74 per cent of the clergy regard themselves as fundamental or conservative in theology—"with slightly more than half preferring to be called 'conservative' rather than 'fundamentalist' "; 12 per cent of the clergy considered themselves neo-orthodox and 14 per cent liberal.[12] No statement was made concerning the sampling techniques used in this poll; in the absence of such a statement the reader can only speculate on the difficulty inherent in the effort to obtain a scientific sample of Protestant clergy-men in the United States and to wonder whether such a sample was in fact obtained.

The Fundamentalists and conservatives are found more in the South than in the North, more in the West than in the East, more in the country than in the city, more in business houses than in colleges, more among the old than among the young.

LIBERALISM

Protestant Liberals more than other Christian groups stress the right of individuals to decide for themselves what is true in religion. The belief in freedom from theological domination by creeds, councils, bishops, pastors amounts to about the most basic religious conviction. Theological divergence among Liberals is, accordingly, great.

Liberals strive to be attuned to modern thought, which is dominated by science. Every contemporary religious thinker must deal with science in some way. A few reject it completely. Others, recognizing its threat to traditional supernaturalism, build compartments in their minds and refuse any commerce between their ideas of science and their religion. The Liberals, however, strive to make science their ally, not their enemy; they take science into the very citadel of religion. Science is a new confederate in the battle for truth; thus Liberals hold that no vital religion can ignore scientific discoveries.

Liberals are frequently criticized for too much ardor in their es-pousal of scientific thought. One gets the impression that in many a sermon "science says" has displaced "the Bible says." One caustic critic observed that after a Liberal is through preaching his hands are usually "red from backslapping physics." Another critic claimed that some ministers he knew considered it "the highest compliment to God that Eddington believed in Him."

Such extreme comments caricature the position of the typical Lib-eral. He is a person who has taken firm hold on the principle that the universe operates through natural laws. Nature's laws, he says, are God's laws. The chemist running his experiments, the astronomer photographing the stars, the psychologist testing his subjects are at-tempting to discover the ways God is revealed in nature and in the affairs of men.

The Liberals do not believe in miracles. All events, they say, are controlled by natural processes. Sometimes Liberals classify the Biblical narratives of miraculous happenings into four categories: (1) narratives with no basis in fact; (2) narratives of actual happenings which became exaggerated in retelling; (3) narratives of actual happenings the character of which was misunderstood; (4) narratives of healings of functional diseases.

Liberals explain these points of view in some such way as the following:

1. Some of the miraculous narratives have not even a kernel of truth in them; they arose in the fancies of a storytelling people. Examples of Biblical miracles of this type are the jar of oil that was constantly replenished and the raising of Lazarus after he had been in the tomb for three days.

2. Frequently, occurrences were exaggerated as they were retold. An example of this type of miracle is the parting of the Red Sea. A hint of what probably happened is given right in the biblical account. We read, "The Lord caused the sea to go back by a strong east wind all that night" (Exodus 14:21). Later the story grew by poetic fancy to "The waters were a wall unto them on their right hand, and on their left" (Exodus 14:22).

3. Actual happenings are frequently misinterpreted even in our more factual day. In ancient times men had even more reason for misinterpretation. They had little understanding of natural law; consequently, they expected miracles and frequently saw them in perfectly natural events. For example, Jesus is said to have walked on the water. A *possible* explanation of what happened is that Jesus may have been walking along the shore. A very low-lying cloud may have obscured the shore line in the early dawn; the disciples had been out in a boat all night and perhaps did not realize that they were close to the shore. Consequently, when they saw Jesus walking toward them, his feet obscured by the cloud, they supposed he was walking on the water.[13]

4. What are thought to be miracles of healing are either erroneous reports or else cases in which no actual physical illness was present. Our emotions frequently block the proper functioning of our bodies. Under stress of religious experience these emotional

blocks sometimes disappear, as the modern faith healing movement amply proves; they also disappear under other types of experience.

Most Liberals intend that these four explanations shall leave no room for the dogma that God sometimes works in ways that are contrary to natural law.

Many of the traditional beliefs about God have been rejected by Liberals. In the past Christianity has stressed God's *transcendence,* that is, His separateness from the world of men and nature. The more simple forms of belief in God's transcendence have pictured Him with a human body seated on a throne high in heaven, surrounded by admiring angels; the more sophisticated forms reject such anthropomorphic notions but nevertheless hold that God is too holy or too spiritual to be linked closely with the processes of the physical universe. Liberals frequently take a tack exactly opposite to the supposition of God's transcendence. They stress the *immanence* of God, that is, His indwelling presence in the universe, His nearness to the processes of nature. According to this conception, God is not an absentee landlord only dimly aware of the struggles going on in his domain; instead he is an active manager, aware of every natural process and every sentient being. Some Liberals deny that God is transcendent in any degree; they hold that he has no existence apart from the physical universe. Probably most Liberals believe that God is both immanent and transcendent.

The majority of Liberals hold to the traditional Christian teaching that God is personal; that is, He is more like a person than like a tree, or a machine, or electricity, or any other thing He might be likened to. Liberals react sharply against the crude anthropomorphism which pictures God with hands and feet and a long white beard; yet most of them do hold that His chief characteristic is mind. He is conscious both of Himself and of us, and cares for us.

Some of the extreme Liberals go as far as to reject belief in a personal God. These men have much in common with those humanists and naturalists who reject the idea of God altogether (see Chapter XVII). The extreme Liberals, however, instead of rejecting the idea, redefine it. Perhaps their beliefs can best be expressed by quoting a passage from Edward Scribner Ames, for many years professor of philosophy at the University of Chicago and pastor of the University Church of the Disciples of Christ, Chicago.

> My idea of God is the idea of the personified, idealized whole of reality. . . . My idea of God is analogous to my idea of my Alma Mater. She is a benign and gracious being toward whom I cherish deep gratitude for her nurture and her continuing good will and affection. . . . In similar fashion God is the personified reality of the world. He is not a mere idea; He has substance, energy, power. He is the common will, the spirit of mankind; He is seen in men, especially in their benevolent corporate life. His image marks the humblest souls, and is more clearly revealed in the great leaders and saviors of the race.[14]

This position is significant not because it represents a movement in Protestantism which is widespread or powerful; it is neither. This position is significant because it indicates the wide range of belief exhibited by people who consider themselves to be within the Protestant tradition and who carry official responsibilities in Protestant churches and institutions of learning.

Some Liberals reject the Trinity and with it the deity of Jesus. Some of them distinguish between the words *deity* and *divinity*. They say there is a spark of divinity in every man, and in that sense Jesus was divine; but he was not the Deity, not God. Liberals of this persuasion consider Jesus to be the supreme religious teacher and the founder of Christianity; but, they contend, only after his death and only gradually did the trinitarian dogma develop.

On the other hand, most Liberals consider belief in the Trinity to be essential. But they wish to define it in modern terms—perhaps in terms like the following: "It is entirely rational to believe in the First Person, the Creator, and in the Third Person, the Spirit of God in the world. Moreover, the basic affirmation of the Christian religion is that God is like Jesus, not that Jesus is like God." A verse by Harry Webb Farrington states this point of view toward Jesus:

> I know not how that Bethlehem's Babe
> Could in the God-head be;
> I only know the Manger Child
> Has brought God's life to me.[15]

Liberals reject the virgin birth and the physical resurrection. Their religion is, they say, the religion *of* Jesus, not the religion *about* Jesus.

Liberals have many different beliefs about prayer. The practice of praying for help of a physical nature has, for the most part, been given up by them. Liberals hold that God would not answer, say, prayers for good weather or for the alteration of the course of physical disease. However, most Liberals believe that prayer for help of a spiritual sort is rational. Thus prayers for wisdom or for patience to endure difficulty are thought to be rational. Liberals stress the value of prayers of "communion." In such prayers, the worshiper seeks to conform himself to God's Will, not to bend God to the worshiper's will. It is the worshiper who must change. One writer said, "The winds of God are always blowing, but we must hoist our sail."

Most of the Liberals believe that man is neither naturally bad nor naturally good. In place of the belief that man has within him evil forces which can be conquered only through the Grace of God, Liberals hold the belief that human beings are plastic and can be molded into either saints or sinners. Liberals look for confirmation of this belief to the social scientists: the psychologists, the sociologists, the anthropologists. Frequently the Liberals cite such discoveries of the social scientists as the following.

Race hatred is not inborn. White and Negro children play together with no feeling of distinction—until they are taught by their elders to dislike each other.[16]

Children are much more apt to develop criminal tendencies in some environments than in others. Sociologists can point to the sections of our cities from which a large portion of the criminals of tomorrow will come.

Competition is not a "law of nature." Co-operation is equally native to humans. Anthropologists who analyzed a dozen primitive societies found that about half of them developed to a marked degree the co-operative tendencies in men.[17]

Many characteristics, supposedly even more firmly fixed in human nature than the tendency to evil, have been shown to be probably the result of training: for example, secondary sex characteristics. A tribe has been found where both men and women exhibit behavior which we would call characteristically masculine; another tribe has been found where both men and women exhibit

behavior which we would call characteristically feminine; still another tribe has been found where the roles of the sexes are just
reversed, where the men have the mannerisms and responsibilities
which we think proper to women and the women have the mannerisms and responsibilities which we think proper for men.[18]

It is from the study of such data that Liberals reject the teaching
concerning original sin and native depravity.

Some of the Protestant Liberals do not believe in immortality; according to them life after death is something about which man can
have no knowledge. A few Liberals would contend that the belief is
unimportant; "One world at a time," they say. But most Liberals hold
that faith in a future life is both important and rational. They think
it is important in order that men may have an adequate perspective
from which to make moral judgments. They think it is rational because a God who conserves the matter and energy of the universe so
carefully that none is destroyed, would not destroy the human mind
which is a much higher manifestation of creative power than either
matter or energy. Liberals have given up belief in hell and in the
Devil; but most of them believe in heaven. However, they profess no
knowledge of what heaven is like.

Out of Liberalism sprang what is called the Social Gospel, the belief that religion must deal more effectively with social injustice. Many
Christians believe that the function of the Church is to develop Christian convictions in the hearts of individual men who in turn will put
these convictions into operation in politics and business. Followers of
the Social Gospel contradict this point of view; they declare that society as well as individuals must be saved; that a healthy individual
cannot develop in a poisoned environment. Consequently, many of the
liberals are much exercised over the problems of war, poverty, and
racial discrimination. The Social Gospel will be more fully dealt
with in a later chapter. (See pages 299 ff.)

Many current writers rather badly misunderstand the liberal orientation. A leading liberal theologian introduced a book defending
Liberalism as follows:

The author of the present volume is not a "liberal" in any meaning of
that term as now most commonly understood in American theological

circles. In the present fashion "liberal theology" is widely used as a whipping boy for all who condemn empty-headed optimism, the substitution of current or past metaphysics for Christian faith, or a rational abstraction for a Christian understanding of man.

The current understanding of liberalism among theologians is less influenced by the actual methods and teachings of theological authors known to their contemporaries as liberals than by Reinhold Niebuhr's critical description of the movement. Niebuhr has rendered valuable service in exposing certain betrayals of the gospel in some popular preaching and teaching in liberal mood. Unfortunately . . . Niebuhr often uses these distorted popular versions of liberal theology as typical of the whole movement. Many Seminarians now practice tilting at these caricatures of liberalism and suppose that they are disposing of the methods and teachings of liberal theology for once and all.[19]

Liberalism in religion is young. It had a brief history two centuries ago in a movement called Deism; but for the most part it is a product of the last hundred years. Many of the implications of this mode of thought are not yet clear, yet the Liberals feel sure that it is to be the theology of tomorrow. It is a modern point of view, they say; it fits the needs of modern man; it keeps abreast of the fast-moving discoveries of modern life. Liberalism does not dispense with faith—no religion or philosophy does—but the Liberal rejects what seems to him to be credulity and puts great trust in reason. Thus Liberals treat their faith with the calmness of men who are sure of themselves. It is not necessary, they think, to be dogmatic or impassioned in defending religion. The facts will speak for themselves. Truth never was worsted in a fair and open encounter. They say with Tennyson:

> There lives more faith in honest doubt,
> Believe me, than in half the creeds.

The proportion of theological Liberals is higher in New England than anywhere else in the nation. Boston is the liberal capital; and it is more or less true that the farther one gets from Boston the less religious Liberalism he finds.

NEO-ORTHODOXY

Liberalism had its innings in the twenties. It was the era of the Coolidge prosperity. World War I had just been won; no one but erudite chart-readers dreamed of a depression; the Republicans were promising two chickens in every pot. But conditions in Europe were a different story. There men were making a slow, bitter fight against the misery which follows actual battle.

European devastation had a profound effect on European religion; theology has ever been sensitive to social conditions. Present-day Liberalism got its initial impetus in the German universities in the heyday of nineteenth-century optimism. Compounded of the wish to retain the privileges wrought by imperialism, and of unlimited confidence in the scientific method, and of a naïve faith in progress, optimistic Liberalism dominated the theological field until the bitter realities of living in a world smashed by war demanded a change. Theologians began to be "realists." They lost faith in the capacities of man to work out his own salvation; they declared man to be naturally evil; they attacked the scientific method as a means of dealing with the basic religious problems; they even lost faith in the power of the reason to guide man to supreme values. European theology turned definitely toward the dogmas of traditional supernaturalism.

In the early years, this kind of thinking won but few disciples in America; the country was too prosperous. But with its own economic "maturing" and the disintegration of its hopes for a world run like a giant American state, there was a rebirth of old-time beliefs and the new supernaturalism began to gain ground. Today its adherents are influential and furnish a significant percentage of the Protestant leadership.

Neo-orthodoxy—a term which makes many of its adherents wince —is too new as a movement to have worked out its concepts in terms which the layman can grasp and understand easily. But today disillusionment has gripped millions of lay minds too. The failure of the United Nations, the threat of atomic warfare, the materialism of our competition with Russia, make men long for a view of life which will give them security in the midst of apparent social disintegration.

This longing does not drive all thinkers in the same direction. Some of our contemporaries look for the answer to politics: to "an American century"; to "protection through American might"; to "our American system of free enterprise." Others try to forget their fears by turning to the pursuit of pleasure: they ignore the news, dedicate all their spare energies to skiing or bridge or poetry, or to drowning their anxieties in alcohol.

But others turn to traditional religion. We are not the first generation, they say, to be confronted by the possibility of disaster. Disaster is a common human experience. And God has not left man in his despair to struggle alone. God has made very clear in the Christian revelation the road which man must travel. Neo-orthodox Christianity is a reaffirmation of the historic Christian doctrines, with such changes as seem mandatory as a result of modern scholarship.

The Neo-orthodox begin by attacking Liberalism. They do this because most of the leaders of the movement were once Liberals. The biography of a typical neo-orthodox leader might read something like this: He was born into an orthodox home and was required by his parents to attend all the established means of Grace—Sunday school, church, and prayer meeting. In his youth he rebelled against the supernaturalism of the church—and against his parents—became enamored of science, and espoused romantic Liberalism. In adulthood he saw that life is not one long junior prom, such as his youthful fancies had pictured; he lost faith in the ability of man to save society; and concluded that only God can bring salvation.

Having reacted against the optimistic view of human capabilities, the neo-orthodox theologian very naturally makes optimism the first object of his attack. Liberalism is naïve, he declares. Man's lot is tragic; his predicament desperate. Man has learned at long last how to produce enough food and shelter to meet the basic physical needs of all the human race; but in his selfishness he won't put his knowledge into practice. He has unlocked some of the doors which lead to the basic secrets of the physical universe; but in his hatred he has used his knowledge to invent weapons which threaten to exterminate half the race. Man is naturally evil. Left to himself he is self-centered, egoistical, tyrannical.

But man's innate depravity, continues the neo-orthodox theologian, is only half the picture. Man also was "made in the image of God." He is a paradoxical creature who is both good and bad at the same time. He reaches out to help his neighbor, but always with one eye on favorable publicity. As Charles Lamb said, the greatest satisfaction in life is to be caught in doing a secret good. Man's tendency to do evil can be conquered and his tendency to do good can be brought to fruition, say the Neo-orthodox, only through God's Grace. Of himself, man is helpless to conquer his tendency to wickedness. By repenting of his arrogant self-assurance and by accepting the judgment and the mercy of God, man can receive God's forgiveness and settle his final relationship with God. Thus sin is overcome in principle, though not in fact; man's earthly problems continue.

Neo-orthodoxy attacks Liberalism's faith in reason. Reason is a human instrument. It can no more settle the problems of basic *religious* faith than could the mind of a chimpanzee settle the problems of physics. Reliance on reason is but another instance of man's arrogance. One neo-orthodox leader writes, "The reason which asks the question whether the God of religious faith is plausible has already implied a negative answer in the question because it has made itself God and naturally cannot tolerate another."[20] Another neo-orthodox leader writes, "Nowhere in the Bible, whether in the Old Testament or the New, is there a philosophical argument for God. There is no evidence that Jesus ever tried even remotely to prove His existence."[21] The basic truths of Christianity are a part of the divine revelation and we either accept them or we don't. "The Christian Faith . . . is not something that is arguable."[22]

Neo-orthodox thinkers accept the results of the modern study of the Bible. They hold that while reason has no power to communicate God's revelation, it has power to deal with nature. The Bible is a natural product; therefore, reason is applicable to biblical problems just as it is applicable to all problems which are of the natural order. Some of the most radical of the biblical scholars are neo-orthodox in their theological point of view, even to the point of rejecting a large part of the biblical account of the life of Jesus as literal historical truth.

The biblical miracles present little difficulty to the neo-orthodox theologian. He by no means denies the possibility of miracles but he exercises much skepticism concerning their actual occurrence. This skepticism would be the natural result of the application of the scientific method to the biblical narratives.

However, the neo-orthodox theologian makes one fertile suggestion concerning our understanding of the miraculous events of the Bible. He says many of the narratives of miraculous events are myths, that is, stories whose actual details never happened but whose central teachings are true to the basic principles of existence. For example, the first chapters of Genesis do not describe the actual creation of the universe; nor was there ever a first man by the name of Adam who was lured by his wife to eat an apple and thereby corrupt the human race. But the first chapters of Genesis are true in the sense that there was a creation, that God was the creator, that man is rebellious against the order that God has established, and that man is estranged from God because of his sins. Many of the greatest truths of the Bible are thus mythologically stated, according to neo-orthodox belief.

Basically Neo-orthodoxy is a reassertion of the judgment of God. It is a reaction against the comfortable theology which stresses God's love at the expense of His majesty. Neo-orthodox thinkers recoil at sermons in which God is pictured as a good fellow, a constant companion, who, "like a real friend," loves us in spite of what we are. This kind of preaching is received favorably by congregations who sit in well-cushioned pews and who will ride home in their own comfortable automobiles. "God is on our side and everything will come out all right." But the stern conscience of the neo-orthodox preacher will not let him forget the tragedy of life in our slums, our "nigger-towns," our hospitals, and on our battlefields. His keen intellect will not let him forget the threat of depression, of war, of race suicide. He sees all these as God's judgment on the sins of man. Man in his conceit thinks he can defy his Creator; but God will not be mocked. Man must humble himself, say the Neo-orthodox, and submit himself to God—and God will forgive. The question is not, "Is God on our side?" but, "Are we on God's side?"

Neo-orthodoxy was for a time the theological rage. It is unfortunately true that a band-wagon psychology shapes the thinking of many Protestant clergymen. Most Protestant communions successfully escape the coercive atmosphere by means of which some conservative religious groups secure uniformity of opinion. But in spite of Protestantism's successful espousal of freedom in religion, many Protestant clergymen, perhaps most, exhibit a tendency to follow in sheeplike fashion the opinions of the strongest personalities which are currently expounding eternal truth. In the forties and early fifties, clergymen who did not espouse Neo-orthodoxy were called "immature," "simple," "laggard," "middle-aged," "arrested in development," "failing to rethink their positions," and even "lazy," while their beliefs were called "outmoded," "frozen," "naïve," and "passé" in contrast to beliefs which were "newer," "up-to-date," "prevalent," and held by "young," "present-day," theologians. In the twenties, Liberals often dismissed "the conservatives" as unscholarly, medieval, not keeping up with the times; Fundamentalism was treated with contempt and Neo-orthodoxy with amusement. In the thirties, the pacifists were a dominant group; at many a conference the man who could not affirm that he would never again bless war felt himself regarded as not Christian and a pariah. Harry Emerson Fosdick in a justly famous remark said that when he was a theological student biblical criticism was the all-important concern of the clergy. Somewhat later, optimistic Liberalism was dominant. Then came the Social Gospel. And now we have Neo-orthodoxy. Fosdick said he hoped he would live long enough to see what would come next. One theologian has written that the new mood has already appeared,[23] and another has written a book which urges a " 'new deal' in theology that will replace the emphasis of the 'neo-orthodox' movement," and which strives to make clear just what are the "new" accents in contemporary theology. No doubt, a new synthesis is forming. Some thinkers have given it the name *Neo-liberalism*. It seems to be a position which puts more trust in reason than does Neo-orthodoxy and which strives to recognize the valid positions of both the neo-orthodox and liberal theologies while seeking to avoid the extremes of both. Will the priests of the new position be as contemptuous of the recent past as were their counterparts in previous decades? One day we will achieve a religious maturity which begins with the assumption that sincere minds can disagree in the realm of theology and still co-operate in efforts to achieve significant spiritual and moral goals.

EVANGELICALISM

A Protestant clergyman once said in my presence, "I love flowers; but I hate botany. I love religion; but I hate theology." He was giving expression to a mood characteristic of Protestants. No one of the three theological positions outlined above—Fundamentalism,

Liberalism, Neo-orthodoxy—is an adequate description of the faith of the average Protestant churchgoer in this country. He distrusts theology; he thinks of the theologian as a person who deals in intellectual subtleties and who has a view of life warped by too much reading and too little contact with "real people." Deep religion, in the view of the average worshiper, is a simple, heart-warming, personal experience. Perhaps the best word to describe this mood is evangelicalism.

The typical evangelical is a mild type of mystic (see pages 276 ff.); he believes that religion is a personal experience, something which sets the spirit aglow. Frequently he is able to point to a time in his life when he had a definite religious experience, when he "got religion." In former years evangelicals insisted that every person, if he is to be sure of salvation, must pass through an emotional conversion. Today the stress on conversion has mitigated but evangelicals still emphasize the necessity for an intimate, personal approach to the religious life.

The typical evangelical is also a mild type of supernaturalist. He no longer holds to the more miraculous aspects of traditional Christianity; but he believes emphatically in the miraculous origin of Christianity and in the Church's obligation to carry the Christian message to the whole world. The evangelical is usually a trinitarian, believes in the unique inspiration of the Bible, takes great comfort in his belief in a future life, and sadly confesses that men are more prone to do evil than to do good. He believes in petitionary prayer and tends in his personal devotions to think of Jesus more as a personal friend and guide than as the metaphysical Second Person of the Trinity.

The evangelical's distrust of the intellect means that usually he has not resolved the contradictions in his faith. He thinks simply that the warmth of his experience verifies his beliefs. In holding this conviction he makes an error which is very common in religion, and in many other types of thought, the error of thinking that because belief gives tremendous personal satisfaction, the belief must be true. Persons of every type of persuasion and practice defend their faith and conduct in the same fashion.

Protestant Worship

Most Protestants believe the heights of worship can be achieved anywhere. The soul cannot come closer to God in the church building than it can in the home or in the factory; God is no more present, say most Protestants, in the Communion Service than he is in private prayer. It is the inner disposition not the outward form that counts. Thus in their worship, Protestants strive to get away from reliance on physical things. Because of this fact, many Protestants believe that their worship is more spiritual than Catholic worship; Protestants say they put less emphasis on physical objects and ceremonies and more reliance on mental development and spiritual renewal. The typically Protestant attitude toward worship might be likened to that attitude toward poetry which insists that poetry is essentially an inner experience, an emphasis, a way of viewing life rather than a pattern for arranging words according to specified schemes of meter and rhyme. Catholic worship centers in the Mass to which physical elements are essential; Protestant worship (in most churches) centers in the teachings of the Bible.

During the Reformation, Protestants deliberately abandoned many of the rites and symbols of the Catholic religion. Protestants dropped five of the seven Catholic sacraments; they discontinued the elaborate vesting of the clergy; they no longer genuflected before the altar, or crossed themselves, or bowed the head at the name of Christ. In a large majority of the Protestant churches, worshipers even ceased to kneel during prayer; this practice was discontinued because it was thought to be an act of veneration of the Host. Distrust of forms in religion went so far at one time that some Protestants even refused to celebrate Christmas, declaring that it is a "popish festival," and to repeat the Lord's Prayer, contending that it is a "vain repetition."

Today Protestants still cherish the belief that no physical object is essential to the highest form of worship. Yet they are coming increasingly to believe that symbolism plays an important part in worship. Some of the older customs are creeping back into use: many Protestant churches today observe the Church Year, vest their choirs, follow a prepared ritual. The use of the cross is an interesting example of the trend toward the increased reliance on symbols. Churches which formerly would no more have thought of displaying the cross in their services than they would of using statues or incense, now place the cross on the altar as the central object in the sanctuary. The continued willingness of Protestants to make innovations in their worship is illustrated by experiments which have put jazz into the service. A "jazz mass" has been conducted under Protestant Episcopal auspices[24] and a Texas music teacher wrote a jazz setting for the Methodist "Order for Morning Prayer."[25] It has been presented a number of times, including on television.

The very architecture of a Protestant church building testifies to the effort to rely in worship on an inner movement of the mind and spirit. In a majority of Protestant churches the pulpit and Bible are the central objects of attention, having displaced the altar. Today in Protestantism there is a strong movement back to the altar-centered church; however, the plans for these church buildings always include a pulpit and a reading desk—and on the reading desk is an open Bible.

Forms of worship in most Protestant churches are not fixed; the local church leaders determine the order of service. (The Episcopal Church is an exception, as will be explained in a later chapter.) If a stranger attends a service of this kind, his experiences probably run something like the following. He is met at the door by an usher and is conducted to a pew. He would notice that in this type of church there are no kneeling cushions and that the pews are set too close together for worshipers to kneel comfortably. Later on he would discover that it is customary during prayer simply to bow the head. The observant stranger would note also that as the parishioners come into the sanctuary and sit down they make no devotional movements—they do not genuflect and usually they do not bow the head in prayer. They simply sit down in silence and compose themselves.

The service begins with the playing on the organ of some well-known classic. After this music is over, the congregation sings a hymn, perhaps the familiar

> Holy, Holy, Holy! Lord God Almighty!
> Early in the morning our song shall rise to Thee;

Holy, Holy, Holy! Merciful and Mighty!
God in Three Persons, blessed Trinity!

The congregation stands during the singing of this triumphant song and the choir and minister march in and take their places. The minister then offers prayer, invoking God's presence in the hearts of all who are in attendance, and all repeat the Lord's Prayer. The choir then sings an anthem. Many a Protestant will testify that for him the most meaningful expression of the Christian message is the devout singing of laymen.

At the conclusion of the anthem a passage from the Scripture—usually a Psalm—is read responsively, the minister and the congregation reading verses alternately. The practice of reading responsively comes from a desire to give the congregation as active a part in the service as possible.

Then the minister reads another passage from the Scripture. If the stranger is a student of the Bible, he may recognize the verses chosen. Perhaps they will be the beloved 13th Chapter of I Corinthians, which begins

Though I speak with the tongues of men and of angels, and have not charity, I am become as sounding brass, or a tinkling cymbal.

Then comes the pastoral prayer, in former times usually called "the long prayer." Until recent decades, prayers in nonliturgical Protestantism were practically never read from a prayer book; they were spoken extempore. Prayer books were thought to be barriers to deep religious expression and a vestige of pre-Reformation practice. Since religion comes from within, written prayers, according to this thesis, can never express the living emotions of people worshiping here and now; expecting people to gain spiritual nourishment from reading a prayer is like expecting them to gain physical nourishment from reading a cookbook. Vital spirituality is a living, present experience. Spontaneity in religion was formerly so emphasized that "one man even objected to hymn books on the ground that singing should be from the heart and not from the book."[26] Just how this idea would be carried out in congregational practice is a bit hard to imagine.

Today the trend is back toward carefully prepared prayers. The present-day clergyman would agree that genuine prayer must come from the heart. But he contends that spiritual and literary giants are more able to express the depths of our devotional experiences than we are ourselves; and in any case, he says, a prayer spoken in the hearing of worshipers is not their own heartfelt prayer and is only designed to stimulate their own prayers. Under these conditions, the better the spoken word can be, the more probability there is that it will be able to stimulate the devotional attitude.

At the conclusion of the pastoral prayer, the offering is taken: the ushers pass through the congregation, carrying collection plates. After each worshiper has had an opportunity to symbolize his own inner dedication by presenting a gift, the ushers march to the front bearing the plates, and the congregation rises and sings a doxology.

> Praise God, from whom all blessings flow;
> Praise Him, all creatures here below;
> Praise Him above, ye heav'nly host;
> Praise Father, Son, and Holy Ghost.

After the offering has been taken, another hymn is sung, and then the minister begins his sermon. The sermon is the most prominent feature of a Protestant church service. It is also the one most frequently criticized. A former ambassador to the United States from Denmark wrote that if the Protestant church should be found dead, the sermon would be the dagger in her breast. Dean Inge of St. Paul's, London, once said that preaching a sermon is like throwing a bucket of water at narrow-necked bottles. The average Protestant would dissent from these opinions. When the Protestant churches no longer emphasize the sermon, he would say, they will cease to be Protestant —they will cease to stress individual judgment and to consider that worship is chiefly a movement in the inner dispositions. The sermon is the agency whereby Protestant church members are lifted out of their spiritual complacency, whereby they gain new insights into religious problems, and make applications of their faith to the conditions of everyday living. So important is the sermon in Protestant thought that almost every devout worshiper has stored in his mind the memory of great sermons that have all but revolutionized his life.

After the sermon is concluded, another hymn is sung, and all bow while the minister pronounces a benediction. Then the members of the church gather in the aisles and vestibule, greet one another cordially, and shake hands with the minister who has made his way to the church door. The stranger senses during these experiences after the benediction that he is in the midst of a community of friends, persons who find joy in one another's companionship and who seek to prolong it. If he continues to attend Protestant services, he will discover that the heights of the worship experience are reserved for those who come to feel that they are part of a church family, a beloved community, which shares its common burdens and perplexities, triumphs and joys.

In addition to services of the type just described, Protestant churches conduct from time to time the service of Holy Communion (Lord's Supper). This service differs sharply from the Catholic sacrament at several points. In the first place, Protestant laymen receive both bread and wine. Thus Protestants follow their understanding of the injunction of Jesus at the Last Supper, "Drink ye all of it." In the second place, Protestants reject the dogma that the Mass duplicates in an unbloody manner the sacrifice on Calvary. In the third place, Protestants, except for the Lutherans and a few of the Episcopalians, do not believe that any sort of physical miracle is wrought in or on the elements of bread and wine.

Several doctrines are current among Protestants concerning what happens during the Communion Service. The following is a brief outline of some of them:

The Physical Presence. This dogma holds that the substances bread and wine remain bread and wine but that with them are the *physical* body and blood of Christ. Luther held this belief. He described it by saying that when a bar of iron is heated red hot, the bar remains iron; but something of great power is added. Thus he asserted that Christ is physically present, though the bread and wine are present also. Lutheran groups hold this theory. Frequently they describe their belief by saying that the body and blood of Christ are "with, in, and under" the elements.

The Spiritual Presence. This dogma holds that nothing physical happens in the Communion Service. Christ is truly present in the elements, but after a *spiritual manner* and is received by the believer as a spiritual blessing. Many Presbyterians, Episcopalians, and Methodists believe this doctrine, as do also members of other denominations.

The doctrine of the symbolic nature of the Communion Service is the position that the elements are but symbols of the presence of God, in somewhat the same sense that the flag is the symbol of the United States. Persons who hold this doctrine consider that the Communion Service is a memorial service commemorating the life and death of Jesus Christ. Many Congregationalists, Baptists, and Christians (Disciples) believe this doctrine, as do also members of other denominations.

The doctrine that physical elements should not be used in the Service of Communion is maintained by the Christian Scientists. They have a Communion Service, but it is conducted without the elements.

The doctrine that no special service of Communion is desirable is held by the Quakers. They contend that every service of worship is a Communion Service and that worship will not be aided by the presence of bread and wine.

The elements are distributed in Protestant churches in two ways. In some of them worshipers are served at the altar rail. This method has the weight of centuries of tradition behind it. In other churches worshipers are served in the pews by laymen who bring the elements from the Communion table. This method is defended by asserting that it makes of Communion a group experience and that it is more like the original Lord's Supper than the traditional method.

A majority of Protestant churches no longer use wine but have substituted grape juice. The Mormons even use water. The purpose here, of course, is to avoid serving alcohol.

Baptism in most of the Protestant communions does not assume the essential role that it does in Catholicism. The baptismal experience is considered by few Protestants to be essential to salvation. Salvation is considered to be dependent on the inner disposition. Baptism is, however, made the first formal step toward becoming a church member.

A matter of debate among Protestants concerns the proper method of baptizing, whether by pouring, sprinkling or immersion. The Baptists and Disciples were once unanimous in affirming that immersion was the New Testament method and should be retained by the modern Church. Most Baptist and Christian churches still insist on immersion, but some of them join churches of other denominations in asserting that the question is unimportant. Most of the Protestant churches use the method of sprinkling, contending that sprinkling is easier to administer and puts less emphasis on the physical aspect of the sacrament.

How Is Protestantism Governed?

Edmund Burke in his great defense of the American Colonies before the English Parliament said:

> The people are Protestants; and of that kind which is the most adverse to all implicit submission of mind and opinion. This is a persuasion not only favorable to liberty, but built upon it.[27]

One historian wrote:

> Democracy did not arise out of eighteenth century political and industrial conflicts, as a momentarily popular view misconceives. Its roots are to be found in the attempted revival of primitive Christianity by the radical lower-class sects of the Protestant Reformation, those peasants and yeomen who were our own ancestors, and who initiated the Reformation and eventually carried out its basic principles—especially in America—to conclusions undreamt of in the beginning. The ideal of local self-government was brought to America by the Pilgrims; the separation of church and state was derived from the Baptists; the right to free speech was a development of the right of freedom of conscience established by Roger Williams and William Penn; the equality spoken of in the Declaration of Independence was an outgrowth of the equality practiced by the Quakers. Democracy was envisaged in religious terms long before it assumed a political terminology.[28]

The reader should not conclude from these striking statements that Protestants have everywhere and always practiced a pure type of democracy. On the contrary, on many occasions they have acted most despotically; for example, the most democratic aspect of early Protestantism, the Anabaptist movement, was bitterly persecuted by Protestants as well as by Catholics; frequently Protestant churches have been ruled from the top.

Yet the basic democratic principles inhere in the Protestant dogma of the right of private judgment; when the common man wins the privilege of judging for himself in religion, sooner or later he demands a similar right in other fields. The Protestant contribution to the development of democratic institutions is thus a very distinguished one.

Protestant governmental arrangements can be classified as episcopal, presbyterian and congregational. In the simon-pure episcopal type of government, authority rests with a bishop; in the presbyterian type, authority rests with elected representatives; in the congregational type, authority rests with an open church meeting.

No Protestant denomination in America has a strict episcopal government, as far as I know. In former decades, some of the denominations were autocratically run; but ecclesiastical organization has so developed that today no Protestant bishop or executive secretary is given anything approaching final authority. On the other hand, few denominations have a strict congregational polity. Even in denominations whose legal framework and traditions provide for complete local autonomy, actual practice has greatly modified the extent to which local congregations act independently—though their power to do so, even to the extent of changing denominational affiliations, is still unquestioned and is sometimes exercised. The growth of ecclesiastical machinery has meant that more and more the denominational organizations are run as representative democracies (at least in theory), and the local churches are run as face-to-face democracies.

On the whole, American Protestants run their institutions in a fashion similar to the way Americans run their governments, and tend to be democratic in the same degree that Americans generally are democratic—a degree not always gratifying. Democracy is a very complex skill and ecclesiastical democracy is beset by many of the problems that confront secular democracy; a large percentage of church members do not exercise their church franchise, local churches frequently are run by a clique, sometimes they are even run by a boss, occasionally the majority in a church meeting laughs down the ideas of a minority, frequently leaders make decisions (supposedly in the interests of efficiency) which should be entrusted to the will of the majority.

This situation should not lead the reader to conclude that the Protestant churches are a drag on the American democratic tradition. Many of America's secular institutions are much less democratic. Her factories and stores, for example, have been little affected by democratic principles; few of her colleges are operated democratically. But Protestant churches, on the whole, are institutions where the popular will determines major policies, where power is widely distributed, where leadership is frequently changed, where property and funds are democratically controlled, where ability is generally recognized and frequently rewarded.

Some of the churches in their national and regional organizations put representative democracy into practice. Other churches run their national and regional organizations by a kind of oligarchy. This situation will be discussed in Chapter VIII.

Unfortunately, it does not appear that the present Protestant groups, like some of their forerunners, will lay the foundations for bold, new democratic adventures. Today, Protestants try simply to put into practice the present popular American understanding of the nature of democratic institutions. The churches, especially at the local level, are chief among the voluntary organizations—lodges, service clubs, sports clubs, co-operatives, labor unions—in which Americans learn that democracy is something more than voting in public elections once a year and thereafter waiting patiently for the decisions of the higher-ups. If Americans have the habit of democracy, it is because such voluntary organizations give individuals the opportunity for much democratic practice.

Protestant democracy means that final power rests generally in the hands of laymen. In a large percentage of the local churches laymen hire and fire the minister, determine the requirements for Church membership, hold the deeds to the church property. Clergymen, of course, exercise great influence; they are not only specialists in church administration but they are also the executives who carry out democratically determined policies. The relationship between the Protestant clergyman and the membership of his church might be likened roughly to the relationship between the executive and legislative branches of a government which operates on the basis of well-defined constitutional principles. In this type of organization the layman has the last word, if he wants to say it.

This placing of power in the hands of laymen is considered by some observers to be a serious weakness in Protestantism. An English bishop visiting in the United States declared that the greatest threat to

American and Canadian church life is the dependence of the minister
on the good will of the people. A writer in the *Atlantic* stated that
clergymen "find themselves merely hirelings of groups they are compelled to please and whom they dare not stimulate or rebuke."[29]

Probably it is true that a considerable percentage of Protestant
clergymen find lay authority irksome and stultifying to moral and
spiritual development. The majority, however, are happy to work
under the Protestant system, because they believe that power is less
apt to be abused if it is in the hands of a whole church group than if
it is concentrated in the hands of a skilled leader. Protestant clergymen
usually think of themselves as teachers, and they know that while
church opinion moves ponderously it also moves persistently, and in
the direction of broad understanding, when it is under the influence
of preaching that is both patient and idealistic. Democratic organization often seems slow and inefficient when contrasted with the speed
and efficiency of centralized control. Most Protestant clergymen,
however, believe that the final event will demonstrate the wisdom of
entrusting power in religion to common people, even as the recent war
demonstrated the initial slowness but the ultimate strength of the
democratically controlled nations.

Perhaps the most disturbing criticism which can be made of the
Protestant churches as democratic institutions is that they frequently
minister to different social classes. This fact can be most clearly seen
in the racial divisions which characterize religious life in the United
States. The color line is about as tightly drawn in the local church
as it is in institutions which never refer to Paul's statement, "There is
neither Greek nor Jew, . . . Barbarian, . . . bond nor free"
(Colossians 3:11). Eleven o'clock on Sunday morning is sometimes
said to be the most segregated hour of the week. One clergyman wrote
that he hoped the church he served might become as racially inclusive
as the local bus line. A study of Congregational Christian, northern
Presbyterian, and United Lutheran churches, reported in 1954,
"permits an estimate that 9.8 percent of all the churches in these
communions have constituents from more than one racial group."
However, the actual number of persons of a nondominant race
integrated into the "congregations or organizations" of these denominations was quite small; even among the integrated churches of

these denominations only 2.7 per cent of the constituents were members of the nondominant race.[30] Moreover, a study sponsored by the National Council of Churches reported that

Opportunity for membership in a church can in no way be equated with the opportunity to become an active participant in its life and fellowship. The survey would seem to indicate that in the average church with non-white members the social life moves along a fairly separate pathway and out of their touch. Membership becomes a sort of "front room" type of fellowship for the non-white person who must get along as best he can with no real voice in forming the programs or policies of the church to which he gives his allegiance.[31]

Among Roman Catholics integration has proceeded further than among Protestants. A third of the Negro Catholics worship in integrated churches; often, however, "there is a tacit understanding within a church that some masses on a Sunday are for Negroes, while others are for whites."[32] For more material on segregation in the churches see pages 296 f.

Division along economic lines also characterizes the churches, especially in cities. In his book *The Status Seekers,* Vance Packard has a chapter he calls "The Long Road from Pentecostal to Episcopal," and presents data to show that "the upper class in most United States communities is drawn more powerfully" to some denominations than to others.[33] One writer asserted he could verify this fact by taking a tour in his city on a Sunday morning and observing the cars parked outside the various church buildings. The older models and the ones needing paint would be parked near the buildings on the wrong side of the tracks; the new models and the big ones would be found outside the buildings in the "best sections." A number of careful studies of social stratification in American communities have verified this writer's assertions.

Some students have explained these economic and social divisions in the church by contending that social divisions in religion are a natural consequence of the need in worship for fellowship with kindred spirits; worshipers of similar backgrounds naturally gravitate together in a free society. Moreover, other things being equal, people will go to the churches located near their homes. Since most American communities are stratified along economic lines, stratification in the churches simply reflects the class structure of American society.

Against such observations as these some students stress the point of view that the big churches are run by the same people who run the big stores and the big factories. Persons of this type are accustomed to exercise authority and influence, and naturally assume leadership in ecclesiastic affairs; the poor man in the rich church is expected to be a follower just as he is expected to be a follower in business. Thus the poor man strikes out for himself in religion; many a struggling church was brought into existence when superficial observation could detect little reason why existing institutions could not meet the need. As one Negro said to me, "At least we can run our churches."

There is a tendency for denominations as well as local churches to divide along racial and economic lines. Booker T. Washington once said that if you find a Negro who is anything but a Methodist or a Baptist you know that someone has been tampering with his religion. The Office of Public Opinion Research at Princeton University conducted in 1945–46 four opinion polls from which data were obtained on the extent to which our various denominations minister to voters with *incomes* in the "upper," "middle" or "lower" brackets. Table I gives these data.[34]

It is obvious that the full implications for church government of the basic Protestant principles have not yet been worked out. The sixteenth-century reformers were not aware of many of the democratic implications of their basic teachings. But through the centuries the implications have become clearer and Protestant institutions have become less and less authoritarian and more and more trustful of the intellectual and spiritual capacities of the common man. Neither the American churches nor the American nation has achieved a democracy that is in any sense ideal. Yet many Protestants are firmly convinced that when a more ideal type of democracy shall be achieved, Protestant principles will have greatly influenced its formation and will be basic to its spiritual structure.

THE ECUMENICAL MOVEMENT

As was noted at the beginning of this chapter, there is a growing conviction among Protestant leaders that the divisions in the Church should be decreased in number. A united Church would be a strengthened Church, it would have greater influence in society, it would be

Table I
Distribution of Denominations by "Class"

Denomination	Number of Cases	Percentage Whose Income Was in the Upper Bracket	Percentage Whose Income Was in the Middle Bracket	Percentage Whose Income Was in the Lower Bracket
Roman Catholic	2,390	8.7	24.7	66.6
Methodist	2,100	12.7	35.6	51.7
Baptist	1,381	8.0	24.0	68.0
Presbyterian	961	21.9	40.0	38.1
Protestant, smaller bodies	888	10.0	27.3	62.7
Lutheran	723	10.9	36.1	53.0
Episcopal	590	24.1	33.7	42.2
Jewish	537	21.8	32.0	46.2
Christian [Disciples]	370	10.0	35.4	54.6
Congregational	376	23.9	42.6	33.5
Latter-Day Saints (Mormon)	175	5.1	28.6	66.3
Christian Scientist	137	24.8	36.5	38.7
Reformed	131	19.1	31.3	49.6
Protestant, un-designated	460	12.4	24.1	63.5
Entire Sample (Including 800 additional cases)	12,019	13.1	30.7	56.2

much less expensive to maintain, and it would be in accord with the biblical injunction to "keep the unity of the Spirit. . . . There is one body, and one Spirit, . . . one Lord, one faith, one baptism, one God and Father of all" (Ephesians 4:3–6).

Several actual unions have taken place. Three Lutheran bodies merged in 1918 to form the United Lutheran Church. In 1931, the Congregational Churches joined with the Christian Church to form the Congregational Christian Churches. In 1934, the Evangelical Synod united with the Reformed Church to form the Evangelical and Reformed Church. In 1939, three Methodist bodies, which had separated prior to the Civil War, were reunited. In 1946, the United Brethren and the Evangelical Church joined to form the Evangelical United Brethren Church. In the fifties, two churches with differing types of government, the Congregational Christian and the Evangelical and Reformed, created the United Church of Christ. In 1958, two Presbyterian bodies created the United Presbyterian Church in the U.S.A. In 1961, the Unitarians and the Universalists formed the

Unitarian Universalist Association. And in that same year, three Lutheran bodies, the American, the Evangelical, and the United Evangelical, formed the American Lutheran Church. In prospect for 1962 is a second Lutheran merger, this time of four bodies: United, Augustana, Finnish Evangelical (Suomi Synod), and American Evangelical; the name of this body will be the Lutheran Church in America. At this writing, no other unions are in immediate prospect.

Some efforts at union have failed. A long negotiation in the forties between the Episcopalians and the northern Presbyterians was abruptly broken off by the Episcopalians. A warm discussion between the American Baptists and the Disciples cooled. But the desire for unity is increasing. One highly publicized proposal is for a merger of four of the largest Protestant churches: The Methodist, the Protestant Episcopal, the United Presbyterian, and the United Church of Christ. The result cannot at this writing be foreseen; nevertheless, the probability is high that the next decades will see further changes in the organization of American Protestantism.

Another fruit of the ecumenical temper is the marked growth of the Council of Churches movement. The churches of many large cities and counties, and the denominations of most states have banded together to form confederations. There are now nearly a thousand such councils. Also twenty-five Protestant denominations (including all but two of the large ones), along with eight Eastern Orthodox denominations form the National Council of Churches. The National Council was organized in 1950 through a merger of the Federal Council of Churches, the Foreign Missions Conference, the Home Missions Council, the International Council of Religious Education, and other Protestant agencies. These various confederations exert a great deal of influence, even though their strength suffers from the fact that they can exercise no authority over their member denominations, and from the fact that the funds given them by the churches are meager.

The formation, in 1948, of a World Council of Churches greatly encouraged Protestant leaders. After long and careful preparation, 140 denominations throughout the world united in the creation of a body through which they can speak and by means of which they can give one another mutual counsel and assistance. The World Council's headquarters are in Geneva, Switzerland. It operates on a budget of less than a half million dollars, "a sum that is reached by the budgets of some local churches in the U.S.A." American churches now contribute about 75 per cent of this amount.[35] The number of denominations in the World Council is now about 170.

Both the National Council and the World Council are weakened by the fact that the difficult spadework in local communities necessary to make them effective has been neglected. Most laymen are ignorant of the ecumenical movement and some clergymen are indifferent to it. In some localities two or more churches (usually the weaker ones) have been federated, but on the whole the number of competing local groups has not been much reduced. As one astute observer commented:

The Church . . . has become overextended. While . . . ecumenical co-operation, a World Council are essential goals, they have been developed at the expense of not mastering the problems of the home base. It is not that these things are not noble achievements; it is that they have no roots, they give a false sense of strength.[36]

A notable factor in the Ecumenical Movement is the refusal of the most orthodox denominations to co-operate with the most liberal. The simple formula for admission is confession of "our Lord Jesus Christ as God and Savior." As a consequence, the most liberal denominations are kept out of both the National Council and the World Council, though the individual of pronounced liberal tendencies is admitted to fellowship, provided he is a member in good standing of a denomination more conservative than he is. This development is a result of the realization on the part of the leaders of the movement that co-operation with many conservative groups is possible only on conservative terms. The theory is that an imperfect advance toward unity is better than no advance.

Action on this theory causes some writers to raise the question whether the ecumenical leaders are "guided by principle, or by church politics." The liberal Universalists were denied membership in 1945 in the Federal Council of Churches although the Russian Orthodox Church (in America) was admitted in that year. When the Universalists were again denied membership in the Federal Council in 1947, *The Lutheran* commented:

Reason for keeping out the Universalists is entirely practical. . . . Charles P. Taft, Federal Council president, had written that "if we let in the Unitarians [or Universalists], we let out the Lutherans."[37]

An American secretary on the World Council was asked why the Universalists were not included; he replied, "Just think how many conservatives we would lose." George A. Coe wrote:

The formula [for membership in the World Council of Churches] would admit the most authoritarian church of all, the Roman Catholic, but it would exclude churches that, however devoted they might be to the practice of love towards God and men, do not toe the mark with respect to a dogma that can be held by unloving minds.[38]

The ecumenical movement is criticized also for its threat to freedom. The larger the body the further removed in general is decision making from the grass roots. The concern among the ecumenical leaders seems to be for size; larger and larger units are urged upon the Protestant constituency. There is no corresponding concern for the development of democratic techniques whereby authority in these units of large size can flow naturally from the bottom to the top. Policies are set at the top; the same personalities continue in power year after year.

The ecumenical movement is criticized also for its threat to freedom of religious expression. One critic of the movement wrote:

. . . the division of Christians into the major families of theological tradition—denominations—is not in itself scandalous. On the contrary, it is seen to be a necessary condition of the fullest understanding of an infinite and inexhaustible gospel which defies containment within any single form of life and expression. Without these families of tradition, or something like them, the Christian movement, past and future, would be permanently impoverished and distorted. . . .

If our present separateness often prevents our being adequately and effectively heard, togetherness may so limit our mobility, reduce our creative adaptability, and dissipate the vitality of our witness as to reduce even further our impact on the variegated life of the modern world.[39]

The difficulties faced by the ecumenical movement are even broader. Large numbers of conservative Protestants will have nothing to do with it. The movement would be happy to admit the Fundamentalists—provided of course they came in a co-operative spirit; but most of the Fundamentalists claim that the ecumenical leaders are in fact little more than self-appointed promoters of a new denominational effort. Some Fundamentalists, stung by competition, have set up their own organization, the American Council of Christian Churches.

Thus all over Christendom we have groups which regret Christian division, regret it profoundly. But unfortunately too many of them are in the position of affirming dogmatically that Christian unity is possible only on their terms. For additional material on the ecumenical movement, see pages 190 ff., 324 ff.

RELIGIOUS EDUCATION

Rabbi Mordecai M. Kaplan writes:

The real test of any religion, movement or civilization is in the process of education to which it gives rise. If that process is rich in content, colorful and stirring, the movement or religion which it transmits is answering a real need, and is bound to live.[40]

Early Protestantism certainly met this test, and nowhere more significantly than in America. Colonial Protestants in Massachusetts, in order to foil the "project of ye old deluder, Satan, to keep men from ye knowledge of ye Scriptures," ordered each town of fifty householders to establish a public school. Massachusetts Protestants also founded Harvard College in order "to advance Learning, and perpetuate it to Posterity, dreading to leave an illiterate Ministry to the Churches, when our present Ministers shall lie in the Dust." These educational efforts are but the best-known of a long series of provisions made by American colonists for the teaching of religion, and other subjects. In those days the term *religious education* had not yet come into use; all education was religious. It would have been as unthinkable then to set up a public school that ignored religion as it would be today to set up a public school that ignored science.

Nor was this situation changed by the adoption of the Constitution of the United States. For the first fifty years of our national life all the public schools taught religion. However, beginning in about 1840, conflicts among the religious groups over which religion should be indoctrinated led to the eventual banishing of all formal religious instruction from the public schools. Catholics saw in this development a serious threat to their Church; consequently they began the creation of their system of parochial schools.

Most Protestants saw in the secularization of the public schools no serious threat to religion. They believed in the promise of the public schools and in what they could do for America. Protestant support made possible the striking success of the American public school system. The national achievements in public education are part of the glory of American democracy—and of American Protestantism. If Protestants had chosen to establish denominational school systems, their churches might now be stronger institutions. But the nation would almost certainly have within it greater diversity, more intolerance, less consciousness of unity of purpose than it now has. The public schools, for all their faults, have done for America what a series of competing parochial systems could never have done.

In teaching religion, Protestants have relied on instruction in the home and in Sunday schools. The home has proved to be highly effective in the teaching of morals; parents are deeply concerned that their children should grow up to be honest, co-operative, benevolent. However, in teaching reverence for the Church and its ritual, in passing on the Christian tradition, the home has proved unco-operative. Charles C. Morrison, former editor of *The Christian Century,* wrote, "What the home has done about it [education in the religious tradition] is too notorious to require comment."

Protestants have put their major reliance for the training of children on Sunday schools. These schools have a tremendous enrollment—about forty million (of all ages)—and they are to be found in practically every American hamlet. They are in session about one hour a week and are manned by laymen. One traveler from the continent of Europe said that the most surprising thing he found in America was that everywhere adults gather around them little groups of children and tell them Bible stories. The education which is provided in the Sunday schools is often criticized. The critics contend that the shortness of the sessions, the poorness of the equipment, and the lack of skill on the part of voluntary teachers mean that little genuine education can take place. The critics also assert that half of the children of Protestant background are not touched by the Sunday schools.

Weekday instruction in religion, on time released by the public schools, has received much Protestant attention in recent decades. Such instruction is usually conducted for one hour a week. This method of teaching is said to reach about three million pupils of all denominations—Catholic, Jew, and Protestant—in three thousand communities.

Protestant concern for higher education has declined. In colonial times most of the colleges were under some type of Protestant control; a large percentage of the colleges and universities founded since colonial times received their stimulus directly from the Protestant churches. Between 1830 and 1860, Methodists founded thirty-four colleges and Baptists founded twenty-one. However, today Protestants sponsor the education of but a fraction of the students now in higher education. The colleges and universities now under Protestant control are usually not the strong and influential institutions.

Protestantism's relative neglect of religious education is strange in view of its original concern and in view of the belief in individual judgment. A strictly authoritarian religion, one which teaches by indoctrination, can survive with much less educational effort than can a religion which gives freedom for individual choice. The free man can make wise choices only after he has abundant information. But we are confronted by the strange fact that it is authoritarian Catholicism which makes the more serious religious educational effort and free Protestantism which neglects religious education, except at the highest level of theological training. Perhaps the slow-moving democratic processes will one day awaken the Protestant conscience to the realization that in the long run the Church cannot be stronger than the education its children, and adults, receive.

In former generations the Protestant press dominated the journalism of the country. One student estimates that a century ago, a time when the percentage of non-Protestants was very small, three-quarters of all reading done by the American people was on religious subjects.[41] For many years Protestant weeklies were the leading magazines of opinion in the country. The *Independent,* beginning as a small antislavery organ, grew until it had an undeniable influence on national affairs. The *Outlook* attained a similar position, counting at one time an ex-President of the United States as a contributing editor. No religious publication today holds such a position. In fact, between 1910 and 1930, a period when the circulation of secular magazines increased greatly, the circulation of Protestant journals decreased steadily. Comparable statistics are hard to secure, but there seems to be good evidence to show that by the end of the period the circulation of Protestant journals represented somewhat less than 1 per cent of the total newspaper and magazine circulation of the country.[42] In later years the circulation of Protestant journals rose, a 40 per cent increase being reported between 1940 and 1944.[43]

Regarding Protestant journals, one writer wrote:

[They are, on the whole,] a welter of thin-blooded Sunday-school papers, educational quarterlies, and miscellaneous publications ranging from collections of conventional devotional literature to handsome scholarly journals carrying abstract articles in the jargon of the theologian . . . The religious journalist is making his principal appeal, not to the masses or the classes but to the spirit of denominationalism and sectionalism.[44]

Some Protestant magazines have of late shown great improvement.

FOREIGN MISSIONS

Christianity conquered the Roman Empire through the indomitable spirit of early missionaries. During the Middle Ages, this crusading temper was less in evidence; and during the hundred years prior to the Reformation, Christian missions had been all but suspended; the Church was prosperous and self-satisfied. For many decades after the Reformation, the Protestant churches had energy for little besides the effort to maintain their own newly won liberties. But the Roman Catholic Church was spurred by the Reformation to renewed missionary zeal. In that age, Catholic nations were the leading maritime powers of the world: they had discovered America, and the route around Africa to India, and had gained footholds in these lands. Catholic missions naturally followed.

In the eighteenth century, the Catholic missionary pace slowed down to a walk. But about the beginning of the nineteenth century, both Protestants and Catholics began such a spurt in missionary activity as has never been equalled anywhere. It continues till the present day. As a result of this activity, Christianity became in the nineteenth century for the first time a truly world-wide movement. Catholic missions during this period have been so effective that a Protestant student of missions could write: "No other one institution that the human race has ever seen has equalled it [the Roman Catholic

Church] in the numbers of countries and peoples among which it is found."[45]

During the past one hundred and fifty years, Protestant missionary efforts have surpassed Catholic efforts.[46] The Protestant movement gave notice of its coming achievements in the remarkable career of William Carey. Carey was an English cobbler who in 1792 was sent out to India by the newly formed Baptist Missionary Society. He was scorned and attacked by the Indians and also by his own countrymen, who by that time were in control of India. A later Bishop of Bombay wrote:

If ever a heaven-sent genius wrought a conquest over obstacles and disabilities, it was . . . this humbly born Englishman. . . . he received hardly any education. . . . [Yet] this man before he died took part in translating the Bible into some forty languages or dialects, Chinese among the number! He started in life as a cobbler—would never let anyone claim for him the more dignified title of shoemaker—he died professor of Sanskrit, the honored friend and adviser of the government whose earliest greeting, when he landed on the shores of the country, had been to prohibit him from preaching.[47]

In America, the dynamic nineteenth-century missionary movement began with a society formed in 1808 among five young men at Williams College. The idea spread; the denominations created foreign missions boards; soon non-Christian lands were being invaded by Protestant missionaries. They worked under the greatest difficulties; not only were the local residents suspicious, but many Occidentals also opposed missionary work. Commercial interests in Africa and Asia saw an end to their system of exploitation, if enlightenment came to non-Christian peoples.

In spite of indifference at home and danger abroad, the missionary movement spread. It is found today in practically every non-Christian land. In 1960, Protestant churches throughout the world were sending over forty-two thousand missionaries to lands other than their own, 64 per cent going from the United States and Canada. During the decade of the fifties, the North American missionary force increased 81 per cent. In 1960 Roman Catholics had almost sixty-eight hundred persons from the United States serving as foreign missionaries.[48] The percentages of the Protestant missionaries engaged in various types of work were: evangelistic, 61 per cent; educational, 19 per cent; medical, 12 per cent; administrative, 4 per cent; agricultural and other, 4 per cent.[49] These percentages may change considerably in the future. An editor of the *World Christian Handbook, 1957* wrote that "in very many countries there are signs that the particular forms of work which have been done in the past will not be possible in the future."[50]

In spite of increases in the number of missionaries and in the *number* of persons who are Christians, the percentage of Christians in the population of the world is dropping. The percentage in 1900 was 32.2; in 1960 it was 30.3.[51] In 1948, the total of all Protestants in mission lands was reported as twenty-five million; the total for all Catholics was reported as one hundred and thirty million.[52]

Among Protestants of recent years the missionary movement has come in for much criticism. The rise of liberal theology has undercut the older motivation for making converts—to prevent the "heathen" from spending eternity in hell. The present attitude of many moderns is that Christianity is a Western religion, that the missionaries work chiefly in countries whose cultures are in many ways superior to our own, and that the religions of these countries are more fitted than is Christianity to meet the needs of the people who live there. Why should we disturb their faith as long as they find it satisfying? What we Christians tend to do, so runs the argument, is to compare our own exalted ideals, not with Oriental ideals, but with Oriental practice.

The reply to this position would vary according to the theological position of the one who gives answer. A large percentage of the missionaries still maintain that salvation is impossible outside their Church. Another large percentage hold that only Christianity is of divine origin and, therefore, Christians must labor unceasingly until all the world is converted. Liberals—decidedly in the minority among active missionaries—take the position that Christianity has much to learn from other religions, and these religions much to learn from Christianity. The proper attitude, say Liberals, for the missionary to take is one of sharing; exchange of spiritual and cultural treasures is vital if we are ever to achieve a world culture—and a world culture is the only possible basis for world peace. Moreover, the liberal missionary would contend that the alternative to Christianity for the overwhelming majority of the peoples of the world is not one of the age-old, noble philosophies of the East, but rather some type of demonology or devil worship.

The missionaries have taught Christian ethics as well as Christian theology. Moreover, they have given practical examples of what this ethics means. Over much of Africa, Asia, and the islands of the Pacific, the missionaries are responsible for demonstrating the work of hospitals, institutions for the insane, and public health programs. Missionaries have brought in scientific agriculture, thereby greatly increasing the yield of crops, the productivity of forests, and the quality of livestock. The missionaries have contended that education is for humble people as well as for aristocrats; when universal, free education comes to the Orient much of the credit will belong to the missionaries. They have struggled to raise the status of women, to abolish infanticide, to outlaw the opium traffic, to smash the slave trade, to care for the blind, to introduce modern methods of famine relief. Millions of the earth's common people have little hope of rising above disease, ignorance, want, and superstition except through the activity of Christian missionaries.

The percentage of Christians in missionary lands is still very small; but the size of the Christian groups is not a good indication of their influence on public affairs. In the Chinese *Who's Who,* before the Revolution, one man in six was a Christian, and one in two was educated in Christian schools. Like the Socialist party in the United States, which had an influence out of all proportion to its size, Christian teachings in the East are manifest in the ideas of a multitude of persons who would deny any connection with Christianity. The present challenge to the caste system in India had its inception in Christian idealism, even as the present challenge in America to the Negro-White line, now taken up by so many secular groups, also had its inception in Christian idealism.

Frequently it has been said by skeptics that missions are simply the religious phase of Western imperialism. Surely that accusation is shown to be false by the tension which exists between missionaries and business interests, by the attitude of friendship rather than of exploitation with which the missionaries have treated nationals, and by the numbers of missionaries who in time of war have stuck stubbornly to their posts in spite of the warnings of their home governments. Now that the era of Western imperialism seems to be drawing to a close, we need not look for the demise of foreign missions.

PROTESTANT STRENGTH

Many Catholics and nonchurchgoers are sure that Protestantism is on its way out. Hilaire Belloc, a leading Catholic author, declared that in the sixteenth century the Protestant "attack upon the Church seemed overwhelming. It continued to succeed from that day onwards almost into our own time. It was like a great flood which runs at first most violently, [then] gradually slackens."[53] But "the hegemony of the Protestant culture in Europe has crumbled."[54] A discouraged Protestant clergyman wrote, "Protestantism is disintegrating and is doomed. It may outlast your life and mine, but ultimately America will see it no more."[55]

These discouraging forecasts are sharply contradicted. For example, Henry P. Van Dusen wrote:

By any reasonable test the nineteenth century was by far the greatest in Christian history. . . . In terms of influence upon the whole life of humanity Christian ideals and spirit had effected greater reforms and improvements in the lot of all sorts and conditions of men than had ever been wrought by any single influence in any previous epoch of history.[56]

Kenneth S. Latourette wrote:

From Luther until this day Protestantism has been gaining momentum. For nearly three centuries it was a minority movement within the universal Church of Christ. The main stream of Christianity seemed to be flowing through other channels, notably the Roman Catholic Church. In the nineteenth century, however, it was becoming obvious that Protestantism had become the main current of the Christian stream. The currents represented by the Eastern churches have been dwindling. That represented by the Roman Catholic Church continues strong and has had something of a quickening in the present century. However, measured by inward vitality as displayed in new movements and the effects upon mankind as a whole, increasingly Protestantism has become the chief channel for the life of Christ in the world.[57]

We have here a sharp division of opinion. Which point of view is correct?

Statistically the Protestant situation in the United States is encouraging. At the beginning of our national life about one person in twenty in the population was connected with the churches, practically all of them Protestant. Today the proportion of Protestant church members in the population, exclusive of persons with Catholic or Jewish backgrounds, is probably ten times what it was at the end of the colonial period.[58] The proportion continues to increase.

Judging statistically, there is reason to think, as has been indicated, that Protestant strength is considerably greater than its membership rolls would indicate. Protestant methods of maintaining church rolls are such that adherents are frequently "lost." Unquestionably a large number of the sixty-five million Americans who are usually classified as "unchurched" consider that they are part of some denomination's constituency. In Canada, where the Protestant situation is presumably very similar to that in the United States, the government census takers ask the individual citizen to name the denomination with which he is affiliated. In 1941, over twice as many Canadians claimed affiliation with the United Church of Canada (the leading Protestant communion) as the Church itself reported as listed on its membership rolls.[59] Similarly a sampling poll taken by the United States Bureau of the Census in 1957 indicated that the number of persons in the United States who considered themselves to be Protestants was over 80 per cent larger than the number of members reported by the Protestant churches. (See page 77.) In 1960, the Protestant churches reported a membership of sixty-three and a half million. Using a Catholic definition of membership, they had eighty to eighty-five million. According to the percentage (66.2) derived from the Bureau of the Census poll, about one hundred and nineteen million Americans (including children) in 1960 considered themselves to be Protestants.

These data indicate that, statistically speaking, American Protestantism is by no means a dying movement. But the strength of Protestantism cannot be appraised merely by reviewing institutional arrangements and statistics. We must appraise also the spiritual health of individual Protestants. How loyal are they to their churches? And how seriously do they take religious teachings?

At this point we must distinguish rather sharply between morals and worship. A very large portion of Protestants take the ethical obligations taught by their religion with great seriousness; a smaller portion take their ceremonial obligations seriously.

The emphasis in Protestant ethics is on personal morality. In their personal conduct most Protestants strive seriously to live up to the moral requirements of the Ten Commandments; they are honest, neighborly, faithful to their wives, and they try to treat their fellows decently, just as they expect to be treated. Protestant ethics have pretty much outgrown the "don't chew, don't swear, don't play cards, don't go to the theater" stage. This former emphasis is understandable in view of the Puritan tradition out of which Protestant ethics grew and in view of the crudities of frontier life. The change from this emphasis is most clearly seen in the Protestant attitude toward the Sabbath. In former times Sunday was a day which Protestants kept free from all labor and frivolity. In my own childhood, spent in a Methodist parsonage, strict instructions were issued not to engage in sports on the Sabbath. My parents were Liberals and would not have objected to a bit of amusement on a dull afternoon; but the long arm of the leading layman made its way into our family life. However, Sabbatarianism is being abandoned by most Protestants; the continental or Catholic Sabbath is more and more observed by them. Many clergymen teach that after attendance at church the day should be used for recreation; a few clergymen even hold early services for families who wish to go on all-day picnics or for sportsmen who hope to get in a full day on the golf course.

Protestantism's influence among the "unchurched" is strong, and primarily ethical. Among Americans nominally Protestant is a multitude—just how many is anybody's guess—who, though they seldom go near a church, conduct their lives on the Christian *ethical* pattern. This situation prevails, of course, in all denominations, but probably to a less extent among Catholics than among Protestants and Jews. These ethically minded persons have adopted Judeo-Christian moral standards, but insist that they can lead just as good a life outside the Church as they can inside. They see no useful function served by their personal church attendance. A large percentage of them do consider churches to be essential in the community and think attendance at religious services to be an excellent thing for young people. They are glad that other people busy themselves in maintaining religious organizations and in providing religious opportunities for the children. But according to this position, the purpose served by maintaining churches and by sending children to Sunday school is an ethical one. These anti-church-attendance Protestants are a testimony to the effectiveness of the Protestant churches in the realm of personal ethics.

Though Protestants as a whole are not nearly as loyal in following the ceremony prescribed by the Church as they are in following its ethical teachings, it would be a mistake to conclude that *esprit de corps* in the Protestant churches is in a serious state. On the contrary, morale is high. Millions of Americans attend the churches regularly, and reverently carry out Protestantism's ceremonial prescriptions. Public and private worship give them their life's orientation, and furnish the vision by means of which they live. This large group of loyal Protestants consider that they are part of a vital enterprise, one that is essential to their own well-being and one without which the nation would perish. Actual observation of a large number of Protestant services does not give the impression that the worshipers are discouraged or disgruntled. Pessimistic statements concerning the internal morale of Protestantism come chiefly from those who attend other churches, from those who attend Protestant churches irregularly, or from those who attend no church at all.

Pessimism concerning the external influence of the Protestant churches has much justification. The long-range trend seems to point in the direction of a declining Protestant influence in our national affairs; secular concerns more and more determine the trend of public events. A student of the sociology of religion puts the situation strongly: "Recent sociological investigations all point to one depressing conclusion: that American Protestantism, as institutionally organized, is virtually impotent in terms of affecting the policy-making processes of our sociey."[60] If Catholics in most communities labor under the handicaps of a minority psychology, Protestants in most communities are handicapped by the psychology of the majority. In the past they have had things pretty much their own way in the United States. They set up the kind of schools they wanted, they had the allegiance of the nation's economic and professional leaders, they usually had control of the political agencies. This situation no longer obtains, but the psychology endures. Protestants tend to be complacent.

Will Herberg states another view:

American Protestantism . . . is having to adapt itself to what is, in effect, a minority status in a pluralistic culture that has become post-Protestant, and it is not doing this very well. It is an anxious, downward-heading minority, bristling with an aggressive defensiveness character-istic of such groups.[61]

Herberg's judgment applies, I think, only to some Protestant leaders. They view with alarm. But the great majority goes on its confident way, insisting on the right to maintain many sectarian divisions and refusing to alter the traditional policy of sharp sectarian competition.

In former times the Protestant churches were the centers of Ameri-can community life. From them came the impulses which created many of the social institutions of the community: schools, hospitals, charitable agencies, reform movements. In them many an American found his social contacts, his opinions, his recreation, and even the best music he could find. Today many of these functions have been taken over by radio, newspapers, movies, television, automobiles, and professional social agencies. The church is no longer the hub of com-munity life. Yet churches continue to do pioneer social thinking. In the future as in the past, we may expect the churches to sense new needs, to provide agencies to care for these needs, and finally after the agencies have won a firm hold on the community conscience, to turn them over to other hands to run.

Before concluding, as so many do today, that the Protestant churches need not be considered in appraising America's future, it would be well to ponder the fact that Protestantism exhibits both the strengths and the weaknesses of democratic institutions generally. Democracies move slowly; but once aroused, they exhibit tremendous power. There is today a latent loyalty to Protestantism on the part of millions of Americans. This loyalty is founded not so much on knowl-edge or experience as it is on a substratum of conviction that religion is an essential thing in a community and that the ethical code of Christianity is the ideal toward which society should strive. The right social situation plus inspired leadership could tap this deep-seated conviction.

Another factor which brightens Protestantism's prospects is the compatibility of much of its theological thinking with science. More and more, American thought will be dominated by scientific points of view. Democratic control, with its attendant ease of theological change, means that Protestant dogma can be altered to meet changed intellectual demands. In the long run the church or synagogue which insists on maintaining an antiscientific view of the universe will be found to be working at a disadvantage.

The trouble with this confident view of Protestantism's future is that it assumes a stable social situation. Can Protestantism deal with crisis? Can a democratically controlled religion act courageously and with sufficient intelligence to deal with totalitarian religions? Above all, can it act in time? The recent history of Europe is not encouraging. If worse comes to worst and America turns to some type of totalitarianism in an effort to maintain its security, democratic religion would probably be lost along with other democratic institutions.

CHAPTER IV

LUTHERAN CHURCHES

"I beg that my name be not mentioned," said Martin Luther, "and that people be called, not Lutherans, but Christians. What is Luther? The doctrine is not mine, nor have I been crucified for any one."

In spite of this vigorous declaration a large number of Protestants are now known as Lutherans. The term was first used in ridicule; the Pope put it into his bull excommunicating Luther. Thus again was exhibited the common tendency to deride whatever religion differs from one's own—especially new movements in religion. Methodists, Quakers, Baptists got their names from attempts at ridicule. The term *Christian* itself probably was first used in the same way, much as present-day undergraduates sometimes use the term *Christer*.

For many years after the beginning of the Reformation, the followers of Martin Luther honored his wish and did not use the term *Lutheran Church*. But today "no loyal son of hers has any desire . . . to drop the time-honored name around which cluster memories of a most gallant fight which the Church militant has had to wage for the heritage of the faith once delivered to the saints."[1]

Luther was a Catholic priest, a widely known professor in the University of Wittenberg. His conflict with the hierarchy was precipitated by an abuse of indulgences (see pages 48 ff.). The monk Tetzel assured the faithful that an offering of money was all that was necessary to release the souls of the dead from purgatory. This unorthodox teaching aroused Luther to a formal challenge of debate—he nailed *Ninety-Five Theses* on the door of the Castle Church in Wittenberg, the University bulletin board. Within a month they were read in

Western Europe "as if they had been circulated by angelic messengers." It was like starting a fire on a dry prairie. The demand for reform was in the air. Rulers were restive under the enormous sums of money which annually went to Rome and scholars were questioning the validity of many papal claims.

Luther was condemned, excommunicated and commanded by the Emperor to journey to Worms, there to appear before the German Diet. With the Emperor sat the great princes of the Empire and also the learned theologians of the Church. It was as though in our day a professor in a small college were called to Washington, D. C., brought before the President, the Congress and the Supreme Court, and sternly commanded to recant his radical ideas about government. Luther, a simple monk, the son of a poor peasant, stood before the Diet, disregarded the possible consequences, denied that he was a heretic, accused the Church of error, and refused to recant. It was an act of such courage as comes only from a profound sense of moral indignation. Luther said:

It is impossible for me to recant unless I am proved to be wrong by the testimony of Scripture or by evident reasoning; I cannot trust either the decisions of Councils or of Popes, for it is plain that they have not only erred, but have contradicted each other. My conscience is captive to the Word of God, and it is neither safe nor honest to act against one's conscience. God help me! Amen!

Luther lived to see the Protestant movement firmly established in Germany, Switzerland, England and the Scandinavian countries.

Today the Lutheran churches are the largest Protestant group in the world, claiming a total of about seventy millions. This figure, however, includes all baptized members. The number of actual communicants is considerably smaller since Lutheranism is, or was, the established (state) church in a number of European countries where the overwhelming majority of the population is baptized, but only a minority attend church with any regularity. In the United States the Lutheran denomination is the fourth largest, being exceeded only by the Catholics, the Baptists and the Methodists. There are over eight million Lutherans in the country; they are divided among fifteen groups. Most of them are small. Four, however, are large: The American Lutheran Church (formed by a merger in 1961) with two and one-quarter million members; the Lutheran Church, Missouri Synod, with two and one-third million members; the United Lutheran Church with two and one-third million members; and the Augustana Evangelical Lutheran Church with six hundred thousand members. The latter two plan to join in 1962 with two other Lutheran bodies (which have a combined membership of about sixty thousand) to form the Lutheran Church in America. Lutherans are found all over the United States, but they are weak in New England and in the South. Their great strength lies in the Middle Atlantic and in the North Central States.

Lutheranism was founded in the United States in the earliest colonial times. Germans, Dutchmen, Swedes, Norwegians, Finns, Danes as they colonized in the seventeenth and eighteenth centuries naturally brought with them the dominant religious ideas and institutions of their homelands. The predominantly English character of the colonies meant that these "foreign" churches were not given the supervision that churches based in England received. Thus for many years the organization of Lutheran churches in America was chiefly the result of the spontaneous effort of pious laymen; and the resulting congregations were only incidentally related to one another.

The task of forming synods, and thus of establishing American Lutheranism on a really firm foundation, was the work of Henry Melchior Muhlenberg, the "patriarch" of the Lutheran churches in America.

Henry Melchior Muhlenberg

Muhlenberg was born in Germany in the second decade of the eighteenth century. When he was twelve years old his father died. As a result the son was forced to spend the next decade of his life at hard manual labor. Not until he was twenty-one was he able to begin his education; he studied Latin and Greek in the evenings after work. His progress was so rapid that at age twenty-four he was elected by the Council of his home town to receive a scholarship to the University of Göttingen. After the completion of his university studies, he took a position teaching in an orphanage in Halle. There he was moved by an appeal for aid sent by three pastorless congregations in Pennsylvania and agreed to go out as a missionary on a three-year trial. He stayed in America the rest of his life.

The physical discomforts and dangers of his voyage are hard for

us to imagine in our time when life on an ocean liner is looked upon
as a special kind of luxury. A contemporary of Muhlenberg estimated
that in one year two thousand people died during passage from
Europe to America, most of them as a consequence of inhuman treatment and the overcrowding of ships. The voyage lasted one hundred
and two days, during much of which time Muhlenberg was overcome
by seasickness. The ship was still in sight of England when he wrote in
his Journal:

> I was very sick, so could not rise the whole day. . . . It is much more
> irksome to be sick on shipboard than on land. Our beer was already
> sour; the water foul; the daily fare for healthy persons was peas, pork,
> stockfish, and salt beef, half-cooked in the English fashion. . . . It is
> difficult to recover from sickness on it.[2]

Weeks later, nearing land but with the ship's water almost gone, he
wrote:

> Today we had no wind at all, only calm. We took counsel together as
> to how to keep alive for several days without water. The captain said
> that he still had a small quantity of olive oil, of which each might drink
> a little every day and keep alive when the water was gone. He also had
> left a few bottles of vinegar, of which we took a little occasionally.[3]

But it was not till seven days later that they received a new supply of
water from a passing warship.

When Muhlenberg arrived in America, Philadelphia was a town of
unpaved streets, provincial manners, and about fifteen thousand
inhabitants; only the eastern edge of Pennsylvania was fully settled.
Muhlenberg found the Lutheran churches throughout the colonies in a
sad condition. They had been unable to find competent pastors and
thus had been preyed upon by "unscrupulous ecclesiastical tramps,"
"spiritual adventurers" who had palmed themselves off as having
genuine Lutheran ordination. He had a brief struggle at the beginning
with these men, who refused peaceably to give up their positions; but
within a few weeks he was acknowledged as the pastor of three Pennsylvania congregations, including the one in Philadelphia; a fourth
congregation was added within a year.

At the beginning, his congregations were very poor. They had not
been able to erect adequate church buildings—the group in Philadelphia worshiped in a butcher (some accounts say carpenter) shop—
and they could give him but a very small salary—the first year one

congregation paid him nothing and another paid less than enough to settle his rent. Consequently, he was forced into debt; yet so great was the need for adequate houses for worship that Muhlenberg neglected his own finances and urged the people to erect church buildings. Within a few years, new houses of worship were built by all four of his congregations. Twenty-five years later the Philadelphia congregation began the erection of a building which for a number of years was the largest (and many thought the finest) church building in North America. It was to this church that the Congress adjourned for a service of thanksgiving after the surrender of Cornwallis at Yorktown.

Muhlenberg's parish was large and transportation was very difficult. Yet he was indefatigable in attending to the many spiritual needs of his parishioners. He conducted services, baptized infants, confirmed youths, administered Communion, visited the sick, buried the dead, founded and taught schools. He wrote the fathers at Halle:

Here are thousands, who, by birth, education, and confirmation, ought to belong to our church but they are scattered to the four winds of heaven. The spiritual state of our people is so wretched as to cause us to shed tears in abundance. The young people have grown up without any knowledge of religion, and are fast running into heathenism.[4]

Sickness and death among his parishioners demanded long, hard riding; sometimes it seemed that he was almost constantly on horseback. His journeys often involved danger. In his Journal for February, 1748, he wrote as follows concerning a visit to two small churches outside the confines of his own parish:

In this month I made another trip to the little congregations in Upper Milfort and Saccum. . . . It was night when I got between the mountains into an unusually deep valley where there are deep swamps and holes and the snow lay very deep. I could not very well go back and it was still six miles farther to my quarters; there was no road and I could not see the snow-covered holes. First I rode two miles in the wrong direction toward the left and had to work my way laboriously back again. After that I kept to the road pretty well, but several times I fell suddenly with the poor horse through the snow and soft ice into the swamp and had to work my way out again with God's help. The horse became weary and reluctant to go through the unbeaten tracks of deep snow, so I was obliged to walk ahead on foot and make a track for the horse, which exhausted me greatly, and I still had three miles to go. I would have been glad to sit down in sheer weariness, but it was so bitterly cold and I was perspiring so profusely that I did not dare to rest and risk a sleep of death. I once more summoned up my remaining energies in the name of the Lord and finally reached my lodgings safely that same night.[5]

Throughout his life Muhlenberg remained in the eyes of his ecclesiastical superiors in Germany a simple pastor; but he was in fact an itinerant bishop. His influence extended all the way from Nova Scotia to Georgia. He ministered to pastorless congregations, organized parishes, regularized orders of worship, wrote parish constitutions, advised young clergymen, persuaded the Halle fathers to send out more pastors, and acted as judge in the settlement of disputes. His correspondence was large; and the number and extent of his trips were almost incredible in view of the difficulties of travel, his meager financial base, and his lack of formal authority. One of his biographers, with perhaps some exaggeration, could write, "There was probably not a Lutheran Church, in his day, in this country in which he did not officiate."[6] Another student of his life has said, "He possessed in an extraordinary degree the grace of finding favor with men."[7]

The services of worship which Muhlenberg conducted were usually long—very long, according to our modern standards. Frequently they began with catechizing the children, and sometimes even catechizing the adults. Then would follow prayers, Scripture lessons, and the singing of hymns. Occasionally the congregation would possess but one hymn book and it would be necessary for Muhlenberg, who was an excellent musician, to sing a line and for the congregation to sing after him. Then would follow a sermon. If the people to whom he was ministering had not received the sacraments recently, the service would include also baptism and the Lord's Supper. Muhlenberg conducted worship in both German and English; he even learned enough of the Dutch language to minister to Dutch congregations which were without pastors.

After he had been in America six years, Muhlenberg was responsible for the organization of the first Lutheran synod in the New World. He and five other Lutheran clergymen, along with delegated elders from ten congregations, assembled in Philadelphia, drew up a constitution, ordained a young man to the ministry, reminded themselves of the faith expressed in the "Unaltered Augsburg Confession," and prayerfully laid plans for the future of Lutheranism in the colonies.

Two and a half years after his arrival in America, Muhlenberg married Mary Weiser, the eighteen-year-old daughter of a well-known Indian agent. They lived together for forty-two years and founded a distinguished American family. They had eleven children, four of whom died in infancy. The eldest son had a notable military career in the Revolutionary War and later served as a United States Senator. Another son became a well-known botanist and the first president of Franklin and Marshall College; a third son was a member of the Continental Congress and was the first Speaker of the national House of Representatives. As this paragraph is written another Muhlenberg sits in the House of Representatives, the sixth of his family to serve in the national Congress.

The relations of the Muhlenberg family to the Revolution are heart-warming to American patriots. At the beginning of the war the eldest son, Peter Muhlenberg, was serving as rector of an Anglican parish in Virginia. He was appointed by George Washington to be Colonel of the Eighth Virginia Regiment. In the middle of January, 1776, Rev. Peter Muhlenberg preached to his congregation a farewell sermon, concluding, "In the language of Holy Writ, there is a time for all things. There is a time to preach and a time to fight; and now is the time to fight." Then, according to a frequently repeated story, he threw off his clerical robe and in the uniform of a continental colonel marched to the church door where he commanded the drums to be beaten for volunteers. His regiment saw action in all the major southern battles of the war and bore the brunt of the fighting at Yorktown. After two years he was promoted to the rank of Brigadier General and later to the rank of Major General.

The elder Muhlenberg's Journal comments on the progress of the Revolution. On July 3, 1776, at home in Philadelphia he wrote:

It is said that the Continental Congress resolved to declare the thirteen united colonies free and independent.

A day later he wrote:

Today the Continental Congress openly declared the united provinces of North America to be free and independent states. This has caused some thoughtful and far-seeing *melancholici* to be down in the mouth; on the other hand, it has caused some sanguine *miopes* to exult and shout with joy. . . . There is One who sits at the rudder, who has the plan of the whole before Him, to whom all power in heaven and on earth is given, and who has never yet made a mistake in His government. He it is who neither sleeps nor slumbers.[8]

Muhlenberg died in 1787; he had lived in America for forty-five years. During this period he had personally been responsible for placing the Lutheran churches in this country on a permanent foundation, and had witnessed the stirring events which brought the United States to birth. For all his loyalty to the new nation, he did not succeed in making Lutheran people desirous of developing a unified culture in America—at any rate not if it meant giving up the culture of the lands from which they had come. As a consequence, Lutheran churches long continued to conduct worship in "foreign" tongues and to be oriented theologically toward Europe, a fact which has had a marked influence on the institutional development of American Lutheranism.

THE CONFESSIONS

There is more conformity of belief among Lutherans in the United States today than among any other large Protestant group; and oddly enough, for such self-conscious followers of a reformation movement, most of the Lutherans are proud of it. To be sure, there are among United and Augustana Synod Lutherans "a goodly number" of Liberals, and "incipient liberal movements" can be found in even the most conservative bodies. But in general, Lutheran clergymen seldom get very far off the theological reservation. Occasionally, one runs into a Lutheran who bemoans the lack of independent thinking among his fellow churchmen. But for the most part, Lutherans are sure that the only infallible standard of faith and practice is the Bible, and that their Confessions (creeds) are in harmony with the "one . . . pure Scriptural faith," preserving it in its purity.

Lutherans believe that their Church is the Church of Christ and the Apostles, the medieval catholic Church purged of false doctrines. Lutheran logic at this point is simple: the Bible is God's revelation; the Bible reveals one religion; the meaning of the Bible is clear to anyone who approaches it from the proper point of view; any Church which deviates from the Bible is in error; the teachings of the Lutheran Confessions did not originate in the minds of Lutherans but in the mind of God; Lutheran Confessions faithfully reflect Biblical teachings; practically all non-Lutheran churches fall short of being truly Christian because their beliefs deviate from basic truth as revealed in the Bible.

The Lutheran Church is not a sect, in the opinion of many Lutheran scholars; it is the Christian religion. Moreover, as one clergyman writes:

Christianity occupies a solitary place and demands an evaluation apart from all other religions. It does not belong in a Parliament of Religions; it is *the* religion. There is no substitute for it, for there is no equivalent in other religions for what Christianity offers. It cannot be improved upon; it is the last word in religion; for it is, solely and alone, *the religion of a real salvation.*[9]

Lutheranism emphasizes its Confessions as does no other Christian group. It inherits from the early centuries the three ecumenical creeds: The Apostles', the Nicene and the Athanasian. These creeds, used widely by orthodox Christians, are not a complete statement of the faith of orthodox Christianity—Protestant or Catholic. They do not say enough. For example, they do not include Christian ethical standards. The creeds were hammered out in the heat of early conflicts in the Christian Church and consequently dealt only with the matters which were then under dispute. But even though all the items of conservative Christian faith are not stated in the ecumenical creeds, they are important because they express a considerable portion of the positive beliefs of orthodox Christians.

The most familiar of the ecumenical creeds is the Apostles' Creed. Since it is used in the services of many Christian groups, we will use it here as an illustration of the teachings of the ecumenical creeds in general.

I believe in God, the Father Almighty, Maker of heaven and earth: And in Jesus Christ, his only Son, our Lord,
Who was conceived by the Holy Ghost,
Born of the Virgin Mary, suffered under Pontius Pilate, was crucified, dead, and buried:
He descended into Hell, the third day he rose again from the dead, He ascended into heaven, and sitteth on the right hand of God, the Father Almighty;
From thence he shall come to judge the quick and the dead.
I believe in the Holy Ghost; the holy Catholic Church, the Communion of Saints; the forgiveness of sins; the resurrection of the body, and the life everlasting. Amen.

Lutherans accept the plain meaning of these words. The Apostles' Creed is used by some non-Lutheran churches where a considerable percentage of the members no longer interpret literally the language of the Creed. Usage under such circumstances is sometimes defended by clergymen as an effort to awaken in the minds of worshipers a feeling of unity with the Christians of earlier centuries. Clergymen of this persuasion contend that where religion is really free, all members of a large church could no more agree on a creedal statement than they could agree on politics or on how to raise their children. Therefore, worshipers should not expect that everything said in a church service must agree with their private beliefs. Such clergymen also frequently assert that present-day Christians need not take the creeds literally in order to maintain fellowship with the early Christians. The growth of language is such that the early Christians did not believe what the creeds say literally to us. Moreover, continue these clergymen, most present-day Christians do believe what the early Christians really meant by the creeds. The creeds should be taken as poetic, "mythological" expressions of deep religious truth.

Some clergymen follow the lead of the learned and brilliant churchman who said, "I do not believe all the dogmas in the Apostles' Creed, but I accept them. If I were more spiritual, I would believe them."

But among Lutherans (and many other groups) such rationalizations of the creeds need not be employed. It is taken for granted that clergymen and laymen alike believe just what the Creed says. They believe that Jesus was born of a virgin, will come again to earth, will raise the physical bodies of the dead.

In addition to the ecumenical creeds, Lutherans subscribe to a number of other specifically Lutheran Confessions. The most important of these is the Augsburg Confession. This creedal statement was formulated in 1530, at the very beginning of the Reformation, but it is still the firmly held faith of American Lutherans, because they believe it states accurately the teachings of the Bible. The paragraphs printed below are brief excerpts from the Augsburg Confession.

Since the Fall of Adam, all men begotten according to nature, are born with sin, that is, without the fear of God, without trust in God, and with concupiscence.

Men cannot be justified before God by their own strength, merits or works, but are freely justified for Christ's sake through faith.

Through the Word and Sacraments . . . the Holy Ghost is given, who worketh faith where and when it pleaseth God.

Baptism . . . is necessary to salvation.

The Body and Blood of Christ are truly present, and are distributed to those who eat in the Supper of the Lord.

Vows and traditions concerning meats and days, etc., instituted to merit grace and to make satisfaction for sins, are useless and contrary to the Gospel.

[Our adversaries] urged only childish and needless works, as particular holydays, particular fasts, brotherhoods, pilgrimages, services in honor of saints, the use of rosaries, monasticism, and such like. . . . Our works cannot reconcile God or merit forgiveness of sins, grace and justification, but that we obtain this only by faith, when we believe that we are received into favor for Christ's sake, who alone has been set forth the Mediator and Propitiation.

It is taught on our part that it is necessary to do good works, not that we should trust to merit grace by them, but because it is the will of God.

To the laity are given both kinds in the Sacrament of the Lord's Supper, because this usage has the commandment of the Lord: "Drink ye all of it."

Civil authority must be distinguished from ecclesiastical jurisdiction. . . . According to divine law, to the bishops as bishops . . . no jurisdiction belongs, except to forgive sins, to discern doctrine, to reject doctrines contrary to the Gospel, and to exclude from the communion of the Church wicked men, whose wickedness is known, and this without human force, simply by the Word.

The meaning of some of these statements needs explanation.

At the beginning of the Reformation practically all Protestants believed that we can do nothing to aid our salvation, our getting to heaven. They said that salvation is the gift of God. Men are universally and fundamentally depraved, prone to do evil. They are so sinful that they all deserve damnation. So deep is our degradation and so exalted is being in the presence of God, that man is as helpless to gain heaven as he is to jump to the moon.

The Catholics, on the other hand, taught the doctrine of "works." It is possible, they said, to increase the effect of God's Grace in us and to aid our salvation by doing good works—performing the spiritual offices of the Church and doing acts of benevolence and charity.

In reacting against the Catholic emphasis on the importance of outward observance, the early Reformation Protestants went so far as to deny that works of any kind have an effect on our salvation; even a whole lifetime spent in unselfish service has no effect. Salvation is strictly a gift of God. Why then, argued the Roman Catholics, lead a moral life? Because "it is the will of God," says the Augsburg Confession.

The teaching that salvation is of God and not of man seems to be contradicted by the dogma that we are saved by faith. But according to the Lutherans, faith is simply the channel through which God offers salvation; faith, too, is the direct gift of God. Luther wrote:

I cannot by my own reason or strength believe in Jesus Christ my Lord, or come to him; but the Holy Ghost has called me through the Gospel, enlightened me by his gifts, and sanctified and preserved me in the true faith.

To the non-Lutheran eye, these teachings read very like predestination—the doctrine that God determines from eternity the destiny of the soul. Calvin, the leading Reformation theologian, taught this doctrine, saying God arbitrarily chooses some souls for heaven and some for hell. The Lutherans vigorously deny belief in this "double" predestination. They say God damns no man to hell; He merely chooses some souls for heaven. This is a puzzling position. On the one hand, Lutherans believe God calls all men to faith, and offers all men redemption in Jesus Christ. On the other hand, they teach that no man can have the faith which is essential to salvation except God gives it to him.

All liberal Protestants today, and most conservative Protestants as well, have abandoned belief in any theory of predestination. They now put as much emphasis as do Catholics on the necessity for salvation of doing good works—works not of a ritualistic but of an ethical kind. Salvation by character is at present the typical Protestant point of view. But this statement is not true of the Lutherans. They still hold to the "unaltered" Augsburg Confession. Salvation, they say, comes from on high and good deeds will get no one to heaven. Lutherans hold "mere morality" in special horror.

Thomas Arnold of Rugby wrote, "The distinction between Christianity and all other systems of religion consists largely in this, that in these other, men are found seeking after God, while Christianity is God seeking after men." Lutherans applaud this sentiment. They are up in arms doing battle against the idea that religion is an achievement of mankind. The priesthood of all believers does not mean, for them, that each man must decide for himself what is good and bad in religion. God forbid! The priesthood of all believers means simply that there must be no human barrier placed between the individual soul and the Word and that every Christian must strive to be a mediator of God to his fellows.

LUTHERAN WORSHIP

Lutheran worship is free; that is, Lutheran congregations are at liberty to follow any order of service they think best. As a consequence, quite a number of Lutheran churches, influenced by the simple tastes of the frontier, came in earlier times to follow an informal order of service, one quite like that of their nonliturgical Protestant neighbors. Today, however, Lutheran churches are choosing more and more to follow a fixed form of worship. It is a liturgical service, based on the Mass of the medieval Church; it was promulgated in the early days of the Reformation. Luther discarded those portions of the Mass which he considered superstitious and added some elements of earlier Christian worship which had fallen into disuse. He also directed that the service be conducted in German rather than in Latin, and he emphasized the importance of the sermon.

The service begins by invoking God's presence by prayer and a hymn. After these devotions comes the Confession of Sin; this Confession is a series of prayers in which minister and congregation declare to God, "We are by nature sinful and unclean and . . . we have sinned against Thee by thought, word, and deed." Then comes the Declaration of Grace on all those who have sincerely acknowledged their sins and who accept the Christian faith.

Theoretically the service really begins with the Introit, a word whose original meaning was the "entering in." In the early days the Introit referred to the entrance of the priest into the chancel; during this entrance a Psalm was chanted. Today the name *Introit* is given to the Psalm itself.

After the Introit, the moving Gloria in Excelsis is sung and the pastor leads in prayer. Then come readings from the Bible, unison recitation of one of the creeds, another hymn, and the sermon. After the sermon come the offering and the General Prayer. The service closes with a hymn and a benediction.

The minister faces the altar during those portions of this service when he leads the people in their communication with God (the sacrificial portions); he faces the congregation when he represents God communicating with the people (the sacramental portions).

This brief description no more mediates the moving experience of worship under Lutheran auspices than a college catalogue mediates the enthusiasm and excitement of life on a college campus. The Lutheran service is a living thing to those who have been born into the Lutheran Church and who believe all its teachings. The service, for them, is the most solemn of all life's experiences, it is the very voice of God speaking through long centuries of Christian devotion. Joining in worship with a Lutheran congregation has always been for me a moving experience. The Lutheran liturgy, employing as do also the Catholic and Episcopal liturgies, the most ancient and beautiful literary expressions of the Christian Church, leaves the worshiper with a sense of majesty. The music of the service is often of unsurpassed spiritual import. Johann Sebastian Bach was a devout Lutheran; some of his greatest works were composed as an aid to Christian worship. Such music sung by a Lutheran choir—a choir composed of musicians who are chosen primarily for their faith and only secondarily for their artistic skill—and such a liturgy read by a Lutheran pastor—a pastor so conscious of its truth that he thinks any tricks of elocutionary emphasis unworthy—have the power to awaken, to challenge, to exalt a congregation, and to move it to Christian action.

The combination of the highest aesthetic forms and of genuine religious faith produces on worshipers an effect which is seldom attained elsewhere in human experience. Would that some genius could capture for the nonliturgical tradition the power which comes from setting forth the deepest religious convictions in art forms which are more capable of touching the emotions. For the average religiously literate American, the problem of finding an intellectually tenable faith is easy compared to the problem of finding a worship form capable of keeping this faith alive and growing. How alert men can seriously contend that religion can reach adequate power through mere communion with nature or solitary meditation, with no provision for periodic revitalization through the symbolism and fellowship of congregational worship, is beyond understanding.

Lutherans are the singing Church par excellence. Prior to the Reformation the Christian Church had resisted the singing of hymns during the Mass; but Protestantism emphasized singing. A church historian writes:

> The peculiar genius of the Protestant religion—the free and joyous spirit inspired by the doctrine of gratuitous forgiveness, and by the part which the laity assumed in worship, and in the management of Church affairs—was manifested in the "outburst of poetry and music," that was especially characteristic of Germany. . . . [Hymns] were sung not only in the church, but also in the household, the workshop, the market-place, and by armies on their march. The gospel was carried on the wings of song, and in this way spread abroad almost as much as by the voice of the preacher.[10]

Luther himself wrote many hymns, among them "A Mighty Fortress Is Our God," which Heine called "the Marseillaise of the Reformation." It has been estimated that there are no less than a hundred thousand hymns in Lutheran literature.

Lutherans really sing when they are in church. In a good many American churches the singing is timid, inhibited, the members of the congregation reminding one of adolescents at their first dance, afraid of doing the wrong thing and therefore doing as little as possible. But in Lutheran churches enthusiasm in singing is good form and the result is a volume of tone that testifies to genuine spiritual vitality.

Lutheran preaching gets as far as possible away from any hint of being merely the wisdom of the clergyman who is speaking. Just as the Lutheran Confessions are supposed to derive wholly from the Scriptures, so a Lutheran sermon is supposed to be based on a Scriptural or Confessional foundation. One Lutheran writer puts it this way:

> The Lutheran pastor as a rule, therefore, is more concerned to cultivate deep inner piety within the souls committed to his charge than to sparkle in the pulpit, attract publicity, and be known as a "great preacher."[11]

The emphasis in Lutheran preaching on the Biblical doctrines means that many Lutherans have been opposed to the discussion of contemporary social problems in the pulpit. For many people, religion is an escape from the cruel uncertainties of modern society into a world of supernatural security. One prominent Lutheran clergyman wrote of "the whole crowd who believe in the social gospel, in which we do not believe one whit."[12] A theological professor wrote, "If one wants to stigmatize a preacher one need only say of him that he preaches a social gospel."[13] In 1934, an investigation of the social opinions of American clergymen demonstrated that Lutherans were easily the most conservative of the Protestant denominational groups.[14] Today there are indications that a different spirit reigns in some quarters. One author wrote: "Lutherans have been aroused to a greater social consciousness. . . . Some Lutheran bodies have not hesitated to issue pronouncements on questions related to war, labor, family life, birth control, and divorce."[15] The executive director of the National Lutheran Council wrote in the *National Lutheran:*[16]

The place to begin the establishment of moral principle as a basis of government policy and procedure is at the local polling booth. . . . We Lutherans do not believe that the organized church should add itself as another pressure group in the capitals of state and nation. But we do believe that the church has two inescapable duties: 1) to urge the intelligent practice of Christian citizenship, and 2) to encourage some of its best qualified young people to undertake government service as their Christian vocation.

LUTHERAN POLITY

Most of the Lutheran churches of the world are governed episcopally; but in the United States the form of government is congregational and presbyterian. In all matters, the local congregation could, since it holds title to the property, exercise complete authority, provided continuance of relations with other Lutherans were not an object. In actual practice, the local congregations delegate much of their power to the synod, a body composed of representatives from the local congregations. The synods in turn send representatives to national conferences, where the policies of the Lutheran bodies are really determined.

Lutheran churches are run by males, like most American institutions. In 1936 the United Lutherans finally voted, after a long debate in which biblical quotations were the determining factor, to admit women to membership in synodical conferences; but it was 1946 before the first women appeared on the floor of the United Lutheran biennial convention. In 1948, this denomination sent a woman as one of its four delegates to the first assembly of the World Council of Churches meeting in Amsterdam. However, in 1959, the Missouri Synod "reaffirmed its historic position that the right to vote in congregational matters is reserved for the male members of the voter's assembly."[17] The ordination of women is not practiced by Lutheran bodies in the United States, except for the six-thousand-member Slovak Lutheran group. *Christianity Today* noted in 1959 that a woman had been "placed in full charge of a Slovak Lutheran congregation."

The Lutheran method of education rests on a thorough system of indoctrination. Again it must be said that Lutherans believe the right of individual judgment in religion is merely the right to interpret correctly what the Bible teaches—and its teachings are known. Most Lutherans have about the same attitude as other Protestants on the type of arrangements necessary for religious education—a Sunday school is considered adequate. The Missouri Synod, however, believes in the necessity for parochial schools; about a third of its children are studying in schools of this type.[18]

Lutherans, like the Catholics, bemoan the divisions in Christendom; but just like the Catholics their plan for union usually is, "Come over and join us." A denominational magazine of the Missouri Synod called the leaders of the ecumenical movement "ecumaniacs." Clergymen of this body frequently call the Papacy the "antichrist," and they usually refuse to stay on a public platform while a clergyman of another denomination offers prayer, unless they are convinced that he believes the faith "in its purity." But the orthodoxy even of the Missouri Synod is suspected by some of the Lutheran groups. The Evangelical Lutheran Synod (fourteen thousand members) in 1955 suspended relations with the Missouri Synod charging it "with unscriptural practice in regard to praying and working with Lutheran bodies with which it does not have doctrinal agreement."[19] In Rib Lake, Wisconsin, in 1947, twenty-five Wisconsin Synod Lutherans threatened to resign from the American Legion unless the practice of opening meetings with prayer was discontinued. They said the practice was "a denial of our faith."

In spite of its claim to infallibility of doctrine, the Missouri Synod does not slam its doors on all co-operative ventures. It would not "forswear its growing participation in the Boy Scout program, though . . . the smaller synods damn Boy Scouting as involving at least incidentally that private Lutheran sin, 'unionism.' "[20] The Missouri Synod has also taken up membership in one division of the National Council of Churches and has participated in discussions of co-operation with the National Lutheran Council, a confederation of the more liberal Lutheran bodies. One clergyman said that ultimate "Missouri membership in the [National Lutheran] Council is as sure as death and taxes." As a result of the Missouri Synod's efforts toward co-operation with denominations of doubtful orthodoxy, the Wisconsin Synod in 1961 severed its former close ties with the Missouri Synod.

The United and Augustana Lutheran groups are much less restrictive in their attitudes. They follow, generally, a policy of co-operation with other denominations; they have joined the National Council of Churches. However, many of their corporate decisions are dictated by a desire to conciliate the conservative Lutheran groups. The most liberal of the Lutheran bodies, the United Lutheran Church, had for years only a "consultative" relationship with the National Council's predecessor, the Federal Council of Churches; that is, its delegates attended sessions and shared in the discussions but did not accept the responsibility of voting.

The theological sternness which characterizes Lutheranism was evidenced by one of the synods of the United Lutheran Church when it tried three young clergymen from suburban Milwaukee for heresy and found two of them guilty and defrocked them. After his trial one of the deposed clergymen said:

In all the investigations and trials, the synod has insisted that the Bible must be taken as literally true in all its parts, whether it happens to be talking about history, biology, geology or faith. . . . I insist that the synod constitution means what it says, namely, that these books are the "only infallible rule of faith" and that we are not called upon to consider them infallible in matters of history, science, etc.[21]

An editor of the *Christian Century* in his report of the synod meeting which tried these clergymen indicated that a mood prevailed which too often characterizes the church meetings of all denominations, a mood which gives honest consideration to but one side of an issue. The editor noted

. . . the unquestioned assumption that one familiar interpretation of a justly revered formula has beartrapped the Truth, so that nothing that needs discussion can even be talked about; the desolating near-unanimity and hearty enthusiasm on the votes to condemn. . . .

Other elements in the picture probably can't ever be filled in; you would have to have seen them to believe them. Once you start on compulsions and personality distortions, where do you stop? . . . But there was more involved than the documentable legalities. You could hear that in misplaced levity, in bitter vindictiveness, see it in faces. . . .

The pastors are conscientious, diligent Christians . . . huddling in the safety of unison voice votes. . . .[22]

The internal strength of Lutheranism is beyond all cavil. Lutheran doctrine is infinitely precious to a multitude of people. Lutheran institutions would be defended in America, if need ever arose, with the same utter consecration with which they were defended in Nazi Germany. The Lutherans are a force to be reckoned with. No one can dictate their religion. They live in a spiritual castle: massive, static, an ever sure defense against the world.

But a major religious task of America is to achieve in a free society spiritual values held in common. The Lutherans, behind their castle wall of theological doctrine, are sure they are right and are sure they have made safe their spiritual liberties. But the question most poignantly presented to our day is: Can a free society survive the spiritual anarchy which follows in the wake of religious isolationism? Are there not spiritual levels on which co-operation among all sects is mandatory if democracy is successfully to weather totalitarian storms?

CHAPTER V

THE PROTESTANT EPISCOPAL CHURCH

NOTHING RAISES the blood pressure of an Episcopal clergyman like the oft-heard assertion, "The Episcopal Church was founded because Henry VIII, King of England, wanted a divorce." It is true, say Episcopalians, that the most important reformation movement in the Church of England (mother of the Episcopal Church) began in the sixteenth century during the reign of Henry VIII; but to assert that the Church was founded then is like asserting that when a woman does her spring cleaning she builds a new house. The Christian Church in England began more than a thousand years before the time of Henry VIII.

According to tradition, Joseph of Arimathea, who laid the body of Jesus in the tomb after the crucifixion, went to England in about A.D. 63 bringing with him the chalice which Jesus used at the Last Supper; this chalice is the Holy Grail of the Arthurian legends. Joseph also, according to the tradition, founded the Christian Church in the British Isles. This tradition is taken seriously by almost no one at the present time; however, church historians generally acknowledge that there were Christians in Britain as early as the second century, probably Roman soldiers sent there to guard the frontiers of the Roman Empire. But not until the end of the sixth century did the Bishop of Rome make an effort to bring the British Christians under his jurisdiction. By the time of the Reformation, the Pope's authority had been firmly established for many years in England, bishops and crown both acknowledging him as the head of Christendom.

THE REFORMATION IN ENGLAND

During the first part of his reign, Henry VIII was a faithful Roman Catholic; the Pope after a time even bestowed on Henry, for a book he wrote attacking his contemporary, Luther, the designation "Defender of the Faith," a title still proudly assumed by the Kings of England at their coronation. Henry married a Spanish princess, Catherine of Aragon. Though they had eight children, only one, a girl, survived more than a few days. The King and the people of England consequently became concerned over the lack of a male heir to the throne; in addition, the King had fallen in love with Anne Boleyn. Thus he applied to the Pope for an annulment of his marriage.

Under ordinary circumstances getting an annulment of the marriage would have been easy; the Popes had proved obliging on similar occasions. But it happened that the Spanish armies were at that time overrunning much of Southern Europe; thus the Pope, Clement VII, did not feel he could risk the anger of the Spanish ruler by freeing Henry from his Spanish queen. The Pope avoided a decision and at one stage even suggested bigamy as a possible solution. After years of waiting, Henry finally took matters into his own hands, broke the ties which bound England to Rome, had himself declared "The Protector and Supreme Head of the Church and Clergy of England," and got his annulment from the Archbishop of Canterbury, then as now the leading English cleric. And the Church of England issued a "declaration of independence," declaring, "The Bishop of Rome hath not by Scripture any greater authority in England than any other foreign bishop."

The Church of England did not become Protestant by this break. Henry continued throughout his long reign to believe firmly in the traditional religion and to follow traditional forms in worship. Only his "unwieldy corpulence prevented him from creeping to the cross on his last Good Friday, as he had been wont to do."[1] However, the rule of Rome in England was definitely at an end. The English ambassador to the Vatican was recalled, all papal revenues were stopped, and the English clergy were required to take an oath repudiating the Pope's leadership. In addition, Henry suppressed the monasteries and con-

fiscated their property; only in the nineteenth century did religious
orders again appear in the Anglican Communion (the Church of England and the twelve churches related to it, of which the Protestant
Episcopal Church is one). Henry also provided for the publication of
the English Bible and ordered that it be made available in the churches.

England now had "the open Bible"; and in the churches groups might
be seen clustered about the lecterns, while anyone who could do so read
for them the ancient and mysterious book . . . [that in centuries had not]
been open to the common people. Here was something by which plain
men would judge the words and deeds of prelates.[2]

After Henry's death the Reformation in the Church of England
gathered momentum; the liturgy was revised according to Protestant
doctrines, it was translated into English, and it was made available to
worshipers in *The Book of Common Prayer;* the marriage of priests
was permitted; images were removed from the churches; the elevation
of the Host for purposes of worship was forbidden; priests began to
study the art of preaching; a general congregational confession was
substituted for private confession; the dogma of transubstantiation was
denied; purgatory, indulgences, Communion in but one kind, the invocation of saints, the use in worship of a language not understood by
the people were all declared to have no warranty in Scripture and to
be "repugnant to the Word of God." Under Henry's three children
there were marked religious contrasts: the short reign of Edward VI
was definitely pro-Protestant, the short reign of Mary was pro-papal,
and the long reign of Elizabeth solidified many of the distinctive characteristics of the Anglican Communion. In the years which followed,
conflict continued between those who wanted to purify the worship
of the Church of England of all forms which had been "tainted by
popery" and those who wished to retain many of the traditional usages.
This conflict, in greatly reduced temper, has characterized much of
the history of Anglicanism in America.

ANGLICANS IN AMERICA

During colonial times the Church of England was in a favored position since it was the state Church of the ruling power. In seven of the
Southern colonies, the Church was supported by taxation, it was

"established." During the Revolutionary War, however, it had very hard sledding, since many of its clergymen were Tories. But at the end of the War clergy were sent to Scotland and England for consecration as bishops (in order to maintain the Apostolic Succession), all jurisdictional ties with England were severed, and an American jurisdiction was established. Since that time growth has been steady. Today the Protestant Episcopal Church is one of our largest and most influential Protestant bodies.

Episcopalians generally consider their Church a "bridge" between Roman Catholicism and Protestantism. In spite of its indubitable Protestant character, the Episcopal Church has retained many features frequently characterized by Protestants as Catholic. The Episcopate has been meticulously maintained and belief in the Apostolic Succession determines many corporate actions. The architecture of Episcopal churches follows the traditional model. Episcopalians observe the traditional Christian Year. Episcopal clergymen are vested— if not as elaborately as Roman Catholic clergymen, at least more elaborately than most other Protestant clergymen. Confirmation by the bishop is retained. Episcopal clergymen are referred to as priests. And, most important of all, in the eyes of the average Protestant, the Episcopal service puts a great emphasis on ceremonial.

To non-Episcopalians, the Church makes a definite impression of solidarity and unity. However, a very little information destroys this idea. The old Protestant-Catholic conflict, under new names, endures. The majority of Episcopal priests are what might be called "Prayer Book Churchmen"; but they stand in the middle between two opposed groups. One group is usually referred to as the Liberal Evangelicals; the term *Low Church,* formerly the designation of the Protestant party, is also sometimes used. The other group is usually called Anglo-Catholic, though the term *High Church* continues. Conflict between these two parties sometimes reaches embarrassing proportions. One Episcopal clergyman wrote:

> We shall probably be obliged, for a long time, to permit a rumpus room in the basement where the extreme "Liberals" can smash up the furniture; and an attic where the lunatic fringe of the "Catholic" party can play at Church—"Let's pretend we are Roman Catholics!" Most of us love children so we shall not mind the noise, but most of us will prefer to live in other parts of the house.[3]

CENTRAL CHURCHMANSHIP

Let us begin our study of these parties by describing the central group, the Prayer Book Churchmen. Most of these clergymen would join with one of the early bishops of New York who said, "My banner is Evangelical Truth and Apostolic Order." Another slogan frequently heard is: "Catholic for every truth of God; Protestant against every error of man." The Episcopal clergyman, according to this conception, is neither a vested Congregationalist nor a disguised Roman Catholic, but both a priest of the ancient Christian Church and a transmitter of its truth in language and forms which "have no superior."

The Book of Common Prayer is widely considered to contain the most precious forms of Christian worship and belief which have been developed by the Christian Church, Protestant and Catholic. These forms were composed in, or have been translated into, "incomparable English" and have been so revised and related as to speak meaningfully to modern minds. However, respect for the Prayer Book springs not so much from its literary qualities as from the fact that it is both the vehicle through which the faith is transmitted and also the standard for public worship.

In leading services of worship, ministers are required to make use of the worship forms provided in the Prayer Book; it is a matter of law. One of the first statements in the Prayer Book reads: "The Order for Holy Communion, the Order for Morning Prayer, the Order for Evening Prayer, and the Litany, as set forth in this Book, are the regular Services appointed for Public Worship in this Church, and *shall be used accordingly.*" (Italics mine.)

These orders of worship prescribe the sentences with which public services are to be opened, the passages of Scripture to be read, the prayers to be offered, the words by which the congregation makes General Confession and receives Absolution, the form which the Benediction shall take. The minister has freedom in the selection of hymns and of anthems, in choosing the topic for his sermon, in determining the ceremonial which accompanies the reading of the service. But the essentials of the service have been provided by the Church.

Episcopalians are sure their worship is more meaningful because its form is prescribed. For them, the Prayer Book frees a service from the idiosyncrasies of the individual priest—his emotional states, his clichés, his periods of spiritual dryness. The Prayer Book contains the writings of saints who agonized in prayer and gloried in praise. The readings have for Episcopalians the sound of authority; they come from the whole Church, ancient and modern. Consider the collect which is set at the beginning of the service of Holy Communion. This prayer has a grandeur which rises to the height of sublimity; it impresses the worshiper as a Grand March of the faithful into the presence of God.

> Almighty God, unto whom all hearts are open, all desires known, and from whom no secrets are hid; Cleanse the thoughts of our hearts by the inspiration of thy Holy Spirit, that we may perfectly love thee, and worthily magnify thy holy Name; through Christ our Lord. Amen.

This prayer is said by the priest whenever and wherever Holy Communion is administered in the Anglican Communion; this prayer is also said by every Roman Catholic clergyman every morning as he prepares to celebrate Mass, and is used widely throughout the rest of Christendom. Because this petition is repeated over and over, and on such solemn occasions, and because it has such definite links with Christians of many times and places, it has for Episcopalians a meaning and spirit which a prayer composed the previous evening in the rector's study, and said for the first time, could never have.

Strangers in an Episcopal Church are often confused by what they consider an excess of formality; they cannot find and keep the place in the Prayer Book, and are confused by so much standing and sitting, even though they may have been told, "We stand to praise, kneel to pray, and sit to be instructed." Any person planning to attend a service with which he is not familiar would do well to remind himself that all services of worship are easily followed by those who love them, and all services seem strange until they have been attended often enough to become very familiar. The spiritual effect of a service should not be judged after one or two experiences with it.

The Prayer Book furnishes the doctrinal standard for the Church; the recitation by minister and people of either the Apostles' Creed or the Nicene Creed is required at all major services of worship. In addition, all members of the Church declare at Confirmation their belief in the Apostles' Creed. This Creed is printed on page 160; the Nicene

Creed, normally used at every service of Holy Communion, is printed below. The relation believed by Episcopalians to exist between these two can be seen in the statement made in the twenties by the House of Bishops: "The shorter Apostles' Creed is to be interpreted in the light of the fuller Nicene Creed. The more elaborate statements of the latter safeguard the sense in which the simpler language of the former is to be understood."

I believe in one God the Father Almighty, Maker of heaven and earth, And of all things visible and invisible:

And in one Lord Jesus Christ, the only-begotten Son of God; Begotten of his Father before all worlds, God of God, Light of Light, Very God of very God; Begotten, not made; Being of one substance with the Father; By whom all things were made: Who for us men and for our salvation came down from heaven, And was incarnate by the Holy Ghost of the Virgin Mary, And was made man: And was crucified also for us under Pontius Pilate; He suffered and was buried: And the third day he rose again according to the Scriptures: And ascended into heaven, And sitteth on the right hand of the Father: And he shall come again, with glory, to judge both the quick and the dead; Whose kingdom shall have no end.

And I believe in the Holy Ghost, The Lord, and Giver of Life, Who proceedeth from the Father and the Son; Who with the Father and the Son together is worshipped and glorified; Who spake by the Prophets: And I believe one Catholic and Apostolic Church: I acknowledge one Baptism for the remission of sins: And I look for the Resurrection of the dead: And the Life of the world to come. Amen.

The question arises immediately, "In what sense do Episcopalians believe in such dogmas as the Virgin Birth and the bodily Resurrection of Jesus Christ?" In no one sense. All Episcopalians do not think alike. To be sure, they are bound to accept the wording of the Apostles' Creed; they declared their faith in it when they were confirmed. But they have wide liberty to interpret the Creed in the manner which seems to them to accord with Scripture. And this liberty is exercised. This latitude of interpretation is tolerated in order to avoid conflict and to maintain religious liberty. It would be safe to say, however, that the overwhelming majority of Episcopal clergymen make conservative interpretations of the Creed, though most of them would not want to be bound to accept its literal meaning, certainly not as interpreted by a person who was unacquainted with the history of the times and tensions which gave the Creed birth. Sampling polls have verified the conservative nature of the theological opinions held by most Episcopal clergymen. In one poll of every seventh priest on the Church's list of clergy, 87 per cent said they believed that "Jesus was conceived in the womb of the Virgin Mary without a human father." A poll conducted by *Christianity Today* found that 83 per cent of the Episcopal clergymen polled thought the virgin birth is important as a basis for church union, and 90 per cent thought the unique deity of Christ as the son of God is important as a basis for church union.[4] On the other hand, one of the bishops asserted that "the biblical evidence and the theological implications seem to be in favor of assuming that Joseph was the human father of Jesus" and wrote, "I prefer the creed to be sung."[5] Some Episcopal clergymen brought charges of heresy against him for these statements,[6] but the Church took no action.

Thirty-nine Articles of Religion, all but one inherited from the Church of England, are printed at the end of the Prayer Book. These Articles are not the standard of belief in American Episcopalianism; no affirmation of belief in them is required of either priests or laymen.

ANGLO-CATHOLICS

Services in some Episcopal churches are scarcely to be distinguished from services said by Roman Catholics, except that the language used is English. All the richness and solemnity of Roman Catholicism is to be found there; also its spiritual intensity and its authoritarianism.

Anglo-Catholics emphasize the Mass; they believe that it, rather than Morning Prayer, should be the most frequent office in the Church. One priest wrote:

The most superficial examination [of the Prayer Book] will show that the Holy Communion . . . was intended to be the principal service of the day, at which the largest congregation could be expected. It is the only service which contains within itself a provision for the taking up of an offering, for the giving out of notices, or the preaching of sermons. Only one place-finding is necessary, as against the numerous page turnings of Morning Prayer.[7] .

Anglo-Catholics—they would rather be called just *Catholics*—practice close Communion, that is, they decline to administer the Sacrament to any person who has not received in confirmation the laying on of hands by some bishop in the Apostolic Succession. The earmark of a Catholic Church, in their opinion, is the possession of clerical orders which go back in unbroken line to the apostles; in other words, the Apostolic Succession is a crucial doctrine. Anglo-Catholics hear private confessions, administer extreme unction, urge the laity to call priests "Father," urge frequent Communion, emphasize the desirability of diligent preparation before and thoughtful thanksgiving after receiving Holy Communion, enjoin making the daily meditation. They urge development of a "rule of life," and as aids in keeping this rule observe fast days, make a systematic use of retreats and quiet days, and "consecrate beauty to the service of God." They have sponsored the re-establishment of monastic orders in the Episcopal Church; there are at present in the United States eleven orders for men and fourteen for women.

The critics of Anglo-Catholicism find in it a considerable amount of ceremony for ceremony's sake. High Churchmen sometimes make the impression of being unnecessarily fussy about small points of external observance.

They perform an overelaborated and sometimes ill-conceived ceremony of kissings, crossings, genuflections, and bowings with great meticulosity, [while they] often read the service so fast and so inaudibly that it might just as well be in Latin.[8]

An English clergyman wrote concerning a visit to the United States:

I failed to discover why it is that when receiving Holy Communion an American high Anglican genuflects four times, where an English high Anglican does so only once or twice, and an Irish Roman Catholic not at all.[9]

EVANGELICALS

The Evangelical party has much in common with such Protestant groups as the Presbyterians and the Methodists. Services conducted by the Evangelicals are sometimes indistinguishable from the run of Protestant services. Of course, the worship orders prescribed in the Prayer Book are followed; but the temper and mood are Protestant. Attendants at Evangelical services are fairly sure, for example, to hear the office read impressively, though there is no guarantee that it will be; Evangelical leanings furnish no necessary guard against a voice that reads the liturgy "like the minutes of the previous meeting."

Evangelicals make comparatively little use of symbolic acts.

They [often] do not use the sign of the cross or bow at the name of Jesus or turn toward the altar for the *Gloria*. They [often] do not raise the alms or communion elements when they offer them at the altar. They [often] seem almost as obstinately Puritan as the Congregationalist who will not under any circumstances kneel to pray.[10]

Evangelicals tend to put the same emphasis on preaching as do other Protestants. True, the most frequent office in Evangelical churches is Morning Prayer, and it makes no provision for a sermon. The Prayer Book, however, nowhere requires that only the forms printed in it shall be used in worship; the minister has liberty to *add* to the prescribed service any material which seems to him to be important and fitting, a liberty exercised by the priests of all parties.

Evangelicals stand for the Priesthood of all Believers. A leading Evangelical says:

[This doctrine saves the Church] from the foolish pretension that one tradition of official ministry is in some intrinsic way superior to the traditions of other bodies. . . . Particularly is the Eucharist [Holy Communion] saved from being a sacrificial rite offered by a superior individual on behalf of others. . . . [Instead] it is the sacrifice of "our selves, our souls and our bodies."[11]

Bishop Angus Dun doubtless expressed the views of many Prayer Book Churchmen as well as of the Evangelicals, when he wrote:

The only apostolic succession I shall claim . . . is that implied in the confession that I know I have not attained, but press towards the mark of our unity in Christ.[12]

The Evangelical believes in open Communion; he feels that the Lord's Table must not be set only for those who have had episcopal confirmation. He believes it will never be possible for Christendom to be united unless all Christians can share their deepest spiritual experiences. The Evangelical is also willing upon occasion to invite into his pulpit members of the clergy of other denominations, an act which scandalizes the Anglo-Catholics.

On the other hand, Evangelicals join in the general Episcopal practice of refusing to give "letters of dismissal" to other Protestant churches. The clergymen of most Protestant denominations are willing to write letters stating that church members are in good standing and commending their reception into the churches of other denominations. But not the Episcopalians; for, there is "no provision in canon

law for giving letters of dismissal, except to an Episcopal Church."
Moreover, priests of the Church, with a few exceptions, will not receive members from other denominations by letter.* The Church will
accept the baptism of other churches, provided the formula used in
the ceremony was trinitarian; but ordinarily a non-Episcopal Protestant can become a communicant member of the Episcopal Church only
through receiving confirmation from a Bishop. However, a Roman
Catholic can become a communicant member of the Episcopal Church
without episcopal confirmation if he has been confirmed in the Roman
Catholic Church.

Unfortunately, the conflict between Evangelicals and Anglo-Catholics is more in evidence today than it was some years ago. In 1948
an English observer made a judgment which is probably still correct.

At present it is bogged in party feuds much as the C. of E. [Church
of England] was twenty years ago. An alarming feature of the P.E.C.
[Protestant Episcopal Church] is the tendency to develop monochrome
dioceses: that is, dioceses of one ecclesiastical color. There is an area in
the Middle West known to the sophisticated churchman as the "biretta**
belt."[13]

Many of the Evangelicals consider that the Anglo-Catholic has a certain "snobbish exclusiveness" and are offended by the "apparent attitude that there are two sorts of parishes—one of which is 'first class,'
and the other only 'second class.' "[14]

On the other hand, the Anglo-Catholics fear what they consider the
easy manner in which Evangelicals throw open Holy Communion,
with few safeguards, to "all those who love our Lord." Many Episcopal clergymen, in fact, practice close Communion. As a consequence,
at conferences gathered to foster the ecumenical movement the Episcopalians, and also the Lutherans, sometimes feel that they cannot join
in Communion with other Protestants. Occasionally at these conferences there are three services: one Episcopalian; one Lutheran, and one
Presbyterian-Methodist-Baptist-etc. At one conference of the Student
Christian Movement in New England, college students rebelled at this
unfraternal conduct, refused to have separate services, and argued and

* A few rectors do "accept the members of other churches by letter and
then prepare them for future confirmation."
** A biretta is a small square cap frequently worn by Roman Catholic and
Anglo-Catholic priests.

pleaded for a joint service. But the Episcopal clergy at the conference were adamant; as a result the conference had no Communion Service. Such actions, however, must be contrasted with actions of a more liberal nature, that of Presiding Bishop Henry Knox Sherrill, for example, at the Evanston meeting of the World Council of Churches when he threw open the Protestant Episcopal service to all who wished to commune. But that service was picketed by some of the Anglo-Catholics.

Anglo-Catholics feel also that they can make no compromise with persons who, though they have had no episcopal ordination, claim to be Christian ministers; such persons are not considered to be priests of Christ no matter how sincere or high-minded they might be, and no matter what type of ordination they may have received from Protestant churches. However, persons ordained by Roman Catholics are considered to be priests.

The effects on Church life of the conflict between the Evangelicals and the Anglo-Catholics have been widespread. Efforts on the part of Anglo-Catholics to take the term *Protestant* out of the name of the Church have been defeated. (At this writing the Anglo-Catholics are making another effort to accomplish this result.) The recent holders of the office of Presiding Bishop, the leading official, if not the candidates of the Evangelicals, were certainly not the candidates of the Anglo-Catholics. Anglo-Catholics spearheaded the movement which scuttled proposals for union with the Presbyterian Church; (more concerning this development later). Anglo-Catholics also defeated an Evangelical proposal to take the Thirty-nine Articles out of the Prayer Book. But on another occasion, the Evangelicals defeated an Anglo-Catholic proposal to do the same thing.

GOVERNMENT

The Episcopal Church is essentially a constitutional democracy. Its bishops are a long way from being dictators; rather, they are officials who carry out policies which have been democratically determined. One bishop said to me, with a bit of exaggeration, "I have only what authority I can win for myself." The carefully defined structure of the Church is designed as much for the control of power as for its exercise.

One rector said to me, "We are the most democratic of all the churches because we recognize the realities of power." Most of the denominations which emphasize congregational government because of its supposed greater liberty are considerably less democratic in their *regional* and *national* organizations than is the highly structured Episcopal Church. The editor of an Episcopal journal could write, "An Anglican is particularly impressed, not to say dismayed, by the organizational complexity and power-concentration that seems to spring so naturally from the Protestant ethos."[15] This surprising judgment springs from the failure of most congregationally organized bodies to define clearly the limits of the power regional and national executives can exercise (except over local churches), and from a failure to provide for natural channels whereby local groups can influence regional and national policy. These propositions will be discussed more fully later; the point being made here is that the Episcopal Church is high among American churches in the quality of the democratic processes it practices.

The spiritual affairs of the local church are in the hands of the pastor, usually called the *rector,* unless the parish is supported by missionary funds; then the title is usually *vicar.* The rector has control of the spiritual ministrations of the church: its services of worship, the music provided for worship, the administration of the sacraments, the running of the church school, the uses to which the church building may be put.

The financial affairs of the local church and the title to church property are ordinarily in the hands of an elected body of men called the *vestry.* The vestry is elected by parish meetings and has the power to select the rector, though it must secure the bishop's approval of the selection. The power to select vicars rests with the bishop. Canon law provides that the rector's tenure is for life and is not to be terminated by the vestry without his consent. The rector can even force in the civil courts the full payment of the salary originally agreed on. But if the pastor cannot be discharged, neither can he resign legally, without the consent of the vestry. However, in case of a dispute over tenure, either the rector or the vestry may appeal to the bishop, who then has power to settle the matter. Of course, such extreme regulations are very seldom invoked; no parish prospers without good will on all sides. The clear intent of giving tenure to the rector is to free him from dependence on the whims of the parish, the bane of clergymen in most denominations. These regulations were invoked in the Melish case. The rector of Brooklyn's Holy Trinity Episcopal Church refused to resign when requested by the vestry. The bishop was called in and after due consideration ordered the rector to quit his post; he again refused since a large majority of the members of the parish wished him to remain. The case was then taken to the civil courts where the bishop's action was confirmed.

Continental United States is divided into seventy-eight dioceses and nine missionary districts. Each diocese and missionary district is headed by a bishop; but he shares authority with a diocesan convention, a legislative body composed of the clergy and elected lay representatives from each parish. The bishop of a diocese is chosen by the diocesan convention, though his election to the office must be approved both by a majority of the Standing Committees of the other dioceses and by a majority of the bishops of the Church.

Final authority in the Protestant Episcopal Church rests with the General Convention, a national body which meets every three years. It is composed of the House of Bishops and the House of Deputies. The latter contains four priests and four laymen from each diocese, and one priest and one layman from each missionary district.

The General Convention chooses a Presiding Bishop whose major function is to serve as the executive head of the National Council, a body charged with carrying out the decisions of the General Convention. The Presiding Bishop occupies an office of great dignity, but he is not the administrative superior of the other bishops. He has no authority in the local dioceses, except as the General Convention may require his direct supervision of specific projects. He does, however, preside over all sessions of the House of Bishops and "takes order" for the consecration of all new bishops.

The Protestant Episcopal Church has fraternal relations with the other churches of the Anglican Communion through the Lambeth Conference, a meeting held approximately every ten years. The Conference is called by the Archbishop of Canterbury, is held in Lambeth Palace, London, and is open to all active bishops of the Anglican

Communion. The decisions of the Conference have great weight but do not have the force of law.

The position of women in the Episcopal Church is traditional. The House of Deputies in 1946 permitted, for the first time, a woman to be seated as a Deputy. But the same Convention refused the earnest request of the Woman's Auxiliary of the Church to interpret the word *layman* in church laws to include women as well as men; the interpretation would have permitted women to hold many church offices. "Male voices shouted down the resolution," commented *The Lutheran*.[16] In the Diocese of Rochester a proposal to send a woman as a Deputy to the General Convention was defeated, one argument against the proposal being, "Women are already doing more than their share of church work." In 1949, the House of Deputies declined to seat four women; the Convention did, however, accept two million dollars which Episcopal women had raised for the work of the Church. The House of Deputies in the Convention of 1961 refused by an overwhelming vote to approve a constitutional amendment which would have changed the word *layman* to *lay person*.

The Protestant Episcopal Church numbers about three and one-half million members. It is roughly one-twelfth the size of the Roman Catholic Church and one-third the size of the Methodist Church. Nevertheless, the Episcopalians are a powerful body. Man for man they probably have more influence on national affairs than any other religious group; they are probably more often the heads and directors of our great corporations, and are probably more often mentioned in the social, educational, and financial columns of the newspapers than are an equivalent proportion of any other denomination. The Episcopal Church is found all over the United States, but its real strength is on the eastern seaboard.

As a denomination Episcopalians have taken a more active part in the ecumenical movement than has any other American group. Episcopal interest in the movement and, indeed, the movement itself are more the result of the efforts of Bishop Charles Henry Brent than of any other one person. Telling briefly the story of his life will give an opportunity to indicate some of the facts about this important aspect of twentieth-century Christianity and also to indicate something of the kind of life led by a great, modern, American churchman.

BISHOP BRENT

Brent was born in Canada in an Episcopal rectory. He attended the elementary and high schools in Newcastle, Ontario, and then went to Trinity College School, where he displayed keen interests in organ playing and Rugby football, making the school team as fullback. After two years, he enrolled in the University of Toronto, made a good academic record, and graduated in 1882. Then he returned to Trinity College School as a teacher.

Brent's choice of the ministry as a vocation was apparently made early, and without much struggle. During his two years of teaching at Trinity he prepared for the examinations which precede Holy Orders. He was ordained in 1886 by the Bishop of Toronto; but there was no opening in the Bishop's diocese. Consequently, Brent took a position in Buffalo, New York, for what he thought would be a short time. But he was destined to have the United States as the base of his operations throughout the rest of his life. In Buffalo, he was appointed curate and organist in St. John's Church; apparently for a time he considered devoting his life to church music. Soon, however, it became apparent that his vocation should be the pastoral ministry and he was placed in charge of a small mission church, St. Andrew's.

There he came into conflict with his bishop over—of all things—the use of candles on the altar. Both the bishop and Brent were high churchmen, but the bishop was very suspicious of any appearance of leanings toward Rome. At this stage in the development of Episcopal ceremony (it has been greatly elaborated in recent decades), the use of candles connoted a Roman tendency. Thus Brent was ordered—somewhat summarily—to take the candles off the altar. He resisted vigorously, since he knew that candles were in use in other parts of the diocese. However, since St. Andrew's was a mission, the bishop had control there, even though he did not have control in the self-supporting churches of the diocese. In the end, Brent had to obey instructions. But he began to look around for another post.

In the summer of 1887, he attended a retreat for clergymen conducted by Rev. A. C. A. Hall, Superior of the Boston house of the Cowley Fathers, a monastic order with headquarters in England. Brent was profoundly impressed by this retreat, established a close friendship with Hall, and landed before many months as a member of the Cowley household in Boston. He continued there for three years, living by the simple, ordered, spiritual pattern of the monks. Years later he wrote:

I can conceive of nothing more admirable or productive of good results in the character and efficiency of a young priest than life in such an environment as I found myself in [in] the Mission House. . . . Simplicity of living, close attention to duty, and punctilious regularity are amiss at no time of one's career, but they are a whole education before a man's character is finally set.[17]

Brent was on the point of becoming a full-fledged monk, taking the vows of poverty, chastity and obedience, when he was outraged by an act of the head of the Cowley Fathers in England. Hall, decidedly High Church, had refused to vote for Phillips Brooks, a Liberal, as Bishop of Massachusetts; but once Brooks had been elected, Hall as a member of the diocesan Standing Committee voted to ratify the election. For this compromise with Liberalism, Hall was called to England by the head of the Order. Brent was so incensed at this highhanded procedure that he also went to England and gave his opinion of the action directly to the Father Superior. This interview ended Brent's formal connection with monasticism. He was no "yes" man.

Bishop Brooks then asked Brent to become a member of the staff of St. Stephen's Church in the poorest section of Boston. He accepted, and for ten years, worked in the midst of the sordidness and squalor of the great city, trying to bring light into dark alleys, love into desolate lives, vision into narrow minds. Father Brent was a genuine pastor; he cared about people. He counseled individuals and helped them solve their problems. He was concerned about social conditions and supported movements directed toward social change. But his greatest ministry was a devotional one. He felt it was his mission to prepare a spiritual haven in the services of the Church where frustrated, beaten men and women could find solace and beauty. Brent conducted worship with the intentness and simplicity of one who personally is receiving great spiritual blessing. He reacted against the fussiness of some High Churchmen; yet his services were Anglo-Catholic, striving to plumb devotional depths through the pageantry of rich vestments, lights, symbols, and a sacerdotal decorum on the part of the clergy.

Distressed by the contrast between the luxury of the rich and the penury of the poor, and noticing the emptiness of the great houses on Beacon Street during the hot weather, he persuaded some of his wealthy friends to open their homes during the summer months to selected South Boston families. In view of American mores, the suggestion was extreme and the results were not happy; but the story shows the courage and magnetism of the man. On another occasion he took passage from England to the United States in the steerage in an effort to learn at first hand the mind of the immigrants who then were flooding American shores.

Two years after the close of the Spanish-American War and the acquisition of the Philippines by the United States, Brent was suddenly elected missionary bishop of the Philippine Islands. His work there made him a national figure. He came to know and have influence on the leaders in many fields, including Theodore Roosevelt and William Howard Taft. He traveled all over the United States soliciting funds for his missions. He helped shape missionary policy. He wrote many books (twenty in his lifetime, the majority while serving in the Philippines). But the activity which first brought him into real prominence was the part he played in the movement against the opium traffic.

Under the Spaniards, opium had been a government monopoly; under the Americans, there was passed in the Islands a bill (later vetoed by Roosevelt) again making opium a government monopoly, and devoting the revenue to education. Bishop Brent fought this proposal, declaring, "We would be educating men in vice in order that we might educate their children intellectually." Soon after, Taft, then governor, appointed the Bishop to a committee whose function was to study control of the drug. Brent became convinced as a result of this study that the problem would yield only to international action; therefore, he wrote to President Roosevelt urging him to take leadership in forming an International Opium Commission. Roosevelt saw the possibilities in the proposal, called the Commission, and appointed Brent to the American delegation. Promptly after the convening of the Commission, Brent was surprised to be elected its president. Two years later, in 1911, another international conference on the opium traffic was held at The Hague; again Brent was a member of the American delegation, and again he was elected president of the conference. His activities against the opium traffic continued until his death, in 1929; one of the last acts of his life was to send a memorandum on the opium traffic to President Hoover.

Twice during his tenure in the Philippines, Brent was elected Bishop of Washington, D. C., a position for which his political acquaintance fitted him. He was also elected Bishop of New Jersey. But he declined all these invitations. "If I should some day cease to be Bishop of the Philippine Islands," he said to the Quill Club in Manila, "it will not be because of extra persuasion from prominent personages . . . but because the body refuses to behave properly in the Tropics." In 1917, his doctors made clear that he would not survive another two years in the Islands. Consequently, he accepted election as the Bishop of Western New York, but with the proviso that he would take up his duties only after he had served his country, then at war, in the American Expeditionary Force.

He went to France simply as a special representative of the Y.M.C.A.; but in a short time, such were his powers of leadership, he was established as Senior Chaplain at Pershing's headquarters, "in effect, if not in name, the chief of Army chaplains."[18] He set up the chaplains' organization, established a chaplains' school, and shaped policies which had a lasting effect on the reorganization of the Army Chaplains Corps after the war. One of his policies, however, was not continued: he was so much concerned that clergymen serving in the Army should establish the pastoral relationship quickly with all kinds of men that he refused to permit the chaplains under him to wear insignia of military rank.

As Bishop of Western New York, Brent was not a great administrator; he had too many national and international problems bidding for his attention. But he did furnish a spiritual leadership that, according to his successor, practically "raised the diocese from the dead." To his lasting regret he was not able in the diocese to exercise widely his pastoral functions. But he did inspire and vitalize many groups of people through addresses, sermons and the leadership of conferences. The clergy and the laity of Western New York came to think of him as the greatest American churchman since Phillips Brooks. They dubbed him "Everybody's Bishop."

At Edinburgh in 1910, attending the World Missionary Conference, Brent came to the conviction that church unity, long considered unattainable, could be achieved within a century; at first he even included Roman Catholicism in his ecumenical vision, though later he changed his mind on that possibility.

Soon after leaving the Conference at Edinburgh, he attended the General Convention of the Episcopal Church, meeting in Cincinnati. He had been asked to speak there to a mass meeting of the Bishops, Deputies and Woman's Auxiliary. The result of that address was the establishment by the Episcopal Church of a Commission to promote a world conference where the differences which prevent unity would be honestly faced and an effort made to resolve them. World War I broke out before the conference could be held. After the war, in 1920, the Lambeth Conference issued a stirring appeal for the unity of Christian people, and in the same year seventy denominations sent representatives to a preliminary Conference which was held at Geneva. At this Conference, initiative for the ecumenical movement passed from the Episcopal Church Commission to an extremely able Continuation Committee, of which Brent was made chairman. This Committee arranged to hold a World Conference on Faith and Order at Lausanne, Switzerland, in 1927. This Conference was recognized by Christian groups all over the world, no less than 127 denominations sending representatives. A large portion of the leadership of the American and European churches was there.

At the opening session, this cosmopolitan company repeated, each in his own tongue, the Lord's Prayer and the Apostles' Creed, sang "Now Thank We All Our God," and Bishop Brent preached. "We are here," he began, "at the urgent behest of Jesus Christ"; and then he pleaded that the delegates ponder the moral qualities essential to Christian unity: humility, patience, charity, faith. Then he was elected presiding officer. "His position as the pivotal person of the Conference was plain," wrote Archbishop Temple, "and his quiet, firm and often humorous control of the discussions was most effective."[19]

The formal accomplishments of Lausanne were not great; but the informal ones began a new era. Lausanne proved that officials representing many diverse Christian groups could sit down in the spirit of conference and make long strides toward understanding. At the end, Bishop Brent said, "God has enlarged our horizons, quickened our understanding, enlivened our hope."

Brent's leadership at Lausanne was his last major contribution to the ecumenical movement; in view of a serious heart condition, he had accepted leadership even there only at considerable risk. After the Conference, his lifework was for all practical purposes ended. But the movement which he had done so much to bring into being went forward, under the leadership of his famous colleagues, through conference after conference until at Amsterdam, in 1948, the World Council of Churches was brought to birth.

The attitudes of most of the present leaders of the ecumenical movement were well expressed by Brent.

Christ's agile feet journey to the human heart along many and diverse paths.[20]

You and I must put ourselves in the right relation to God. I am as strongly convinced on many subjects as the rest of you, but I am anxious to get rid of prejudice and ignorance, and it is for us, in a way that perhaps we have never done before, to put ourselves at the disposal of God, to give our minds and our hearts and our judgments into His hands that He may sway us whither he will.[21]

What is needed more than anything else is courage to try God's way. . . . I reaffirm my belief that the Christian Church, if it be so minded, can, in the name of Christianity, rule out war and rule in peace in a generation. I may be a fool, but if so I am God's fool.[22]

Episcopalians and the Ecumenical Movement

The results of some other efforts at church unity have not been so happy. The Episcopal General Convention in 1937, in a burst of ecumenical enthusiasm generated by two world conferences held that same year at Oxford and Edinburgh, made overtures to the Presbyterian Church in the U.S.A. looking toward the union of the two bodies. The Presbyterians showed genuine interest and enthusiasm. But after nine years of study and conference by a Joint Commission of Presbyterians and Episcopalians, the General Convention refused to accept even for "study" in the local churches a *Proposed Basis of Union*. This action was perhaps the major setback the ecumenical movement has received—and the blow was delivered by the Church which has carried the banner for church union. The action came through a coalition of Southern and Anglo-Catholic votes, the Bishop of Chicago even threatening secession if the union were consummated.[23]

Presbyterians hit the ceiling. Henry P. Van Dusen, an Episcopal layman as well as a Presbyterian clergyman, said:

Is it any wonder that throughout American Protestantism, the Episcopal Church is increasingly likened to an adolescent school-girl who proposes marriage in leap-year, and then, when her offer is accepted, searches frantically for some escape from her pledged commitment?[24]

Another Presbyterian clergyman stated the case still more bluntly in a statement wryly reprinted by an Episcopal journal:

Among the Protestant clergy in any community the ministers of what Church almost always (there *are* rare exceptions) can be counted on *not* to be counted on in joint religious services? What denomination keeps its pulpits closed to ministers of other evangelical Churches? [There are exceptions to this generalization also.] What denomination refuses to grant letters of dismission to all other evangelical denominations? . . . Pastors who eagerly try to promote union enterprises in their communities . . . receive neither encouragement nor cooperation from their Episcopalian brethren in the ministry.[25]

Many Episcopal bishops and priests were greatly embarrassed and disappointed by the action of the General Convention rejecting the *Proposed Basis of Union.* Yet the leaders of the ecumenical movement in the denomination have not lost hope. Canon Theodore O. Wedel, a member of the Joint Commission, wrote:

Our Church has now to wrestle with the central issues of Church re-union. We should have faced these issues and settled our own internal dilemma of conscience before we burdened another communion with the embarrassments of our disunity.[26]

Episcopalians have developed churchmanship to a far higher degree than most Protestants. As a result, they have a greater concern for achieving a united Christendom than other Protestants and frequently are tempted to speak beyond their willingness to act. At the founding of the World Council of Churches in Amsterdam, no group wanted to talk as much; yet earlier that same summer at Lambeth, the bishops refused full communion to the newly formed and united Church of South India, even as Anglicans in Canada a quarter of a century earlier had refused to become an organic part of the United Church of Canada. Only in 1940 did the Protestant Episcopal Church become a full-fledged member of the Federal Council of Churches.

The issue responsible for the greatest disagreement between the Episcopalians and the Presbyterians, and probably in the ecumenical movement generally, is that of the validity of the various types of ordination. The Joint Commission made a fertile suggestion at this point, proposing that Presbyterian clergymen be ordained priests in the Episcopal Church and Episcopal clergymen and bishops be ordained presbyters in the Presbyterian Church. The acceptance of some such device will doubtless be essential if the ecumenical movement is ever to become strong. Such acceptance is probable in many communions. Increasingly, Protestant clergymen would agree with the Methodist Bishop G. Bromley Oxnam, who said:

I would be proud to kneel at any altar and have the hands of Harry Emerson Fosdick placed upon my head. . . . Similarly, I would rejoice in receiving from Henry Sloane Coffin and from Rufus Jones the treasures of their traditions . . . [and] to have the hands of Bishop Henry Knox Sherrill laid upon my head, symbolizing the unbroken tradition of the centuries.[27]

The reception of Episcopal orders by a Methodist minister in good standing was a step in this direction. George Hedley, chaplain at Mills College, Oakland, California, was ordained by the Bishop of California, James A. Pike, in order that Episcopal students might "feel free" to attend Holy Communion when it is celebrated by Dr. Hedley. Bishop Pike was rebuked by some Episcopalians (he took a "short cut to chaos") and Dr. Hedley was rebuked by some Methodists (he is "neither fish nor fowl; he has impugned his ordination as a Methodist"). An Episcopal editor stated the situation when he wrote that

virtually all Protestant Churches would accept Anglican clergy into their ministry without reordination, but Anglican Churches will not accept Protestant ministers without ordination to the diaconate and priesthood.[28]

Subsequent to the ordination of Dr. Hedley, the Evangelical and Reform Church instructed its clergymen not to accept Episcopal ordination if they wished to continue as Evangelical and Reform ministers: "it could be construed as an implied acknowledgement of a deficiency in our own ministerial orders."[29] And a Presbyterian clergyman wrote, "No proposal can be supported which even by inference implies that my present ordination is not to the Holy Catholic Church. . . . such terms as 'extension of ordination' or 'reordination to a new church' have no reality for there is no way of broadening an ordination that is already to the church universal." Late in 1961 both the Methodist Council of Bishops and the Protestant Episcopal General Convention passed recommendations designed to prevent any further dual ordinations of the Hedley type.[30]

Perhaps the next move needed is for Episcopal clergymen to give clear evidence that they truly respect and are anxious to receive the orders of other Protestant churches. None have as yet received such orders, as far as I know, but Bishop Pike has written, "I am ready to kneel down for the purpose."[31] The Episcopal Church has made very clear its conviction that clergymen out of the Apostolic Succession are not qualified to serve as Episcopal clergymen. If it will take the next step and affirm that persons in the Apostolic Succession are not qualified by that fact to serve as Methodist, Presbyterian, or other Protestant clergymen, a large step toward union would be taken.

Surely it is but a matter of time until many of the non-Roman churches find a way of lessening our confusion of tongues, and at the same time of preserving the preciousness of each church's distinctive contribution to our religious life. When this end is achieved the impact of Christian idealism on civilization will have been greatly increased and the spiritual stature of our churches mightily enhanced.

But the person who is concerned about the spiritual stature of the whole of society wonders whether the attainment of Christian unity will not be so difficult as to consume all the surplus energies of most denominational leaders and whether the negotiations will not be so delicate that the ecumenical leaders will be tempted to reduce their message to a kind of sweetness and light vagueness. Moreover, is the uniting of the old-line Protestant and Eastern Orthodox churches, those which "confess our Lord Jesus Christ as God and Savior," the most important task confronting enlightened churchmanship? What is the answer of ecumenical leaders to the thesis that the United

States confronts a clear and present spiritual danger, the danger of an inadequate commitment to the ethical values central in our civilization? And can this danger be met without the wholehearted co-operation of the skilled and devoted churchmen who today spend most of their spare energies in promoting a movement which at best can minister to less than a third of the American people?

CHAPTER VI

PRESBYTERIAN CHURCHES

SOME PRESBYTERIANS would like to believe that the first General Assembly, the highest body in the Presbyterian governmental system, was held in Jerusalem in the middle of the first century and is described in the 15th Chapter of the book of Acts. On that memorable occasion, Paul and Barnabas came from Antioch, conferred with the "apostles and elders," and matters of great moment were settled. A few contemporary Presbyterians would even agree with the Puritan professor in Elizabethan times who contended that the Bible prescribes not only the doctrine of the Church, but also its governmental form—and that that form is Presbyterianism.

These extreme positions are seldom the belief of the present-day Presbyterian churches; Presbyterians do hold, however, as do also many non-Presbyterian scholars, that the early church was ruled by its weightiest members, its elders, called in the Greek language *presbyters*. Presbyterians believe that with the Reformation, the Church finally *returned* to its original faith and form of government. The major figure in that phase of the Reformation which resulted in the re-establishment of the Presbyterian system was one of the most influential Christians who ever lived, John Calvin. He is the outstanding figure among the second generation of the Reformers.

JOHN CALVIN

Calvin was a brilliant French student of law who was converted to Protestantism by reading Luther and the New Testament. Forced to flee from France, he settled in Geneva, Switzerland, and succeeded, according to the Scotsman, John Knox, in turning that city into "the most perfect school of Christ that ever was on the earth since the days of the Apostles."

Knox's judgment resulted from the central role which religion played in the municipal life and from the puritanical conduct on which Calvin insisted. His study of the New Testament led him to believe that Christ required of his followers earnest striving for moral perfection. Moreover, Calvin considered it the duty of ministers to point the road toward purity, and when necessary to force travel along that road. Consequently, the citizens of Geneva were punished not only for such sins as dishonesty, violence and lewdness, but also for dancing, staying away from church, criticizing ministers and speaking well of the Pope. One man received a three-months' banishment for observing when he heard an ass bray, "He chants a fine Psalm." This stern view of morals had subsequently a pronounced influence on the moral ideals of the Puritans, both in England and in America, as we shall see.

Calvin set in motion forces which resulted in the modern evangelical form of worship, the form most widely used in the United States today. His influence on theology was also very great. He wrote a best seller, *The Institutes of the Christian Religion,* a work which furnished for many years the doctrinal basis for the majority of the Protestant churches. Calvin was sure, however, that he had not originated a theology; he and his followers believed that he merely expounded what was clearly taught in the Bible. A modern student of Calvin has declared, "No writings of the Reformation era were more feared by the Roman Church, more zealously fought against, more hostilely pursued, than Calvin's 'Institutes.' "[1]

Calvin's influence on politics was also great and furnishes an excellent illustration of the carrying power of ideas which achieve the deepest religious sanctions. He contended that, according to the New Testament, ecclesiastical government, instead of resting in the hands of one man, should be in the hands of public representatives, persons chosen to protect the interest of the Church as a whole. This ecclesiastical theory spread widely over the Western world and was adopted by secular as well as religious thinkers. Calvinism undoubtedly played a leading part in the development of democratic institutions.

PRESBYTERIANISM IN SCOTLAND AND ENGLAND

For the past three centuries, the center of Presbyterianism in Great Britain has been Scotland. The leader in bringing this nation to Protestantism was John Knox. Knox, a devout Roman Catholic priest, was converted to Protestantism about the middle of the sixteenth century. He joined those in his country who had taken arms against the Catholics, was captured, and spent nineteen months as a rower in the galleys of France, the nation which came to the assistance of the Catholic rulers of Scotland. After his release he spent twelve years in exile, first in England, and then as pastor of the English refugees at Geneva; Calvin was then at the height of his powers and influenced Knox greatly. Knox returned to his native land in 1559; a year later the Scottish Parliament, supposing it could legislate belief, abolished Catholicism, and declared the Reformed faith to be the religion of the state and *all* the people.

Naturally this action did not accomplish the Reformation in Scotland. A long and dramatic struggle ensued between the Catholic party, led by Queen Mary, and the Protestant party, led by Knox. In the end, Knox, aided by the misdemeanors of the Queen and the power of the English government, brought his party to victory. He established Calvinist theology as the doctrine of the Church and Calvinist polity as the government of the Church. The Scotch have succeeded, sometimes in spite of great difficulty, in maintaining Presbyterianism as the dominant religion of their country down to the present day.

Calvinist beliefs exerted a tremendous influence on English politics in the decades between the death of Henry VIII, in the middle of the sixteenth century, and the passage of the famous Act of Toleration, in 1689. This period of European history was dominated by the celebrated principle, *Cujus regio ejus religio,* the religion of a country and of all its people should be that of its sovereign. The sovereign of England during the last forty years of the sixteenth century was Elizabeth, a woman whose primary interest was politics, not religion. Nevertheless, she was the daughter of Anne Boleyn and, therefore, was forced to maintain England's Protestantism, or else have herself declared illegitimate and unlawfully Queen. Yet, Elizabeth had no wish to alienate unnecessarily her subjects who retained Catholic sympathies. Consequently, she and her bishops steered a middle course; under their direction the Church of England adopted, generally, a Calvinist theology while it retained many of the traditional rites and ceremonies.

The political nature of this arrangement meant that religion was often attended to as an official duty. The clergy, dependent on the favor of the Queen, frequently went about their tasks in a mechanical, ear-to-the-ground sort of way. Soon most Englishmen of genuine piety were intent on change. Two movements resulted: one was called Separatism and the other Puritanism. The Separatists were the most radical; they believed the difficulty of getting reform in the Church of England was so great that they must withdraw, must separate, from the existing Church in order to re-establish the Church of Christ and the Apostles. The story of their experiences will be told in the chapters on the Congregationalists and the Baptists.

The Puritans, on the other hand, insisted that the Church of England must be "purified" of Roman Catholicism, rid of the "rags of Popery." The Puritan (originally a term used in contempt) found the basis for his faith in the Bible, a book which he studied with great care. All over England, by the end of Elizabeth's reign, were to be found individual men seated by their firesides, spending long evenings poring over the Scriptures, searching for the will of God. Thus Bible study became fixed in the habits of many generations of English-speaking Protestants.

The Puritans are responsible also for other customs which had widespread influence in England and America: the possession of a family Bible, family prayers, faithful self-examination as a substitute for the confessional, the undecorated church building, extempore prayer in worship, the omission of the sign of the cross and of kneeling before the altar in worship, the wearing of lay apparel by the clergy, belief in the correctness of public worship in unconsecrated buildings, and the "English" (solemn) as against the "Continental" (gay) Sabbath.

The conflict between the Anglicans and the Puritans grew so serious during the reigns of Elizabeth's successors, the Stuarts, as to become the all-absorbing concern of British politics. The Stuart kings, especially Charles I, were bent on exercising the royal authority over the Church. The Puritans were in high disfavor. "Such as could not flee," as a writer of that day declared, "were tormented in the bishop's courts, fined, whipped, pilloried, imprisoned, and suffered to enjoy no rest." Finally, rebellion broke out—the King and the Anglicans on one side, the Puritans and the bourgeois on the other. The latter were victorious, and eventually Charles I was dethroned and beheaded. The Puritan armies of this period had a morale such as comes only from deep-seated conviction. Their victories gave rise to the quip, "There is nothing as dangerous as a Presbyterian just off his knees."

During the Puritan ascendancy, Parliament, in an effort to establish standards in religion, called to Westminster a group of leading clerygmen, most of them Presbyterians. This Westminster Assembly prepared the famous Westminster Confession of Faith, a creed which in general summarized the teachings of Calvin. Parliament abolished the *Prayerbook,* adopted the Presbyterian *Directory of Worship,* and adopted the Presbyterian form of church government.

Then came a reaction under Cromwell. He was an Independent who favored tolerance for all persons who held to the Protestant fundamentals; he had as little use for compulsory Presbyterianism as for compulsory Anglicanism. During the ten years of his rule, religious toleration became the guiding principle of the government.

But after Cromwell, the Stuarts again required religious conformity. Finally, under William and Mary a new dynasty was established, the old idea of conformity was abandoned, and the legal foundations of religious toleration were laid down. The Toleration Act (1689) is a great landmark in religious history. This Act was not perfect; it failed to grant freedom to all groups, notably to non-trinitarians and to Catholics. But it did establish toleration as a principle, both in England and in her colonies.

PRESBYTERIANISM COMES TO AMERICA

All Protestant churches in America show markedly the influence of Calvin's genius. This influence is most evident today in two of our denominations. One is called Reformed. On the continent of Europe—chiefly in Switzerland, France, Holland, Hungary and Germany—most of the churches which adopted the Calvinist teachings took this name. When members of these bodies migrated to America, they naturally brought their churches with them. There are in the country about a half million persons in churches now using the name "Reformed."

The second American denomination in which Calvin's influence is especially evident is, of course, the Presbyterian. The first presbytery was organized in 1705 and the first synod in 1716. In these early decades Presbyterian churches made steady progress, chiefly in the middle colonies, New Jersey to South Carolina.

During the Revolutionary War, Presbyterians were active in the struggle for independence; one writer claims that "more than any other single group, [they] were responsible for the development and success of the American Revolution."[2] Some English leaders called the Revolution "a Presbyterian rebellion," and Horace Walpole said in Parliament, "Cousin America has run off with a Presbyterian parson." These statements no doubt put too much stress on religion as a factor in the immediate causes of the American Revolution. Yet there can be little doubt that the contribution of the Presbyterians to the war was very great. Many of them lived in those colonies where the Church of England was the religion of the state and where proposals had been made to bring over from England a bishop who would be supported by taxation.

JOHN WITHERSPOON

One Presbyterian clergyman played such a prominent role in the struggle for independence that a review of his career will be illuminating. He was also the leading figure in the Presbyterian Church through two formative decades.

John Witherspoon was born in Scotland, fifty-three years before the signing of the Declaration of Independence. His father was a clergyman of substantial though not outstanding gifts. But the son proved to have exceptional abilities. It is said that he could read the Bible at age four, and that subsequently he was forced to memorize so much of it that at one time "he could repeat nearly all the New Testament."[3] Later in life, he once declared that after writing a sermon he would engage to repeat it word for word after reading it over three times. At thirteen he was pronounced ready for the University of Edinburgh, an early age even for that day (the philosopher, David Hume, Witherspoon's elder contemporary, entered at eleven). Three weeks after celebrating his sixteenth birthday, Witherspoon became a Master of Arts, no doubt a comment on the kind of education provided by the University as well as on the brilliance of the student's mind. He continued at the University for four more years, studying theology. At age twenty-two he was called to a substantial parish and two years later first became a member of the General Assembly, the ruling court of the Church. At age twenty-eight, he was selected to preach at the General Assembly the "annual sermon before the Lord High Commissioner."

At age forty-five, after a successful and stormy career in Scotland, Witherspoon was asked to become president of Princeton, then called the College of New Jersey. The College in that day was a tiny institution of four professors (including the president), two or three tutors, and about a hundred students. It was a Presbyterian stronghold, supplying most of the Presbyterian clergymen in all the middle colonies. Witherspoon came to America as a missionary venture and at considerable financial sacrifice. The College was a young, struggling institution, badly in need of vigorous leadership. Moreover, American Presbyterianism, rent by controversy and feeling the attacks of eighteenth-century secularism, needed guidance from the old country. Benjamin Rush, later to become a famous physician and a signer of the Declaration of Independence, enthusiastically assured Witherspoon that he could become "a bishop among the Churches."

In America he soon came into prominence; at his first appearance at the annual meeting of the Synod of New York and Philadelphia, he was made a member of no less than eight separate committees. Never a great preacher and a rather ugly man ("an intolerably homely old Scotchman"), he nevertheless, according to one observer, had more "presence" than any public character save George Washington. One Presbyterian clergyman wrote that Witherspoon was "as plain an old man as I ever saw and as free from any assumption of dignity"; yet at General Assembly he "evinced such an intuitive clearness of apprehension and correctness of judgment, that his pointed remarks commonly put an end to the discussion."[4]

He served as president of the College of New Jersey until the end of his life, twenty-six years after his arrival in America. He proved to be an able administrator; and the College prospered—until the Revolutionary War took away its students, damaged its buildings, and dissipated its funds. But his administrative powers were outweighed by his abilities as a teacher. He liked and understood boys; and he believed in them. To one graduating class he said:

It is a common saying that men do not know their own weaknesses; but it is as true, and a truth more important, that they do not know their own strength. . . . Multitudes of moderate capacity have been useful in their generation, respected by the public, and successful in life; while those of superior talents from nature, by mere slothfulness and idle habits, or self-indulgence, have lived useless and died contemptible.[5]

Here is expressed an understanding worthy of the best modern educational psychology; the powers of most students will blossom in a climate so full of encouragement and confidence. Witherspoon's students made a distinguished record in the early decades of our national life. Less than five hundred graduated from the College during his tenure. But of these, one, James Madison, became President of the United States. Another, Aaron Burr, became Vice-President. Ten became members of the President's cabinet; three, justices of the Supreme Court; twenty-one, members of the Senate; thirty-nine, members of the House of Representatives; twelve, governors of states. But it was in the fields of religion and education that his students made their most notable contributions. Twenty-three per cent of his students became clergymen, and of these 11 per cent became presidents of colleges, including such institutions as Princeton, University of North Carolina, University of Nashville, University of Georgia, and Oglethorpe. In addition, his students founded academies and elementary schools in community after community. In that day, as in this, Presbyterian interest in education was high; the Presbyterian minister on the frontier was half schoolteacher.

After Witherspoon had been on this side of the Atlantic scarcely half a decade, he began to take a prominent part in the struggle against England. Prior to this time he had held firmly to the conviction that clergymen should not meddle in politics. He feared state control of religion and defended the theory that the Church can best serve society by being above social conflict. But he found eventually, as will those who hold a similar thesis today, that the theory is good only when it concerns relatively unimportant matters. When men are struggling for survival, or for liberty, or for justice, the Church will either be a part of the struggle or the Church will be abandoned.

Witherspoon arrived from Scotland a loyal subject of the Crown; but soon his democratic nature was greatly impressed by the relative absence of class distinction in America, by the assured and straight-forward demeanor of common men, by the extent to which prosperity was shared by all groups, by the absence of beggars, by the possibility of traveling from town to town without armed escort, and by the climate which, as one Scotsman put it, is "no such black, foul weather as at home." Like many a later immigrant, Witherspoon soon gave to America his unstinted loyalty. When England made clear that she intended to rule the colonies in the interests of British economic imperialism, there was no question where his sympathies lay.

On July 4, 1774, he was made one of nine members of the leading revolutionary group in Somerset County, New Jersey, the Committee of Correspondence. Two weeks later he represented his County at a New Jersey congress, met to consider the developing situation. Later that same summer, he published his first political writing, a vigorous pamphlet entitled *Thoughts on American Liberty*. In August, John Adams went through Princeton on his way to Philadelphia to attend the first session of the Continental Congress; there he drank a glass of wine with the president of the College of New Jersey and pronounced him "as high a son of liberty as any man in America."

The news of the Battle of Lexington and Concord, eight months later, galvanized the Presbyterian Synod of New York and Philadelphia into political action; Witherspoon was made chairman of a committee to draw up a pastoral letter in which the Synod urged clergymen to avoid politics in the pulpit no longer, and to remember that the most courageous soldier is the pious man who has been filled with righteous anger.

In June, 1776, Witherspoon and four others were elected to represent New Jersey in the Continental Congress and were empowered, ". . . if you shall judge it necessary or expedient," to join with the delegates of other colonies ". . . in declaring the United

Colonies independent of Great Britain." The New Jersey representatives arrived at the Congress on June 28; six days later they joined in signing the Declaration of Independence.

Then began for Witherspoon a period of strenuous activity and great harassment. Service in the Continental Congress proved very burdensome not only because of the magnitude of the problems with which the Congress had to deal, but also because of the personal difficulties in the midst of which delegates had to work. The states refused to provide properly for the support of their representatives; in addition, this was a time of great inflation. At one point conditions became so bad that "it began to look as if some morning soon members of Congress would not be able to pay for their breakfasts."[6] Attendance at meetings was very irregular, delegates often yielding to the temptation to mount their horses and ride home. Moreover, Congress had upon occasion to flee before the British and even before the wrath of its own unpaid troops.

In the midst of this situation, Witherspoon proved to be one of the most valuable members of the Continental Congress and one of its most consistent attenders. He served for four straight years, and then, after a year's interval, served another year—and all at great personal sacrifice, since he had of necessity to neglect his personal affairs. Once when Robert Morris needed a sum of money for the use of the Congress, Witherspoon signed a bond for the amount; after his death his estate had to make the bond good.[7] He served in all on more than one hundred and twenty committees and was a member of two standing committees of supreme importance—the Board of War and the Committee on Foreign Affairs. So prominent was his role that a rumor—a false one—circulated widely to the effect that he was a member of a junto which was supposed to control congressional action. His biographer sums up his contribution in the Congress as follows:

He . . . took an active part in the . . . debates on the Articles of Confederation; shared in the formation of the new government's foreign alliances; witnessed that government floundering in a bankruptcy of which he had given it plain warning; assisted in organizing the executive departments that superseded the earlier committee plan; was a leader in the discussion of the perplexing problem of western lands; . . . when peace with Great Britain was impending he was conspicuously prominent in directing the preliminaries in selecting the American commissioners, and actually dictated himself their most important instructions. In addition to these major concerns he was occupied with a multitude of lesser activities which may be classed under the head of humanitarian endeavors —such as the kindlier treatment of prisoners, the checking of cruelty in warfare, the better administration of military hospitals, the improvement of health and morals and therefore of discipline, in the army.[8]

When the surrender of Cornwallis at Yorktown ended the Revolution, Congress promulgated a Thanksgiving Proclamation which it had asked Witherspoon to write.

In spite of his consuming revolutionary activities, Witherspoon continued his educational and religious leadership. He struggled to keep the College from disintegrating; all through the war, he managed never to miss commencement or a trustees' meeting and he usually was on hand at the beginning of the school year. In 1779, at his suggestion, his salary was cut in half and the amount saved was applied to the salary of one of the professors; four years later this arrangement was made permanent. For several years, he permitted himself to work under a rule which provided that he was personally responsible for student arrears in tuition. It was necessary for him to put forth heroic efforts to repair the campus buildings after their habitation by both British and American troops.

He also continued throughout the war his leadership of American Presbyterianism. Often he was found in the pulpit, preaching vigorously, as his duties took him about the country. In 1771 he had been elected treasurer of the Synod of New York and Philadelphia, an office he filled for a decade and a half. When he was free at the close of hostilities to devote himself again to the Church, he was placed by his clerical colleagues on many important committees. The most significant of these assignments was the chairmanship of a committee which framed a new constitution for the Church. Thus he had a prominent part in shaping the present structure of American Presbyterianism. At the first meeting of the General Assembly of the newly organized Presbyterian Church in the United States of America, Witherspoon preached the opening sermon and presided until a moderator was elected.

Two years later at the meeting of the Assembly, Witherspoon is reported to have said to a delegate, "You can scarcely imagine the pleasure it has given me in taking a survey of this Assembly to be-lieve that a decided majority of all the ministerial members have not only been sons of our college, but my own pupils." By actual count, of the thirty-six ministers present, twenty-eight were graduates of the College of New Jersey and sixteen Witherspoon's own students.[9] What better evidence could be given of the worth of the religious train-ing he had given at Princeton?

THE DEVELOPMENT OF THE CHURCH

The activities of the Presbyterians during the Revolution—only one of their ministers was Tory and he was expelled early in the war —placed the Church in a comparatively strong position at the close of hostilities. The conflict had made the religious climate rigorous for all church groups; but public opinion reacted so favorably to the indomitable temper of Presbyterianism that it was placed in a strong position from which to attack the problems of the expanding West.

In the decades that have intervened between that time and our own, Presbyterian churches have had a steady growth; they have today a total membership of four and one-third million and a total constituency much larger. Except for New England where their cousins the Congregationalists are exceptionally strong, the Presby-terians are well-represented throughout the nation. They are especially strong in the Middle Atlantic area.

They are divided among ten bodies. Most of them are small. Two, however, are large institutions. The United Presbyterian Church in the U.S.A., often called Northern Presbyterian, has three and one-quarter million members; it was formed in 1958 by the joining of the large Presbyterian Church in the U.S.A. and the much smaller United Presbyterian Church. The Presbyterian Church in the U.S., often called Southern Presbyterian, has nine hundred thousand members. The Northern Presbyterians and the Southern Presbyterians formed one Church prior to the Civil War.

PRESBYTERIAN THEOLOGY

Although a large percentage of Presbyterian laymen would be sur-prised to hear it, the seventeenth century Westminster Confession of Faith is still the nominal standard of Presbyterian doctrine, in both the

North and the South. This three-centuries-old document contains the familiar Christian teachings—such dogmas as are found in the Apostles' Creed. Present-day Presbyterianism holds to the sovereignty of God, the Trinity, the Incarnation, the Virgin Birth, the Resurrection, the Grace of God, the Revelation of God in the Bible, Original Sin, Eternal Life. Presbyterianism is conservative in its theology and evangelical in its mood.

But in addition to these traditional teachings, the Westminster Confession contains other dogmas, to which few contemporary Christians adhere; among the most noteworthy of these is predestination. The following are examples of the statements on predestination which are still printed in the Constitution of the Northern Church:

> By the decree of God, for the manifestation of his glory, some men and angels are predestinated unto everlasting life, and others foreordained to everlasting death.

> These angels and men, thus predestinated and foreordained, are particularly and unchangeably designed; and their number is so certain and definite that it cannot be either increased or diminished. . . .

> The rest of mankind, God was pleased . . . to pass by, and to ordain them to dishonor and wrath for their sin, to the praise of his glorious justice. . . .

> We are utterly indisposed, disabled, and made opposite to all good, and wholly inclined to all evil. . . .

> Elect infants, dying in infancy, are regenerated and saved by Christ through the Spirit, who worketh when, and where, and how he pleaseth. So also are all other elect persons, who are incapable of being outwardly called by the ministry of the Word.

> Others, not elected, although they may be called by the ministry of the Word, and may have some common operations of the Spirit, yet they never truly come to Christ, and therefore cannot be saved: much less can men, not professing the Christian religion, be saved in any other way whatsoever than by Christ, be they never so diligent to frame their lives according to the light of nature, and the law of that religion they do profess. . . .

> There is no sin so small but it deserves damnation.

Does the average Presbyterian believe all these dogmas? Emphatically No! In the classic era of Presbyterianism, they were believed. At one time all Presbyterian (and most Anglican, Congregational and Baptist) ministers preached that all men deserve damnation, but

God chooses some men for heaven and others for hell. As late as two or three generations ago, some Presbyterian clergymen were still preaching that some infants, though they could not possibly have sinned before their death, were to spend eternity in the tortures of hell. Some morbid preachers, driving the point home in their sermons, cried, "There are babies in hell not a span long." Nor was the blow softened by the insistence that infant damnation was but half the story, the other half being the salvation of other infants.

Such doctrines have been vigorously denied by Presbyterians for decades. One clergyman was quoted as saying in 1884 concerning the extreme statements of the Westminster Confession, "Need I assure you, that we reject every one of these revolting ideas, with as much sincerity as any of those who charge us with them."[10] The Constitution of the Presbyterian Church, U.S.A. verified his position. In 1903, a Declaratory Statement was adopted by this Church; the Statement continues as part of the Constitution of the United Presbyterian Church. A portion of this Statement follows:

Concerning those who perish, the doctrine of God's eternal decree is held in harmony with the doctrine that God desires not the death of any sinner, but has provided in Christ a salvation sufficient for all, adapted to all, and freely offered in the Gospel to all; . . . The Confession of Faith . . . is not to be regarded as teaching that any who die in infancy are lost.

This Statement, of course, directly contradicts parts of the Westminster Confession of Faith, which still *begins* the Constitution. The contradiction is not explained or resolved. Subscription to the Confession is not required of persons who seek ordination. Rather they are asked, "Do you sincerely receive and adopt the Confession of Faith of this Church, as containing the system of doctrine taught in the Holy Scriptures?" An affirmative answer to this question, as some Presbyterians are at pains to point out, is not equivalent to asserting personal faith in the Confession. Also, laymen who seek to join the Church are asked merely to make a public profession of their faith. One who is not a Presbyterian wonders why a document which so definitely misrepresents the actual beliefs of most contemporary Presbyterians, is permitted to occupy so prominent a place in Presbyterianism. One Presbyterian clergyman wrote, "Presbyterians have a built-in invitation to dishonesty as long as the Westminster Confession of Faith and the Catechisms are subscribed to as the authoritative statements of Christian Doctrine. These documents no longer say what they mean and when we repeat them, we don't mean what the say."[11] Other denominations also retain creedal statements which are not representative of current beliefs, as will be noted later.

Among the Northern Presbyterians, theological self-consciousness had led to a number of embarrassing church fights. One was the heresy trial of Professor Charles A. Briggs. This scholar was expelled in 1893 from the Presbyterian ministry for teaching propositions which the overwhelming majority of Presbyterian clergymen in the North today would accept.

In 1924, Harry Emerson Fosdick, a Baptist employed as preacher by a Presbyterian church, was "invited" by the Presbytery of New York to become a Presbyterian minister; thus the Presbytery made clear that Fosdick could not continue in one of its pulpits unless he conformed doctrinally. He chose to resign.

In 1936, about one hundred ultraconservative ministers, headed by J. Gresham Machen, were expelled because they established an Independent Board of Presbyterian Foreign Missions; they felt that the Presbyterian Board of Foreign Missions followed policies which were not in accord with a strict interpretation of the Confession of Faith.

In 1959, a group of Fundamentalists tried to persuade the General Assembly not to confirm Dr. Theodore A. Gill as president of San Francisco Theological Seminary because he had written concerning the virgin birth, "What of us who make the Virgin Birth no part of our personal confession . . . ?" and had added that this miracle is "something to worry about." Some members of the Assembly demanded that Dr. Gill be asked simply to say, "I believe in the virgin birth."[12] The Assembly, however, voted overwhelmingly to confirm him without requiring any such declaration. One estimate put the fundamentalist vote at about twenty out of nearly one thousand Assembly members.[13]

These controversies were a sad experience for all concerned. Today there is apparent among most Presbyterians a desire to permit real latitude of theological expression—but at the same time to hold firmly to the Christian tradition. They hope for doctrinal unity without uniformity. Most Presbyterians appear determined to drive down the middle of the theological road—veering neither toward Fundamentalism nor toward Liberalism.

REPRESENTATIVE GOVERNMENT

Presbyterian polity can be illustrated by outlining the government of the United Presbyterian Church, U.S.A.

A local church is governed by the *Session,* a body composed of the minister and a small group of ruling elders. A ruling elder is a layman who has been elected by the members of the church, and then ordained to eldership. In the old days elders held power for life. At present their term is limited to six years. The Session has complete oversight of the spiritual life of the Church. It is the body which supervises worship, controls church buildings, plans educational programs, admits to membership; the Session can even excommunicate. But it does not have the power to choose the minister; that power is exercised by the congregation.

The Sessions of a district—usually ten to thirty—are organized into *Presbyteries.* The Presbytery consists of all ministers in the district and of as many elders from each church as the church has pastors. The powers of this body are extensive: the Presbytery must veto or approve a local church's selection of minister, must arrange for his formal installation, must approve his resignation, must give consent to his dismissal, and may require his dismissal. The Presbytery examines and ordains candidates for the ministry, has oversight of vacant churches, founds churches, unites them, divides them. These powers are, of course, used with discretion; Presbyterians will not be pushed around any more than will the rest of Americans. Yet a Presbytery can show its teeth on occasion.

The authority of the Presbytery restricts the powers of the local churches and increases the security of the pastor. Thus Presbyterian clergymen are in a much less vulnerable position than are the clergymen of congregationally controlled groups, but are in a more vulnerable position than are Episcopal clergymen, whose dismissal cannot be arbitrary, and whose salaries cannot suddenly be lowered by disaffected vestries or administrators.

The Presbyteries (not less than three) are organized into *Synods.* The Presbyteries are also organized into the *General Assembly.* This body is the highest authority in the Church and is vested with legislative, executive and judicial powers.

In all these governing bodies, every member stands on an equal footing. Elders have the same voting power as the clergy and one clergyman's vote is as good as another's. There is no House of Bishops in Presbyterianism. Yet each minister is officially a "bishop"; he joins with his fellow bishops in exercising the episcopal function whenever his Presbytery acts in a supervisory capacity over a group of churches. Another title held by the minister is "teaching elder."

Women may be elected as ruling elders and as representatives to the higher courts. But they were not permitted to be ordained as ministers until recently. A vigorous effort in the late forties to secure the passage of legislation permitting the ordination of women went down to ignominious defeat. The mood of the debate was frequently on a level beneath the seriousness of the problem. One clergyman asserted that with women in the pulpit a new item would have to be introduced into the worship service: an organ interlude so the pastor could powder her nose. Another declared ordination "would increase the alarming tendency to throw the whole burden and responsibility of church work upon women." A pastor in New York City said:

> [The proposal to ordain women] is absolutely contrary to the Bible and to common sense. . . . Women are not temperamentally fitted to be ministers. . . . Women are not especially good at keeping other folks' secrets and that is one of the things a minister must do. Women are apt to be influenced by their feelings in matters of belief rather than by sound judgment. . . . Women are usually too kind and sympathetic with other women.[14]

In 1956, however, the Church declared officially:

> The Bible does not prescribe a permanent and specific social structure for the Church or society; it is proper to speak of equality of status for men and women both in terms of their creation and redemption; and there is no theological ground for denying ordination to women simply because they are women.

For a long period Presbyterian efforts at church union had disappointing results. In 1904, three-fifths of the Cumberland Presbyterian Church (a body which had broken away a century earlier) went back to the Northern Presbyterians; but the remaining two-fifths refused to reunite and insisted on maintaining the Cumberland name and organization. A recurrent flirtation between the Northern and Southern churches has shown the Southern Presbyterians to be filled with a spirit of denominational complacency, even of self-righteousness. In 1957 the vote by Southerners not to merge with Northerners was influenced by some men who objected to being "unequally yoked together with unbelievers." The unfortunate experiences of the Northern Church in dealings with the Episcopalians have already been recounted. Recent experiences have been more encouraging. In 1958, the Presbyterian Church in the U.S.A. and the United Presbyterian Church joined to form the United Presbyterian Church in the U.S.A. And in 1960, the leading official of this denomination preached a sermon which precipitated a nationwide discussion of union by the Methodist Church, the Protestant Episcopal Church, the United Presbyterian Church, and the United Church of Christ.

As has been noted, one of the things Presbyterians are famous for is their high educational standards for the clergy. The Presbyteries examine the academic qualifications of candidates for ordination with great care. The standing of Presbyterian theological seminaries, if not the highest of any denomination, is certainly as high as any. The Cumberland secession sprang in part from opposition to ordaining none but educated men. This insistence was a severe handicap to the growth of the Church in the pioneer period, since not enough men could qualify to meet the demands of the growing frontier. Baptists and Methodists, whose educational standards were much lower, grew much faster. A Southern Presbyterian comments as follows on the contemporary educational emphasis:

We think we have a very scholarly ministry. One can scarcely lift a leaf in this part of the world without uncovering a doctor of divinity. Almost exactly half of the men who died in our ministerial ranks last year were D.D.s. That means that everybody thinks of himself as an expert in the mysteries of God with the result that each is inclined to look on everybody else as wrong if he does not agree. "I've been to school and I know," might be written over the study door of every Presbyterian pastor.[15]

MIDDLE-OF-THE-ROAD RELIGION

A traveler trying to find a characteristic expression of American religious life could do no better than to investigate Presbyterianism.

It is at the religious center. Theologically, Presbyterians are moderate supernaturalists. Governmentally, they are conservative democrats. Liturgically, they are tamed evangelicals. Politically, their clergymen fall slightly on the conservative side of center.[16]

Such facts do not mean that Presbyterians are merely lukewarm about the Church; in many ways the Presbyterian record is outstanding. In missionary giving, in concern for religious education, in religious journalism, in support of colleges, in church extension, Presbyterians are at or near the top among Protestant denominations. Yet one Presbyterian near the close of World War II could write about his Church:

Nothing—no great cause nor any master irritant—is deeply disturbing the Presbyterian Church at the present time. . . . This is, of course, exactly what disturbs those who love it—that in a day of such indescribable catastrophe our church should be relatively undisturbed, characterized by the same old comfortable spiritual lethargy, continuing to sit on well cushioned pews and listen to sermons preached primarily to please the parish palate, . . . and never dreaming of using its full power to implement its principles. . . .

One of our greatest troubles is that, as Spinoza observed, "We confine ourselves to the possible, and so we have no problem." . . .

Stanley Jones was disturbingly close to the truth when he remarked that the modern church is more of a field for evangelism than a force for it.[17]

Fifteen years later another Presbyterian could write that

at our best we have tried to transform the world more into accordance with the will of God, by the Spirit and in the might of God. In recent generations it has taken quite a spasm of historical imagination to remember this about our placid, settled selves. But it is true, nonetheless: we Presbyterians were a revolutionary party, we were a radical bunch when our bright reputation was burned into the history books. Presbyterians in their shining hour knew that the Church is not here for its own sake, that the Church is here to help with all the wrenching and razing that is necessary if the world is ever to be reoriented to the will of God who made it and loved it well enough to live in it and die for it—to save it. We Presbyterians came boiling up out of Geneva, spreading out over the world, a rambunctious, disputatious, bookish outfit, resisting the massive status quo wherever we went, suffering martyrdom, leading revolutions, raising rebels, crossing oceans, founding schools and hospitals and libraries, inventing new forms of government, even monkeying with the inherited economic system. . . .

That day is largely over now. Running as hard as we can, we cannot be sure that we will even catch up with the world, much less take the lead again. Nowadays world leaders lead, events happen, and the panting Church far back in the ruck gasps its consternation and feebly pipes its well-meant counsel, knowing all the while in its despairing bones that leaders will continue to lead by less avoidable lights than ours, and that events will continue to happen by another quite oblivious logic. How could we expect more when so few for whom the church seeks to speak understand any more what she says or why she says it, and when so few to whom the Church speaks have any idea of her reference or her meaning? . . .

What ails us as a Church? It seems to me that we Presbyterians are relying too much on our vast machinery to move us. And the farther out we push our glittering rim, the hollower we seem to go at the heart.[18]

These judgments apply with equal cogency to most of our religious organizations. Can the spiritual problems of our time be met by middle-of-the-road policies sponsored without passion? Revolutionary occasions demand revolutionary measures. No lesson of history is plainer than the fact that in time of crisis the voice of moderation is drowned by the shouts of a chorus of extremists. This principle is as true of religion as it is of politics or economics.

CHAPTER VII

THE UNITED CHURCH OF CHRIST

THE BOLDEST and in many ways the most interesting effort to practice the ideals of the ecumenical movement is the formation of The United Church of Christ. Prior to its emergence, only bodies with similar types of church government had united: episcopally governed bodies had united, and congregationally governed bodies had united. But in the United Church of Christ a presbyterially governed body, the Evangelical and Reformed Church, and a congregationally governed body, the Congregational Christian Churches, formed a union into which went aspects of both systems of government.

An understanding of this newly formed denomination can be had only on the basis of a description of the two bodies which joined together.

THE CONGREGATIONALISTS

"To belong to a Baptist church," wrote one authority on Congregational polity, "one must be a Baptist. . . . To belong to an Episcopal church one must be an Episcopalian." But "a Congregational church is not a church of Congregationalists, but a church of Christians in which the congregation governs. It has absolutely no sectarian tests."[1] Another Congregational writer contended that Congregationalism is the pristine type of Christianity; "the Churches of the apostolic age were *Congregational* Churches."[2]

The claims of these writers are greater than would be made by most students of Congregationalism. Nevertheless, these claims point the direction of thinking in this group. Like the members of the other

216

Christian bodies we are studying, Congregationalists are firmly convinced that they are the true inheritors of the spirit and message of early Christianity. The modern phase of Congregationalism—out of which grew Unitarianism—began with the conflicts in Elizabethan England. But Congregationalists would no more date the beginning of their movement in the sixteenth century than Roman Catholics would date the beginning of theirs with the Council of Trent or Episcopalians the beginning of theirs with Henry VIII.

The Elizabethans who were given the name *Congregationalists* found in the New Testament a clear command to seek perfection, to which end they felt compelled to hurry on the Reformation "without tarrying for any." They believed in a "gathered" church, a church composed only of True Christians—as they called themselves. They lived in a time when it was generally assumed that a person became a Christian simply by being born into a Christian society and receiving the rite of baptism. But the True Christians contended that Christ's Church consists only of persons who give clear evidence of their Christian character and thus are "proved saints." These Congregationalists also contended that the local church is not subject to the outside control of either bishop or presbytery but is properly governed in ecclesiastical matters by the members who are united together by a covenant.

The Congregationalists differed among themselves over their views of the Church of England. Some of them were Separatists; they believed the Church of England was not genuinely Christian and that they must withdraw, must separate, from it. Others believed that the Church of England was the true Church because, they affirmed, there had always been in it some genuine Christians who had practiced Congregational principles. Both of these groups played a prominent part in the establishment of Congregationalism in New England as will be indicated later. The story of the Separatist Congregationalists will be highlighted here because they founded the first successful colony in New England and because they espoused ideas somewhat more like ideas which came eventually to dominate the New England churches.

The Separatists met in little private groups for instruction and worship after what they considered to be the manner of the New

Testament. Of course, such practices were in direct opposition to the clearly stated policy of the government. It considered conformity to the Church of England essential to the nation's safety, and legislated all manner of penalties against nonconformity. An ecclesiastical Commission, a kind of Inquisition, was established with the purpose of enforcing uniformity: the Commission was commanded to prevent private prayers if more than the immediate family were present, to forbid preaching by all persons except men ordained by the Church of England, to require the use of *The Book of Common Prayer* in the conduct of services of worship. In 1592, the death penalty was prescribed "for the Punishment of Persons obstinately refusing to come to church."

But the True Christians had seen a vision and were not to be dissuaded from following it by persecution. They gathered secretly in homes, on board ships, out in the forests, in abandoned gravel pits. Many were thrown into prison and some were hanged. But the meetings continued. Men are not easily deterred from a course of action on which they believe rests their eternal salvation.   ·

Separatism had a foothold in London, where the very size of the city gave protection. But for our story, the chief center of the movement was one hundred and fifty miles north of London on the great road that ran to Edinburgh. There, near Sherwood Forest, were two villages, Gainsborough and Scrooby, in which lived a group of people whose fortunes were destined to become well known in America. To this rural section, in the reign of James I, came men who had been trained in the Puritan atmosphere of Cambridge University. They had moved on in their thinking to Separatist convictions, and before long had a considerable following among the countryfolk.

The King resolved to make these sectaries conform or to "harry them out of the land." Soon their position became intolerable and they decided to flee. William Bradford, later governor of Plymouth Colony in America, wrote concerning their situation:

Having kept their meeting for the worship of God every Sabbath in one place or another, notwithstanding the diligence and malice of their adversaries, seeing that they could no longer continue under such circumstances, they resolved to get over to Holland as soon as they could— which was in the years 1607 and 1608.[3]

Thus began the story of the Pilgrim Fathers. The record of their experiences will furnish another illustration of the extremes to which people are willing to go when they are in the grip of deep religious conviction.

THE PILGRIMS

The Scrooby Separatists moved to Holland because they knew that there they would be granted religious liberty. They were not the first group of Separatists to migrate to that country; groups from London and Gainsborough had already gone over. The Dutch were unique in that era—and in most others—in the extent to which they had learned from their own sufferings the lesson of tolerance.

In preparing for the trip, the Scrooby group had to work by stealth; for, leaving England without permission was as unlawful as was remaining at home without conforming to the Church of England. The crossing was completed only at the expense of many trials, dangers and conflicts with the law.

Settling in Leyden, the Pilgrims found it difficult to make a living; so difficult that some "preferred to submit to bondage, with danger to their conscience, rather than endure these privations. Some even preferred prisons in England to this liberty in Holland, with such hardships."[4] Moreover, they saw their children adopting the customs and speech of a strange land. And in addition, King James, by bringing pressure on the Dutch government, was finding ways to continue his harassment. And there was a strong probability of the renewal of war between Holland and Spain. Thus, these simple folk began again to ponder the wisdom of moving their homes, and their Church, to a foreign land, this time to America, a place as far from the reach of the King's molestation as earth afforded them.

Even the contemplation of an attempt to colonize in America seemed foolhardy to many. Their resources were so meager that one authority on American colonization could write, "No enterprise in overseas settlement thus far undertaken can compare with this desperate project of the Leyden Separatists."[5] But in spite of the hazards, a portion of the Church determined to make the attempt. After months of negotiation, planning and anxious conference, they finally made an agreement with the Plymouth branch of the Virginia Company to settle in the lands under its control. The prospects were so unfavorable that only about a sixth of the members of the Church decided to make the first voyage.

John Robinson, their pastor, remained behind with the majority of the Church. Before the departure he preached a sermon in which he stated principles still revered by Congregationalists:

> I charge you that you follow me no farther than you have seen me follow the Lord Jesus Christ. . . . I am verily persuaded, the Lord has more Truth yet to break forth out of His Holy Word. . . . I beseech you, remember, 'tis an Article of your Church Covenant, that you be ready to receive whatever Truth shall be made known to you from the written Word of God. . . . It is not possible . . . that perfection of Knowledge should break forth at once.

Sailing to England they were joined by about eighty more passengers, only one or two of whom were avowed Separatists; the rest had been hired by the Plymouth Company to aid in the work of colonization. The colonists set out in two ships; unfortunately, three hundred miles beyond Lands End they had to turn around and make for port, one ship proving "leaky as a sieve." They abandoned her and transferred passengers and goods to the *Mayflower*—except that a score of the passengers, including one of the Separatist families, abandoned the project. "Thus, like Gideon's army," wrote Bradford, "this small number was divided, as if the Lord thought these few too many for the great work He had to do."[6]

On board was a company of just over one hundred passengers, thirty-five from the Leyden Church, crowded on a tiny ship of one hundred eighty tons. In midocean, a storm was encountered during which one of the main beams amidships bent and cracked alarmingly. Some persons counseled turning back. But temporary repairs were made, and the ship sailed on. During another storm, one of the passengers, venturing out on deck, was washed out into the sea; but providentially he caught hold of a rope that was trailing in the water and was dragged back on board. Finally, at the beginning of the winter of 1620, the Pilgrims sighted Cape Cod.

Before they landed, they drew up and signed the famous Mayflower Compact: "solemnly and mutually in the presence of God and of one another [we] covenant and combine ourselves together into a civil body politic." This type of covenant was familiar to the Separatists; for, the covenant idea inheres in the very nature of the gathered church. The Mayflower Compact extended the covenant idea to civil government and was a long step down the road toward democracy. It was signed by the adult males without distinction of class or religious conviction.

The rigors of the succeeding months cannot even be imagined by persons who have never at any time of year spent a single night out-of-doors with no covering and in wet clothing. That needs to be experienced to be understood. The Pilgrims were forced to anchor a mile and a half out in the harbor, and could come ashore only at high tide. On December 25, the first building was begun, and by the middle of January most of the group were living ashore. But in the following weeks they faced the most terrible ordeal of all their experiences. A sickness which they called "galloping consumption" came upon them. Six died in December, eight in January, seventeen in February, and thirteen in March.

In the time of worst distress, there were but six or seven sound persons, who, to their great commendation be it spoken, spared no pains night or day, but with great toil and at the risk of their own health, fetched wood, made fires, prepared food for the sick, made their beds, washed their infected clothes, dressed and undressed them; in a word did all the homely and necessary services for them which dainty and queasy stomachs cannot endure to hear mentioned; and all this they did willingly and cheerfully, without the least grudging, showing their love to the friends and brethren.[7]

Fear of the Indians and wild animals caused much anxiety. Most of the colonists were so innocent of the true nature of life on the frontier that they had not mastered the use of firearms and received instruction in this essential skill *after* the plantation was reached. Fortunately, before many months the colonists made a treaty of peace with the nearest Indian chieftain, a treaty which remained in force for many years and which furnished a continuing source of security and satisfaction.

During the second and third years the colonists faced the constant threat of starvation. According to their agreement with the Plymouth Company, they were to receive supplies from England; but the supplies never came. In the spring after the arrival, the twenty-one men and six boys who had not succumbed to the epidemic began to prepare the land for planting. Twenty-seven acres were cleared and sown to corn and small grains. The latter crop failed but the corn was a success. By fall, prospects were brighter.

Unexpectedly, in November, a ship arrived bringing thirty-five new colonists, most of them healthy young men. "But they brought not so much as a biscuit-cake, or any other victuals with them." The colony had then to take "an exact account of all their provisions in store, and proportioned the same to the number of persons, and found that it would not hold out above six months at half allowance, and hardly that."[8] In the months ahead, except when the fish were running, hunger was always near.

That spring more land was cleared and planted. But in June, the representative of the Plymouth Company sent seven men whom he asked the colonists to accept as guests until he could establish a settlement for them up the coast. A little later, sixty more arrived for a stay of six weeks. This group brought their own food, but they were quartered on shore. Many of them were of doubtful character and, learning that green corn roasted in the ear furnished a welcome relief from the monotony of their fare, they began foraging in the Pilgrims' fields. Such as were caught were whipped soundly; nevertheless they did much damage. These depredations, along with insufficient cultivation due to the Pilgrims' own weakness, seriously reduced the autumn yield and food again was short. In the months before the next harvest (the third), there was a considerable period when the colony was without bread or any kind of corn; it had to depend on nuts, clams, and what fish could be caught with inadequate tackle. And in June, there set in a drought of seven weeks' duration.

During this trying period, the Pilgrims' reliance on religion never wavered; they continued to put their trust in the providence of God. We are told that Elder Brewster, sitting down to a meal of boiled clams and a pot of water, thanked God that he was permitted to "suck of the abundance of the seas, and of the treasures hid in the sand." As the drought wore on the Pilgrims began to examine their own lives to discover secret sins which might be the cause of God's disfavor. Finally, they decided to meet publicly for prayer and confession. For the space of eight hours they humbled themselves before God; they recalled His promises, acknowledged their sins, exhorted one another, and prayed for the end of the drought. Even as they left the place of meeting, the clouds were already gathering; the rain began to fall and it continued until "it was hard to say whether our withered corn or drooping affections were most quickened or revived."[9]

The subsequent crop proved to be a good one. "God gave them plenty, and the face of things changed, to the rejoicing of the hearts of many for which they blessed God." The desperate part of their struggle was over.

Twenty years later, Bradford, reminiscing on their trials, wrote as follows concerning the source of their strength and the reasons they had been put to the test:

What was it then that upheld them? It was God's visitation that preserved their spirits. . . . God, it seems, would have all men behold and observe such mercies and works of His providence as towards His people, that they in like cases might be encouraged to depend upon God in their trials, and also bless His name when they see His goodness towards others. Man lives not by bread alone. It is not by good and dainty fare, by peace and rest and heart's ease, in enjoying the contentment and good things of this world only, that health is preserved and life prolonged. God in such examples would have the world see and behold that He can do it without them; and if the world will shut its eyes and take no notice of it, yet He would have his people see and consider it.[10]

NEW ENGLAND RELIGION

The territory to the north of Plymouth, the Massachusetts Bay Colony, was settled by Congregationalists who did not wish to withdraw from the Church of England; they agreed with Plymouth on many important ecclesiastical questions, but did not approve of separation. Massachusetts was more advantageously situated commercially than was Plymouth—it was better located, was backed by adequate resources, and was under the supervision of influential leaders. It prospered from the beginning and soon outshone Plymouth and wielded more influence. Both Massachusetts and Plymouth favored congregational principles: they believed in the gathered church, practiced the covenant relationship, insisted on the right of the local congregation to choose its own pastor, opposed the use of the Prayerbook. Their polity became normative for the early New England churches.

The movement toward ecclesiastical democracy undoubtedly had a marked influence on the development of political democracy. Most of the early colonial leaders did not believe in popular political rule. Although the governor at Plymouth was elected annually by democratic vote, once elected his rule was arbitrary. John Winthrop, a leader of the Massachusetts Colony, declared that the few are the wise "and the fewer the wiser." Seats in church and in schools were assigned on the basis of social status. In 1631 Massachusetts passed a law limiting the privilege of voting to church members and for decades church membership was a jealously guarded privilege, a privilege tied closely to the attempt to restrict the ownership of land. "Up to 1643, out of fifteen thousand inhabitants [of Massachusetts] only seventeen hundred and eight had been permitted to become freemen."[11] Two-thirds of the adult males of Boston were disfranchised even at the time of the signing of the Declaration of Independence.

Yet a drift toward political democracy set in soon after the beginning of colonization. This movement—still in process—was no doubt greatly influenced by secular factors, chiefly the freedom of economic opportunity afforded by the developing country; but of large importance also were the widespread convictions that local control of church affairs is prescribed by the New Testament and the habits engrained by long practice in church meetings.

The Congregational churches in colonial New England were "established," that is, they were supported by taxation. This system continued even after the adoption of the United States Constitution and its First Amendment, which provides that "Congress shall make no law respecting an establishment of religion." Establishment was not abandoned in Connecticut until 1818, in New Hampshire until 1819, and in Massachusetts until 1833.

New England clergymen played an important role in achieving American Independence. Before the era of newspapers, sermons had a far more general significance than they have today. Often they were the means of broadcasting news, and clergymen did not shun editorial comment. They were not afraid of politics as are their successors today. Some careful scholars contend that in New England the clergy were among the chief agitators for the American Revolution.[12]

In the first quarter of the nineteenth century, the Congregationalists split into two groups, the Trinitarians and the Unitarians. The Unitarian movement will be considered in a later chapter.

CONGREGATIONAL DEVELOPMENT

At the beginning of the nineteenth century, the Congregationalists were the largest and most influential religious group in America. Today they are far from that. When they joined in creating the United Church, they had but one million four hundred thousand members, being outnumbered by the Roman Catholics, Baptists, Methodists, Presbyterians, Lutherans, Jews, Episcopalians and Christians (Disciples). How can we account for the failure of Congregationalism to increase as rapidly as other denominations?

One factor was the loss of the Unitarians. Another factor was the very sense of power and leadership which Congregationalists had. A feeling of security, if it leads to complacency, is a disadvantage at any time, but especially so during a time of expansion such as the nineteenth century in America. Another factor was the conviction on the part of many of the denomination's leaders that the Congregational type of government was ill-suited to the frontier; they felt that the more closely knit Presbyterian government would furnish infant churches more careful supervision and make for a more effective religious life in the pioneer communities. A Plan of Union was adopted between Congregationalists and Presbyterians whereby the two groups agreed not to compete in the western communities, both communions supplying men and money. The Plan "worked in almost every instance to the advantage of the Presbyterians."[13] Congregational historians estimate that the Plan lost the denomination over two thousand churches. These statistics are disputed, but there can be little doubt that through the arrangement with the Presbyterians "Congregationalism lost the momentum of an always-westerning frontier."[14] The denomination finally abandoned the Plan of Union and undertook sectarian home missions; but the end result was poor Congregational representation in many states. Two-thirds of the Congregationalists are to be found in New England and five other states: New York, Ohio, Illinois, Michigan and California. The representation in most of the other states of the North and West is thin and is especially small in the South.

Throughout most of their history, Congregationalists have had a deep concern for education. This fact can be seen by simply listing some of the colleges whose beginnings were determined or greatly influenced by Congregationalists: Harvard, Yale, Dartmouth, Williams, Amherst, Mount Holyoke, Wellesley, Smith, Oberlin, Grinnell, Carleton, Pomona. All these institutions, in accordance with the Congregational theory, were placed under the charge of independent boards of trustees. As a result the denomination has control over none of them today.

In 1931, the Congregationalists joined with The Christian Church to form The Congregational Christian Churches. The Christian Church had about one hundred thousand members and was itself the result of a merger of some Baptists in Vermont, some Methodists in North Carolina and Virginia, and some Presbyterians in Kentucky. This body was one of the denominations influenced by the Campbells, whose work will be discussed later, in the section on the Christian Churches (Disciples of Christ). The Christian Church was congregational in its polity.

Congregationalists, like a number of other groups in America, wield more influence than their numbers would seem to warrant. Part of the explanation is social. On the whole, Congregationalists have more wealth than the members of most other American churches (see page 135); and with wealth comes privilege and prestige. Also, the Congregationalists are well educated. They live in the sections of the country which provide the best educational advantages. Moreover, the Church's system of government gives a clergyman the freedom which is essential to the development of creative ideas. Congregationalism thus tends to attract independent minds from other denominations. The result is a leadership which is intellectually alert and influential in the religious councils of the nation. It is my impression, however, that in those sections where Congregationalists are prominent, clergymen exert less influence in their communities than they do in the South and in some parts of the West.

CONGREGATIONAL GOVERNMENT

An English priest once said that from the viewpoint of Europeans, all American churches are congregational in their polity. This obvious

exaggeration points to the fact that nowhere in American Protestantism is to be found that "fine, old ecclesiastical arrogance" which
characterizes so much of Europe's church government. In many
sections of Europe, the churches are governmental or semigovern-
mental institutions, their leaders holding office through the sufferance
of public officials. Such leaders are freed from dependence on popular
approval to a degree that an American clergyman is not.

But in the formal sense, only about two-fifths of American Protestants have a Congregational polity; among this number are the Baptists, the Disciples, the Quakers, the Unitarians, and, of course, the
Congregationalists.

In this polity, the power of the local church is theoretically unlimited. It has complete control of its own affairs, both financial and
doctrinal. It hires its minister. It holds title to its property. It maintains such relations with other churches as it itself determines. It co-
operates or declines to co-operate with denominational agencies. It can
even ordain men to the ministry, though in actual practice this power
is almost never exercised, the local congregation preferring to join
with the other congregations of its neighborhood in judging the fitness
of clerical candidates.

The denominational executive has nothing but an advisory relationship to the local church, and frequently not even that. He wields a
large influence in some churches; in others he is completely ignored.
He has no power to discipline or to command local groups, except in
missionary churches where he holds the purse strings. About the ultimate in disciplinary action is for the denomination to vote to "drop the
name" of a local church "from the *Yearbook*."

However, these theoretical propositions concerning congregational
government overstate the real situation. Regarding churches of the
Congregational type, a former editor of *The Christian Century* contended:

[They] are not a whit more independent than Presbyterian or Episcopalian congregations. If they exercise their imagined independence beyond a certain point they are dealt with just as similar variations are
dealt with by the other denominations, the only difference being in the
technique, not in its effectiveness.

The editor amplified this remark, saying:

[There is] no suggestion that the *kind* of irregularities or variations which would cause them to be "dealt with" were the same in all denominations. . . . But the principle of conformity obtains no less in one group than in the other. A Congregational, Baptist, or Disciples local church . . . conforms to an organic whole which is larger than itself, a whole of which it is a subordinate and cooperating part.[15]

The position of the local church in the larger denominational fellowship can be likened to the position of a college student; a student will put up with a great deal of what he thinks is autocratic interference with his personal liberty in order to stay in college. Similarly, the average local church will go to considerable lengths to avoid expulsion from the larger fellowship. Thus, being "dropped from the *Yearbook*" is not as ineffective a penalty as it might appear to be.

The Congregationalists federate their local churches into three types of organization. Next above the local church is the Association; then comes the State Conference; and finally a national organization called by the Congregationalists the National Council and in the newly formed United Church called the General Synod. These bodies are composed of representatives from local churches. The state and national organizations have paid executives, usually called Secretaries, Superintendents, or Ministers.

The Associations control ordination, a departure from the practice of the early New England Congregationalists who ordained men through the agency of the local church. The power of the Association in this matter is illustrated by the case of one well-known New England clergyman who was defrocked because the Association of which he was a member disapproved of his divorce and remarriage. On the other hand, the power of the local church is illustrated by the fact that he was called to a Congregational parish and serves as its pastor. The local church is not disciplined. Although the clergyman is under many professional disabilities because of the disfavor of his fellow ministers, his only clerical disability in ministering to the local parish is lack of authority to solemnize marriage; and this authority comes not from the church but from the state government which looks to the Associations to determine who are bona fide Congregational clergymen.

Though Congregational officials have little authority over the local church, they have great authority over the specific organization they head. One Congregational clergyman wrote, "The development of the state superintendency in personnel and prerogative in the last decade evidently presages a sort of pragmatic diocesan episcopacy."[16] This increase in power on the part of Congregational executives comes from the fact that much of the work of the denomination is performed by boards which have an independent corporate existence. The best-known boards of this type are the American Board of Commissioners for Foreign Missions and the Board of Home Missions; the General Council is itself incorporated, as are most of the State Conferences. The power of election to membership in these corporate bodies usually rests with the representatives of the local churches. But once the boards are elected, they have great power. They choose the denomination's leaders and determine denominational policy. Most of the actual decisions of the boards are as far removed from the consent of the local church member and pastor as the decisions of the State Department are removed from the consent of the average citizen. Even the board members themselves—meeting but briefly and sporadically—are usually under the necessity of accepting the advice of their paid executives. Many board actions simply rubber-stamp executive recommendations. This situation obtains, of course, in all the denominations; it is inherent in the American approach to democracy.

Control by administrators so disturbed Roger Babson, the investment counselor, during his period of office as moderator of the General Council, that he called the denominational secretariat "the Vatican." One pastor complained: "Our leaders draw down the curtains on their inside negotiations." "Whether it's in business, the army, or in church life," he wrote, ". . . the most dangerous thing for an ambitious young man to do is to oppose the powers-that-be."[17]

That there is truth in these allegations can scarcely be denied. Yet a false picture is drawn if the reader concludes from these statements that the Congregational clergyman is under the control of denominational machinery to the same extent that clergymen in the presbyterian or episcopal systems are. Some of the problems which confront congregational government will receive further discussion in the next chapter.

One phrase quoted on the previous page from *The Christian Century* editorial needs amplification. The editor pointed out that while congregationally organized denominations have in practice genuine power over the local churches if they want to exercise it, the *"kind* of irregularities" which are tolerated vary greatly. To this comment should be added another: the *amount* of irregularity which is tolerated also varies greatly. The Southern Baptists, for example, are much less tolerant of deviation than are the Congregationalists. Congregational churches tolerate a great deal of deviation. In the leading Congregational state, Massachusetts, no local church has been expelled from the fellowship in the current century.

This willingness to tolerate deviation accounts not only for the theological variety which is found in the denomination, a situation which will be discussed later, but also for the number of ecclesiastical prima donnas who find shelter under the denomination's wings. The amount of freedom permitted for personal idiosyncrasy is so much greater among Congregationalists than in most other groups (much greater certainly than the amount granted by the episcopal or presbyterian systems) that many a personality who wants to play a stellar role unhampered by the restrictions inherent in group effort naturally gravitates into Congregationalism. Almost every Congregational clergyman knows half a dozen such men. If they are asked to speak at church gatherings, they accept with alacrity, and their participation makes the occasion in their eyes important. But they do not attend such meetings while other leaders speak. The churches served by such men, though often large and rich, are usually near the bottom in the amount they give to denominational causes. One pastor, author of many books, widely known lecturer, much loved by his own people, wrote in answer to a denominational appeal for a gift from his unusually wealthy church for help with church extension: "We have just undertaken to raise a hundred thousand dollars for the installation of a new heating plant and have found it necessary to discontinue all outside giving." The clergymen of this man's neighborhood are unanimous in condemning this and similar actions on his part. But he is not disciplined. He will continue to make merry with the ecclesiastical machinery and to push only those enterprises from which he hopes to get front-page coverage; he will also continue to have his name printed in the *Yearbook*.

Theoretically there is no sex segregation in the Congregational ministry. Women are ordained, and have been elected to the highest honorary office in the Church, the Moderatorship of the General Council. Practically, however, there is discrimination against women, just as there is in most American institutions. "Of the approximately six thousand parsons listed in our 1954 *Year Book,* only 4 per cent or about 240 are female and less than one-third of these—about eighty—are actually ministers of congregations."[18] The churches which these women serve are generally small. They are also generally rural. One woman wrote concerning the status of ordained women in the churches:

There is often small difference between the denominations that do and do not ordain women. Ordination of women really means very little when the same churches that ordain these women balk at accepting them as ministers of their churches. Ordination thus becomes the end rather than the beginning for many women seeking service in the pastoral ministry.

Many women ministers I have talked with believe that most churches would not even consider calling a woman minister if they could afford a man. But even this is not accurate. Many churches with a small budget prefer having a student preacher rather than a fully qualified woman minister. And it is all too sadly true that the women in our local churches are often stanchest against accepting their sisters in the pulpit.

If ordination of women is to mean anything more than a pat on the back for the denominations that are broad-minded enough to follow this practice, then the local churches must come to recognize that a woman who is called of God and who has diligently prepared herself by college and seminary training has both the desire and the ability to serve as a faithful minister.[19]

A former Director for Town and Country Work for the denomination wrote concerning the woman minister:

While women have proved that they can serve any type of church, whether rural or urban, it is true that some smaller churches take women pastors, particularly single women, mainly because they can live on a smaller income than a man, especially a family man.

What kind of work does a woman pastor do? They do better work than men in rural churches, not quite as well as men in city churches. This latter fact may be partly because city churches which women serve are weaker fields to begin with. However, the rural churches served by women are smaller than our average rural church and still they do better than the denominational average in quality of rural work.

Women pastors have a higher than average rating in members received by confession of faith and for total members received. Women excel among rural pastors very decidedly in Sunday school and young people's work if comparative enrollments mean anything. Their greatest single achievement in which they exceed the average performance for all our rural churches is in contributions of their churches to apportionate benevolences. They also get their town and country churches to do better . . . in contributions to home expenses and they preach in better than average rural church buildings.[20]

Congregational churches frequently invite clergymen from other denominations to serve as their pastors; many of these men transfer their ministerial affiliation to Congregationalism. But "often as many as five hundred churches have continued under the pastoral care of ministers who belong to other denominations."[21] A great deal is learned about Congregationalism when the full implications of this fact are pondered.

CONGREGATIONAL THEOLOGY

The early New Englanders were vigorous Calvinists. The Westminster Confession, after its formulation, became the standard of doctrine and remained so for many decades. However, congregational polity, unlike presbyterian or episcopal polity, permits of easy change. Whenever a majority in a local Congregational church came to feel that the Westminster Confession no longer represented its living faith, the Confession was abandoned and another affirmation of faith put in its place.

Today Congregationalism is less traditional in its doctrine than is any other large, old-line denomination. A few of the churches are very conservative; a few also are very radical. But most of them adopt a solid Liberal position. Few of them use the ecumenical creeds in services of worship. Many Congregationalists would agree with the clergyman who said he accepted the Apostles' Creed as an "adequate refutation of the heresies it was intended to combat." Probably the statement which would fit the beliefs of the large majority in the Church is the Kansas City Statement, so named because it was adopted by the General Council which met in that city in 1913:

We believe in God the Father, infinite in wisdom, goodness and love; and in Jesus Christ, his Son, our Lord and Savior, who for us and our salvation lived and died and rose again and liveth evermore; and in the Holy Spirit, who taketh of the things of Christ and revealeth them to us, renewing, comforting, and inspiring the souls of men. We are united in striving to know the will of God as taught in the Holy Scriptures, and in our purpose to walk in the ways of the Lord, made known or to be made known to us. We hold it to be the mission of the Church of Christ to proclaim the gospel to all mankind, exalting the worship of the one true God and laboring for the progress of knowledge, the promotion of justice, the reign of peace, and the realization of human brotherhood. Depending, as did our fathers, upon the continued guidance of the Holy Spirit to lead us into all truth, we work and pray for the transformation of the world into the kingdom of God; and we look with faith for the triumph of righteousness and the life everlasting.

Most local churches do not require before admission to membership formal subscription to any statement of faith; most of them ask instead subscription to a covenant or to a statement of purpose.

THE EVANGELICAL AND REFORMED CHURCH

The Congregational Christian Churches formed but one of the sources of the United Church of Christ. The other source was a body which itself was the result of a merger, consummated in 1934, of two church bodies. These bodies were The Reformed Church in the United States and The Evangelical Synod of North America.

The term commonly used to describe the churches on the Continent of Europe which owed their reformation impulse to the work of John Calvin is *reformed*. His influence was so pervasive that churches following reformed doctrine and polity were found even in Germany where most of the churches were Lutheran. The German reformed churches, especially influential in the Palatinate, held a theology similar to that already described in the discussion of the earlier phase of Presbyterianism; this theology was formulated in the middle of the sixteenth century into the Heidelberg Catechism. At this time also was formulated "a plan of modified Presbyterian church government."

In the late colonial period Germans who held reformed convictions migrated to America settling for the most part in the middle colonies. Before the American Revolution the churches organized by these persons were under the guidance and supervision of the reformed churches of Holland. After the ties with Europe were broken by the Revolution these churches became self-supporting and autonomous. They grew in number, expanding especially into Ohio and Wisconsin. Eventually the German Reformed Church in the United States was organized; very shortly thereafter the word "German" was dropped from the name. In the early years the German language was the major means of communication. Dropping from the name of the Church this reference to a foreign country symbolizes a development which characterizes the experiences of most American church organizations which have had their origin in non-English-speaking lands. At first these churches have used the foreign tongue; but as their communicants became more and more at home in this country and became more and more skilled in the use of English, the pressure mounted for a shift to English as the language of worship and of publication. The change-over has often been resisted, especially by the older members of the churches. Bilingual services have often been resorted to. In the end the English language has been generally adopted. Churches now struggling with this language problem are the Eastern Orthodox, as will be noted later.

The persons who formed The Evangelical Synod of North America also came from Germany, but a century later than those who organized the Reformed Church. By this time in Germany, many of the Lutheran and Reformed groups had been merged into churches to which the name "Evangelical" was often applied. Most of these churches acknowledged the validity of both Lutheran and Reformed ordination, and while using the Augsburg Confession also permitted the use of the Heidelberg Catechism. Immigrants from these churches arriving in the United States at the time of the great nineteenth-century westward expansion often settled in the Midwest "with St. Louis as the main point of distribution." In the early years the churches formed by these immigrants received missionary support from the homeland; support also came from the American Home Missionary Society, an organization largely supported by eastern Congregationalists and Presbyterians. In 1872, after thirty-five years of grouping into smaller organizations, these churches united into one body, which shortly afterward was given the name, The Evangelical Synod of North America. In 1922, this body began discussions of merger with the Reformed Church; together they created in 1934 The Evangelical and Reformed Church.

The Constitution of this body provided that the doctrinal standards of the Church were to be the Heidelberg Catechism, Luther's Catechism, and the Augsburg Confession; however, the President of the Church wrote that the Constitution "hastens on to provide" that:

Wherever these doctrinal standards differ, ministers, members and congregations, in accordance with the liberty of conscience inherent in the gospel, are allowed to adhere to the interpretation of one of these confessions. However, in each case the final norm is the Word of God.

The type of government provided was a "modified presbyterianism." The President of the Church wrote:

Powers of the synod and general synod and their respective officers are carefully defined in the denomination's Constitution and By-Laws, but the rights and freedoms of ministers, laity and congregations are preserved with equal care; so much so that Evangelical and Reformed people feel that a denominational constitution and by-laws serve much more as a guarantee of freedom than as a threat to it.[22]

The Evangelical and Reformed Church had about eight hundred thousand members; "95 per cent of its congregations and its communicant membership" were "found east of Montana, Wyoming and Colorado, and northeast of Texas and the Mississippi Delta States."[23]

THE MERGER

The effort to bring the Congregational and the Evangelical and Reformed groups together was long and stormy. Negotiations began in 1942. Strong opposition developed immediately in the Congregational Churches; it arose because of a fear that the liberty and authority of the local church would be abridged. A poll of Congregational churches on the proposed merger was taken in 1948; prior to the poll, the announcement was made that approval by 75 per cent of the churches would be considered sufficient warrant to proceed. At the deadline date only 65.5 per cent of the churches had voted affirmatively. After an extension of seven months 72.2 per cent had approved. The General Council voted to proceed on this basis. A lawsuit was brought by one part of the minority group in an effort to prevent the merger. The suit was lost but it delayed the merger for several years.

Much bitterness remains as a result of the conflict. The Constitution of the new denomination is attacked on several grounds. It is attacked because it is alleged to take away the freedom of the local church. Yet it provides that the "autonomy of the local church is inherent and modifiable only by its own action. Nothing in this Constitution . . . shall destroy or limit the right of each local church to continue to operate in a way customary to it."

The Constitution is attacked because "what a document grants it can later retract," "even to the deleting of the 'guaranteed' paragraphs." Yet it provides that "Nothing in this Constitution . . . shall be construed as giving to the General Synod, or to any Conference or Association, now or at any future time, the power to abridge or impair the autonomy of any local church. . . ."

The Constitution is attacked because it threatens the ultimate control of church property by a "centralized hierarchy." Yet it provides that no authority above the local church shall "now or at any future time" be able to limit the power of the local church "to acquire, own, manage and dispose of property and funds," and further provides that the local church can "withdraw by its own decision from the United Church of Christ at any time without forfeiture of ownership or control of any real or personal property owned by it."

The Constitution is attacked because certificates of ordination must bear the signature of the President of the Church; some President at some "future time" might withhold his signature against a man whose theology or politics he disapproved. However, ordination is definitely fixed as the prerogative of the Associations, and the By-Laws clearly provide that "after ordination or in anticipation of it a certificate is issued bearing the signatures of the proper officers of the Association and the President of the United Church of Christ."

The Constitution protects the rights of the local church as adequately as any document which defines a relationship of trust should. Where the Constitution does fail is in not providing adequate control by the local churches of the Church's superstructure.

The officials in the two uniting denominations have believed deeply that the merger is the will of God; one of them said, "This is probably the most important step our churches have been called upon to take, or will be called upon to take, in this century." Convictions have run so high that the debate has often been characterized by stubbornness, meanness, and inaccuracy. A writer in the official Congregational magazine characterized those who opposed the merger as "a recalcitrant and divisive group,"[24] and one clergyman wrote, "They are a group opposed to the vigorous expression of the Gospel in *any* sphere."[25] The term *antimergerite* has become in most of Congregationalism a stigma which affects profoundly a clergyman's professional status. But the minority also has used sharp words. Some of its members have said that records have been "juggled," that "contractual agreements" have not been lived up to, and that "the whole thing seems more like a power-centered political movement than a devout, spiritually motivated call for united witness and work 'in Christ.' "[26]

Focusing attention on the bitterness which has arisen fails to indicate the deep spiritual motivations which have prompted action on both sides. One distinguished clergyman, a member of the minority, wrote:

We are interested in a drawing together of Protestants to affirm a basic devotion to Jesus Christ, but we cannot support a movement aimed at exalting an institutional church as the necessary expression of that devotion.

The Pilgrims came to America to get away from superstructures. They affirmed the local congregation as a fully adequate "outcropping" of the invisible fellowship of Christ's people. Each local Church is to ordain its own pastor, govern its own affairs, make its own witness, and look to the presence of the Eternal Christ in its midst. It is Jesus who guides and inspires. No other organizational structure is Scripturally recognized.

But whatever safeguards the United Church may propose to guarantee local freedoms, these are secondary to its intent to subordinate local congregations to a real, visible, ecclesiastical superstructure called "The United Church."[27]

Another member of the minority wrote:

The sin of ecumenicity is that it puts right organization in the place reserved for a right spirit, obedience in the place reserved for free response, and possession of power in the place reserved for possession of love. Ecumenicity is, indeed, one of the crop of modern heresies which corrupt the gospel, and which lead men to believe in something very much less than the good news of our Lord Jesus Christ. . . .

The goal of one "Great Church"—one big world-embracing organized Church body—*was* achieved in most of Western Europe in the days of Pope Hildebrand. The visible, worldly authority of "that Church" was virtually complete. But, at the same time unprecedented corruption came into the ecclesiastical courts. What happened was not that "The Church" conquered the world. What happened was that the world conquered "The Church!"[28]

On the other hand, the leader among Congregationalists in the fight to create the United Church wrote:

The uniting groups do not expect the beams of the millennial sun to break out through the new relationships; but neither do they expect to be separated from that light which illumines every achievement planned for God's glory and carried out with devotion. If nothing else of reward comes to them than to have the sense of having felt a call of God in this age and having responded, it will be enough. If their plans carry, there will be one less denomination among the many, the too many, in America. One step will have been taken away from the scandal of denominationalism toward that one world which the church should at once prophesy and demonstrate.[29]

No doctrinal norm has been set for the United Church. The Constitution affirms that the Church

claims as its own the faith of the historic Church expressed in the ancient creeds and reclaimed in the basic insights of the Protestant Reformers. It affirms the responsibility of the Church in each generation to make this faith its own in reality of worship, in honesty of thought and expression, and in purity of heart before God.

In 1959, the following Statement of Faith was adopted by the United Church as a "testimony not a test."

We believe in God, the Eternal Spirit, Father of our Lord Jesus Christ and our Father, to whose deeds we testify:
He calls the worlds into being, creates man in His own image and sets before him the ways of life and death.
He seeks in holy love to save all people from aimlessness and sin.
He judges men and nations by His righteous will declared through prophets and apostles.
In Jesus Christ, the man of Nazareth, our crucified and risen Lord, He has come to us and shared our common lot, conquering sin and death and reconciling the world to Himself.
He bestows upon us His Holy Spirit, creating and renewing the Church of Jesus Christ, binding in covenant faithful people of all ages, tongues, and races.
He calls us into His Church to accept the cost and joy of discipleship, to be His servants in the service of men, to proclaim the gospel to all the world and resist the powers of evil, to share in Christ's baptism and eat at His table, to join Him in His passion and victory.
He promises to all who trust Him forgiveness of sins and fullness of grace, courage in the struggle for justice and peace, His presence in trial and rejoicing, and eternal life in His kingdom which has no end.
Blessing and honor, glory and power be unto Him. Amen.

CHAPTER VIII

BAPTIST CHURCHES AND CHRISTIAN (DISCIPLES) CHURCHES

THE SIXTEENTH century Reformation was a half-done job, according to most Baptists and Christians (Disciples). To be sure, it freed the Church of the usurpation of Rome and "went back of the distinctly medieval dogmas, such as transubstantiation and exaltation of the Virgin Mary and the saints."[1] But the Reformation did not go "back of the Nicene dogmas of the fourth century."[1] Genuine New Testament Christianity, according to this view, lapsed soon after the Apostolic Age and was not again given a real chance until a century after Luther; moreover, in the opinion of highly placed Baptists, it was "in America, not in Germany, that the genuine Reformation culminated."[2]

On its negative side this "genuine Reformation" consisted of the rejection of restrictive creeds, of episcopal or presbyterial government, of a stated ritual, of infant baptism. On its positive side, this reformation consisted of the "adoption of a spiritual and New Testament Christianity." The Baptists and the Christians (a Baptistlike group here in America, sometimes called Disciples, which became a separate denomination about a century and a quarter ago) find little in the New Testament to support "the elaborate forms" found in most other types of Christianity. According to the Baptist view, the Roman Catholic religion is dominated by externals; the Lutheran and Episcopalian churches moved a step or two away from this domination; the Presbyterians moved considerably farther away; the Congregationalists farther still; but it was with the Baptists that true freedom from unscriptural, medieval formalism was achieved.

Baptists hold that no person can be a Christian until he has received *regenerating grace*. As they read the New Testament, they find that no one in the Apostolic Age was admitted into the Church by baptism until he was old enough to understand and accept Christ's message and until he had been "born again," until his moral and spiritual life had been regenerated. The Baptist churches, therefore, are composed exclusively of "adults," that is, persons who can choose for themselves. Baptists hold that their organization and doctrine rest solely on New Testament teaching and not on "human opinion" (creed). They consider a local church or a denomination to be only an external body, the Universal Church being wholly invisible.

THE ANABAPTISTS AND THE MENNONITES

Most Baptist historians place their denominational beginnings in biblical times, but modern Baptist history begins with the Reformation. One important aspect of this beginning was a group called the Anabaptists. These biblical literalists were committed to the development of a strict ethical religion. They did their best to follow scrupulously the life described in the Bible—as they read the Bible. For example, they held their property in common, and, usually, refused military service—these actions find sanction in Holy Writ. But in addition to such practices the Anabaptists were also the first modern Christians to practice "believer's baptism"; like the present-day Baptist groups, they denied the validity of infant baptism and baptized only believers ("adults"). Catholics and Protestants alike were horrified by this practice and called it "ana" or "re" baptism, since these adults had all received "baptism" in infancy. For a time the Anabaptists flourished, but their combination of radical social and radical religious ideas brought upon them persecution of the bitterest kind, including death by drowning, a fate considered by their enemies to be particularly appropriate. Soon their movement died leaving no direct institutional progeny—except the Mennonites, and they rejected many of the radical Anabaptist practices.

The Mennonites were named for their founder, a converted Catholic priest, Menno Simons. Their movement began in the sixteenth century, in freedom-granting Holland. From Holland, the Mennonites spread in small numbers over much of Europe. Persecution drove many of them to America. Today there are about one hundred and sixty thousand in this country; they are divided into more than a dozen different sects.

They adopt on the whole a very conservative theology; most of them emphasize the Second Coming, the necessity for conversion, baptism of believers, baptism by pouring, opposition to taking oaths, simplicity of worship, and foot washing (in addition to baptism and the Lord's Supper) as an ordinance instituted by Jesus. Pacifism is the doctrine for which the Mennonites are best known; during World War II, 40 per cent of the conscientious objectors were members of this small denomination. The Amish Mennonites (numbering about twenty thousand, in three groups) are the ultraconservative wing of the denomination; they strive valiantly to maintain the simplicity of their way of life, fastening their garments with hooks and eyes instead of with buttons in order to avoid the ostentatious display condemned by the Bible. The Old Order Amish will not use carpets, curtains or wall pictures in their homes; and they worship in private houses, thinking church buildings improper. This group is opposed also to centralized schools.

English Baptists

Some historians begin Baptist history with English Separatism. (See pages 199, 217 ff.) In the time of James I, John Smyth, an Anglican clergyman, set out to convert the Separatists in Gainsborough, near Scrooby. But instead of turning them to the Church of England, he was himself converted to Separatism and soon became their teacher. Under his leadership, a group from Gainsborough moved in 1608 to Holland, migrating ahead of the Pilgrims, as has been noted. There Smyth, following out the logical implications of his Separatist principles, became convinced of the correctness of believer's baptism, and in 1609, along with thirty-seven other Englishmen, established (according to one opinion) the first modern Baptist church. Smyth baptized himself by applying water to his own head, and then baptized his followers. To persons who question the validity of such a baptism, Baptists point to their contention that the Church is spiritual and not mechanical in nature. They contend that no person can truly be a member of the Christian Church until he has been "born again." As the group in Amsterdam put it, "the churches of the apostolic constitution consisted of saints only." After a few years most of the Amsterdam Baptists concluded that it was their duty to return to England, there if necessary to face persecution in the effort to re-establish true Christianity.

The earliest English Baptists were not concerned about *how* a person should be baptized; they were concerned about *who* should be baptized. Baptism by complete immersion of the body in water, rather than by pouring or sprinkling the water, is supposed by many people to be the distinctive Baptist principle. But it was not until nearly a half century after John Smyth's time that the mode of baptism came into serious controversy. Then careful study of the New Testament convinced the Baptists that immersion, which was the general practice of the Church until the thirteenth century, was the method used in apostolic times; immersion accordingly was adopted and is retained by most Baptist and Disciples churches.

The term *Baptist* was coined, like the names of most of the other Christian groups, by the opponents of the movement. But its members preferred the names "Baptized Believers," or "Christians Baptized on Profession of Their Faith." Soon, however, these terms proved too cumbersome and the popular usage was adopted.

It was through the Baptists that the singing of English hymns of original composition became popular; formerly, English Protestants had sung rhymed versions of the Psalms, a practice still followed in America by a few Presbyterians. The new hymns were long considered by non-Baptist Christians in England to be a bold and improper innovation. A powerful weapon in the controversy over the new hymns was *Pilgrim's Progress;* many of the songs Bunyan put on the lips of Christian came to enjoy great popularity.

The English Baptist churches grew slowly during the eighteenth and nineteenth centuries. Today there are somewhat more than a million Baptists in Europe. It was in America, as has been intimated, that the Baptist movement had its major growth; twenty-five million persons are reported as baptized members of Baptist, Christian or Churches of Christ groups in this country. If they counted children in their membership totals they would no doubt number considerably above thirty million.*

ROGER WILLIAMS

It is probably not true that Roger Williams founded the first Baptist church in America, as is commonly supposed. He was in fact a very poor Baptist, denominationally speaking. But there is no question about the immense contribution which Roger Williams made to the development of Baptist tenets, nor about his immense contribution to American democracy. His career is outlined here because by common acknowledgment his life embodied Baptist principles to a greater extent than any other American leader. He was the first to found a modern commonwealth in which full religious liberty and the separation of church and state were put into practice; moreover, he created in this commonwealth a simple but vital political democracy.

Williams was the son of a London shopkeeper and was born, probably, in the year that Queen Elizabeth died. In his youth, by his ability to use shorthand, then a new and uncommon art, he attracted the attention of Sir Edward Coke, a famous jurist who was electrifying England by his opposition to King James' claim to rule by divine right. Sir Edward, attracted also by the keenness of the young man's mind and by the charm of his personality, offered to help him through the University of Cambridge. As an undergraduate, Williams gained a high degree of scholarship, refinement of manners, Puritan sympathies, and a keen sense of the injustice of the University's class structure which permitted young aristocrats to lord it over the sons

* The result obtained by increasing the total reported for these churches by 38.7 per cent is above thirty-four million. This is the percentage of persons under fourteen years of age in Protestant households according to a poll of thirty-five thousand households by the Bureau of the Census; see the Bureau's report Series P–20, No. 79, Feb. 2, 1958. However, some of the churches, in the Southern Convention, for example, are said to admit to membership a significant number of persons who are under thirteen years of age. Thus the thirty-four million figure is probably too high. According to Bureau of the Census figures, extrapolated from the poll noted above, the number of persons in the country fourteen years of age and over who said they were Baptist was 23.5 million; the number who said they were Roman Catholic was 30.7 million. Thus the Baptist total (without the Christian Churches and the Churches of Christ) is about 77 per cent of the Catholic total. These figures are subject to sampling errors.

of the commonalty. This sense of injustice is the key to much of his later behavior.

After graduation, he received ordination and served for a short time in a minor post in the Church of England. But Williams saw clearly that he could not continue in the Church unless he was willing to manhandle his conscience and pretend to revere what in fact he despised. He might have had a distinguished ecclesiastical career; certainly he had most of the qualities which make for preferment in the Church: piety, intelligence, wit, enthusiasm, leadership, winsomeness. But he had also a stern conscience, so stern that often in later years he seemed to scruple at trivia.

Becoming convinced that the Christianity of the New Testament was Separatist, he gave up his position with the Church, left England —though the move was to him "bitter as death"—and migrated with his bride to Massachusetts, two years after the founding of that colony. In Boston, he made such a good impression that the Puritans invited him to become their teacher in the First Church; but after a careful survey he declined because, as he wrote later, "I durst not officiate to an unseparated people." I might "have run the road to preferment, as well in Old as New England." He startled the people of Boston by saying that he could become their teacher only if they made public repentance for their sin of having been members in the old country of the Church of England. He also told the magistrates that they had no right to punish persons who broke the Sabbath; he said civil officers should have power over civil conduct only, not over religious practice, belief or expression.

After these unpleasantries at Boston, he was offered the position of teacher in the church at Salem. This position he accepted. But the Massachusetts magistrates, smarting under his censures, brought their influence to bear and in a short time he was dismissed from the Salem church. He then settled for a time in Plymouth. The Pilgrims recognized his great abilities. Bradford described him as "a godly and zealous man, with many rare abilities," and wrote that his teaching was "highly approved, and for its benefit I still bless God, and am thankful to him even for his sharpest admonitions and reproofs, so far as they agree with the truth." Bradford as well as Williams could read the New Testament and try his hand at interpreting it.

After about two years in Plymouth, conditions in Massachusetts permitted Williams' return to Salem as assistant pastor of the Church there. His years of experience had not taught him caution, nor had they silenced his conscience. Soon he was again in conflict with the Massachusetts officials. He continued to teach that magistrates have no right to interfere in religious matters. He also taught that "a magistrate ought not tender an oath to an unregenerate man"; and, most significant of all, that the colonists had no right to their land, the rightful owners being the Indians who had never been properly compensated. "Christian Kings (so-called) are [not] invested with right by virtue of their Christianity, to take and give away the lands and countries of other men."

These teachings brought Williams into immediate trouble. The members of the oligarchy which had control of Massachusetts were intent on perpetuating the inequality inherent in the English social structure. They forbade the commoner to ape the gentleman in his manner of dress. They arranged that the punishment imposed on a poor man for a crime should be more severe than the punishment imposed on a member of the aristocracy for the same crime. They silenced, or fined, or deprived of political privilege any person they considered too bold in speaking out against their power. They controlled the distribution of land in their own interest, granting small parcels to some of the more energetic and tractable colonists, but assigning themselves large and impressive holdings.[3]

Roger Williams came into direct conflict with this undemocratic system. Like the other clergymen, he might easily have become a member of the ruling clique. Instead, he chose to stand for a more democratic way. After he had been back in Salem for about a year, the magistrates, thinking to further intrench themselves, passed a regulation requiring all residents to take an oath swearing their loyalty to the magistrates and promising to give speedy notice of any sedition plotted or intended against the government. Persons who persisted in refusing to take this oath were to be banished. Williams refused to make this obeisance to the local squirearchy. Instead, he declared his opposition and persuaded others to join him. He said an oath requiring the use of God's name is an act of worship and the taking of an oath by an unregenerate man is taking God's name in vain. Williams believed the separation of church and state should be complete and considered that the clergy had entered into an unholy alliance with the magistrates to bolster an unjust social order.

Such defiance brought prompt action. He was summoned to appear before the General Court and roundly lectured. But he remained unconvinced and obdurate. No sentence was passed at that time and in the weeks that followed he continued to oppose the oath and won so many to his stand that "the Court was forced to desist from that proceeding." The Salem church then voted to promote him from assistant pastor to pastor.

But the will of the Massachusetts rulers was not to be balked. They had previously decreed a uniform church discipline throughout the colony, a ruling which Williams of course opposed. Accordingly, he was called before the Court a second time and charged with teaching freedom of religion. The leading ministers pointed out during the proceedings that under his system "a church might run into heresy, apostasy, or tyranny, and yet the civil magistrate could not intermeddle." The Court also censured the Salem church for calling as pastor a man who was "under question"; if such actions and attitudes spread, a free pulpit and an independent Church would surely follow. Salem and Williams were granted time in which to ponder the situation, and Williams was instructed to return to the next session of the Court and give satisfaction or expect sentence. Then, as the crowning act of that session, the Court rejected a petition of the Town of Salem to lands in Marblehead, because the Town had chosen Williams as its pastor.

When news of the Court's action reached Salem, indignation ran high. The colonists had but recently fled from England to escape the Stuart tyranny; now they were running into another tyranny, one imposed by some of their fellow religionists. A public meeting was held and a letter was drafted which explained the Salem situation and appealed for the support of the populace. Copies were sent to the churches of the colony. But the oligarchy, getting wind of the enterprise, warned its representatives in the various communities; the letters were quietly pocketed and never were read before the people.

Before Williams could take effective further action, he became seriously ill of an infection. During his convalescence, he wrote a letter to the Salem church in which he reaffirmed his Separatist opinions and urged severance from the rest of the colony in spiritual matters. But now the oligarchy struck hard. It unseated the Town's deputies from the General Court and sent them home. Moreover, it continued to refuse to grant the Marblehead lands and made clear that dismissing Williams was a condition of receiving them.

The "cooler" heads in the Town now began to waver in their conviction and soon concluded that the time had come to submit. The Church members held a meeting and voted both to continue communion with other churches and to accept the conditions dictated by the Court. (Six months later the Marblehead lands were granted to Salem.) Williams immediately resigned from the Church. The Court now had him completely at its mercy; he was summoned again for trial.

In that day, the General Court of Massachusetts was a body which held final legislative, executive and judicial powers, and it operated in inquisitorial fashion. Williams was presented with no formal indictment; he had no attorney; no jury was chosen; and Governor Haynes acted as both presiding judge and leader of the prosecution. Still weak from illness, Williams was confronted by fifty of the best minds in New—or old—England, including ex-Governor John Winthrop; Allerton Hough, former Lord Mayor of Boston, England; John Cotton, Puritan apologist; and Thomas Hooker who would subsequently found Hartford and of whom it was said his majesty was so great he could put a king in his pocket. Against this group Williams was forced to stand alone, no member of the Court daring to come to his defense.

His conviction was a foregone conclusion, but the leaders shrank from passing sentence. Consequently, they thought by argument and by weight of dignity to persuade him of the error of his way. Hour after hour they grilled him. But he would not recant. The Court wearied, and still he stood his ground. Finally, the Court adjourned in order to give him time to ponder the consequences of forcing a sentence. But when they convened again they found him ready "not only to be bound and banished, but to die also." Accordingly, he was ordered to leave Massachusetts. However, in view of his recent illness and the imminent birth of his second daughter (who was to receive the name "Freeborne"), they permitted him "to stay till spring," but enjoined him "not to go about to draw others to his opinions."

Williams refrained from further public agitation but did continue to expound his views in private. The General Court hearing of this activity ordered his arrest, thinking to send him to England. Friends apprised him of this threat and he fled, on foot, in the depth of winter, into the grim forest.

Historians seeking a full view of Roger Williams' banishment have pointed out that his years in Massachusetts were "a desperate time in the history of the colony,"[4] and that he would probably have received rougher treatment in most of the nations of the seventeenth century. Even today we have not learned the full implications of freedom; we still deal sternly with pioneer thinkers. Three centuries after Roger Williams' trial, the General Court (legislature) of Massachusetts sought to make amends by rescinding his sentence of banishment; the Governor of Massachusetts personally carried the writ of acquittal down to Rhode Island, where the chief executive dressed himself up in colonial costume to receive the document. But it was in that same year that Massachusetts (aping Rhode Island) established a teacher's oath. Robert M. Hutchins has observed, "We do not throw people into jail because they are alleged to differ from official dogma. We throw them out of work."[5] To be sure, we seldom throw people out of work for their *opinions* on *religion*. But opinions in the field of politics and especially economics are quite another matter.

Williams might well have perished in the forest had not friendly Indians taken him in. He lived with them till spring and later that year (1636) joined half a dozen of his followers in founding Providence, so named because he believed God had at last brought him to a safe refuge—a faith which proved correct. After a time, under his leadership, a commonwealth developed which was dedicated to "soul liberty," as he put it, and designed "for those who were destitute especially for conscience's sake." Sanctuary was available in Rhode Island for persons of any type of religious faith—Separatists, Puritans, Anglicans, Roman Catholics, Jews, Moslems.

Such practice of religious freedom brought down scorn and derision. Cotton Mather wrote that Williams had a windmill in his head, and John Cotton called him "the most prodigious minter of exorbitant novelties in New England." Because "all sorts of riff-raff people" collected there, Dutch clergymen in New Amsterdam called Rhode Island "nothing else than the sewer of New England." Because the colony did not meddle with the small matters of private morals, it was habitually called "Rogues' Island." The rest of the colonies were appalled by the fact that a Rhode Island court denied the right of a husband to force his wife to behave in religious matters as he thought proper.

Williams did not stop with mere toleration; he believed in full-fledged equality for all religious groups. He wrote, "I desire not that liberty to myself, which I would not freely and impartially weigh out to all the consciences of the world beside." He disagreed vigorously with the religious ideas of many of his neighbors. Often he argued with them; on one occasion he engaged in a three-day debate with some distinguished Quakers. But always he kept his disagreements in the realm of ideas; governmental authority was never used to bolster religion. Rhode Island, unlike most of the colonies, never had an established church.

Through Williams' influence, the government of Rhode Island was controlled by the democratic vote of all persons who owned land, no matter how small an amount. The right to vote did hinge on the ownership of land; however, he fought for, and to a large extent succeeded in securing, the right of all men to purchase land. He wrote, "I reserved to myself not one foot of land or inch of voice more in any matter than to my servants and strangers." To be sure, his democratic practice did not measure up to our own; it fell still farther short of the democratic ideal. But in its own setting his achievement was truly stupendous. He was nearly two centuries ahead of his time.

Unfortunately, political democracy in Rhode Island did not long survive him. The sons of the founding democrats learned to love the fruits of special privilege and to develop methods of keeping new-comers in their place. Rhode Island grew rich in the eighteenth century on the triangular trade in molasses, rum and slaves. Not until Dorr's Rebellion in 1842 did the State give up land ownership as a qualification for voting.

The sons of the fathers liked better the experiment with religious liberty; it lived on and became a major influence in the development of American politics and religion. The Baptists had a large share in this signal achievement. We do not know just what relationship Roger Williams had to the Baptists. Some students of his life contend that he never joined a Baptist church. Unfortunately, during King Philip's War, the Indians burned most of the buildings in Providence, including Williams' own house; this fire destroyed many valuable documents including some which might have cleared up the question of Williams' relationship to the Baptists. He was, of course, familiar with the teachings of the English Baptists, who stressed democratic church management and the freedom of the church from civil domination. We know that soon after founding Providence, Williams became convinced that Baptist Separatism was nearer New Testament Christianity than Congregational Separatism. It seems probable that he was "rebaptized" by one of his own followers, whom Williams then in turn "rebaptized," along with ten others. But it is certain that he was active in the Baptist group for a period no longer than four months in length. His ever-roving mind finally came to the conclusion that no visible church can demonstrate that it is in the true Apostolic Succession. Consequently, he aligned himself with the Seekers, an unorganized and extremely individualistic group whose members denied the apostolicity of any visible church. In general, he continued to accept and to preach Baptist doctrines; but on the question of the nature of the Church, he concluded that he could not accept even the lowest of low-church conceptions.

Williams' theories were roundly denounced by pamphleteers on both sides of the Atlantic; he wrote many defenses. The most famous was *The Bloudy Tenent of Persecution for Cause of Conscience*. This work, bristling with Biblical quotations and historical allusions, was produced at the beginning of the Cromwellian Revolution, while Williams was in England obtaining a charter for the new colony. The *Bloudy Tenent* touched off an angry debate which reverberated all through English political thinking in the 1640's. In one year, no less than twenty-six pamphlets made direct mention of Williams and his ideas; in another year, at least eighteen did the same, and Parliament ordered the book burned.[6] He was the friend of many leaders of Independency, including John Milton and Sir Henry Vane. Williams belongs among the thinkers who shaped political and religious destiny in England and America. One writer calls him "the true founder of American liberty and democracy."[7]

The following passages from his writings are classic expressions of ideas which eventually secured a firm hold on the thinking of Americans. They need to be pondered against a seventeenth-century setting, a setting filled with such ideas as the divine right of kings, heresy equated with treason, some classes born to privilege and others born to ignominy, the suppression of opinion, and an established church casting over all these dogmas the rosy light of divinity.

The sovereign, original, and foundation of civil power lies in the people. . . . A people may erect and establish what form of government seems to them most meet for their civil condition. . . . Such governments . . . have no more power, nor for no longer time, than the civil power or people consenting and agreeing shall betrust them with.[8]

It hath fallen out sometimes, that both Papists and Protestants, Jews and Turks, may be embarked in one ship; upon which proposal I affirm that all the liberty of conscience that ever I pleaded for, turns upon these two hinges—that none of the Papists, Protestants, Jews or Turks be forced to come to the ship's prayer or worship nor compelled from their own particular prayer or worship, if they practice any.

[A state church is] opposite the very essentials and fundamentals of the nature of a civil commonweal.

The *Christian* church doth not persecute: no more than a lily doth scratch the thorns, or a lamb pursue and tear the wolves.

A believing magistrate is no more a magistrate than an unbelieving.

The spiritual and civil sword cannot be managed by one and the same person.[9]

Highlights in Baptist Development

After the beginning in Rhode Island, Baptist churches grew but slowly. They were under many disabilities in New England, and made no great gains there until the time of the Great Awakening. It was in the middle colonies that the Baptist movement in America really got its start. By 1707 the churches of this section were strong and numerous enough to form the famous Philadelphia Association. This Association, gathering in the churches from southern New York to northern Virginia, came to exert a tremendous influence on the later development of the denomination. One historian even wrote, "Pretty much everything good in our history, from 1700 to 1850, may be traced to its initiative or active co-operation."[10]

During the Revolutionary War, the Baptists came into prominence through their efforts to further religious liberty. They appealed to the Massachusetts authorities for the disestablishment of Congregationalism; taxation without representation, they said, is as bad in religion as in politics. The appeal was turned down. "The Baptists," John Adams is reported to have commented, "might as well expect a change in the solar system as to expect that the Massachusetts authorities would give up their establishment."[11] Nevertheless, a decade later the First Amendment was written into the Constitution of the United States. As a result, religious liberty soon came to exist "in America in a sense and to a degree in which it had never before existed anywhere in Christendom. . . . In England, at this time [the early nineteenth century], a nonconformist could not be a member of Parliament, attend a university, hold a commission in the army or navy, or be legally married except before a minister having 'holy orders.' "[12] Undoubtedly, more credit for the achievement of religious liberty in America is due the Baptists than to any other religious group.

The major growth of the Baptist and Christian denominations came with the nineteenth century and the opening up of the territories beyond the Allegheny Mountains. The freedom, informality and warmth of the Baptist way of doing things appealed to the frontiersman. Moreover, the older denominations could not find and support enough trained clergymen to serve the multiplying frontier communities. This situation offered an opportunity to unlettered but fervent evangelicals to assume leadership. The Baptists and the Methodists were willing to take these men as ministers and thus were able to lay on the frontier the foundations for their tremendous later growth. As a consequence, today these two groups are the largest

Protestant denominations in the country, with a combined membership far exceeding that of the Roman Catholic Church.*

The Baptists "have had a schism about every twenty years of their history"[13] in America—an evidence of vitality as well as of a tendency to disagree. Today the Baptist family includes such groups as the General Six Principle Baptists whose distinctive feature is the laying on of hands at the time of baptism; the Seventh Day Baptists who celebrate Saturday rather than Sunday as the Sabbath; the Two-Seed-in-the-Spirit Predestinarian Baptists who believe that every person contains in his heart two seeds, one of good and the other of evil, one from God and the other from the Devil; the Primitive Baptists who are stern Calvinists; and the Duck River Baptists who are liberal Calvinists and "practise washing the saints' feet."

The bulk of the Baptist membership is in three major groups: the American Baptist Convention (called until 1950 the Northern Baptist Convention) which has a membership of about one and one-half million persons; the Southern Baptist Convention which has almost ten million members; and two bodies of Negro Baptists totaling about seven and one-half million members. The names of these two bodies are National Baptist Convention of America and National Baptist Convention, U.S.A. Inc. Prior to the Civil War these bodies were one. The Northern and Southern groups separated in 1845. The creation of the Negro bodies was a post-Civil War effort on the part of ex-slaves who found in Baptist democracy and simplicity the answer to their religious needs.

Baptists live in large numbers all over the United States, but their concentration is heaviest in the South; they are easily the dominant religious group of that section.

BAPTIST BELIEFS

"We have constructed for our basis no creed," announced the Southern Baptists at the formation of their Convention, "acting in this matter upon a Baptist aversion for all creeds but the Bible."

---

* To obtain this result, membership must be judged by the same standards; i.e., children counted for all groups or only adult communicants counted for all groups.

Yet there are Baptist principles. As we have seen, Baptists emphasize extreme loyalty to the New Testament, believer's baptism, immersion, separation of church and state, simplicity of worship, and have in general an evangelical outlook. One other matter needs to be noted before we have the full picture; the Baptists believe in what they call "soul competence." Every person, according to this teaching, is competent himself to deal directly with God and has the right and need for such dealing. True religion, therefore, cannot be fixed in creeds and handed down from one person to another. True religion is voluntary, dynamic, alive; it is personal and must be experienced. There is no inescapable necessity for a person to belong to a church or to receive the traditional ordinances (baptism and the Lord's Supper). The individual soul itself can enter into the very Holy of Holies.

However, the doctrine of soul competence is not equivalent to saying that an atheist, for example, or even a unitarian could continue as a member of most Baptist churches. "It has never been true that one could be a Baptist and believe as he pleased."[14] Regional organizations have maintained a few unwritten theological norms, and exacted adherence. One way Baptists took to enforce their standards was the practice of having membership open only to persons immersed on profession of their faith. Another way was "close Communion"—Communion refused to any but Baptists. These practices were once universal in Baptist churches; however, by the middle 1940's, one careful observer could write:

Today Northern Baptist churches which practice close communion are rare. But the majority still require the immersion of Christians who come from non-immersion churches before admitting them into full membership. Yet the practice varies all the way from strict application of this rule to a very few churches which will admit into full membership anyone who states that he believes in Christ. The matter of baptism is left optional.

Several years ago there was some agitation among Baptist churches of the north over the adoption of open membership. This movement took two forms. One is complete open membership by which a baptized Christian from an evangelical church was admitted into full fellowship without immersion. All church rights and privileges were granted to such a member. Incidentally, this is the accepted practice among English Baptists and is commonly held without regard to theological views. The other form is that of an "associate membership," which provides for the reception of Christians from other churches into membership in the local church, but with the understanding that they are not to be voting delegates at a Baptist convention.[15]

In 1961, one highly placed American Baptist official estimated that 80 per cent of the American Baptist Churches still require immersion for full membership.

The Baptists and Christians hold that there is no such thing as a sacrament, an act which conveys saving grace to the soul. They call the Lord's Supper and baptism "ordinances," acts prescribed by God, and consider them to be simply the "marks" of discipleship. The ordinances can, if a local church so directs, be administered by unordained persons.

The power to ordain clergymen rests with the local church; but in practice few churches presume to act independently. Thus, ordination is usually performed by the Association. (The structure of denominational organization follows in general the Congregational pattern.) Ordination conveys no sacerdotal powers; clergymen are considered to be simply laymen who carry organizational responsibilities. This doctrine is so vigorously held that among the Christians some of their early ministers "never were ordained, and even today an occasional pastor will grow into his work and rise to national prominence without taking time for this particular formality."[16]

Baptists believe they are even more congregational than the Congregationalists. Some Baptists contend that their movement constitutes "the most extensive experiment in pure democracy in all history."[17] The autonomy of the local unit is carried so far that many Baptists "are highly scandalized"[18] by the phrase Baptist Church; the correct expression is thought to be Baptist churches. The churches of the Southern Convention are so jealous of their prerogatives that they do not send "delegates" to national meetings; delegates might become representatives and assume power to make decisions which could hobble the independence of the local congregation. The Southern churches, therefore, are represented by "messengers."

The Baptist determination to be democratic comes out clearly in the conduct of national assemblies. At the Northern meeting in 1944 at Atlantic City one session was thrown into such a turbulence of expression over a war resolution that an official Baptist paper could say the "session was as wild as the raging sea that surged against the steel pier supporting the Convention Hall."[19] But that same meeting of the Convention patiently went through all the labor of distributing, marking, collecting and counting twenty-six hundred ballots, because one man, exercising his constitutional rights, objected to the casting of one ballot by the secretary.

Baptists continue to place great emphasis on the separation of church and state. This emphasis is the ground of such actions as the refusal of a large portion of the Southern ministers to co-operate with the United States Government in taking the 1936 census of religious bodies (an action which was a major cause of the discontinuance of the census of religious bodies) and the questioning by some Southern leaders of the propriety of chaplaincies in the army and navy.[20]

THE SOUTHERN CONVENTION

A large percentage of Protestant leaders consider the Southern Baptist Convention to be the "problem child of American Protestantism." To be sure, these churches show evidence of vigorous health. They possess a very high degree of devotion to Christianity, and the statistics of their growth are impressive. Their rate of membership gains has been well above the national average. Their Sunday school enrollment steadily increased in a period when a declining birth rate produced a decrease in most denominations. In 1960, they reported a total of 1,406,326 tithers. And in the 1956–60 period "nearly 10,000 new churches and missions" were established.[21]

But the credit of such a record is not enough, in the opinion of many observers, to blot out the debit of an intransigent insistence that only the ways of Southern Baptists are scriptural. Most Southern Baptists insist that their faith and it only is ordained of God; their plan for a united Church is for all Christians to become Southern Baptists. "Southern Baptists have the only message to save lost men."[22] The Southern Convention refuses to join the National Council of Churches; the President of the Convention in 1960 said there are no possibilities that it will join the National Council. The Convention of 1948 "shouted down" a committee recommendation that an unofficial observer be sent to the founding of the World Council of

Churches at Amsterdam. In at least one Convention "almost all references from the platform to the Northern Baptists were unfriendly in tone, and several of them were applauded."[23] The Southern Convention has deliberately embarked on a program of expansion into territory formerly thought of as belonging to the American ("Northern") Baptists, even though the Convention refused to admit Canadian Baptist churches to convention fellowship on the grounds that "it would encroach upon territory Canadian Baptists have occupied for years." The Convention has twice held its annual meeting in Chicago.[24] A Southern journal characterized Chicago as "the greatest challenge of any area in America for Southern Baptist witness"; a Southern official predicted three thousand Southern Baptist churches in the Great Lakes area by 1975;[25] a Southern report noted that the number of Southern Baptist churches in Oregon-Washington increased from 19 in 1948 to 155 in 1960.[26] Spurred no doubt by these developments, the ecumenically-minded American Baptists have recently appointed a "general missionary in the South" and are establishing churches there.[27]

The Southern Baptists reiterate the conviction that their denomination has no peer. "The leadership of the Christian world today is largely a Southern Baptist problem, because Baptists are the most rapidly growing denomination in the world."[28] One embarrassed member of the denomination wrote: "We are forever talking about ourselves. . . . Take as an example the 1955 meeting of the convention in Miami, which blew the Southern Baptist horn long and loudly. . . . Perhaps we feel that by constantly speaking about ourselves and to ourselves we can keep the 'evil spirits' away."[29]

A major cause of such sectarian behavior is the constitutional setup of the Convention. It has adopted a "mass meeting" approach to the solution of its problems. Since no less than ten thousand messengers attend the annual meetings and since they have a "propensity for feudin', fussin' an' fightin'," according to ex-Congressman Brooks Hays who served a term as Convention President,[30] anything like genuine deliberation is out of the question. Carefully considered committee recommendations are pushed aside by opinionated majorities who carefully shield themselves from exposure to any religious ideas except Southern Baptist ones and who are fond of listening to Bryan-esque orations which extol Baptist virtues and excoriate the weaknesses of other groups. It was such a rejection of a committee report which led to the vote to "invade" the "Northern" states; the vote was accompanied by a "volume of cheers and applause [which] was deafening."[31] In the midst of World War II the president of the Convention, a former Governor of Texas, opened the sessions with an address the substance of which was reported by Harold E. Fey of *The Christian Century* to run as follows:

Evil men have arisen . . . to challenge the continuance of Christian civilization. So God has commissioned Christian America to free the world of its chains. The invasion is about to begin in Europe. "George Washington will be there. Thomas Jefferson will be there. The Unknown Soldier will be there. He will stand by Woodrow Wilson and in the hour of certain triumph they will stand and sing, 'My Country, 'tis of thee.' " The last time America defaulted on her responsibility to the rest of the world. "But if this great country will look to the southland, where democracy is purest, where Anglo-Saxon blood is purest, and take its direction from the Southern Baptist Church, it will not default again. Our first responsibility therefore is to bring about a revival of the old-time religion, the religion of heaven and hellfire and damnation, to keep the church bells ringing, to inoculate our youth with faith in the principles of our forefathers and in God, the same yesterday, today and forever. Let us rise and sing 'Onward Christian Soldiers.' " The Convention rose and sang lustily.[32]

Time reported the 1953 meeting in these words: "They clapped, rumbled their approval and cried an occasional 'Amen' as this year's president . . . made a metaphorical speech ridiculing the idea that Southern Baptists might affiliate with the National Council of Churches."[33]

An editor of *The Christian Century* commented as follows on the situation inside the denomination:

Some leaders talk of "the miracle of Southern Baptist unity," but others know that underneath the surface appearance of unanimity presented by a controlled and undemocratic convention there is serious and growing disaffection. Large numbers of Southern Baptists are disturbed by the fact that at present all avenues to peaceful, democratic change seem to be blocked. . . . They confess that . . . the only way in which dissenters can bring about change is through backdoor methods of personal influence, backstairs intrigue at conventions, "deals" between persons in power, and currying the favor of officeholders and patronage dispensers. . . .

Meanwhile the boards and agencies of the church go their own ways, following the policies which are sometimes dictated by nothing more than the exigencies of maintaining and extending their own power.[34]

To infer that such generalizations characterize the attitudes of all Southern Baptists would be very far from correct. Southern Baptists are no more all alike than are all Jews, or all Congregationalists or all Negroes. One writer declared: "Southern Baptists probably have the widest range of belief and practice, culture and ignorance, to be found in any denomination."[35] Moreover, changes are occurring in denominational moods. Two decades ago one estimate had it that there were "a number" of open-membership churches and that close communion was "not nearly so common as many would suppose." Today according to the estimate of one denominational official "possibly ten per cent but probably not more than five per cent of the churches practice open membership." This official furthermore writes: "Without a great deal of fanfare about the matter, I would say that most Southern Baptist pastors make no attempt to limit participation in communion beyond the fact that it is an observance for believers who are members of churches."

Changes are occurring in other areas also. One observer wrote, "The day when demagogues could sway the great Southern Baptist Convention may be over, although the fear of the demagogue is still strong in official circles."[36] Another observer wrote that Southern Baptist churches are entering councils of churches "in increasing numbers all across the south."[37] There are signs also of a quickening social conscience. In 1948, the Messengers refused to send congratulations to their fellow member, Harry S. Truman; many observers felt this refusal stemmed from Mr. Truman's stand on civil rights. However, in 1954, soon after the Supreme Court's decision on segregation in schools, the Convention issued a statement urging compliance with the law. Prior to this time, the Convention's theological seminaries (although not its colleges) had been opened to Negroes, and the denomination's Social Service Commission had issued a "solemn warning" to all churches against accepting "bribe money" from the Ku-Klux Klan. One prominent clergyman, when laymen refused at Communion to serve Negroes, defied Southern mores and served them himself.

THE CHRISTIAN CHURCHES

During a Communion service held in Glasgow in 1809, a young Presbyterian sat thinking intently. He had attended the preparatory service, had satisfied the elders as to his fitness to commune, and had received a metal token which, according to the custom of that day, permitted him to receive the sacrament. As he sat in the quiet of the church, group after group of communicants went forward; but he delayed. Finally, as the last group began to commune, he rose, marched to the front, deposited his token on the communion table, and stalked out of the church.

This young man was Alexander Campbell. At that time he was twenty-one years of age, had just spent a year in the University of Glasgow, and was soon to embark on a voyage to America where he would settle on the frontier (West Virginia) and join with his father and some other Presbyterians in rejecting Calvinism and in re-establishing the "Christian Church." Young Campbell had come to the conclusion that the Church of which he was a member in Glasgow—the Anti-burgher Seceder Presbyterian Church—was not established after the New Testament pattern. He came to believe that each local group must be independent, that the New Testament is a perfect constitution for the doctrine and discipline of a New Testament Church, that clerical privileges and dignities have no place, that creeds are nonscriptural, that laymen have the right and duty to preach, that the Lord's Supper should be observed every Sunday, that immersion of regenerated believers was the apostolic practice.

After Campbell reached America, he found under the leadership of a Presbyterian minister, Barton W. Stone, a body of persons who held similar ideas. He joined with them in focusing the attention of many people on these doctrines and they became the locus of religious practice in a widening area. Their affinity to Baptist doctrines was soon noted and for a time the new group was numbered among the members of that group. But finally the Campbells and their colleagues came to consider that the Baptists were a "denomination" while the New Testament provided simply for "Christians." Consequently a new body was born—ironically enough through an effort to avoid sectarianism.

There was considerable debate concerning a name for the new group. One faction favored the term *Christian,* quoting Acts 11:26, "The disciples were called Christians." Another faction favored the term *Disciples.* Alexander Campbell believed this name to be more distinctive and less invidious than "Christian." The name *Disciples of Christ* was finally adopted. In 1957, however, the official name was changed to *International Convention of Christian Churches (Disciples of Christ).* But the terminology is not exact. Local churches of this group (they do not like to be called a denomination or a sect) are called "Christian," "Disciples of Christ," and "Church of Christ" (though they have no connection with the Churches of Christ which are described later in this chapter. (See also page 226 for a discussion of another body which used the term *Christian.*) The Christian Churches have frequently been called Campbellite, a practice which for a number of years has been obsolescent.

The early history of the Disciples was marred by violent controversies. The members accepted a literal view of the Bible, as was customary in that day, and sometimes acted on the belief that no religious practice is proper unless it has explicit warrant in the New Testament. Some of the societies accordingly adopted resolutions opposing preaching from notes, paying salaries to the clergy, and using the title *Reverend.* "Imagine," ran the argument, "saying 'The Rev. Simon Peter!' " The most violent controversy concerned the correctness of missionary societies. (There was no question about missionary *work.*) The antisociety group contended that the New Testament sanctions no organization but the Church itself, that a missionary society has its beginning in *human* thoughts, that it has a "money basis," that it "joins churches together" and is, therefore, a dangerous "temporizing with ecclesiasticism" and the "beginning of apostasy."[38] This controversy continued for many years and was finally settled only by the withdrawal from the Disciples of the antisociety group.

Another controversy had to do with the propriety of using organs in the churches. One spokesman for the antiorgan group declared that since organs were not used in the worship of the first century, they were "an insulter of the authority of Christ, and . . . a defiant and impious innovator on the simplicity and purity of the ancient order. Let no one who takes a letter from one church ever unite with another using an organ. *Rather let him live out of a church than go into such a den.*" (Italics his.)[39] One church in Indiana put up a building with windows and doors so narrow that the customary organs of that day could never be moved in. In 1960, a decision in the Kentucky courts was reported in which a group in a local church who favored the use of instrumental music was declared entitled to exclusive use of the church property.[40]

The bulk of the Christian churches believed organs should be used. But a stubborn minority continued to hold otherwise. They were the same group which opposed missionary societies. As a result, a new sect was born, an antisociety, antiorgan sect. It has the name *Churches of Christ* and reports a membership of over two million adults. However, in 1947 it reported but one million and in 1936 but three hundred thousand. This rate of increase seems improbable. Texas has more members of this group than any other state; other states where the Churches of Christ are prominent are Tennessee, Kentucky, Alabama, Arkansas, and Oklahoma.

Some of the members of the Churches of Christ share with the members of many other sects a conviction that there is but one road to salvation—their own. The stubbornness with which this conviction is sometimes held is made clear by a letter which came from a local minister to a woman who asked for the transfer of her church membership to a Congregational church. The letter ran in part as follows:

We do not issue church letters for anyone to enter into some denomination. Perhaps you didn't understand that according to the Scriptures when you accepted Christ and became obedient to Him, He added you to *His* Church (the church of Christ). You have no Scriptural authority to go join any man-made organization that is called a church. . . . Congregationalism with human names and false teachings will cause your soul to be lost. . . . Don't listen to what men have to say but be sure you have a "Thus saith the Lord" for what you do religiously.

The spirit of this letter does not characterize the Christian churches. They began as an effort to get rid of sectarianism and achieve Christian unity; their plan was for all Christians to get back to pristine Christianity. To date, this body has succeeded in being merely another denomination. They have never been party to the actual union of two denominations. They deplore this fact. Nevertheless, a serious recent discussion of merger with the American Baptists was discontinued.

The Christian Churches are now discussing with the United Church of Christ the possibilities of merger with that body.

The Christian Churches celebrate Communion each Sunday morning and practice open Communion. Open membership, however, is another matter. In 1929, only nineteen churches out of about six thousand "openly and avowedly" admitted to membership persons whose baptism had not taken the form of immersion. By 1948, one hundred seventeen churches were "known to practice open membership," but probably four hundred more actually practiced it "quietly" and said "nothing about it." A still larger number who do not practice open membership "are restrained from it by considerations of expediency only, not by conviction."[41] In 1957, a well-informed observer wrote that "not one-fourth" of the eight thousand churches practice open membership.[42]

The Christian Churches number about one million eight hundred thousand adults. Their chief strength is in the North-Central States with Indiana, Illinois, Missouri, Ohio and Kentucky leading the procession. Texas also has a large Christian contingent. These six states contain about half of the Christians in the country.

THE DISCREPANCY BETWEEN DEMOCRATIC THEORY AND PRACTICE IN THE FREE CHURCHES

The claim so often made by the congregationally organized churches that they practice an especially purified brand of democracy is far from accurate. A careful investigation by Paul M. Harrison of the government of the American Baptist Convention led him to write in his impressive *Authority and Power in the Free Church Tradition* that "the Baptist denomination has been no more successful in establishing a 'democratic polity' than many other Protestant denominations which do not place a primary emphasis on this goal."[43]

The power of American Baptist leaders he said is "oftentimes considerably greater than the official ecclesiastical authority of the Episcopalian or Methodist bishop."[44]

In all of the "free" churches—Baptist, Christian, Unitarian-Universalist, United Church of Christ—the local church has a high degree of control over its own affairs, much greater control than in the episcopal or presbyterian systems. The democratic failure occurs at the state and national levels of denominational life. If democracy is a system in which power flows from the bottom to the top, in which the persons at the top are chosen by the persons at the bottom, and in which policy is determined by persons at the bottom or by their elected representatives, then free-church government as ordinarily practiced in the United States is not outstanding for its democratic procedures. Usually the national denominational leaders are not democratically chosen, they serve for indefinite periods (commonly until retirement—an average of about twenty years), and they habitually determine as well as administer denominational policy.

On paper, the free churches operate democratically. The newly adopted Constitution of the United Church of Christ, for example, provides for a General Synod composed of delegates from the grass roots who have power to elect officers and to determine policies. This theory is general. In practice, however, on most policies and in most elections the national assemblies are "bypassed" by denominational "elites," to use Harrison's words, who use the national assemblies as means for promoting previously determined policies and for electing previously chosen persons.[45]

The composition of the average national assembly is such as to force this *de facto* disavowal of democratic processes if denominational life is to move onward at a pace which most informed observers consider to be wise and necessary; the average national assembly is incapable by itself of making wise and fully considered judgments concerning denominational policy and personnel. A large number of the delegates are attending for the first time. Frequently they have been chosen as a reward for yeoman service in the local church and not because they are prepared by experience and study to take an active part in the assembly's deliberations. They do not know the ropes: they do not know how to present motions so that they will be considered, how to gain admittance to the ecclesiastical equivalent of the smoke-filled room, how to get the floor at times of debate. They are intimidated by the onward rush of an overfull agenda, bewildered by the complexity of the problems on which they are asked to vote, and know that the kind of free debate and discussion which would clarify the issues would lengthen the time the assembly must sit to very expensive lengths. Accordingly the assembly tends except for an occasional issue to rubber-stamp committee and board recommendations, thus transferring real authority in the denomination from itself to the committees and boards.

As has been noted, some of the denominations have a mass-meeting type of national assembly. "If all [American Baptist] churches were fully represented there would be a minimum of 12,744 voting delegates at annual meetings."[46] Over ten thousand delegates attended the 1960 assembly of the International Convention of Christian Churches. Over eleven thousand Messengers attended the 1959 Southern Baptist Convention. The new Constitution of the United Church provides for a Synod of about seven hundred, a size considerably larger than that of the national House of Representatives, a size which many political scientists feel is unwieldy. "In the present stage of their development, the annual conventions bear a singular resemblance to county fairs."[47] Confronted by such a situation, men who have the responsibility for national denominational leadership must choose among anarchy, inaction, submission to demagoguery,[48] and efforts to "bypass" the assemblies. Effective power both for the making of policy and for executing it rests in fact with the leaders. Thus the basic separation-of-powers principle does not operate.

In some of the assemblies, discussion is discouraged. One leader said, "Conditions are such at the present time that we cannot risk it." Another leader said, "If we started an open discussion of theological issues we'd blow the lid off a boiling pot."[49] One reporter wrote concerning the 1959 General Synod of the United Church of Christ:

Dissatisfaction erupted on the last evening. Speakers rose and asked an honest airing of issues which had been troubling delegates. After some veiled efforts toward this end, it developed that Congregationalists were unhappy over the constitutional provision for what they look upon as fragmented or compartmentalized boards which report directly to General Synod, as distinguished from the old inclusive Congregational boards. . . . But apparently there had been a gentlemen's agreement to confine discussion of such matters to committees, and E & R President James E. Wagner . . . was deeply cut. With a grim smile he spoke of this "unfortunate sequence of events" which had violated the hope that such matters would be wrestled with "in smaller meetings."[50]

Harrison's study indicates that less than three hundred persons "actually run" the American Baptist Convention.[51] Some such figure would probably result from an equally careful study of the leadership of other congregationally organized bodies. These leaders tend to be a self-perpetuating group since they often determine nominations for board and committee memberships and for salaried positions. The assembly often merely rubber-stamps these nominations. Concerning the United Church, one Congregational clergyman contended:

The proposed Constitution and By-laws lend themselves to the making of a benevolent oligarchy.

. . . There is no limit to the number of years a staff executive can serve in the same or a related position. It is possible for a man to become an executive or staff member of an instrumentality at the age of 30 or so and continue in office until he retires at 70.

. . . The present by-laws would entrench a system that makes the selection of instrumentality or board members the right of the present board members and staff. Prolonged over decades this would make the United Church the virtual property and private domain of a very few individuals. . . .[52]

The situation here described has resulted from an uncompromising determination to preserve the freedom of the local church. The theory in some of the denominations is that persons who go to regional or national assemblies must not go as delegates or as representatives. Such persons could take action which might bind the local church. Consequently a power vacuum developed which pragmatically minded leaders filled as best they could. In the Congregational Christian Churches, a theory developed which if put into operation would have denied representative government. The theory claimed for the General Council the same freedom and independence that was claimed for the local church. In the midst of the merger struggle, the chief executive officer of the denomination wrote a book in which he asserted that "a council is a kind of congregation,"[53] and the editor of the denominational magazine wrote that " 'Power' no more rises from the bottom in our fellowship than it descends from the top."[54]

Representative democracy is surely the best means yet devised for determining the will of large numbers of people, for distributing power in large organizations, and for preserving the freedom of local organizational units. The free churches do not practice representative democracy at the national level. They have not yet faced up to the realities of power. Their national organizations are comparatively new, creations of the past century for the most part. These self-proclaimed archdemocrats will surely one day provide in their governmental structures for the separation of powers, for smaller assemblies, for assemblies so financed that they can meet long enough for true deliberation, for genuine elections of delegates and board members and salaried officials, and for limited tenure in office. Until the time when reforms such as these are made by the congregationally organized churches, claims that they practice a superior type of democracy are in error at the regional and national levels. The government of the local churches is another matter; there the forms of democracy obtain and are exercised according to the concern and skill of the local members.

CHURCH AND STATE

As a denominational family the Baptists and Christians make their greatest contribution to our national life through their insistent pressure for the preservation of religious liberty and the separation of church and state. These churches can always be counted on to take action opposing an ambassadorship to the Vatican, or disapproving the granting of public funds to sectarian schools, or condemning the State Department for adopting Roman Catholic policy in South America. It is no accident that the first executive of the organization "Protestants and Other Americans United for the Separation of Church and State" was a Baptist clergyman. Baptists in Alabama and in North Carolina turned down large federal grants for hospital buildings at a time when Methodists and Roman Catholics in Washington, D. C., sought such aid.[55] On the other hand, Baptists seem on the whole to agree with most of the denominations in opposing the move to take away tax exemption from religious property—and thus are willing to accept police and fire protection as well as other services from the state. "In one shockingly casual vote the [Southern] convention ordained all of its foreign missionaries just to satisfy a provision of the Social Security law."[56]

Belief in the separation of church and state has become the firm belief of the overwhelming majority of Americans. Unfortunately, a considerable percentage assumes that the doctrine means the separation of religion and state, that all phases of religion are outside the province of the state. But, if the position taken in the first chapter of this book is correct, a state is possible only because a group of people share a set of religious values; the heart of a state is religious faith, faith on the part of citizens that their common ideals are infinitely precious and worthy of supreme devotion.

It may be, therefore, that unthinking efforts to keep the state from all commerce with religion will be self-destructive. The thesis is that if a state cannot preserve its spiritual center, it will perforce be supplanted by a kind of society which is more spiritually aware; and this type of spiritual awareness need not, indeed usually has not, led to religious freedom. We will return to this thesis in the last chapter.

CHAPTER IX

THE QUAKERS

THE SOCIETY of Friends was born in seventeenth-century England, the same England which produced the Separatists, Cromwell, the Baptists, Roger Williams, the Seekers and the Act of Toleration. The Society was founded by George Fox. He was the son of a poor weaver. Fox wrote in his *Journal* that at nineteen years of age, in 1643,

... at the command of God, ... I left my relations, and brake off all familiarity or fellowship with old or young ... a strong temptation to despair came upon me. ... I continued in that condition some years, in great troubles, and ... went to many a priest to look for comfort, but found no comfort from them. [Then I] ... looked more after the Dissenting people. ... But as I had forsaken the priests, so I left the Separate preachers also. ... And when all my hopes in them and in all men were gone, so that I had nothing outwardly to help me, nor could I tell what to do; then, oh! then I heard a voice which said, "There is one, even Christ Jesus, that can speak to thy condition"; and when I heard it, my heart did leap for joy. ... Jesus Christ ... enlightens, and gives grace and faith and power ... this I knew experimentally. My desires after the Lord grew stronger, and zeal in the pure knowledge of God, and of Christ alone, without the help of any man, book, or writing. For though I read the Scriptures that spake of Christ and of God, yet I knew Him not, but by revelation. ... Then the Lord gently led me along, and let me see His love, which was endless and eternal, surpassing all the knowledge that men have in the natural state, or can get by history or books."[1]

Fox's spiritual victory brought with it important convictions: such as that an Oxford training in divinity is not of itself sufficient evidence of a man's fitness for the Christian ministry, that religious services need not be held in consecrated buildings, that within the heart of every man is a divine spark which will leap into flame unless it is smothered by sin and indifference, that speaking in religious services should be under the immediate guidance of the Holy Spirit, that true religion must be personally experienced, that every plowman is as precious as any peer, that all men should be treated with love and respect.

These were radical and creative ideas. But Fox's true significance lies not as much in the nobility of his vision—at any time in history there is a multitude of people who dream of how the world can be made better—as in the extent to which he incorporated his ideals into actual living, his own and that of other people. Without a university training and without ordination by the Church, he dared to become a minister of Christ. His presumption brought him into conflict with the authorities, and he was thrown into prisons—foul places—eight times, for a total of six years. He was mobbed and beaten; but he never became bitter or counseled violence. And he insisted on treating all men as equals.

In carrying out his belief in equality, Fox declined to adopt the growing custom of showing honor to highly placed persons by using the pronoun "you" (then used only in a plural sense) in addressing a single person. Commoners were addressed by the singular "thou" or "thee." Fox insisted on using the singular in addressing any one person of whatever class. Some Quakers still continue the practice among themselves, even though it has ceased long since to be a protest. Most of them also drop all titles when addressing one another; they use simply first and last names: Herbert Hoover, Rufus Jones, Richard Nixon.

Fox also, like Roger Williams, refused to take oaths. An oath, he said, is in direct contradiction to the Sermon on the Mount which reads, "Swear not at all." Moreover, an oath assumes a double standard of truthfulness; Fox asserted that there are no special occasions for which scrupulous truthfulness should be reserved.

The Quakers also adopted the practice of numbering the days of the week—First Day, Second Day, etc.; they adopted a similar practice with the months. These names were chosen because the names commonly given are not Christian, but "pagan" in origin.

Fox had no intention of establishing a sect; he wanted simply to help stumbling, unhappy men and women. He found in town after town a multitude of people who flocked eagerly to hear him; the soil had been prepared by the sectaries of that day: Separatists, Baptists, Seekers, Ranters, Levellers. He made many converts from these groups. Soon a considerable number, especially in the north of England, came "under convincement," a rudimentary organization appeared, and about three score of persons—some of them women!—began to emulate Fox in his itinerant ministry.

These traveling preachers were youths, practically all of them in their twenties. It is instructive to ponder the fact that the majority of the epoch-making contributions in the field of religion have been initiated by men under forty, as is evidenced by the careers of such persons as Jesus, Gautama, Loyola, Luther, Calvin, Wesley and Fox.

At first the new group called themselves "Children of Light" or "Children of Truth" or "Friends of Truth." Finally, the name "Religious Society of Friends" was chosen. The term *Quaker* was first used in derision, probably because some of the members of the Society trembled under the urgency of their emotions.

Thirty years after the beginning of the Society of Friends, there were about sixty-five thousand Quakers in England. The number in that country today is about a third of what it was then; thus the proportion of Quakers in the population has greatly decreased, since the population has steadily increased. Today there are about two hundred thousand Quakers throughout the world, 60 per cent of them in the United States.

QUAKERISM COMES TO AMERICA

In 1667, William Penn, an English gentleman, twenty-two years of age, son of a famous admiral, fresh from a successful term in the army and a season of polishing in France, came under convincement. He adopted Quaker ways, wrote, preached, came into conflict with his father, offended the bishops, and like many another Quaker contemporary spent weary months in prison.

At the death of Admiral Penn, his son came into possession of a debt of sixteen thousand pounds owed his father by King Charles II. Penn proposed that the debt be settled by a grant of land in America. The King, hard pressed for funds to keep his elaborate ménage going, was pleased to get rid of his obligation by giving feudal rights to a section of the American wilderness—over which he nevertheless retained the right of taxation. In 1681, the charter was signed and the King named the region Pennsylvania, in honor of the Admiral, not of the son.

The Friends had already gained a strong foothold in America, especially in Rhode Island, New Jersey, Maryland and the Carolinas. Penn wished not only to establish a refuge for persecuted Quakers, but also to undertake a "holy experiment" in applied Quakerism. The government he founded was remarkably progressive for its time, his relations with the Indians proving especially happy; he was father to the famous Indian treaty of Shackamaxon, "the only treaty," said Voltaire, "never sworn to and never broken."

By 1700, claims one Quaker historian, the "Friends were the greatest single religious organization in the English colonies as a whole, both in their influence and in their promise."[2] A century later, the Friends were still a large body, ranking in number of local groups below the Congregationalists, Presbyterians, Baptists and Episcopalians, in that order. Below the Friends, came the Reformed, Lutherans, Roman Catholics and Methodists, again in that order.[3] Today the Quakers have dropped in numbers far behind all of these groups. Only one hundred and twenty-five thousand Americans are Quakers today. "Two centuries ago, 10 per cent of the population of Pennsylvania and New Jersey were Friends. Today they are one tenth of 1 per cent there."[4]

The major reasons for this decline were probably two. One was the cooling of the spiritual fires which had provided power for the founding of the movement. A time came "when American Quakerism became so quietistic and ingrown that concern for non-Friends almost died."[5] One blunt observer "likened the Society of Friends in Philadelphia to a fragile antique: exquisite, quaint, otherworldly."[6] The quietistic mood lingers on. A contemporary Quaker wrote in the *Friends Journal* of her "great disappointment and sorrow" that another writer in the *Journal* had urged Quaker membership in the National Council of Churches thus showing "little conception of going the lone road of individual responsibility."[7] A member of the meeting where I worship replied to some proposals for increasing attendance by saying, "I have no interest in numbers." The birth of another temper may be signaled by the announcement that "in 1962 the first three-year Quaker seminary is expected to come into being and to develop over a period of five years into an institution fully accredited under the American Association of Theological Schools."[8]

Another reason for the Quaker failure to grow was the type of life the Quakers insisted on their members living; they became a "peculiar people." They wore broad-brimmed hats, said "thee" and "thy," and refused to take oaths. They disowned Quakers who married non-Quakers, permitted women to be "ministers," conducted their worship in silence, and refused military service. Such sectarian standards naturally repelled many people. Any religious group whose mores are markedly different from those of the general population, is bound to be small. Conversely, no religious group can become large unless the standard of behavior on which it insists is fairly similar to that of the general population.

There was a revival of spiritual fervor among many Quaker groups about the middle of the nineteenth century. Beginning then some of the distinctive "nonessentials" began to be abandoned. Today, Quakers no longer wear the plain dress or disown their members for "marrying out of meeting." And they date their mail just as do other Americans. But the spiritual and ethical essentials remain.

A number of divisions have occurred in American Quaker history. The best known was the split in the third decade of the last century between the Orthodox and the Hicksite groups. The Hicksites felt that only through separation could they preserve their liberty, and "witness" to an increasing liberalism in theology. A marked tendency toward the healing of these divisions is evident in the twentieth century, influenced no doubt by the widespread ecumenical temper of American Protestantism.

WORSHIP AND THEOLOGY

Quakers take satisfaction in the fact that they have no creed; in fact, one observer wrote that modern Quaker theology is "nebulous almost to the point of invisibility."[9] Consequently, Quaker thinking roams all over the theological landscape. Most Friends are evangelical in their viewpoint; but some are practically Fundamentalists and a very few practically Humanists.

Nevertheless, there are some teachings on which all Quakers agree.

One is the nature of the Church; all Quakers believe that the true Church is spiritual and not formal or ecclesiastical in nature. Another teaching is the necessity for religion to be personally experienced. Not even the Bible, according to most Friends, can be put in the place of personal adventures with the Spirit. Fox wrote: "You will say, Christ saith this, and the apostles say this; but what canst thou say? Art thou a child of the Light, and hast thou walked in the Light, and what thou speakest, is it inwardly from God?"[10]

The distinctive Quaker teaching is the doctrine of the Inner Light. All men, say the Friends, have within themselves the true source of their religious inspiration. God is known directly. He is truly immanent and thus is present in *every* human heart. (He is also transcendent and thus "extends infinitely out beyond and above all human life.")[11] Many names have been given to this inner experience of God: the Light, the Light of Christ, the Light Within, the Spirit, the Seed, the Root, the Truth, That of God in Every Man. Today the name almost universally used is the Inner Light.

This teaching accounts for the dignity and confidence with which early Quakers treated all sorts of persons: criminals, Indians, jailers —and women. The Quakers early gave women full spiritual equality with men. They continue to rank above all other old-line religious groups in the opportunities they give women.

At times in Quaker history, the doctrine of the Inner Light has been debased by some Quakers to a kind of superstition. They came to believe that the inner voice must be trusted hour by hour for all the little decisions of ordinary living; and more than one leader came to the conviction that he had a wire connecting him directly with God, and that his opinions, therefore, were infallible.[12] Present-day Friends hold that the promptings of the inner voices are not always or all of God, nor their apprehension of spiritual realities always correct and must, therefore, be checked by experience and be presented for the judgment of other Christians.

The most notable application of the doctrine of the Inner Light has been to the traditional service of public worship; see pages 1 f. for a brief description of such a service. This historic type of Quaker worship (still practiced by about a third of American Friends) is built on the assumption that God speaks directly to the human heart. It is a misconception to think of this service chiefly in terms of silence. Silence is but a means for quieting the tumult of life and letting the Inner Light shine. Traditional Quakers insist that though their worship is "unprogramed," it is silent only as long as no person present is moved of the Spirit to witness to the Truth that is in him. The Quaker silence is alive, not dead; it is the silence of a greenhouse, not of a graveyard. And there are no special persons who because of ordination are thought to be especially fitted to preach or teach. Consequently, there is no professional leader in the *traditional* meeting. "Since we have no [professional] minister"* in the unprogramed meeting, writes one Quaker, "all of us have a responsibility—it is not the abolition of ministry, but the abolition of the passive laity that the Society of Friends has ever striven for."[13]

Worship in the ideal meeting is a wondrous experience. It can take place only among friends, kindred spirits who trust and love each other, who have risen above the necessity of saving face, and who are willing to share their spiritual travail. When a true friend stands simply and humbly in his place and speaks of the joy which follows after selfless devotion, or of the insight that is gained from honest prayer, many a bewildered and lonely seeker feels as though a hand has clasped his in the dark; he ceases to wander alone and becomes part of a spiritual brotherhood which moves onward together.

Of course, many a meeting fails to reach the ideal. Meetings are sometimes attended by strangers whose chief interest is curiosity; they sit the silence out, doggedly waiting for something to happen. Some meetings are disturbed by the habitual speaking of persons whose "opening" seems to come not from God but from the "will of the creature." Cranks and neurotics also occasionally put in their appearance. Such disturbers are "eldered," that is, they are visited and gently but nevertheless firmly instructed that their inspiration is "not of God." But of all the ills which beset traditional Quaker worship the most common is unwillingness to speak. Many a Friend goes to meeting with the firm intention of keeping silent, no matter what vistas may open to him. In most meetings all the witnessing is done by a small

---

* The term *minister* is used by traditional Friends to designate a person who has been singled out as one whose public witness is especially notable. Other denominations would use the term *lay preacher*.

percentage of the membership, and consequently during some sessions no one speaks.

The traditional pattern of worship has been abandoned by the majority of American Quakers. This development resulted from a revival movement which began shortly after the Civil War. The revival brought into the Society a considerable number of persons who were not birthright Quakers. Thus, in some meetings the number of persons who were familiar with the traditional pattern dropped to almost zero. These meetings often employed a pastor. Soon he began to perform the functions ordinarily performed by pastors in other denominations; soon also services of worship followed a more or less set pattern and became in general like the services conducted in neighboring Methodist or Baptist churches. Programed (pastoral) meetings center in the Middle West and unprogramed ones on the Atlantic seaboard. In other parts of the world, except for a few meetings in Canada and some in Africa, Friends meetings are unprogramed.[14] About half of the pastoral meetings have Fundamentalist leanings.[15]

MYSTICISM

The silent-meeting Friends are the best-known American exponents of that type of religious experience known as mysticism, an experience which is found among all traditional religious groups. Students of mysticism disagree sharply concerning its nature, characteristics, and definition. For purposes of this exposition, a religious mystic can be defined as a person who feels that he has "an immediate, intuitive, experimental knowledge of God."[16] The following are some mystics whose religious experiences are well known: the prophet Isaiah (see Isaiah 6), the Apostle Paul (see Acts 9), Francis of Assisi, Martin Luther, Ignatius Loyola, Bernadette Soubiroux and George Fox.

To such sensitive persons come moments when they feel they are enveloped in God's power, and earth becomes the very vestibule of heaven. Fox describes one such experience as follows:

Now was I come up in spirit through the flaming sword, into the paradise of God. All things were new; and all the creation gave another smell unto me than before, beyond what words can utter. I knew nothing but pureness, and innocency, and righteousness. . . . The creation was opened to me; . . . And the Lord shewed me that such as were faithful to Him, in the power and light of Christ, should come up into the state in which Adam was before he fell.[17]

Wordsworth in *Tintern Abbey* describes a mystical experience:

> That serene and blessed mood
> In which the breath of this corporeal frame,
> And even the motion of our human blood,
> Almost suspended, we are laid asleep
> In body, and become a living soul:
> While, with an eye made quiet by the power,
> We see into the life of things.

Students of mysticism sharply disagree concerning its source and nature. Some of them contend that all mystical extremes are to some degree pathological. Contemporary psychologists especially tend to dismiss the extreme forms by calling them self-hypnosis, hallucination, hysteria, endocrine imbalance, and point out that similar experiences can be produced in almost anyone by the administration of certain drugs.[18] The mystic himself often will admit that such factors play a part in his experiences.[19] But he will not admit that they account for the whole of it; on occasion he may even insist that God works through such psychological mechanisms, just as He works through the other laws of nature. The extreme mystic has no doubt whatever that something real has happened to him; he "knows" he has communed with God and not merely with his own unconscious self; the testimonies of the extreme mystics are unanimous at this point. The person who wishes to understand religion will ponder this fact; perhaps the mystical capacity is a sixth sense, slowly emerging in the long evolutionary process.

Extreme forms of mysticism are rare, very rare in pragmatic America. But milder forms are common, even usual. Any person who has the *emotion* of worship has a type of mystical experience, one whose most significant difference from the experience of the extremist is the degree of intensity. It is this mild sort of mysticism that the Friends cultivate. No visions, or apparitions, or dazzling lights appear to Quakers in meeting. Silently they seek the presence of God; they feel the warmth of His presence, gain a sense of wholeness, review their actions, renew their dedication. This same experience is, of course, the fruitage of every successful service of worship, held under whatever auspices.

Most mystics who do more than just have a pleasant journey in a dream world ordinarily find it necessary to follow some system or order in their contemplation. Quakers sometimes say they need pegs on which to hang the silence. One Quaker in a widely read article described his own very personal method as follows:

The first thing that I do is to close my eyes and then to still my body in order to get it as far out of the way as I can. Then I still my mind and let it open to God. . . . I thank God inwardly for this occasion, for the week's happenings, for what I have learned at His hand, for my family, for the work there is to do, for Himself. And I often pause to enjoy Him. Under His gaze I search the week and feel the piercing twinge of remorse that comes at this, and this, and this, and at the absence of this, and this, and this. Under His eyes I see again—for I have often been aware of it at the time—the right way. I ask His forgiveness for my faithlessness and ask for strength to meet this matter when it arises again. There have been times when I had to reweave a part of my life under this auspice.

I hold up persons before God in intercession, loving them under His eyes—seeing them with Him, longing for His healing and redeeming power to course through their lives. I hold up certain social situations, certain projects. At such a time I often see things that I may do in company with or that are related to this person or to this situation. I hold up the persons in the meeting and their needs, as I know them, to God. . . .

When I have finished these inward prayers I quietly resign myself to complete listening—letting go in the intimacy of this friendly company and in the intimacy of the Great Friend who is always near.[20]

QUAKER POLITY

Quaker business at the level of the local congregation is transacted through what is called a monthly meeting. Above the monthly meetings are quarterly meetings and yearly meetings, each rising in significance and in area of jurisdiction. The yearly meeting is the chief conference in any particular branch of Quakerism. Theoretically, these regional meetings are gatherings of the whole group; any member can attend and give voice to his concerns. Most of the yearly meetings send delegates to the Five Years Meeting or to the Friends General Conference. The Five Years Meeting is the larger having about 65 per cent of the total number of American Friends; the General Conference has about 25 per cent. A half dozen other Quaker bodies are listed in the *Yearbook of American Churches*.

A Quaker business meeting—at its best—is a highly democratic procedure. The Quakers aim to arrive at a general agreement rather than for one faction to gain a victory by out-talking or out-voting other factions. When a problem is under discussion, the Friends seem to be very leisurely, every member receiving full opportunity to speak his mind. At the end of the discussion, no vote is taken. The "recording clerk," who has listened carefully to the discussion, writes out a "minute," a short statement, in which he records the "sense of the meeting." This minute is then read aloud; if it is generally approved, it stands as the action of the meeting.

Ordinarily the clerk is careful not to record the sense of the meeting unless there is practical unanimity. Most Quakers are very anxious to do complete justice to minority opinion; unless they are in the midst of a crisis, they usually prefer to let a matter go unsettled, even for months or years, rather than to force a small but sincere group to act against its convictions. Many Friends think that this process is the most democratic form of doing business yet devised.

One factor which militates against the democratic character of the Quaker system is the possibility that clerks can take more seriously the remarks of some Friends than of other Friends. The most serious split in Quaker history—the Orthodox-Hicksite division—was occasioned in large measure by the fact that the clerk of the Philadelphia Yearly Meeting was accustomed to give "no weight" to the opinions of "practically all" the men who later became the leaders of the Hicksite group.[21] That such tendencies have not been entirely overcome is shown by the following excerpt from a letter printed in the correspondence columns of the *Friends Intelligencer*.[22]

What are Quakers offering the seekers who fill half the benches especially in the city meeting houses? . . . Many convinced [as opposed to birthright] Friends have been given Meeting responsibilities. But how many times are those the ones who have formed matrimonial or economic ties, or both, with the old Quaker families? . . . The Religious Society of Friends must use all the abilities of all its members.

The tendency to bureaucracy also plagues the higher reaches of Quaker administration. A leading Quaker notes:

[There is evidence of an] unintentional transfer to a few persons of both the labor and the authority that formerly large committees or the whole community shared. In spite of appearances to the contrary, often a few secretaries and their associates formulate policies and in effect make decisions, and often the rank and file are only too willing to have it so.[23]

Status plays an important role in Quakerism, just as in other American institutions. Indeed, the consciousness of status—consciousness of a person's position in a social structure—plays a very prominent part in the psychology of all the religious groups with which I am familiar. A priori, one might suppose that professed followers of Jesus would lean to humility. But many leaders do not. The motive of service is strong among churchmen, but not that of self-effacement. A man who is in a position to make personal observation says that at one denominational headquarters the thickness of the carpets on the floors of the offices corresponds to the ranks of the various executives. Among ecclesiastics there is at least an average amount of pride, rivalry, competition, and comparing position with position. This opinion is expressed here because striving to achieve status is probably less characteristic of the Quakers than of the other groups discussed in this book. Yet self-magnifying motives are still clearly evident in many a Quaker meeting. One Quaker leader in his personal actions reminds me of the story of the monk who is said to have declared about his order, "We do not have as much influence, or learning, or zeal as some other people; but in humility we beat the world."

SOCIAL SERVICE

Most Quakers have not been content merely to enjoy their religion; they have harnessed it to the task of alleviating the ills of society. Their record in the field of social service is consequently very distinguished. All through the colonial period their friendship with the Indians was notable. At the time of the Revolution, the Quakers initiated a great reform in the treatment of criminals. One historian of the social sciences writes:

The reaction against the brutality of corporal punishment seems to have been due almost entirely to the Quakers. . . . This group took its Christianity very literally. In the last quarter of the seventeenth century, in the two Quaker colonies of West Jersey and Pennsylvania, the savage European and Puritan criminal law was repudiated, corporal punishment abandoned, and imprisonment substituted as the normal method of treating the violator of the law.[24]

The Friends also pioneered in the abolition movement, in the agitation to secure equal rights for women, and in the effort to secure humane treatment of the insane. Religious institutions often take the lead in meeting human need and then after the feasibility of a given approach has been demonstrated step aside while secular institutions take over. An interesting example is the Student Y.W.C.A.; it pioneered in such firmly established academic procedures as student employment services, vocational counseling programs, foreign-student advisory programs, freshman orientation sessions, and even physical education classes for women.[25]

The best-known contribution of the Quakers to public morals has been in the peace movement. They rejected the use of force as a substitute for fair dealing. They believed that all men have the capacity to respond to love. In their Christian experience, they saw no occasion for war, not even "to do God service." By the time of the American Revolution, they had adopted a doctrinaire pacifist position, and most of them refused to serve in the army; those who did not refuse—including Generals Nathanael Greene and Thomas Mifflin—were read out of meeting. The Orthodox repeated this pattern during the Civil War, one Quaker periodical of that period declaring that to speak of a "fighting Quaker" was like speaking of a "blunt sharpness, a jet black whiteness, or a sinful godliness."[26]

The strictness of the Quaker opposition to participation in war slackened considerably in the subsequent decades; by World War I, Orthodox meetings no longer disowned members who enlisted. The chief cause of the change was that the nineteenth century revival emphasized a type of religious experience which did not necessarily include the removal of all military service from the "saved believer's" character. One historian estimates (on the basis of incomplete statistics):

. . . of the young men drafted or liable to the draft possibly 350 stood against any service under military direction as straight out C.O.'s [conscientious objectors]; about 600 accepted some form of non-combatant service, and about 2300 went into combatant service. . . . The "non-pastoral" yearly meetings and the students and alumni of Quaker schools furnished a larger proportion of C.O.'s than other groups.[27]

By the time of World War II, a large percentage of the Quakers agreed with a resolution of the Indiana Yearly Meeting, passed in 1943:

> We do not censure Friends who conscientiously take part in the war effort any more than ourselves for failure to create conditions that might have done much to have avoided this war.[28]

Many Quakers who personally were absolute pacifists favored this resolution. According to one estimate "roughly 1,000 men were in C.P.S. [Civilian Public Service], 1,000 in non-combatant service, 1,000 deferred, and 100 in prison; while 8,000 were in the armed forces."[29] Some Quaker families had one son in the army and another in the conscientious objector's camp.

During the cold war nonpacifist attitudes among Quakers have continued. Elton Trueblood, a widely known Quaker, after a term of leadership in the United States Information Agency, wrote as follows in answer to the question: "Is the old pattern of conscientious objection to war enough?" "Assuredly it is not. . . . The Quaker, providing he is sincere about wanting peace, will not try to undermine the deterrent power of the West."[30] Trueblood was reported in the press as having attacked presidential candidate Adlai Stevenson for being "irresponsible" in proposing an end to nuclear tests and for "talking nonsense" in suggesting the end of the draft.[31]

Some indication of the extent to which pacifist attitudes still obtain among Friends is given by a poll of 287 Friends and nonmember attenders in the Philadelphia Yearly Meeting "between" the ages of sixteen and twenty-one. Some type of pacifist position was taken by 46 per cent, while the position favoring regular military service was taken by 32 per cent; the rest were uncertain.[32]

Quakers are probably more prominent than any other denomination in promoting dramatic efforts to bring pacifism to public attention. An Illinois group sponsored a movement to promote voluntary taxation for the benefit of the United Nations. The Pacific Yearly Meeting proposed legislation permitting persons "with sincere convictions against military preparations" to pay their income tax to the United Nations International Children's Emergency Fund—provided that as a proof of sincerity they paid 5 per cent more than their normal tax liability. The New England Friends delivered to President Eisenhower a petition signed by ten thousand persons urging the banning of nuclear bomb tests. Quakers organize peace walks and peace vigils. One vigil in which Quakers participated largely was held at Fort Detrick, Maryland, a center for research in germ warfare. The vigil lasted twenty-one months; ten hours a day, two to over one hundred persons stood quietly opposite the main gate displaying a sign which explained that they were making "an appeal to stop preparation for germ warfare."[33]

Quakers with pacifist beliefs have not been content to sit the wars out. On the contrary, they have considered that the suffering and tragedy of war imposed on them an obligation to serve. The best-known of the Quaker efforts in this direction has been the American Friends Service Committee, founded in the middle of World War I; the founders declared:

We are united in expressing our love for our country and our desire to serve her loyally. We offer our services to the Government of the United States in any constructive way in which we can conscientiously serve humanity.

They applied their principles even more effectively outside than inside the United States—the need was greater there. After the lifting of the food blockade, the Quakers went to great lengths to feed the starving children of enemy nations. In 1921, a million children were fed; in 1922, half a million; and again in 1923, a million. At one point the Quakers were the largest distributors of milk in the city of Vienna.

These humanitarian activities, along with the assurance by the Quakers that they had no ideological goods to peddle but were trying merely to alleviate suffering, won a wide measure of confidence. So much so, that the Nazi government even permitted them in 1938 to give help to the Jews who were victims of the great pogrom touched off by the shooting of the Nazi leader von Rath.

After thirty years of existence, the American Friends Service Committee had dispensed about seventy million dollars (much of it given by non-Quakers) in goods and services to about six million people.[34] In its fortieth year of existence, the Committee dispensed more than six million dollars and had four hundred and twenty members on its world-wide staff.[35] And in 1947, the Committee received (along with the Friends Service Council, London) the Nobel Peace Prize.

The Friends have insisted that their service activities should be non-political in nature. This stand—and the Quaker philosophy in general —has recently come in for much admiration. That this praise is by no means universal is shown by the following pointed paragraphs:

The Quakers are by and large social conservatives. They are not so "non-political" as they like to think. Typical of a group which has known persecution (even if in the historical past) and has finally gained social acceptance, they have a deep desire for conformity. Economically, too, they have a large stake in the preservation of the status quo—witness the large percentage of strong anti-New Dealers. Wealth is not a sin to most Pennsylvania Quakers, and the system through which they got their wealth must of necessity be worthy of protection.

But to attribute the Quaker political position entirely to such motives would be superficial. A more subtle reason for stand-pattism among the sincere Quakers is their doctrinaire opposition to "violence" and their advocation of the "friendly way of life." Their position on the war is plain and generally known, but from a similar motive they are apt to be conservative in domestic affairs. By "violence" the Quakers are likely to mean any kind of coercion, not alone military, and opposition to violence might conceivably mean opposition to expropriation of all incomes over $25,000 as well as opposition to one man's killing a brother possessed of the same "inner light. . . ."

He can excuse his political indifference by saying that had he opposed fascism in a Madison Square Garden rally, he would not be allowed to help the Jews in Germany.[36]

A leading Quaker counters these objections by noting the existence of the Friends Committee for National Legislation and by saying, "We do not work through political parties, but we can hardly be said to be unconcerned with governmental policies."

That the Quakers generally are conservative in their politics is probably true; that they are more conservative than the other old-line denominations is surely not true. The probability is that in all American churches economic conservatism and radicalism are functions of the stake in the status quo held by the membership. Genuine economic radicalism is confined to a small minority in the churches.

The Quakers must be reckoned with in any assessment of the spiritual condition of the nation. Their numbers are small; their influence is large. Quaker tolerance, Quaker democracy, Quaker spirituality, and above all, Quaker philanthropy have been much copied by other denominations. During the thirties, Quaker pacifism found many followers among the Protestant clergy, and it may find them again.[37] The chief contribution of this denomination, apart from its ministry to its own people, might be said to be in quickening the consciences of the other denominations. Quakers often pioneer. But Quaker methods place a tremendous weight on the sensitivity, initiative and concern of the individual church member. Thus, it would seem clear that in our present religious climate the Quaker denomination is destined to continue to be small and the Quaker method can hardly be expected to play a dominant role in solving the major spiritual problem which immediately faces the American people as a whole, the problem of achieving a spiritual integration of American culture.

CHAPTER X

THE METHODIST CHURCH

THEODORE ROOSEVELT once said:

I would rather address a Methodist audience than any other audience in America. . . . The Methodists represent the great middle class and in consequence are the most representative church in America.[1]

The magazine *Life* commented on this denomination as follows:

In many ways it is our most characteristic church. It is short on theology, long on good works, brilliantly organized, primarily middle-class, frequently bigoted, incurably optimistic, zealously missionary and touchingly confident of the essential goodness of the man next door.[2]

In number of members, counting children, the sixteen million persons in the Methodist family of churches rank third—behind the Roman Catholics and the Baptists. However, one organization, The Methodist Church, containing "99 per cent of all the white Methodists in the United States"[3] probably continues to be our largest Protestant body, although the Southern Baptists may soon overtake it and may even surpass it in size. It has about twelve and one-half million members (counting children). There are a score of other Methodist bodies. The large influence of this denomination in national affairs is seen by the fact that in 1961 it had a representation in the national Congress almost as large as that of the Roman Catholic Church, ninety-five senators and representatives listed themselves as Methodists while ninety-eight listed themselves as Roman Catholics.*

* See *Christianity Today,* Dec. 5, 1960, p. 205 and Jan. 2, 1961, pp. 287 f. Using the data presented by *Christianity Today,* I have computed congressional membership in comparison to the reported size of the denomination. This computation shows that the groups below the average for the denominations are the Roman Catholic, Baptist, Lutheran, Jewish; the groups near the

Yet one hundred and fifty years ago the Methodists were hardly noticed whenever the ecclesiastical great of that period gathered to appraise the religious status of the nation. Eighty-three preachers and about fifteen thousand members comprised the whole of the Methodist Episcopal Church at its organization in Baltimore in 1784; it was a young, upstart organization, often treated with contempt; yet in a little over fifty years it had become the largest Protestant body in America.[4] The story of this remarkable growth begins in England, a century after the Puritan revolution.

JOHN WESLEY

The founding of Methodism was the work of an Anglican priest, John Wesley, perhaps the greatest religious leader the Anglo-Saxon peoples have produced. He was born in 1703 in a country parsonage, just fourteen miles from Scrooby, original home of the Pilgrims. During his long life, he lived through the revolt of the American colonies, saw the industrial revolution develop strongly, and got news, a year and a half before his death, of the storming of the Bastille.

Wesley was greatly concerned over the religious condition of his time. The Puritan revolution of the previous century had provided for religious freedom but had not abolished the Establishment. Sheltered behind the wall of tax support, many Church of England leaders became complacent. They catered to the aristocracy and all but ignored the needs of the common people, who were being forced by the industrial revolution into new localities and into new ways of living. "Enthusiasm" in religion was particularly despised; all throughout polite society in eighteenth-century England, visible concern for the

---

average are the Methodist, Christian, and Mormon; and the groups above the average (in a rough order of increasing distance from the average) are United Church of Christ, Presbyterian, Friends, Christian Scientist, Episcopalian, Unitarian. This listing compares one denomination with another. All the denominations listed have their share of representation according to their ratio of the total population. This result is possible because according to the data only three members of Congress are "not listed" according to denomination while the churches reported as members less than two-thirds of the population. Relationships among the denominations correspond roughly to the relationships which obtained in 1948 (see Jacob S. Payton, "The Church in Congress," *The Christian Advocate*, Apr. 15, 1948, p. 495). The chief difference between 1948 and 1961 is that both the Jews and the Roman Catholics rose in the computation to the point where they became roughly equal to the Baptists and the Lutherans.

spiritual welfare of either oneself or one's neighbors was definitely bad form. Wesley's movement was the most successful of a number of eighteenth-century spiritual movements which challenged this moral and spiritual indifference.

Fifteenth among nineteen children, his pietistic and methodical rearing developed within him both a remarkable religious sensitivity, and a remarkable capacity for scholarship. At Oxford, he had a distinguished record, and at graduation was made a Fellow of Lincoln College. There, he became the center of the Holy Club, a group of young men who banded together for the purpose of cultivating personal piety. The Club was strongly ascetic, adopting austere rules for the guidance of the members. One rule required:

[They should frequently] interrogate themselves whether they have . . . prayed with fervor, Monday, Wednesday, Friday, and on Saturday noon; if they have used a collect at nine, twelve, and three o'clock; duly meditated on Sunday, from three to four, on Thomas à Kempis; or mused on Wednesday and Friday, from twelve to one, on the passion.[5]

Such religious fervor and, above all, such systematic regularity was completely foreign to the Oxford pattern of life. Consequently, undergraduates derisively called the group *Methodists*.

In his middle thirties, Wesley began a serious effort to revitalize the church life of England and also to carry the message of Christianity to the unconverted. For the rest of his life, he traveled back and forth over England, preaching and forming *Societies*. At first he worked within the parish churches, but soon their pulpits were denied to him. Then he made the momentous step of preaching out-of-doors. This action scandalized his fellow clergymen, who were strongly imbued with the idea that religious services could take place only in a properly consecrated building. Vigorous opposition developed, but in spite of protest, ridicule, and even mob violence, Wesley persisted. He strove particularly to reach the neglected poor.

His magnetic personality soon attracted able leaders; some of them were clergymen, but most were laymen, and before many years the country was being systematically covered by his band of itinerant preachers. Progress was rapid. The year before he died, at the advanced age of eighty-seven, there were over five hundred preachers and one hundred and fifty thousand members in the Methodist Societies.[6]

Wesley always remained a priest of the Church of England. Moreover, he considered that his Societies were an adjunct of the Church of England; and he refused to acknowledge that he had in fact founded a new denomination, even after the logic of events clearly pointed in that direction. Immediately after his death, however, systematic snubbing by the Anglican clergy forced acknowledgment that a new Church had indeed come into being.

The most difficult problem for the early Methodists involved ordination. Wesley twice asked the English bishops to lay their hands on his preachers; but the request was twice refused. Finally, Wesley himself undertook to ordain, affirming in defense of his action that in the first century any full-fledged Christian minister had the right to consecrate clergymen.

METHODISM IN AMERICA

Methodism crossed the Atlantic ten years before the outbreak of the American Revolution. During this first decade, the development was encouraging; but during the Revolutionary War, the Methodists had hard sledding. Their preachers from England were Tories, and consequently all but one thought it wise to leave the country after the beginning of the fighting. Nevertheless, they continued to grow throughout the war years as the result of the work of lay preachers.

After the war was over the ties to England were definitely cut, the Methodist Episcopal Church was organized, and leadership was assumed by an ecclesiastical genius, Francis Asbury. Like Wesley, Asbury was an itinerant, a man on horseback. Tirelessly, he rode from Maine to Georgia and back, and across the mountains and back, in and out of the new frontier, organizing and visiting the Methodists. "He visited practically every State in the Union every year."[7] He was the prototype of the circuit rider, the consecrated and fervent evangel to whom Methodism owes most of its present strength and to whom (along with his rival, the Baptist farmer-preacher) the West owes most of its religion and much of its education.

In that day the frontier was a rough and often immoral place, characterized by gambling, hard drinking, swearing, sexual looseness and violence.

> For at least fifty years following independence a vast struggle was going on between the Alleghenies and the Mississippi River; between civilization and Christian morality on the one hand, and barbarism on the other, and upon the outcome of that struggle hung the fate of the new nation.[8]

The preachers struggled to keep alive the standards of civilized society, to save souls from damnation, and to found churches.

The frontier Methodist minister lived a hard life. He had responsibility for organizing and maintaining churches in a given territory, over which he was required to travel regularly. Some of these territories were heroic in size: in 1804, one man was appointed to "Illinois," and in 1807, another was appointed to "Missouri."[9] Ordinarily, the circuits were of smaller compass, such as the preacher could cover in a month or six weeks of systematic travel. Wrote one man:

> I traversed the mountains and valleys, frequently on foot, with my knapsack on my back, guided by Indian paths in the wilderness, when it was not expedient to take a horse. I had often to wade through morasses, half-leg deep in mud and water; frequently satisfying my hunger with a piece of bread and pork from my knapsack, quenching my thirst from a brook, and resting my weary limbs on the leaves of the trees. Thanks be to God! he compensated me for all my toil; for many precious souls were awakened and converted to God.[10]

The circuit rider planned to preach every day except Monday. Daily he would arrive at some appointed place—usually, in the early years, simply a settler's cabin—and lead the small group which gathered in divine worship; he would speak words of comfort and instruction, administer the sacraments, receive members (on probation and into full membership), adjudicate disputes, expel persons who had proved unworthy.

His financial status was precarious; in 1816, salaries were *raised* to one hundred dollars annually. Asbury wrote that at one conference "the brethren were in want, and could not suit themselves; so I parted with my watch, my coat, and my shirt."[11] So rigorous was the life of the circuit rider that only the most robust could stand it for long. "Of 672 of those first preachers whose records we have in full, two-thirds died before they had been able to render twelve years of service."[12] But their devotion to the cause of religion and morality was deep and they left an indelible impression on the West. In any realistic and fair appraisal, they must be ranked among the creative and formative influences in American history.

In 1844, the Methodists split into two groups over the slavery issue. After the close of the Civil War, efforts began almost immediately to get the two groups back together again. These efforts were not successful until 1939; in that year, the Methodist Episcopal Church, the Methodist Episcopal Church, South, and the Methodist Protestant Church united to form the Methodist Church.

THE METHODIST GOSPEL

The Methodists describe their Church as "a company of men having the form and seeking the power of godliness, united in order to pray together, to receive the word of exhortation, and to watch over one another in love, that they may help each other to work out their salvation."[13]

It is commonly said that there are no distinctive Methodist doctrines. In the first decade of the Methodist movement, Wesley wrote:

The distinguishing marks of a Methodist are not his opinions of any sort. His assenting to this or that scheme of religion, his embracing any particular set of notions, his espousing the judgment of one man or another, are all quite wide of the point. Whosoever, therefore, imagines that a Methodist is a man of such or such an opinion is grossly ignorant of the whole affair.[14]

Yet Wesley had no hesitancy in setting forth his own opinions on certain matters. For example, he attacked vigorously the dogma of predestination, a doctrine which had dominated Protestant thinking ever since the Reformation. He would have no part of arbitrary damnation. Instead, he held that God desires the redemption of all men, that His Grace is free, and that man has an indispensable part to play in his own salvation. The difference between the Calvinist and the Wesleyan teachings at this point can be indicated by an illustration which comes from India; it contrasts theories of salvation by comparing them to the differing ways cats and monkeys carry their young. Kittens are carried without effort on their part; but young monkeys must themselves hang on. Wesley taught that man is himself responsible for being in a position where he can receive the Grace of God.

Salvation can, therefore, come to anyone who has faith and leads a good life. Wesley was not the first Protestant to declare this doctrine, but it was through him that it became dominant throughout American Protestantism. "We are all Methodist now," said one clergyman. Thus in point of sober fact, Wesley made one of the greatest contributions to modern theological thought.

Wesley was heir to the Pietist-Baptist-Quaker teaching that religion is primarily an experience, an emotion, an attitude rather than primarily a theology. Present-day Methodism continues to be deeply concerned that the presence of Jesus be felt warmly in every Christian heart and that his teachings be taken seriously. Borden P. Bowne, famous Methodist teacher of the last generation, overheard a scholar from another church reproaching the Methodists for contributing little to theological thinking; Bowne's reply is classic: "They have been too busy providing the data of religious experience, about which other men may generalize."

One result of this refusal to split theological hairs has been a deep sense of theological freedom. The Methodists have never gone in for heresy hunts. As a consequence, there is a wide range of theological opinion in the Church. The majority of the membership is solidly Evangelical; yet a sizable proportion is Fundamentalist and a still larger proportion is Liberal.[15] In the South, the Methodists are the most liberal of the denominations and in the North only two small groups are more liberal: the Congregational part of the United Church of Christ and the Unitarian-Universalists. In a poll of clerical beliefs taken by the Opinion Research Corporation for *Christianity Today,* the order of the denominations from liberal to conservative was: Methodist, Presbyterian, Episcopalian, Baptist, Lutheran.[16] In a poll of Methodist beliefs (including both clergymen and laymen), 37 per cent said they believe that "Jesus Christ is both divine and human"; however, 36 per cent said they believe that "Jesus Christ is a man uniquely endowed and called by God to reveal Him to man."[17]

In spite of this Liberalism, an inquirer who reads the *Doctrines and Discipline of the Methodist Church* gets a decidedly conservative impression, even though the book is revised every four years. Twenty-five Articles of Religion are printed near the beginning. They are a redaction—made by Wesley—of the Thirty-nine Articles of the

Church of England. The Methodist Church still stubbornly clings to these statements even though they contradict the living faith of a substantial proportion of Methodists. In a study made of the opinions of Methodist ministers and official board members living in and near Springfield, Massachusetts, disagreement with some of the essential propositions in the Articles of Religion averaged over one-third.[18]

Moreover, one finds printed in the Methodist *Discipline,* immediately after the Articles of Religion, a series of statements called the General Rules. These directions were framed by Wesley for the guidance of his societies and fitted admirably the moral and spiritual needs of eighteenth-century converts who often came from locales noted for their vice. But the Rules contain certain items which today have a decidedly antiquarian flavor. They proscribe, for example, slaveholding, buying or selling slaves, brawling, railing, the using of many words in buying or selling, the putting on of gold and costly apparel, and laying up treasure upon earth. Few Methodists could be found who would think of these proscriptions as speaking to their spiritual condition. A person who wishes to join the Church is not asked to subscribe to the Rules or to keep them. Yet they are so placed in the Discipline that anyone who undertakes a serious study of Methodism is inevitably confronted by them.

Here again we meet the tendency in religion to conserve and to honor the heritage from the fathers, especially their words. This reverence for the words of the past persists long after religious faith has changed, and after language too has changed. The growth of language is such that the words of yesteryear cease after a time to convey the meanings even of yesteryear. Yet the Church persists in lauding its ancient formularies and often requires the faithful to subscribe to them.[19]

METHODIST ORGANIZATION

The local Methodist church is controlled by a large committee called the Quarterly Conference. This body is commonly self-perpetuating; the low democracy of this method of control seems to be satisfactory to most Methodists.[20]

The local churches are organized into District Conferences. At the head of each District is a clergyman called in earlier years Presiding Elder but now called District Superintendent. This official "travels" among the churches of the District, counseling, directing and encouraging them. But his actual authority is small, except in situations where missionary funds are involved and except in meetings where the local laymen are ignorant of his true powers. ("Did you ever see a District Superintendent put over a program in Quarterly Conference?" asked a layman who read this paragraph.)

Above the District Conferences are the Annual Conferences. These groups, "the fundamental bodies in the church," are composed of "all the traveling preachers in full connection"[21] and of one lay member from each pastoral charge. The bishop who presides over the Annual Conference "appoints," theoretically, the ministers to the various local churches—more of this later. The Annual Conference controls ordination and the relationship of ministers to the Church; in these responsibilities the lay members of the Conference have no vote, and no voice. There are about one hundred Annual Conferences in the country.

The Annual Conferences in the United States are organized into about forty Areas; a Bishop presides over each Area. The Areas are organized into six Jurisdictional Conferences. These conferences elect the bishops. Aside from this responsibility, the functions of the Jurisdictional Conferences vary from the North to the South; in religion as well as in politics southerners tend to a "states rights" view. Accordingly southern Methodists put a considerable organizational emphasis on the two southern Jurisdictional Conferences. But a northern bishop wrote that he "pays little mind to jurisdictional organization."[22]

Five of the Jurisdictions are geographical; they cover the entire nation. One Jurisdiction (the "Central") is racial; it is for Methodism's Negroes. The creation of the Central Jurisdiction was a concession on the part of northern Liberalism to southern Conservatism. Prior to Methodist reunion, there were no Negroes in the Methodist Episcopal Church, South; most southern Negro Methodists had chosen right after the Civil War to withdraw and found their own branch of Methodism. However, Negroes in the northern branch of the Church had continued their membership in the white organization, but on a segregated basis; northern Methodism had nineteen Negro Annual

Conferences, most of them in the South. They continue on in the united Church as the segregated Central Jurisdiction.

The highest body in the church is the General Conference. It is composed of representatives elected by the Annual Conferences; these representatives are one-half clergymen and one-half laymen. Bishops do not vote and are not permitted to speak unless they are invited to do so. The General Conference has "full legislative power over all matters distinctly connectional."[23]

According to Methodist practice, a clergyman who is a member of an Annual Conference in the "effective relation" has a right to employment by the Church; he can demand a pastoral charge. In the old days, the power to appoint the pastors to the churches was exclusively in the hands of the presiding bishop; said one clergyman, "I can remember when the bishop was God."[24] The bishop still determines pastoral appointments in some sections, especially in the South. However, in most of the country, the bishops allocate ministers chiefly to the smaller churches. The laymen in the larger and wealthier churches usually insist on making their own selection of ministers, just as do the laymen in congregationally organized denominations. In some cases the laymen have obtained the clergyman they wanted in spite of the disapproval of the bishop.[25] "The bishop is like the King of England," said one minister; "he has all sorts of power, but he dare not exercise it."[26]

Not until 1956 were women permitted to become full-fledged ministers. Prior to that time they had been ordained "local preachers," and as such were appointed to pastoral charges and were permitted to administer the sacraments. But the "traveling" relationship (membership in the Annual Conference) was denied. During one debate in the General Conference on a motion to permit women full clerical privileges until they married, a clergyman said: "We must keep the pulpit masculine so we can get red-blooded men. Our sacred ministry is a fraternity, not a sorority."[27] Clearly, there is a considerable percentage of the clergy in all the larger denominations who echo the sentiment of Samuel Johnson; he said that a woman preaching is like a dog dancing, the wonder is not that it is done well but that it is done at all. In 1956, a proposal which came to the General Conference from Committee recommended full clerical rights only for unmarried women and widows. But after a stirring debate the Conference gave all women full equality with men in competing for Methodism's pulpits. Less than one-tenth of 1 per cent of the persons serving as ministers are women.[28]

The Central Jurisdiction with its provision for racial segregation has been the object of much discussion. Defenders of the arrangement point out that other denominations have had to accept segregation, and that the climate of opinion in the South has been such that a denomination either segregated Negroes or it did no work in that section of the country among whites. Episcopalians and Roman Catholics usually segregate Negroes at the local level. The Congregationalists organized separate synods for their Southern Negro members. The Northern Presbyterians maintain some segregated synods. Negroes themselves have established separate Baptist and Presbyterian as well as Methodist bodies. Some years ago leaders of the American Unitarian Association prevented (with difficulty) the passage of a resolution by the Association declaring that no church which practiced segregation could be a member of the Unitarian fellowship. The passage of such a resolution would have given effective witness to the overwhelming conviction among Northern churchmen that segregation must be destroyed. But the resolution would have been equally effective in keeping Unitarianism out of an area in which it has great opportunities for expansion. Since the Supreme Court decision calling for the integration of the public schools, practically all of the local Unitarian groups in the South have taken integrationist stands. Thus the churches deal with the problem variously; they all compromise to some degree and many of them, appallingly enough, lag behind other institutions.

In 1956 Methodists set up a heavily financed commission to study the problem presented by the Central Jurisdiction. After exhaustive study the commission recommended that segregation be retained and the recommendation was adopted by the General Conference. The General Conference even refused to set a target date for the abolition of the Central Jurisdiction. This action was taken in spite of "the overwhelming grass-roots willingness"[29] discovered among Negro Methodists to forgo the special privileges conferred by segregation. Two arguments were advanced by the majority in favor of retention of the Central Jurisdiction. One was that segregation was the agreement made between the North and the South prior to the union. Abandoning segregation would not be playing completely fair with the southern members of the Church and might even jeopardize the union. Defending the scheme, one of the northern bishops wrote: "It is very easy . . . to proclaim the evils of compromise. But I remember that Lincoln wrote . . . 'My paramount object in this struggle is to save the Union.' . . . I feel the same way about the Methodist Church."[30]

The second argument advanced in favor of the Central Jurisdiction is that it provides for a marked degree of integration in the upper levels of the Methodist hierarchy. Negroes are less than 4 per cent of the total Methodist membership, "and yet they have on the high echelons of the Church a representation of approximately 16.7 per cent."[31] Four Negroes serve as bishops. (However, "no Negroes have been appointed to the highest administrative posts in general boards and agencies."[32]) No other church has a similar record. The Protestant Episcopal dioceses are integrated; yet no Negro serves as bishop in that denomination. Much public attention has been directed toward the laudable Roman Catholic efforts to effect integration, but the Roman Catholic Church in America has no Negro bishop; and in 1953 it had only fifty-four Negro priests.[33] This latter figure contrasts with the over fifteen hundred Negro Methodist ministers. If the Central Jurisdiction were abandoned, Negroes would become "a hopeless minority" in the white jurisdictions. "They could never hope to elect a Negro bishop."[34]

Moreover, integration at the jurisdictional and conference level will not be very realistic until there is a widespread desire among laymen in the local church to really practice integration. When Bishop Gerald H. Kennedy of Los Angeles appointed a Negro as pastor of an all-white church, two-thirds of the members resigned in protest.[35]

Many persons inside and outside the Methodist Church criticize it for an alleged lack of democracy. Methodist machinery is certainly complex, and many people find it to be overly restrictive. One Methodist wrote that Methodists are "the most supervised people outside the Roman Catholic church."[36] But democracy does not mean the absence of authority and order. All government must deal in these commodities. The important questions to ask in judging the democratic quality of a government are: Who determines policy? Who executes the policy? What means are there for choosing and controlling the executives? The Methodist practice of democracy is a long way from perfect, as the brief foregoing descriptions show; but at the denominational level it compares favorably with that of most American institutions and specifically with most religious denominations. If Methodist supervision seems oppressive, the elected General Conference could change that any time it chose.

The Church is often accused by those who know it best of being a great political machine. Borden P. Bowne once remarked that "undoubtedly the Grace of God is operative everywhere, but it often seems less evident in an Annual Conference."[37] One member of the United States Congress is reported to have said, "All I know about politics I learned at General Conference." Concerning the 1960 General Conference, the editors of an official denominational periodical wrote that there was a "tendency to freeze in lines of force dictated by power blocs. . . . This resulted in battle plans for debate and in many agreements made off the floor, even outside the committee rooms. It was a Conference of compromise. . . . Some said it compromised on everything but tobacco."[38] * One observer wrote in *Christianity and Crisis:* "A General Conference of The Methodist Church resembles nothing so much as the Democratic National Convention. There is the same feverish air of excitement; the same unfortunate tendency to lapse into the more florid forms of rhetoric; the same unpredictability of action."[39]

Church polities, of course, play a major role in determining clerical careers in all the sects. The ambitious man keeps his ear to the ground; the sincere man is fortunate if his convictions coincide with the current opinion.

Methodists, wrote Paul Hutchinson, have a well-developed case of "denominational egotism which breaks out in speeches and articles and which tries to keep alive the idea that there is something

* The Conference refused to lift the ban on the use of tobacco by ministers.

special about the way Methodists preach and pray and behave."[40] One group of theological students, seeking admittance into one of the Annual Conferences, was asked on a written examination by a Conference committee to "Enlarge on the Method, Might, and March of Methodism." Like Americanism, Methodism tends to be insular. The Church is so large and the competition in it so keen that many Methodist ministers are preoccupied with the internal state of the institution and with how they can get ahead in it. When Edward L. Thorndike wished in his study of adult learning to investigate a highly competitive occupation, one in which men quickly find their occupational level, he chose the Methodist ministry. In Methodism the ambitious man keeps his fences mended; that endeavor cuts down the time and interest he has for non-Methodist movements and personalities.

The Methodists are strong all over the United States, but they are strongest in the Middle West and the South. From 1940 to 1948, their rate of membership increase was 5 per cent greater than that of the population as a whole.[41] In 1960, however, the rate of increase was 20 per cent less than that of the population as a whole.[42]

SOCIAL ACTION

Methodists are prominent in the movement to use the churches as agencies for social change. Since this movement has found vigorous protagonists in all of the large denominations, our story of its development will deal with the broader picture.

The "Social Gospel," according to its followers, is as old as the Hebrew prophets; Amos, Micah, Isaiah were fearless in their condemnations of luxury and in their demands for a better deal for the poor. During most of Christian history, the desirability of Church action in social situations was accepted by everyone. But latterly many Christian leaders promulgated the thesis that the Church has responsibility for the salvation of the individual only; reformed individuals will reform society.

This point of view is challenged by the social action effort. A major manifestation of social action in America was the antislavery movement. Another dramatic episode was prohibition; it became a Protestant crusade. Working through the Anti-Saloon League, the

Protestant churches—chiefly Baptist, Methodist, Presbyterian and Congregational—were the major factor in the enactment of temperance laws.[43] During these campaigns, many clergymen did not hesitate to bring politics into their pulpits. For example:

On a single Sunday the pastors in more than 2000 churches in Illinois discussed a pending temperance measure. . . . On the last Sunday in January, 1917, more than 3000 [New York] pastors had engaged in a concerted discussion "of the issues then pending before the legislature."[44]

After the collapse of prohibition, the Protestant churches continued to be the major enemy of the liquor traffic. To be sure, many ministers ceased their open attacks on drinking and within the broad ranks of Protestantism were to be found a few "sophisticated" church members who came to "think it a breach of good manners not to offer the parson 'a short one' when he calls."[45] Nevertheless, in 1950, Protestants still wrote "more letters to their Congressmen on Prohibition than on any other subject."[46] In one carefully planned campaign, Methodists sent no less than two hundred thousand letters to a Congressional committee about an antiliquor bill.[47]

The campaign against liquor seems to have quickened in recent years. *The Christian Century* published an editorial in 1958 entitled, "And Now Repeal Has Failed." The editorial asserted that since repeal, insanity attributable to alcohol has increased three times more than other mental disease cases, that crime directly related to drinking has increased three times more than crimes not "stemming from the use of alcohol," that the nation has spent more than twice as much for liquor as it has spent on schools, that in Massachusetts the "gross alcoholic beverage taxes pay only one-eighth of the expenses the governments and people of Massachusetts incur through the use of intoxicants."[48] A writer connected with the Methodist Board of Temperance asserted the correct figure for "social drinkers who have become alcoholics is not one in 20. It now has increased to one in 12."[49] The Methodist bishops in their Episcopal Address to the 1960 General Conference called beverage alcohol "a beast tearing at the vitals of society." And the editor of an official Baptist magazine attacked a correspondent in the magazine who urged the "biblical distinction between moderate and excessive drinking" by asking whether the writer was "prepared also to justify moderate murder, moderate stealing, moderate pluralistic sex relations, and moderate aggressive war, each of which would seem to be sanctioned in some part of the Bible?"[50]

The complexity of the Protestant picture is shown, however, in the fact that other evidence indicates a continued easing of the opposition to liquor. A survey made by the Boston University School of Theology, reported in 1961, shows that "nearly one-third of all Methodists [according to a sampling poll] see no harm in moderate drinking of alcoholic beverages."[51] And in that same year, the United Presbyterian Church for the first time in its history recognized the right of individual members to drink moderately, saying that those who drink and those who abstain "should respect each other and constructively work together in dealing with the problem of alcohol."[52] *

Many other social problems have received the attention of the churches and synagogues. Most of the major denominations have made ringing pronouncements on social questions. The following are illustrative of the tenor of these statements:

It is the tragic record of humankind that many of those who find comfort in the existing order often fail to apply themselves seriously to the consideration of the ills that plague society. It is part of the great social message of the prophets of our faith that salvation can be achieved only through the salvation of society as a whole. *Central Conference of American Rabbis (1928)*

Ruthless competition must give way to just and reasonable State regulations; sordid selfishness must be superseded by justice and charity. *Archbishops and Bishops of the Administrative Board of the National Catholic Welfare Conference (1940)*

We recommend that new motives besides those of money-making and self-interest be developed in order that we may develop an economic system more consistent with Christian ideals. *General Assembly, Presbyterian Church, U.S.A. (1934)*

We stand for . . . the subordination of the profit motive to the creative and cooperative spirit. *General Conference, The Methodist Church (1940)*

Wage workers in agriculture are denied most of the legal and economic protections long accorded to wage workers in industry. . . . The principles of workmen's and unemployment compensation, minimum wage laws, and the right to organize and bargain collectively under the National Labor Relations Act should be extended to wage workers in agriculture. *General Board of the National Council of Churches (1958)*

* Chapter XVII contains a brief description of Alcoholics Anonymous.

Concentrations of power, controlling the channels of information, beguile masses of people into unthinking conformity and slavish striving for goods that do not satisfy. *General Synod, United Church of Christ (1959)*

We . . . urge Congress to enact legislation that would require as a condition of every contract for federal aid in the housing program . . . that the property involved shall not be restricted against any person, on the grounds of race, color, religion, creed or national origin. *American Baptist Convention (1960)*

We urge that Governments of the United States and Canada recognize the People's Republic of China and . . . work for its admission to the United Nations. *American Unitarian Association (1960)*

The widespread influence of social action in the Churches is indicated by a number of opinion polls. Two of the most informative were conducted three decades ago. One poll asked Protestant ministers and Jewish rabbis whether they favored "drastic limitation" of the "annual income that may be legally retained by an individual." The clergy answered "Yes" by the following percentages: Jewish, 84 per cent; Methodist, 81 per cent; Congregational, 81 per cent; Disciples, 80 per cent; Unitarian and Universalist, 77 per cent; Presbyterian, 75 per cent; Baptist, 72 per cent; Episcopal, 68 per cent; Lutheran, 68 per cent.[53]

Another poll asked the question, "Are you personally prepared to state that it is your present purpose not to sanction any future war or participate as an armed combatant?" Clergymen answered "Yes" to this question in the following percentages: Disciples, 69 per cent; Methodist, 67 per cent; Jewish, 59 per cent; Northern Baptist, 57 per cent; Congregational, 55 per cent; Unitarian and Universalist, 53 per cent; Southern Baptist, 52 per cent; Northern Presbyterian, 46 per cent; Episcopal, 39 per cent; Southern Presbyterian, 38 per cent; Lutheran, 35 per cent.[54]

The conservative journal, *Christianity Today,* expressed the opinion in 1960 that "the Protestant clergy . . . are slowly moving away from their earlier larger commitment to the left toward a more conservative social view." This opinion was based on a poll conducted for *Christianity Today.* The methods used in this poll were not published (see pages 109 f.); nor were the results. The results were summarized by *Christianity Today* as follows: "Whereas a decade ago a ministerial survey indicated that 33 per cent of the Protestant pastors (as attested by their answers to barometer questions) subscribed to the processes by which a socialist economy is effected, the more recent poll narrowed the figure to 25 per cent (in contrast to 40 per cent for the general population average)."[55]

The opinions of Methodist laymen in thirty-eight states were sampled by two members of the faculty at the University of Illinois.[56] The following are among the results:

Statement	Percentage of Methodist laymen who accepted the statement
Politics should call forth the serious and intelligent concern of the conscientious Christian.	91.1
Christians should support the U.N. and try to be informed on the major issues that come before it.	88.6
Racial discrimination and segregation in such areas as education, employment, and religion should be eliminated.	78.6
The church should encourage disarmament among the nations with the United States taking the lead.	71.6
The church is responsible for encouraging better farming and business methods which will lead to better living standards and the possibility for a more wholesome life for all.	74.1
Every person should have the freedom to refuse to serve in the Armed Forces if such service conflicts with his religious convictions.	58.3
Profit, resulting from methodical and well-ordered work, is valued and praised as a profit of God's blessing.	82.7
The church should be responsible for helping attain fair and just relations between labor unions and management.	53.7
In the sight of God, no race or color of man is better than another.	98.0
In advancing economic and technical aid to underdeveloped countries and under-privileged people, the United States should have as their first concern not American interests, but the needs of the people involved.	86.6

A vigorous and vocal section of Protestantism reacts violently to the social action program. This group wants no changes in the present social order. One man, addressing the National Council of Presbyterian Men, wrote that no people "so accurately or successfully as Americans" has "set up institutions in conformity with the true nature of man."[57] Another writer in a journal specifically established to oppose social action in the churches wrote:

Who can deny that in the entire story of God and man as given in His revelation, His Church is built on and in the capitalist system? . . . Those people who are known as the best business minds in history were chosen by God for His special revelation and kingdom building. . . . A wealthy and successful cattleman was called to be the father of the faith and the faithful. A millionaire king was appointed to build Him a house of prayer.[58]

The men who attack the social-action point of view claim that they represent the true attitude of an overwhelming majority of laymen. One former clergyman wrote:

There are tens of thousands of Christian laymen to whom the luxury of writing resolutions is not vouchsafed, who are devoted to the capitalist system—though such devotion, to the minds of the cliché-ridden left, is in itself proof of ignorance or venality—are daily giving courageous evidence that our capitalist economy can be used to advance Christian objectives.[59]

Statistical verification of this point of view comes from a survey of lay Protestant opinion in Baltimore:

The [local] church, as an institution, elects its conservatives to office; and, unless one is the minister, the farther one gets into the administration of local church affairs, the more conservative one is likely to be.[60]

Church members vote more conservatively than citizens in general.[61]*

The opponents of social action have formed many organizations in the effort to prevent the churches from "meddling in secular affairs." Spiritual Mobilization and the Christian Freedom Foundation are two of the most prominent. The well-known industrialist J. Howard Pew has been very active in such organizations. He made an address before a meeting of the National Council of United Presbyterian Men in which he said: "Many men in business and in the

* The various sets of data cited concerning the opinions and actions of church people on social issues are conflicting; the data are not cited in the hope that a clear picture can be given of the status in the churches of given positions on social issues but in order to indicate the amount of attention which they receive from some present-day churchmen.

professions . . . could afford to give far more [to the Church] than they do. Most of them give liberally to charity, moderately to their local churches, and not at all to the corporate Church. . . . They cannot understand how our corporate Church could tolerate such statements and pronouncements on social issues as they have seen in the press."[62] (The Church is "not for sale" sharply rejoined the editor of *Presbyterian Life*.) At one meeting of Presbyterian Men, the delegates after listening to many speeches from conservative businessmen treated with "unforgivable rudeness" the speech of one lone representative of labor.

There were insulting cries from the audience, applause and shouts for the honest admission that labor unionism "is not a perfect organization," calls of "No, no, no" when he proposed to develop a line crediting unions. The moderator of the meeting at one point had to reprimand the delegates with his own little lecture on courtesy. One huddle of Presbyterians . . . proposed a flying wedge to move on the speaker singing "Onward Christian Soldiers."[63]

The ardent social gospeler would agree that wealthy laymen in the churches are indeed of a different mind than are those persons who frame and pass the churches' social resolutions. But the opinion of wealthy laymen, though they may in general control the Protestant churches and the synagogues, is not a good index of the opinion of the man in the pew, any more than the newspapers are a good index of the opinions of the man on the street, especially in an election year. The devotee of social action is "sharply critical of the variety of devices whereby a conservative finds measles as bad as leprosy, regulation the same as socialism, high taxes the same as dictatorship, and Franklin Roosevelt easily linked with Karl Marx."[64] One bishop wrote that "no one is against pronouncements on public questions if the given position taken happens to correspond with his own position. . . . I have often preached against communism . . . and I have received nary a letter of condemnation."[65]

Methodist Bishop G. Bromley Oxnam tells the story of how Thomas E. Dewey, Republican candidate for President in 1948, challenged him by saying that when clergymen talk of economic and social problems they become "fuzzy-minded." Oxnam replied that if so the Republican Platform of 1948 was also fuzzy-minded since that Platform contained most of the proposals urged by the churches forty years earlier in the Social Creed of the Churches, adopted in 1908 by the Federal Council of Churches.[66] Edwin T. Dahlberg, former President of the National Council of Churches, wrote, "It was the vigorous pronouncements Jesus made on controversial matters that sent him to the cross. If he had confined himself to little Mickey Mouse morals, he would never have been heard of."[67]

The average protagonist for social action holds a position much like that of the Methodist who wrote concerning unemployment in a period generally considered to be prosperous:

Nearly five million people, desperately needing and seeking jobs, cannot find them. . . . Physical misery, moral and spiritual breakdown grow. Mass purchasing power declines, effecting more cutbacks, more joblessness, more misery, more degradation. . . .

Let the economists debate the question as to when our unemployment will be "serious" and of depression proportions. It is already alarming for us who seek full, useful employment; who believe with Jesus in the vast worth of *every* individual; and who serve a God who wills not one of the least of His children to perish. For millions of our jobless brothers and sisters depression is already here.[68]

FRANCIS J. McCONNELL

A leading exponent of social action was Francis J. McConnell; he was a bishop of the Methodist Church. The editor of a prominent liberal weekly called him "one of the greatest assets which our nation possesses," but a conservative editor called him "the most dangerous man in the Church." A review of his career will illuminate some Methodist patterns and indicate the concern which the devotee of social action brings to the problems of suffering and injustice.

McConnell was the son of a Methodist minister, the graduate of a Methodist college and theological school, the pastor in rapid succession of five Methodist churches, and for a period of three years the president of DePauw, a Methodist college. At DePauw, he demonstrated his administrative ability by leading a financial campaign which doubled the college's endowment.

In 1912, at forty-one years of age, McConnell was elected bishop. He was assigned to the Denver Area and was also given supervision of Methodist missions in Mexico. In order to fulfill this latter responsibility, he familiarized himself with the problems of Latin America, soon becoming expert and influential—so much so that President Wilson called him into council. Dollar Diplomacy was then at its height, the United States Government interfering in the internal affairs of many Latin American nations, even to the extent of military intervention in Cuba, Haiti, Santo Domingo, Panama, Nicaragua, Honduras, and Mexico.

In Mexico, some American oil companies, headed by the infamous E. L. Doheney, with the help of some United States senators (including a man later to be President of the United States and another later to be sent to prison for accepting a bribe), were using the threat of war in the effort to steal oil rights from the Mexican people. In the end, this effort failed and the Good Neighbor Policy was substituted for Dollar Diplomacy. This shift was possible because of an aroused public opinion. Samuel Guy Inman, a leader in the effort to change the climate of opinion, commented:

> The history of our changing foreign policy . . . during this period shows the influence of moral ideals and the religious spirit. It illustrates how courageous public officials like Woodrow Wilson, Dwight Morrow, and Cordell Hull, when backed by vigorous Christian leaders of public opinion, can successfully challenge the most powerful reactionary forces, industrial and ecclesiastical. At the same time it indicates the notable part that Bishop McConnell played as leader of these religious forces.[69]

In 1920, McConnell was transferred to the Pittsburgh Area of the Methodist Church. There he was catapulted into national prominence by the *Report on the Steel Strike of 1919* issued by the Interchurch World Movement. This Movement was the dramatic and highly publicized effort of some post-World War I church leaders to put interdenominationalism into practice on a wide scale, at home and abroad. For various reasons the Movement failed: because it smacked too much of high pressure methods; because interdenominationalism was too feeble at the grass roots; and because, so it was often said, the *Report on the Steel Strike* offended big business.

In 1919, conditions in the steel industry were deplorable. A fourth of the employees worked twelve hours a day, seven days a week. Hours for all workers averaged 68.7 per week. "The annual earnings of 72% of all workers were, and had been for years, below the level set by government experts as the *minimum of comfort* level for families of five."[70] The industry was not unionized and any workman caught joining or even advocating a union was promptly discharged. The companies owned the steel towns lock, stock and barrel, and systematically suppressed freedom of press, speech, and assembly. The following excerpt from a letter written to McConnell by a working man gives a small indication of conditions:

> I am trying hard to live a Christian life, I love my Church . . . but I must confess that when I leave my home in the morning at 5:15 and do not get home until 6:30 at night, and then eat my supper it is after seven and I am too tired to get ready for prayer meeting, and then I must be in bed at nine if I want to get even seven hours or seven and one-half hours sleep. . . . Now, what time have I to spend with my family under such a working system? . . . My prayer is that you will be successful in seeing that humane beings are treated as humane beings.[71]

In September, 1919, the steel industry boiled over in what was up to that time the largest strike in United States history. The issue was fought out in the arena of public opinion. And the strikers lost, chiefly because the newspapers were almost unanimous in giving only the companies' side of the story and in labeling the newly formed union as an organization of "reds" and "bolsheviks." The newspapers of Pittsburgh, Pennsylvania, the center of the steel industry, were particularly servile in publishing only the interpretations furnished by the steel companies.[72]

In the midst of the strike, the Interchurch World Movement appointed a Commission, headed by McConnell, to investigate. The Commission did a thorough job, concluded that the steel companies were engaged in callous exploitation of the workers, exonerated the Union from any connection with "Bolshevism," and recommended "limiting of the day to not more than ten hours on duty, with not more than a six-day and a fifty-four hour week, with at least a minimum comfort wage." Moreover, the Commission urged organized labor to "democratize" the control of unions, to repudiate "restriction of production as a doctrine," to formulate "contracts which can be lived up to."[73]

The *Report* turned the tide of public opinion. The published document appeared a year and a half after the strike had been broken. Yet within eight months after publication, the United States Steel

Corporation had announced that the seven-day week had "been entirely eliminated" and the twelve-hour day would probably be eliminated "in thirty days or a little more."[74] The power of an aroused public conscience had again been demonstrated.

The reaction to the *Report* was by no means all favorable. The unfavorable reactions can be summarized by quoting from the book which presented the steel industry's answer to the Commission:

From the very nature of their business, ministers of the Christian religion have not the training or the experience to make such an investigation, or even to plan and guide such an investigation. . . . If, in the discussion of a partisan question, he [a minister] confines himself to dwelling upon the importance of the truthful and wise solution of the question, . . . he will usually have accomplished his duty far more effectively than if he attempts to instruct his congregation in the merits of the question itself.[75]

To this point of view, McConnell made an emphatic answer:

There should be unremitting emphasis on human values; we should be willing to accept every new insight into the moral character of God and apply it to human life. . . . It is not the business of the Church to tell in detail how to conduct industry, but to create a public opinion insistent upon the human values which the industrial world must heed.[76]

If you want to know the facts about the human consequences of the modern industrial system, the best place to go is to the pastors of the churches. Employers as a class are hopelessly ignorant of the human values involved. They know their job superbly, but they do not and cannot see the human side. The Church can speak with authority there.[77]

Nobody is pleading that the Church turn itself into a bureau dealing with social and economic problems, but it can do much to build a righteous public opinion.[78]

McConnell was no demagogue. Indeed, his native reticence and brevity of utterance were responsible for a widespread reputation for coldness. But coldness is an improper characterization. He was rather concerned to let facts speak for themselves. He was an intellectual, a creative philosopher and theologian, a man who liked for recreation to study higher mathematics. He once debated with atheist Clarence Darrow, who reported afterward that the experience was like "monkeying with a buzz-saw." To such a person, the appeal to force or excessive emotion in a conflict seems especially inappropriate.

He was the object of much criticism from conservative groups; charges of "bolshevism" and "communism" were legion. At General Conference in 1928, he was accused of "maladministration" and "immorality"; but he was exonerated "with the most tumultuous personal tribute"[79] seen at General Conference in a decade. In spite of the opposition to his social message, McConnell was given many posts of leadership. In 1928, he was moved to the New York Area, then unofficially recognized as the leading administrative post in the Methodist Church. In 1929, he was made President of the Federal Council of Churches. In addition, he headed the Religious Education Association, the American Association for Social Security, The People's Lobby, the North American Committee for the Aid of the Spanish Democracy, the Board of Foreign Missions of the Methodist Church, and (for three decades) the Methodist Federation for Social Action.

McConnell's motivation was always religious. His fearlessness, calmness and patience came from his faith that God has created all men of infinite worth and that the churches can be major instruments in achieving the Kingdom of God. He saw and had a major part in bringing about a widespread change in the Protestant conception of the Church's social function.

The power of religious motivation is commonly overlooked by social analysts. Yet men who really believe they labor for ends which are sanctioned by the very nature of things have a staying power denied to all others. One testimony to this fact comes from the nonclerical secretary to the Commission which made the steel *Report;* writing years later, he said:

At first skeptical of clergy, I wondered why they all went through to the end. . . . A commission of lawyers, or legislators, by my experience, would have weaseled out, and scientists would have qualified. These churchmen, faced by the simple "This is the truth, shall the word be spoken?" all, though with some "Lord help me's," voted aye.[80]

McConnell's religious motivation and moral earnestness are illustrated by the following passages:

We must believe in the God of the fair chance for men here and now —the chance to find those laws of life the adjustment to which means the largest liberty—laws which are the expressions of God's own life—

and mastery of which means communion with Him. Progress in Christian liberty can only mean increasing voluntary assumption of higher and higher laws.[81]

Under the present competitive economic system we have reached a stage which lends warrant to the remark . . . that this earth is the lunatic asylum of the solar system. The [present] social organization . . . allows men to starve because there is too much food, to go without roofs over their heads because there are too many houses, and to do without clothes because there are too many clothes. If this is not social lunacy the word lunacy has no meaning.[82]

What is there so sacred about the profit system that the Church must not call it to account? What is profit? It is what remains in business after wages and salaries, interest and insurance, and all forms of service are paid for. In prosperous times this remainder may be just like "findings," representing no service on the part of the finder.[83]

Jesus did not attack wealth as such, but protested against the injustice and misuse of wealth. One difficulty in which we have come at the present time is that we have not taken material wealth for granted; we think of it in a sordid, materialistic way.[84]

THE ROLE OF RELIGION IN THE RESHAPING OF AMERICAN CULTURE

The major thesis of this volume is that society by its very nature is a religious affair, an affair of sharing a code of behavior, and of believing that code to be part of the very structure of things; the thesis is that such sharing and believing is essential to the preservation and vigor of any society.

But merely sharing a code and believing in it profoundly is no guarantee of its moral quality. Many a society is stable and vigorous without being righteous. The moral quality depends on the nature of the code, not on the amount of reverence given.

In the America of the future, a religious code—a mode of life believed to represent ultimate reality and believed with passion—is bound to play the dominant role. The question is what kind of code will it be? Will it be a code such as the ardent social gospelers espouse, which would deny luxury to any until decency has been made possible for all? Or will it be a code such as our business community wants, which gives the major rewards to the strong, and the intelligent, and the industrious—and to those who inherit wealth? Or will it be a code such as the Communists and Fascists seek which attempts to gain social security at the expense of individual freedom? Or will it be some compromise code? The answer depends on which set of moral convictions come in the future to have the deepest foundation in the religious beliefs of Americans.

In Methodism we can see the moral potential of America. Methodists have over and over again shown their willingness to exhaust themselves and their resources in moral crusades. Today, like the rest of Americans, their indignation has been aroused—by systematic propaganda—to hatred of Communism and of Russians. By an equally deliberate effort, their moral passion could be aroused to the support of a more positive program, to the support of a kind of society more truly democratic than the present political and economic structure of the United States.

CHAPTER XI

THE UNITARIAN UNIVERSALISTS

THE NEWEST denomination to appear on the American scene is the Unitarian Universalist Association. It is the result of a merger between the American Unitarian Association and the Universalist Church of America. Talks between these two groups looking toward union were held as long as a century ago. The present union is the result of active negotiations held during the past dozen years; the last stone in the arch was put in place in 1961. The story of each of these denominations must first be outlined.

UNITARIAN BEGINNINGS

"Unitarianism is simply a return from corrupted doctrines of orthodox Christianity to the pure religion of the New Testament," wrote a Unitarian historian.[1] The claim that the first Christians were unitarian (believed God is a unity and did not believe Jesus was God) has far more basis than the average orthodox Christian would like to think. There is no positive affirmation of the Trinity in the Bible. A definite reference to this doctrine does appear in I John 5:7, the King James Version: "For there are three that bear record in heaven, the Father, the Word, and the Holy Ghost: and these three are one." But doubtless this verse was not written by John, since it does not appear in the oldest copies of John's Letter. Most Liberals contend that the other New Testament passages where the Trinity is supposedly mentioned have, if read honestly, no trinitarian meaning. One Unitarian wrote that in the first three Gospels, the oldest ones, there is "not the remotest suggestion of the doctrine of the Trinity. . . .

In these Gospels we find Jesus simply regarded as the Messiah—a man, sent of God for a high purpose, endowed with superior powers, yet dependent upon God, acknowledging himself not so good as God, and limited in knowledge, authority, and power."[2]

The modern unitarian movement dates from the Reformation. During this period, orthodox Protestants as well as Catholics punished persons who persisted in denying belief in the Trinity. Calvin was instrumental in bringing to death the great antitrinitarian, Servetus. The Catholic Inquisition used every device to ferret out and have punished unitarian tendencies. "Even in England at least ten Protestants were put to death for some form of Unitarianism, and there is no telling how many more died in prison."[3] But in spite of persecution, liberal doctrines continued to be believed. By the seventeenth century unitarian ideas had gained considerable status and were espoused in one form or another by such stalwarts as John Milton, John Locke, William Penn and Sir Isaac Newton.

Modern Liberalism in Christianity got its first great impetus from Deism, a radical religious movement of the eighteenth century. The Deists put their trust in reason rather than in revelation; they contended against the supernatural view of the universe, the miracles, the deity of Jesus, the infallibility of the Bible. For a time such views became the fashion in England; and in America they had a wide influence, both inside and outside the churches. Many of the Founding Fathers held Deistic positions: Franklin, Washington, Jefferson, Madison.

Among the New England clergy the liberalizing tendency was far advanced by the beginning of the nineteenth century—especially in eastern Massachusetts. Ministers in Leominster, Hingham, Quincy, Salem, and Boston attacked Calvinism and espoused a "more reasonable way." In 1785, King's Chapel (Anglican) in Boston adopted unitarian doctrine and worship, and in 1787 this church itself ordained a layman as its minister. In 1805, Harvard appointed an out-and-out Liberal to its Hollis Professorship of Theology.

These developments so alarmed the conservatives that they began a vigorous attack. But the Liberals did not fight back; controversy was far from their wish. They were in a strong position; many were pastors of old and distinguished churches, and had no wish but to exercise the freedom inherent in congregational polity. Yet the conservatives insisted on battle. They denounced Liberalism in sermon and pamphlet. They asserted that the Liberals were not Christian, refused them co-operation, and formed associations from which they were excluded. Finally, in 1825, in order "to diffuse the knowledge and promote the interests of pure Christianity," a small group of laymen and a few of the younger ministers joined in forming the American Unitarian Association.

The adoption of this name repeated a familiar pattern. At the beginning of the controversy, the term *unitarian* had been used by the conservatives in derision. The Liberals had preferred to call themselves rational, or catholic, or liberal Christians. But by 1825 the term *Unitarian* was chosen by the new group; thus again a name used in odium came into honor.

The Unitarians were particularly strong in and near Boston, where the majority of the people of wealth and position were liberal in sympathy. Twenty of the twenty-one oldest churches in Massachusetts went with the Unitarians—including the historic First Parish in Plymouth. The development in that town was typical of the development in many localities. The orthodox members of the Plymouth congregation formed a new organization and claimed that it was the real Church of the Pilgrims. Two church buildings now face each other across the street. On the Congregational building is a tablet which reads in part:

This tablet [honors those who] adhered to the belief of the fathers and . . . perpetuated at great sacrifice . . . the Evangelical faith and fellowship of the church of Scrooby, Leyden, and the Mayflower, organized in England in 1606.

But the Unitarian tablet claims that it honors

The Church of Scrooby, Leyden and the Mayflower, [which]
Gathered on this hillside in 1620, [and]
Has ever since preserved unbroken records,
And maintained a continuous ministry.

In many a Massachusetts town the First Parish is Unitarian, and the Second Parish, Congregational.

The development of Unitarianism belied its early promise. In 1959 the reported membership was a hundred and ten thousand. The reasons for this lack of growth will be discussed later. Boston continues to be the Unitarian stronghold—so much so that the denomination's enemies sometimes taunted it by saying that its creed runs as follows: "I believe in the Fatherhood of God, the Brotherhood of Man, and the Neighborhood of Boston."

UNITARIAN THEOLOGY

Belief in the unity of God rather than in the Trinity is by no means an adequate characterization of Unitarian belief. It is much more complex than that. Unitarians want no part of a restrictive creed; they insist on freedom of belief. Yet they would in general subscribe to the positions outlined under Liberalism in Chapter III. Perhaps the best way to indicate the full range of Unitarian conviction would be to set down some of the Affirmations of Faith which are frequently used in services of worship. One which for a generation was used almost universally in the churches is:

> We believe in
> The Fatherhood of God,
> The Brotherhood of Man,
> The Leadership of Jesus,
> Salvation by Character, and
> The Progress of Mankind Onward and Upward
> Forever.

This statement is seldom used now; belief in inevitable progress is too hard to maintain in a time when man threatens to destroy himself. Of a half dozen Affirmations printed in the Unitarian hymnbook, the following appears to be the most comprehensive:

We believe in God, Father of our spirits, life of all that is; infinite in power, wisdom and goodness, and working everywhere for righteousness and peace and love.

We believe in the ideal of human life which reveals itself in Jesus as love to God and love to man.

We believe that we should be ever growing in knowledge and ever aiming at a higher standard of character.

We believe in the growth of the kingdom of God on earth, and that our loyalty to truth, to righteousness, and to our fellowmen, is the measure of our desire for its coming.

We believe that the living and the dead are in the hands of God; that underneath both are his everlasting arms.

Some understanding of the differences between Unitarian and orthodox thinking can be gained by reading a redaction of the Apostles' Creed which was prepared by Charles Edwards Park, pastor emeritus of First Church in Boston.

I believe in [a single, eternal, all-inclusive, all-pervading Life Principle whose source and perfect embodiment is God, who finds varying degrees of embodiment in all forms of life, who is the prototype of every grace, power, and nobility found in his creation, and whom I call] God, the Father Almighty, Maker of heaven and earth; and in Jesus Christ, [not] his only Son, [for whose son am I? But] our Lord, [because he is a more nearly perfect embodiment of the Life Principle than any one I know;] who was [neither] conceived by the Holy Ghost, [nor] born of the Virgin Mary, [but was conceived and born exactly as we are all conceived and born; and who] suffered under Pontius Pilate, was crucified, dead and buried. He descended into [no] hell, [for, as hell is not a place but a spiritual condition, he never saw the outer door-mat of hell.] The third day [the eager women found his tomb empty, and jumped to the conclusion that in the night] he rose again from the dead; he ascended into [no] heaven, [for as heaven is not a place but a spiritual condition, he never left heaven,] and sitteth on the right hand of God the Father Almighty [if it is any comfort to you.] From thence he shall come [if he is not already here] to judge the quick and the dead. I believe in the Holy Ghost [whom I call Holy Spirit: the spirit in which God works;] the holy catholic Church [so long as it tries to be holy and catholic;] the communion of [what] saints [there are;] the forgiveness of sins; the resurrection of the body [if body means personality: not, if body means this mortal frame, for I am sick to death of my mortal frame and hope to be rid of it soon;] and the life everlasting [meaning a chance to finish out the interrupted opportunities of this life.] Amen.[4]

According to a well-worn quip, "Unitarians believe that there is at most one God." However inadequate this sentence may be when used to summarize Unitarian doctrines, it does clearly indicate a major point of tension in the denomination. Some Unitarians are humanists. This term is used in some quite different ways. When it is used to refer to a type of left-wing Protestantism, *humanism* means a belief that all that exists is natural (as against supernatural) and that human life is the aspect of the universe which religion should be most concerned about. Many humanists would go further and affirm that human life is the highest aspect of the universe; thus they deny the existence of God. One critical Unitarian clergyman wrote:

The humanists are not just activists who think God helps those who help themselves; they are for the most part outright atheists. They have no prayers in their services, the name of God is not mentioned, and they dismiss the people with "closing words" instead of the benediction. While some of them are gentle in their atheism, others are blatant and offensive.[5]

But another clergyman states the issue quite differently:

It is much more in keeping with the spirit of religious humility to admit that whatever divine or creative element is present in life is best apparent on the human level in personalities consecrated to new out-reaches of the human spirit rather than by making unwarranted assertions about the behavior of the divine apart from the human.[6]

With this kind of thinking going on inside the denomination, a good many conservatives have challenged the right of Unitarianism to be called "Christian." An organization was formed within the Association itself, the Unitarian Christian Fellowship, whose purpose was to insure the Christian character of Unitarianism. Ex-Governor Robert F. Bradford of Massachusetts, descendant of William Bradford, spoke before this group saying, "I, for one, will not continue with the label 'Unitarian' if it cannot include Christianity." At the 1949 meeting of the American Unitarian Association, the women of the Church engaged in a heated debate over changing the name of their organization from the General Alliance of Unitarian and Other Liberal Christian Women to the Unitarian Women's Alliance. Many feared the effect of dropping Christian out of the title. This issue is still very much alive, as later pages will show.

UNITARIAN POLITY

Unitarian government is identical in form with that of Congregationalism. Local churches have the same freedom and they are organized into the same kind of superstructure. Unitarian government has been troubled by many of the same problems which trouble Congregational and Baptist government. For example, *The Christian Register,* the former name of the denominational monthly, published an editorial complaining about the concentration of power in the hands of the Executive Committee of the Board of Directors of the American Unitarian Association:

All actions of the Committee were reported to the Board through the medium of mimeographed minutes. But the minutes were elliptical rather than complete and were sometimes so excessively discreet as to carry little meaning to the reader. . . . The effect is that many major policy items are decided by the Executive Committee alone.[7]

There is probably even more freedom in Unitarian government than in Congregational government. The Unitarian desire for freedom amounts to a passion. This fact explains why there are no doctrinal tests in the denomination and why the humanists are so strong in it. On the other hand, there is in the Church a considerable amount of theological smugness, and of intolerance for any but leftish liberal thinking. One well-informed observer who read these pages even wrote, "The kind of liberalism for which a large part of the Unitarian group stands may be the most utterly dogmatic and intolerant."

There have been many explanations of the failure, until the last two decades, of the Unitarian group to grow at a pace corresponding to the pace of many other denominations. One is that Unitarians are "too intellectual." Most Unitarian services leave a large percentage of potential converts with an impression of coldness, of lack of emotional depth. Many observers could say, "When I go to a Unitarian church I usually feel as though I were attending a lecture rather than a service of worship." The effort to achieve a rational faith has frequently taken a negative turn. One minister said, "Many times we have seemed surer about what we don't believe than about what we do believe. People join the Church for the answer to skepticism, not for skepticism."

Perhaps the major reason for the small size of the Unitarian denomination is its lack of missionary zeal. Unitarians have developed an antipathy to missions. Deploring the situation, one minister wrote:

The very word "missionary" was anathema: it was synonymous in our thinking with ecclesiastical inquisitions, religious imperialism and the sight of innocent, morally-erect natives being forced to wear unnecessary clothing and embrace a totally unwanted and even harmful faith.[8]

Shortly after the end of the Spanish-American war, several million Filipino members of the independent Church of the Philippines began a serious flirtation with the American Unitarians. "But our Unitarian principles obliged us to refrain from sending missionaries to the islands to help propagate the faith."[9] The bishops of this

Filipino Church finally secured ordination to the episcopacy from bishops of the Protestant Episcopal Church.

The attitude toward missionary work within the United States was until recently of the same order. In 1949 the American Unitarian Association reprinted a pamphlet in which the pastor of a Portland, Oregon, Church wrote, "We are not actively engaged in proselyting and I think, almost without exception, those who belong to this church . . . belong because they took the initiative."[10] Two years before the merger another clergyman wrote:

> To imagine a world won to Unitarianism has no meaning; a Unitarian world, a mankind persuaded of liberal religion, is from our point of view neither possible nor desirable.
> The varieties of human thought are too numerous, differences in human temperament are too deep, and the needs of people emotionally and intellectually are too divergent to permit the possibility that any one religion ever could satisfy the world.

Yet this writer went on to say:

> But most religious liberals are not yet convinced that anything beyond their local church is really necessary to the advance of their religious outlook in society. . . .
> Yet the truth is that we shall not advance toward enlisting our potential following and achieving our potential influence in the world unless and until we are ready to recognize that an active and adequately supported program on a continental scale is imperative for the advance of Unitarianism. . . .
> It is only lack of money that is delaying the organization of dozens of new Unitarian churches which need assistance at their outset.[11]

One denominational leader affirms that though Unitarians still oppose the idea of a missionary program to "convert the heathen," they no longer think that "they must remain silent and that potential liberals must blunder into a Unitarian Church by accident." He further asserts that during the 1950's the rate of growth was the greatest of any denomination in the country. Growth has been especially large near college and university campuses.

THE UNIVERSALIST CHURCH OF AMERICA

The term *universalism* refers to a belief that in the end all men will be saved. The term arose as a reaction against the orthodox belief that hell is not only a place of punishment, but of eternal punishment.

The Universalists assert that their faith is older than Christianity and was the belief of some of the most distinguished of the early Christians: for example, Clement and Origen, and "probably Chrysostom and Jerome." The assertion is also made that universalism had defenders during a large part of Christian history. The organized form of the movement, however, is American in origin. In the late colonial period, men who reacted against the view that a loving God could condemn men to an eternity of punishment and who were influenced by the mood of Deism began to preach, to write, and to organize churches. In the midst of the Revolutionary War, in 1778, the first association of Universalist societies was formed.

The Winchester Profession, a doctrinal statement, was adopted in 1803. It expresses the general faith which was held in the churches:

We believe that the Holy Scriptures of the Old and New Testaments contain a revelation of the character of God and of duty, interest, and final destination of mankind.

We believe that there is one God, whose nature is Love, revealed in one Lord Jesus Christ, by one Holy Spirit of Grace, who will finally restore the whole family of mankind to holiness and happiness.

We believe that holiness and true happiness are inseparably connected, and that believers ought to be careful to maintain order and practice good works; for these things are good and profitable unto men.

Often the Universalists were attacked for what was alleged to be a belief that a man could do evil without being punished. But in fact they believed that the sinner must pay for his misdeeds. They emphasized God's justice. The teachings which appalled them were those which emphasized God's supposed vindictiveness. The true punishment of God, they believed, brings healing, it is intended to effect repentance, reformation and final salvation.

Universalist theology like Unitarian theology has been in the vanguard of liberal thinking. In general the affinities between the Universalists and the Unitarians have always been close; but these affinities were often ignored in the early years. A Unitarian leader wrote:

Sadly, there was never any enthusiasm among emerging Unitarian leaders for close ties with the embattled Universalists. From 1815 to

1840, as Unitarians were seceding from Congregationalists and Congregationalists were repudiating Unitarians, the Universalists, whose cause was virtually identical on the issues involved, looked longingly for encouragement, cooperation, understanding, and fellowship from Unitarians. Universalism did not possess the social and cultural status of Unitarianism, and reinforcement was desperately needed. It was not forthcoming. Unitarians, for whatever reason, were unable to muster any enthusiasm for the beleaguered religionists who were their closest theological cousins.[12]

The polity adopted by the Universalist churches was congregational. The freedom possible under this system has been made full use of. "Liberty clauses" which guarantee that no doctrinal statements will be used as creedal tests have characterized the denomination almost from its organization. As a result a considerable theological development ensued. In 1899, a statement was issued which reflected influence by the Darwinian theory of evolution and which affirmed belief "in the Bible as containing a revelation from God." By 1935, a change in emphasis was reflected by a statement which avowed "faith in the authority of truth, known or to be known." At one time the idea of salvation held by most Universalists was similar to the idea of most Christians; it centered on the afterlife. In the twentieth century the emphasis has been placed on this life and on its processes of individual fulfillment and social transformation—but not to the necessary exclusion of belief in an afterlife.

In the last century a favorite means of contrasting Universalists and Unitarians was the quip: "Universalists believe God is too good to eternally damn man; Unitarians believe man is too good to be eternally damned." This proposition is much too neat. Both groups believed emphatically in "salvation by character" as the Unitarian phrase put it. The Universalists declared: "We avow our faith in the power of men of good will and sacrificial spirit to overcome all evil and progressively to establish the Kingdom of God."

Freedom in the denomination has meant that persons of all kinds of liberal belief have joined its membership: theists, humanists, mystics. Some wish to be known as Christians; others wish to abandon that name. In 1942, the charter of the Church was changed to read: "To promote harmony among adherents of all religious faiths, whether Christian or otherwise."

New England has been the center of Universalism. In 1960 the Church had seventy thousand members. To the average American Protestant, the usual Universalist church has less to distinguish it from his own church than has the average Unitarian church. The Universalists "know something more of piety, of the emotional components of religion."[13]

THE MERGER

In both denominations the votes to merge were overwhelmingly favorable, more so than were the votes of the Congregational Christians, for example, in the formation of the United Church of Christ. A "ground rule" was set up for the votes which provided that 60 per cent of the local units must vote on the merger and 70 per cent must approve. Over 90 per cent did vote and approval came from 91 per cent of the Unitarian groups, and 79 per cent of the Universalist groups.[14]

The opposition to the merger though small in numbers was spirited in attack. A committee formed by the opposition wrote: "We proclaim that cooperative freedom is more precious than restrictive consolidation. We trust that our brethren will not barter their heritage for the beguiling myth of merger."[15] A central point of the opposition was the alleged lack of a satisfactory name for the merged churches. One opponent wrote: "We have spent many years building up the names Unitarian and Universalist. . . . Now the proposal is that we throw these names out the window." He called this problem "the crux of the matter."[16] Tensions over this problem were no doubt increased by a well-publicized crack made by a delegate at one of the conventions. He proposed that the name be Unitarian; "uni" for the Universalists and "tarian" for the Unitarians.

Of even more import, however, was the relation of the new denomination to Christianity. Long debate at a joint conference of the two churches led to the following paragraph as one of the "corporate purposes":

To cherish and spread the universal truths taught by the great prophets and teachers of humanity in every age and tradition, immemorially summarized in the Judeo-Christian heritage as love to God and love to man.

This paragraph was attacked from both the right and the left. One group published a "remonstrance" in which they wrote, "The name of Jesus has been excluded from the proposed constitution." On the other hand, a well-known clergyman said: "Religion is a bigger word than the word Christian. We look upon truth as universal. . . . There are many good Unitarians who did not grow up in the Judeo-Christian tradition."[17] He hailed an emerging "new world faith." Another clergyman wrote: "We are coming to see that our basic philosophy is making of us a distinctive religious movement that is no longer just a heretical sect within Christianity. Our approach to religion constitutes the birth of a fresh faith in the world of organized religion."[18]

Concerning the theological situation in the denomination a reporter in *The Christian Century* wrote:

> It would be quite inaccurate to see the churches of this new association as simply preserving the modernism of the 1920s; to do so would be to overlook a left wing that shades off almost imperceptibly into nontheistic humanism and Ethical Culture and a right wing of various neoliberals and Tillichians. Within this numerically small denomination is probably to be perceived the greatest diversity that has ever sought and claimed to have found common religious experiences.[19]

One very interesting aspect of the new Constitution is the provision for what appear to be genuine elections for the leadership of the denomination. As has been noted, in some congregationally organized bodies the leadership is actually selected by committees, and these selections are approved by national assemblies in what are called "elections." In the Unitarian Universalist Association, nomination is by committee; but additional candidates may also be nominated by petition. In 1961, a genuine election developed for President of the Association. The terms of office in the Unitarian Universalist Association are for four years; the elected leaders of the denomination may serve no more than three terms.

UNITARIAN UNIVERSALISTS AND
THE ECUMENICAL MOVEMENT

A major force which urges Unitarian Universalists toward an active break with the Christian tradition has no doubt been the treatment they have received from the ecumenical movement. In some instances, they have been refused admittance; in others, they have been expelled. Neither the National Council of Churches nor the World Council of Churches admitted them into membership. Concerning these actions one Unitarian wrote:

I call upon the liberals in other churches to repudiate the two-faced position in which they are placed by the formation of the World Council of Churches with a creedal test. If the World Council has to be formed on this basis—and it is *fait accompli*—then liberals must decide once and for all whether they belong to a liberal or a creedal tradition.

You can see the unenviable position in which the liberals in these other churches are placed. It is the same old position in which a liberal in any creedal church is placed; that of professing creeds which he does not believe that he might remain a part of a great ongoing organization. Or else a man must decide to take the creeds out of their historical context and interpret them allegorically or poetically. . . . Either one must be honest and profess not to believe the creed which makes one suspect, and in this instance would bar one from membership in the World Council of Churches, or else one must accept the creeds and give them an interpretation which historically is dishonest. . . .

I have no quarrel with those who can take this creed and repeat it honestly, straightforwardly, consistent with its history and original interpretation. Men of this mind should form some sort of an orthodox World Council of Churches. My issue is with those who accept this rock bottom basis of unity with tongue in cheek and link themselves with orthodox and traditional Christianity to "get in on the big game" and be on the popular side. They have sacrificed principle for power.[20]

Concerning the exclusion of the Unitarians from the Federal Council of Churches (an ecumenical organization which preceded the National Council), the late Frederick May Eliot, former President of the American Unitarian Association, wrote:

Unitarians may be simple-minded and naïve, but they just cannot imagine Jesus Christ barring the door of any church, or fellowship of churches, that accepts his name and sign, to anyone who asked in sincerity for admittance. By *our* definition of Christianity, such an act would be unchristian in the highest degree, and any definition of Christianity which makes such an act proper seems to us in itself a denial of the essential spirit of the Master.[21]

Evidence that anti-Unitarian Universalist attitudes continue is given by the action in 1959 of the Greater Philadelphia Council of Churches. It expelled two Unitarian and one Universalist churches from full membership and offered "associate membership," a relationship which would require financial support but would deny voting rights. The alleged reason for the action was the refusal of these churches to "accept our Lord Jesus Christ as God and Savior." This allegation is called into question by the fact that in 1957 a Friends Meeting was admitted to the Council even though it explicitly declared its unwillingness to accept the Council's creedal statement; and by the fact that these three churches had been accepted into membership in 1946 when the creedal statement was adopted by the young and struggling Council even though they could not subscribe to the statement. The ministers of the three churches wrote:

> We have grave doubts as to the wisdom, in view of the moral, ethical and spiritual problems which confront the world (and Greater Philadelphia) today, of any action by a group of churches which would indicate that the council is unwilling to accept, on equal terms, the support of all who believe that the ultimate solution of these problems lies in a more widespread acceptance of Christian values as formulated by Jesus of Nazareth. . . .
>
> It is said that this action is necessary to induce other churches, now outside the council, to come in and work with the council. . . . But can they seriously be concerned about the voting rights of three churches in an organization of over 500 churches? Or is it that they will not be embarrassed if we join with them in promoting through the council those spiritual values for which we all stand, as long as we do so as strictly "second class citizens"?
>
> Whatever the reason, we believe that the proposed amendment can be justified only on the ground of expediency, and not on principle. A prophet was moved by God many years ago to say, "And what doth the Lord require of thee, but to do justly, and to love mercy, and to walk humbly with thy God?"[22]

THE FUTURE OF LIBERALISM

The orthodox frequently contend that Liberalism of the kind found in Congregationalism and Unitarianism is bound in the end to fail, that Liberalism is only a bread-and-water religion and lacks the capacity to nourish the deepest loyalties of men, that the denominations most deeply touched by liberal thinking have not grown as fast as the nation generally. Men naturally, the orthodox continue, are trustful supernaturalists and any religion which puts the emphasis on man's responsibility rather than on God's sovereignty is bound in the end to fail.

Liberals, stung by such fighting words, retort that movements far to the theological left of Christian Liberalism have shown abundant capacity to endure: Confucianism, early Buddhism, Marxism. In the struggle against Nazism, the record of Liberals was at least as good as the record of the more conservative religious groups; "the first international treaty of the Nazi regime was the concordat with the Vatican."[23]

Moreover, Liberals contend that the size of the so-called liberal denominations is not a correct indication of the growth of the movement. The number of Liberals outside the Unitarian or Universalist churches far exceeds the number inside; Liberalism has strong representation in all the large Protestant denominations, except Lutheranism. Some indication of how high these percentages are is gained from a review of a study made in the late twenties of the beliefs of five hundred ministers in and near Chicago.[24] The following are some of the dogmas concerning which the ministers indicated their belief or disbelief. The figure following the statement is the percentage of the ministers who were uncertain or who disagreed; these percentages were considerably lower than they would have been if the Lutheran ministers—one-fifth of the total—had not been included.

Dogma	Uncertainty or Disbelief
God is three distinct persons in one.	20%
God occasionally sets aside law, thus performing a miracle.	32%
The devil exists as an actual being.	40%
The New Testament is, and always will remain, the final revelation of the will of God to men.	34%
Jesus was born of a virgin without a human father.	29%
Heaven exists as an actual place or location.	43%
The body will be resurrected.	38%
A visible bodily second coming of Jesus to establish a reign of righteousness on earth will occur.	60%
All men, being sons of Adam, are born with natures wholly perverse, sinful, and depraved.	47%
In order to be a Christian it is necessary and essential to belong to the church.	56%

Another investigation showed in even more emphatic fashion the extent of Liberalism in supposedly orthodox denominations. Thirty-seven Methodist clergymen living in and near Springfield, Massachusetts in 1935–36, indicated as follows their disbelief or uncertainty concerning certain dogmas:[25]

Dogma	Uncertainty or Disbelief
When Christ rose from the dead, he took again his body.	86%
In the unity of the Godhead there are three persons, the Father, the Son, and the Holy Ghost.	27%
The Son took man's nature in the womb of the blessed virgin.	73%
Original sin is the corruption of the nature of every man whereby man is of his own nature inclined to evil, and that continually.	65%
Christ ascended into heaven and there he sitteth until he return to judge all men at the last day.	86%

The reader may remember that in a previous chapter, reference was made to a poll by *Christianity Today,* reported in 1960, which indicated that only 14 per cent of the clergy considered themselves to be Liberals (see pages 109 f.). The discrepancies in the results of these various studies can be accounted for in such ways as the differences in the composition of the groups which were queried, in the phrasing of the questions, in the techniques of sampling, and in the skill of the investigators. These statistics on the beliefs of clergymen must be used with caution.

Obviously, Liberalism is a strong movement. But the question remains: What is the future of Liberalism? What are the prospects of religions of tolerance in competition with religions of dogmatism? Can the effort to maintain an open mind be ultimately successful in an area whose very nature demands an ultimate devotion? Can a church withstand violent social pressures unless its followers believe it to be a divine institution? On a college examination a group of students in a class on "Marriage and the Family" were asked, "In what ways do you want the rearing of your children to differ from your own rearing?" A large percentage answered in substance, "In religion our parents gave us little direction; they told us to choose for ourselves. We want to keep our children from floundering the way we have had to." William James is credited with the remark, "We can't have anything without having too much of it." If Liberalism is to be the religion of future ages, as many adherents believe, will it not have to be a new Liberalism, a Neo-liberalism, whose major orientation comes, not from negation, but from bold and dynamic affirmations?

CHAPTER XII

JUDAISM

ALL THE organized religions of America, with but minor exceptions, are the spiritual children of Judaism. Moreover, the Jewish culture is the oldest mode of living which survives in the West. The fact of its survival, in spite of centuries of attrition and savage persecution, is one of the most amazing and moving stories in history.

Jews debate the question whether Judaism is a religion or a complete civilization. Some hold that Judaism, like Methodism or Christian Science, is a part of culture. Others hold that Judaism is a complete way of life. In order to understand this debate and to get an adequate perspective on the role of Judaism in present-day America, and its potential for the future, we must note some of the major facts of the Jewish historical development.

BIRD'S-EYE VIEW OF JEWISH HISTORY

Our knowledge of the first centuries of Judaism comes chiefly from the Bible.* The story begins in Palestine three and a half millenniums ago with the familiar names of Biblical history, Abraham, Isaac and Jacob, and includes Moses, Joshua, Samuel, Saul, David, Solomon,

---

* What Jews call the Bible, Christians call the Old Testament. From a strictly Jewish point of view, the terms *Old Testament* and *New Testament* are invidious. They imply that the Jewish Scriptures have been superseded. Traditional Jews teach that God made a covenant with the Hebrew people through Moses. Traditional Christians accept this thesis, but contend that centuries later God made a new covenant through Jesus Christ. Jews deny that there was ever a new covenant (testament), reject the New Testament, and contend that their Holy Scriptures are not an "Old" Testament. But, since the term *Bible* in our predominantly Christian culture usually is meant to include the New Testament, Jews in talking to Christians commonly use the term *Old Testament* as a name to designate their Bible, thus illustrating again the tendency of religious groups to give derogatory terms new meanings.

Elijah, Amos, Hosea, Isaiah, and Jeremiah. A notable civilization developed in this era, a civilization characterized by theocratic government, a nonidolatrous religion, a high code of ethics, and the writing of the most influential book of all time. From a military point of view, this civilization was weak. Consequently, during more than a millennium of Hebrew cultural ascendancy in Palestine, the Jews were politically sovereign for only two brief periods; once under David (c. 1000 B.C.) and his immediate successors, and again under the Maccabees, in the second and first centuries B.C.* The rest of the time, they were subject to one mighty empire after another.

In the centuries immediately preceding the Christian Era, there arose in Palestine the sect of the Pharisees. Concerning their importance, a Jewish scholar writes:

The Pharisees constituted a religious Order of singular influence in the history of civilization. Judaism, Christianity, and Mohammedanism all derive from this ancient Palestinian Society; and through their influence in the preservation and advancement of learning, it has become the cornerstone of modern civilization. . . . Fully half the world adheres to Pharisaic faiths.[1]

Most liberal scholars feel that Jesus, though he severely criticized the Pharisees, was greatly influenced by them.

Early in the Christian Era, Palestine ceased to be the center of Jewish culture. Goaded by national pride and the maladministration of the Romans, who at that time were in power, Jewish groups maintained for decades a state of almost constant rebellion. Finally, the emperors determined to destroy the remnants of the Jewish state. A vicious war ensued which ended in A.D. 70 with the destruction of Jerusalem and the issuance of decrees forbidding any Jew to live in the city.

* Instead of the abbreviations A.D. and B.C., some Jews write C.E. and B.C.E. (*Common Era* and *Before the Common Era*). This action is a protest against beginning the calendar with the birth of Jesus.
Traditional Jews have a calendar of their own. In it the year 1 is set, supposedly, at the creation of the world, and the beginning of the fifty-eighth century, the year 5701, occurred in September, 1940 A.D. This Jewish calendar, which is based on the cycles of the moon, comes down from ancient times and consequently does not correspond as accurately as does the Gregorian calendar to solar time. As a result, complex adjustments are a regular necessity if Jewish festivals are to continue to be celebrated in the same seasons. Even with these adjustments, the festivals "migrate" within the Gregorian calendar, much as does Easter.

Following this catastrophe, the cultural center of Jewish life shifted to the provincial towns in Palestine. During these years the so-called Oral Law was developed to a point which has enabled it to serve as the foundation of Judaism down to modern times. The nucleus of that Oral Law is the *Mishnah,* which is supplemented by collects of Scriptural interpretation known as *Midrash.* After about two centuries, the center of Jewish life moved outside Palestine altogether. It is probable that except for a brief period, the Jews were never altogether driven out of the country;[2] but it was not again to be a center of Jewish culture until our own day.

For half a millennium or more residents of Babylon assumed leadership of the widely dispersed Jewish population. There was developed the famous *Babylonian Talmud.* This encyclopedic work is for the most part a record of discussions based on the *Mishnah;* they are interpolated by homilies, legends and ethical maxims. The Old Testament is far more than simply a book of "religion," in the narrow, modern sense of that term. The Old Testament is also a book of history, of poetry, of fable, and especially of law. The Pentateuch (the first five books of the Bible) contains the major legal enactments of the ancient Hebrew civilization. (See, for example, Exodus 20–24, Deuteronomy 12–26, 28.) These laws were believed to have been revealed by God, to be the Torah, the Teaching, the Law of God; (the term *Torah* in the narrow sense is applied to the Pentateuch). But the Pentateuch, according to an ancient rabbinic count, has only 613 laws. They are believed to be fundamental; yet they leave many points of conduct untouched. Thus arose the *Talmud.* It is a vast compendium of writing, over four thousand pages of print, according to one edition. Traditional Judaism is, like every complex culture, a legalism, an attempt to carry standards of conduct down to the level of practical regulations for everyday living. Consequently, in ancient and medieval Judaism, law and religion tended to be one.

In the tenth century, the center of Jewish culture shifted to Moslem Spain. There the Jews were granted economic opportunities, political security and religious liberty—as long as the tax levies were met promptly. Judaism flourished; so much so that this period in Spain is called by some historians the Golden Age.

The first crusade marked the beginning of four centuries of continuous and religiously motivated Jew-hatred. European Jews became outcasts. They were forced to live in ghettos, in segregated and walled sections of the cities into which they were driven at night. They were required to wear a badge, frequently a red or yellow wheel-like piece of cloth, in order that they should never be mistaken for Christians. They were permitted to enter only the meanest of occupations. They were accused of invoking magic to inflict the Black Death. They were expelled from England (in 1290), from France (several times, finally in 1394), and Spain (in 1492).

Driven into the ghetto, the Jew fortified his spirit by religion; he erected out of the Law a wall behind which he retired and to which he gave deep reverence. More and more his conservatism grew. Gripped by fear of any change that might disturb the little security he had, he came to worship tradition.

Yet he also dreamed of the future. He forecast the day when he would again be independent, as he had been in the time of King David. He looked for the Messiah, a heroic leader who would make Jewry the leading people in the world, and Jerusalem the center of the earth's civilizations. The Christians claimed that the Messiah had come in the person of Jesus. They called him Christos, which is the Greek for Messiah. But for the Jews it was unthinkable that the Messiah could have been executed on a cross. They continued to believe that God would send a leader who would lift them out of all their troubles.

The major effort toward Jewish emancipation was initiated by the French Revolution. Little can be said in favor of war as an institution; it is a highly reactionary force, making for spiritual havoc. Yet wars sometimes jar people loose from their outworn ways and make for salutary change. It was so with the attitude of Europeans toward the Jews. The French Revolution leveled the walls of the ghetto and began the movement which gave the Jews in most nations the same privileges, legally, which were enjoyed by other citizens.

But systematic, governmentally sponsored Jew-hatred did not completely disappear. In Russia, throughout most of the nineteenth century and up to the Communist Revolution, fierce persecutions continued. From Germany emanated anti-Semitism, which spread throughout the world and which culminated in the diabolic plan during World War II to exterminate the Jews. "Eighty per cent of all European Jews [exclusive of Russia and England] were starved to death or killed by the Nazis."[3] "Six million were slaughtered in the Jewish Lidice of central and southeastern Europe."[4]

Throughout the centuries, one reaction of the Jews to persecution has been migration; they have spread all over the Western world. In the past hundred years these migrations have had two focal points. Prior to World War I and the subsequent passage of American immigration quotas, Jews came in large numbers to the United States. From the earliest colonial times, Jews were to be found in America; the first synagogue was established as early as 1682. But the bulk of Jewish immigrants arrived between 1880 and 1914. During the Nazi madness, very few, unfortunately, were admitted; but they included some fifteen thousand intellectual leaders.

In recent decades, the bulk of Jewish migration has gone eastward, to Israel. The establishment of a Jewish state in Palestine is a story with which we will deal more fully later.

There are about twelve and one-half million Jews in the world today. Somewhat more than one and one-half million of them live in Israel and about five and one-half million live in the United States. This last figure is often given as the number of Americans allied with synagogues, but is instead an estimate of the number of all American Jews. Many of them have no interest in Judaism as a religion. The number of actual synagogue affiliations is not known with any accuracy. One opinion poll (1952) indicated that half of the American Jews think of themselves as members of a synagogue.[5]

Each of the three major branches of American Judaism claims about a million members. American Jewry is predominantly urban, 87 per cent, according to one estimate, living in cities of over one hundred thousand population.[6] Jews tend to be concentrated on the North Atlantic seaboard, as do other recent, European immigrants.

ORTHODOX JUDAISM

Jews disagree about religion almost as much as do Christians. There are Jewish fundamentalists, middle-of-the-roaders and modernists.

The most conservative group is the Orthodox Jews, or, as they like to refer to themselves, the Torah-true Jews. They maintain as much as possible of the traditional culture, trying to observe scrupulously all the commandments of the Law. According to their belief, Judaism is a revealed religion, Palestine is a sacred land, the Jews are a sacred people, Hebrew is a sacred tongue, the Bible is literally the word of God, the Messianic prophecies will be literally fulfilled, the Jew should take seriously the commands of the Law.

The commands considered to be obligatory in America deal chiefly with ethics and ceremonial. The ethical commands are the familiar moral injunctions of our Judeo-Christian tradition. The best known are, of course, the Ten Commandments. But the Jewish moral code includes also many loftier commandments; for example, Leviticus 19:18, the famous injunction quoted by Jesus, "Thou shalt love thy neighbor as thyself." In that same chapter, Jews are enjoined to love *strangers* as themselves.

The ceremonial commands of the Torah are many and complex. They require, for example, that no work shall be done on the Sabbath day. The devout follower of Orthodoxy goes to extreme lengths to carry out this command. Since the Jewish Sabbath begins at sundown on Friday and ends at sundown on Saturday, places of business belonging to truly Orthodox Jews are closed during some of the best trading hours of the week. On the Sabbath, no food is prepared, no burdens carried, no fires kindled. I once saw an Orthodox rabbi avoid picking up, on the Sabbath, a small object he was describing to a class; if he had picked it up, he would have violated the law against carrying burdens. Orthodox Jews are taught not to turn on the electric lights on the Sabbath; such action is considered a violation of the law against kindling a fire. The lights may be turned on before sundown or a gentile may be hired to turn them on. Orthodox Jews consider that riding on a train or in a car on the Sabbath is contrary to the Torah.

The best known of the Orthodox laws are doubtless those which deal with food. In the Torah, Leviticus 11, is a list of the animals which may not be eaten. Forbidden are animals that have died of themselves, animals which have solid hoofs, animals which do not chew the cud, fish which have no scales or fins. Forbidden also is the

mixing of dairy and meat dishes at the same meal. In order to be
certain that he does not in any degree mix these two kinds of food,
the Orthodox Jew has two sets of dishes, one for dairy meals and the
other for meat meals. During one of the festivals, Passover, he must
also avoid all leaven. In order to make doubly certain that he runs no
risks of eating leaven during Passover, the Orthodox Jew has two
more sets of dishes.

Food that is ceremonially clean is called *kosher*. The manufacture of kosher food is carefully guarded; it must be prepared by or
under the supervision of especially trained and ceremonially appointed functionaries. Consequently, kosher is ordinarily more expensive than is food bought in gentile markets. Yet the number of
Jews who buy kosher is substantial enough that many large manufacturers seek the Jewish trade. The H. J. Heinz Company, for
example, advertises a long list of its products as kosher. And at Passover, this company is at considerable pains to warn the Jewish house-
wife which of its products are not ceremonially clean for that season.
The *Wall Street Journal* reported that the sale of kosher foods is "a
big and growing business."

The inconvenience and difficulty which the Orthodox Jew experiences in attempting to keep the many regulations prescribed by
the Torah are obvious. He clings to them in the difficult environment
of America because he believes they have been commanded by God.

Orthodox veneration for the Scriptures is as great as is to be
found among any sect of Christians. The Bible is studied with the
same absolute belief in its inerrancy that an astronomer would use
in his study of the stars. The orthodox scholar believes that every
word is inspired, the genealogies of the patriarchs being no less
sacred than the 23rd Psalm.

The Orthodox rabbi is firmly convinced that the Law is complete
and final. He feels that Jews who fail to try to keep the whole Law
are renegade. Moreover, he tends to look upon the principles of
Orthodoxy as "beyond reproach," "the source of every Jewish asset."[7]

In the early 1940's an Orthodox rabbi said that Orthodoxy was
the "Judaism of the overwhelming majority of Jews in the United
States."[8] This generalization applied especially to the earlier Jewish
immigrants to this country. But, said this same author, "as the immi-

grant and his children and grandchildren gain a firmer economic and social hold, they regard the demands of Orthodoxy as burdensome and tend to break away. Those who remain are, as a rule, the financially weak."[9] About half the synagogues in the United States are Orthodox. Although this proportion is large, the size of the average individual Orthodox congregation is small when it is compared with the average size of the less traditional congregations.

The Orthodoxy of most of those who try to be faithful to the tradition is a watered-down variety. Living according to all the precepts of ancient Jewry is so difficult in modern America that even most of the Orthodox rabbis do not succeed in it. There is, for example, an injunction in the Torah which reads, "Thou shalt not mar the corners of thy beard." American manners and prejudices are such that practically all Jews disregard this precept. Many keep a kosher home; but many more abide by the Law for only certain items of food or for certain seasons. One writer estimated that in a Midwestern city "no more than 20 per cent of the Jews . . . purchase kosher meats for home use. . . . [And that] the number who insist upon using kosher meat exclusively within their homes and eating only kosher foods outside their homes, is no more than 10 per cent."[10] A large majority are forced by economic necessity to work on the Sabbath. The consciences of a considerable portion are, of course, much troubled by such compromises. But a large number of others are either in a state of open rebellion against all ceremonial laws or demand changes in those laws. Those who rebel are secularists; those who demand changes belong either to the Reform or Conservative wing of Judaism.

REFORM JUDAISM

The most radical of the Jewish religious sects is Reform: in theology, it is very like Christian Liberalism; in ritual, it has abandoned a large portion of the traditional Jewish practice.

Reform Judaism began in Germany in the first decades after the defeat of Napoleon. That was a time of profound reaction, such as is common after a period of war. Most of the privileges granted the Jews under the influence of the French Revolution were then taken away: Jews were forced back behind ghetto walls. The only way of escape from discrimination and ignominy was apostasy. Consequently, a wave of baptisms struck Western Jewry; in Berlin a third of the Jewish population turned Christian.

This alarming situation forced on Jewish leaders a reconsideration of the role of tradition. Some of them concluded that Judaism needed to cast off its ancient dress. By making Judaism "less medieval," they reasoned, they could fortify the spirits of those who were tempted to renounce the faith. Changes in creed and in ceremonial requirement were made—few at first but eventually many.

The Reform temper soon spread to America: since the middle of the nineteenth century American rabbis have had a major influence on the development of the movement. Their most important contribution was in theology. They abandoned belief in an inerrant Bible, and became leading exponents of the liberal approach to Biblical studies. They ceased to believe in the coming of a human Messiah, and adopted instead a belief that mankind can by collective effort achieve a glorious Messianic age. They denied the authority and relevance of the Talmud, and substituted the dictates of rational ethics for the requirements of the Law.

Many ritualistic changes were made by the Reform synagogues. The minutiae of Sabbath observance were abandoned: travel, the conduct of business, the preparation of food on the Sabbath, all became acceptable. The dietary laws were also abandoned; Reform Jews have no theoretical objection to eating pork, for example, although many of them, reared in Orthodox homes, feel uncomfortable in doing so.

In earlier decades Reform Jews believed that for Americans Judaism is a religion and not a civilization. Jews in America, they held, live in one culture and not in two. Their culture is American; their religion, Jewish. Today, however, many Reform Jews hold that Judaism is a culture as well as a religion.

During the last decade or two, many Reform congregations have concluded that they abandoned too much of the Jewish tradition. There is accordingly a moderate movement back toward reinstating the "more meaningful" ceremonies.

This branch of American Judaism has 575 congregations. The Reform synagogues tend to be large, well-financed, and served by able leaders. Moreover, the membership is largely recruited from the well-educated, prosperous, influential and younger members of the community. Consequently, Reform Judaism plays a role in American Jewish life which is out of proportion to its numerical strength.

CONSERVATIVE JUDAISM

Early in the present century, some American rabbis founded a movement which lies between Orthodoxy and Reform. Like Reform rabbis, the leaders of this movement—called Conservative Judaism —contend that any literal following of all the ceremonial requirements of the Torah is not only undesirable but impossible. On the other hand, Conservative rabbis hold that Reform goes so far in discarding the ancient practices as to emasculate religion, and favor retaining all the traditions which can be kept without sacrificing personal efficiency and public esteem. Conservatives retain many of the Sabbath customs, many of the dietary laws, and large amounts of Hebrew in the service of worship.

A wide range of thought and practice characterizes Conservatism; it runs all the way from near Orthodoxy to near Reform. Congregations decide for themselves the extent of their adherence to the tradition.

Conservatism has 660 congregations; it is "the most rapidly growing group of American Jews," receiving "many accessions from the ranks of the former orthodox."[11]

Within the past two decades a small group of Conservative leaders has formed a movement called Reconstructionism. They hold that Judaism is a religious civilization—complete with folkways, law, religion, education, philosophy, literature, art—and urge the enhancement and revitalization of this civilization. They contend that Jews, in order to fulfill their obligation to Judaism, must live in two cultures, one American and the other Jewish. However, Jews "owe their sole civic or political allegiance to the states to which they belong and are nationals of those states alone."[12]

The Reconstructionist view of ceremonial observance is as follows:

A regimen of Jewish religious habits and practices is essential for Jewish religion. But for a great number of Jews in our day the traditional regimen of observance has broken down, in whole or in part, because of changes in conditions or in mental outlook. For them a revised regimen needs to be developed based on a revaluation of traditional observances, one that would satisfy their spiritual needs.

Traditional forms of Jewish ritual observance should be retained, even if their original meaning is no longer valid, provided they have acquired new and valid meanings for us through reinterpretation. Those traditional forms, however, which have no valid meaning for us and do not lend themselves to reinterpretation, need to be modified or replaced by others. Such changes should not be imposed by communal or organizational pressure on those who are not convinced of their need.

New forms of observance giving expression to newly felt needs should be introduced into the ritual of the synagogue and the home.[13]

Reconstructionists deal chiefly with the Jewish culture but they apply their theory to civilization in general. They hold that religion must play a vital role in every worthy culture and consider that the major need of every modern civilization is the revitalization of its spiritual values. Reform as well as Conservative Jews support the movement.

JEWISH BELIEFS

Creeds play no such role among the Jews as they do among the Christians; there is no Apostles' Creed in Judaism. Jews may, indeed, depart widely from the common beliefs of the members of their synagogues and still suffer no disabilities as members. One Jew has written that Judaism is "theologically . . . feeble,"[14] and another that "no religion has since the Middle Ages concerned itself less with philosophy."[15] Judaism is often defined as a system of ethics, a way of living, a way which includes both moral and ceremonial standards.

The famous Rabbi Hillel, who lived in Palestine in the century before Jesus, was once asked to sum up the Law briefly, while he stood on one foot. He replied, "That which is hurtful to thee do not to thy neighbor. This is the whole doctrine. The rest is commentary." It is sometimes said that Judaism is summed up in Psalm 15, another ethical statement.

But it would be a mistake to assume that there is not in Judaism a basic core of theological belief on which a large percentage agrees. Before the emancipation practically all Jews accepted the thirteen principles of Maimonides, the most famous rabbi of medieval times. They are as follows:

1. The belief in God's existence.
2. The belief in His unity.
3. The belief in His incorporeality.
4. The belief in His timelessness.
5. The belief that He is approachable through prayer.
6. The belief in prophecy.
7. The belief in the superiority of Moses to all other prophets.
8. The belief in the revelation of the Law, and that the Law as contained in the Pentateuch is that revealed to Moses.
9. The belief in the immutability of the Law.
10. The belief in Divine providence.
11. The belief in Divine justice.
12. The belief in the coming of the Messiah.
13. The belief in the resurrection and human immortality.

Amplification of some of these statements will indicate the extent to which they are held by modern Jews and also will indicate some points of contrast between Judaism and Christianity.

Jews are strict monotheists and consider that the dogma of the Trinity is a denial of monotheism.

All Jews in the past believed and Orthodox Jews still believe that God spoke in a supernatural manner to mankind through the Old Testament prophets. But the age of prophecy ceased with Old Testament times. Jesus is not considered by traditional Jews to be a prophet. Many liberal Jews do consider that Jesus was in the line of the prophets. But Liberals would deny supernatural powers to any man.

Liberals would deny that the Law as contained in the Scriptures was completely, or even chiefly, revealed to Moses. They would also deny the unchangeability of the Law. And they substitute, as has been noted, belief in the coming of the Messianic age for the belief in the coming of a Messiah in human form.

Jews believe that human nature contains the seeds of both good and evil. They deny "natural" and "total" depravity.

Judaism teaches that man is composed of both body and soul and that the soul is immortal. Jewish conceptions of the future life are much less explicit than are some Christian conceptions.

Judaism teaches that the Jews are a "chosen people"; the Jewish liturgy "constantly refers to the term, 'Thou hast chosen us from all the people.' "[16] This teaching is interpreted variously. For many

Jews, it is undoubtedly a source of great pride; they consider that God set the Jews above all other people and in His own good time will make Jewish superiority manifest. Gentiles are frequently irked by this attitude; for example, George Bernard Shaw wrote, "The fault of the Jew is his enormous arrogance based on his claim to belong to God's chosen race."[17] Chauvinistic as this interpretation of the chosen people doctrine is, it finds clear parallels in most of the sects we have studied. Moreover, the conviction that they were to be the most favored of all the sons of God doubtless made it possible for the Jews to retain their self-respect and maintain their civilization through centuries of persecution.

The dogma of the chosen people is also interpreted to mean that they are chosen to serve mankind or to fulfill a mission. One Jew wrote as follows:

The doctrine of the chosen people offers the Jews no privileges denied to others; on the contrary, it imposes on them a mission, loyalty to which must bring them suffering, humiliation, agonies of pain and death; . . . the doctrine implies no superiority inherent in the Jewish people, apart from the superiority that is attached to one who is charged with the duty to carry an important message. It is the message and not the messenger that is superior.[18]

The Reconstructionists hold that the "idea of Israel as the Chosen People, must . . . be understood as belonging to a thought-world which we no longer inhabit." They consider the idea to be an anachronism, and not susceptible of successful reinterpretation.[19]

It should be observed parenthetically that the Jews are not a "race" in the biological sense. The whole concept that there are significant biological differences among the major human groupings has been pretty thoroughly exploded by modern anthropologists. The only peoples who "breed true to type," apart from such *biologically* unimportant traits as skin color and hair texture, are small groups who have been isolated from their neighbors for many generations. All the large land masses are occupied by hybrid peoples. The Jews are mongrel, like the rest of Europeans and Americans. There is no biological characteristic which identifies a person as a Jew.[20]

On the other hand, there is a considerable "race consciousness" among Jews. Their attitudes toward Negroes, for example, are about the same as those of other whites. Moreover, there is often prejudice within the Jewish group itself, the older immigrants looking askance at the more recent. A student of German ancestry told me that her mother objected strenuously whenever she dated Polish or Russian Jewish boys.

WORSHIP AMONG JEWS

In traditional Judaism the home rivals the synagogue as a place of worship; an elaborate system of prayers and blessings is prescribed both for the family and for its individual members. The ritualistic obligations of the adult male are especially heavy. They begin immediately after he has awakened: with each act of getting up and making his toilet, he recites formal blessings or prayers. Then, before eating, he begins a period during which he gives his full attention to worship; if the entire ritual is completed, this period may run to nearly an hour. Twice more during the day—in the afternoon and at dusk—the observant Orthodox Jew engages in formal worship. These periods are best observed in the synagogue; but today they are frequently, perhaps generally, observed at home or at the place of business. One rabbi wrote:

> Between times he invokes God's name frequently, since the Tradition ordains benedictions for almost every juncture of his life. Should he partake of food between meals, should he don a new garment, taste a fruit just then in season, see a flash of lightning, hear thunder, catch a glimpse of the ocean or of a rainbow or of trees burgeoning in the spring, encounter one learned in Torah or in secular lore, hear good news or be the recipient of bad—for almost every conceivable contingency there exists a brief but appropriate word of blessing.[21]

In the traditional home, prayers are said at the beginning and end of each meal. At sundown on Friday evening, the Sabbath is "ushered in": the mother recites a blessing while lighting the Sabbath candles, the father chants a prayer over a cup of wine, and the family sits down to the most elaborate meal of the week, a meal which is sanctified by prayers, hymns and the recitation of passages from the Scriptures. Throughout Saturday, the members of the family are kept aware of the presence of the Sabbath by ritualistic prescriptions, and just before sundown the Sabbath is "ushered out" by another home service. Many of the annual festivals are celebrated by special prayers, blessings, and meals in the home. A traditional Jewish family, as an evidence of its piety, places on the doorpost of its house a small, pencil-shaped box, a *mezuzah,* in which are placed specified passages of Scripture. If the family moves, a special service of dedication is held for the new dwelling.*

Worship in the traditional synagogue takes place three times daily. It requires the presence of ten adult males, and is conducted in the Hebrew language, except when a sermon is delivered. Then the language used is the one which happens to be the vernacular of the worshipers. No instrumental music is permitted in the traditional synagogue; nor are men and women permitted to sit together. Both are required to wear hats.

The central object in the synagogue building is the Ark, a curtained cabinet which contains one or more copies of the Pentateuch; it is located at the front and center of the assembly hall, the focus of all eyes. Above the Ark is a symbolic representation of the two tablets of the Ten Commandments, and above and in front of the Ark is a light which is supposed never to go out; it symbolizes the eternal light of God. Near the Ark are many-branched candlesticks. In front of it, American synagogues customarily place two desks. From one, the cantor, the leader of the ritual, intones and chants the service, and selected members of the congregation read the Scriptures. From the other desk, the rabbi speaks. In the most Orthodox synagogues, these desks are sometimes placed in the center of the room, the traditional position.

Synagogue worship follows a prescribed order; "extemporaneous services are discouraged in Judaism."[22] "Only divinely-favoured individuals are capable of spontaneous prayer."[23] Thus the exact words to be said during worship are set down in a Prayer Book. The services include prayers, blessings, Scripture readings, hymns, benedictions, sermons. There are no sacraments in Judaism; that is, no

* As a result of an investigation in Minneapolis it is estimated that "less than 50 percent" of the Jews in that city make a religious occasion out of the Sabbath evening meal, and "no more than 10 percent" place mezuzah on their doorposts.—Albert I. Gordon, *Jews in Transition* (University of Minnesota, 1949), pp. 93, 191.

"external" acts which are believed to "mediate the grace of God."
Jews affirm that their relationship with God is solely dependent on
their moral and spiritual condition, and their worship is a spiritual
and not a physical function.

During worship, every person present should read or repeat all
of the service. The function of the cantor is not to say prayers while
others listen, but to set the pace of a congregation, the members of
which are individually offering the prayers and saying the blessings.
If a worshiper arrives late, he is expected to start at the beginning of
the ritual and to catch up with the congregation as soon as he can.
A worshiper may repeat the service silently to himself; but he may
also in moments of fervor break into song. At certain points, unison
and responsive reading are indicated. The cantor does not intone the
entire service aloud; he intones the important sections, such passages as he thinks necessary to guide the congregation, and passages
where his own fervor may break forth. In some Orthodox synagogues, the cantor is assisted by a male choir, a cappella.

Some of the traditional services are very long; the conduct of Sabbath morning prayers, for example, requires about three hours to
complete. Close attention during all this period is not given by the
average worshiper, unless he requires the full time to read the various
passages. Consequently, an air of informality characterizes the
service.

Admiration and reverence for the Prayer Book are very high. The
following are two passages of appreciation.

It breathes a spirit of invincible faith, an earnest desire to be in harmony with God and to understand and do his will. . . . It is a not unworthy sequel to the Psalter from which it has drawn so much of its
inspiration.[24]

When we come to view the half-dozen or so great Liturgies of the
world purely as religious documents, and to weigh their values as devotional classics, the incomparable superiority of the Jewish convincingly
appears.[25]

The most important service of the week in the traditional synagogue is held on Sabbath morning. A brief outline of some of its
details follow; the original is, of course, in Hebrew.

Upon entering the House of God worshipers say:

How goodly are thy tents, O Jacob, thy dwelling places, O Israel! . . .
May my prayer unto thee, O Lord, be in an acceptable time: O God, in
the abundance of thy lovingkindness, answer me with thy sure salvation.

The service proper begins with a statement of some essential principles of Judaism; the first lines of this statement are:

> The living God we praise, exalt, adore!
> He was, He is, He will be evermore!
> No unity like unto His can be:
> Eternal, inconceivable is He.
> No form, or shape has the incorporeal One,
> Most holy He, past all comparison.

Then after a number of prayers and blessings, nine Psalms are
read. After these comes a period of praise; and then is read, after
introductory prayers, the oft-quoted and much-loved *Shema*. It is
composed of passages of Scripture (Deuteronomy 6:4–9; 11:13–21;
Numbers 15:37–41) and holds a place in Judaism corresponding in
prominence to the place of the Lord's Prayer in Christendom. The
best known lines of the Shema read:

Hear, O Israel: the Lord our God is one: And thou shalt love the
Lord thy God with all thine heart, and with all thy soul, and with all thy
might. And these words, which I command thee this day, shall be upon
thine heart: and thou shalt teach them diligently unto thy children, and
shalt talk of them when thou sittest in thine house, and when thou
walkest by the way, and when thou liest down, and when thou risest up.

After reading the Shema and a long passage glorifying it, worshipers recite a series of benedictions—prayers of praise and thanksgiving—called the *Amida,* which is considered to be "the essential
element in all these services"[26] of the synagogue. On weekdays, the
Amida—which must be recited standing—consists of nineteen benedictions; on Sabbaths, of seven.

After the Amida comes the reading of the Bible. The Ark is
opened with great solemnity, a copy of the Torah is ceremonially
brought forth, and it is carried in procession around the synagogue
for all to see. The scroll is then placed on a desk and, after prayer,
the reading begins. The Torah is so divided that in a year the Pentateuch is completed. Traditionally the reading was done by seven men
from the congregation. But today the average man's knowledge of
Hebrew is inadequate for this task; he, therefore, repeats the se-

lection *sotto voce,* or but the first few lines. The passage is read to the congregation by a special official.

After the passage from the Torah, a passage from the Prophets is read. Then come prayers and blessings and the solemn return of the scroll to the Ark.

The rabbi now may, and usually does, preach a sermon. Then follows an additional Amida, and the closing of the service. Just before the end all mourners rise and repeat silently a passage of praise to God. Then, in conclusion, all those present sing a hymn, the beloved "Lord of the World":

> Lord of the world, He reigned alone
> While yet the universe was naught,
> When by His will all things were wrought,
> Then first His sov'ran name was known.
>
> I place my soul within His palm
> Before I sleep as when I wake,
> And though my body I forsake,
> Rest in the Lord in fearless calm.

The services in Reform and Conservative synagogues are revisions of this Orthodox form. The Reform congregations shorten the service, conduct most of it in English, demand decorum of all worshipers, install organs, do not require men to wear hats, permit men and women to sit together. (The novelist Herman Wouk in his defense of Orthodoxy wrote that he was "almost ashamed to record" that seating the sexes together is "the biggest religious issue today in American Jewry.")[27] Some congregations hold a major service on Sunday mornings. Conservative congregations make more use of Hebrew than do Reform congregations and require men to cover their heads. Women may sing in the choirs of Reform and Conservative synagogues.

Another important service in the Reform and Conservative synagogues is held on Friday evening. The holding of this service is an innovation, a concession to our modern schedule of living. The Friday evening service follows the Sabbath morning pattern, except that the service is much shorter, more hymns are sung, and the reading of the Torah does not take place except in some Reform congregations. Hymn singing among the Jews seems to the Christian ear to be very monotonous, the same songs being repeated in service after service.

No collection is ever taken during worship; the primary support of the synagogues comes from annual dues.

Jews have a major conception concerning worship which is foreign to Christian ideas. It is that study of the Torah and the Talmud is itself an act of worship, "a means of communion with God. According to some teachers, this study is the highest form of such communion imaginable."[28] Consequently education and scholarship rank much higher in traditional Jewish opinion than they do in Christian opinion.

Just as Christians observe special seasons throughout the year—such as Christmas, Lent and Easter—so the Jewish calendar provides for the celebration of holy days and festivals. They are for many Jews the most meaningful and colorful aspects of Judaism. Three periods in the Jewish calendar are of especial interest: the New Year, the Day of Atonement and Passover.

The New Year falls, usually, in late September. It is a very awe-inspiring festival and begins the Ten Days of Penitence, a period which ends with the Day of Atonement. On this day observant Jews fast—no food or drink—from sundown to sundown, and spend all of the daylight hours praying in the synagogue. The number who complete this arduous regimen is "still surprisingly great";[29] those who do not complete it feel obliged to find a good excuse. The New Year and the Day of Atonement are called the High Holy Days.

Passover usually comes in early April and commemorates the Exodus from Egypt, the liberation of the ancient Hebrews from bondage. Passover is pre-eminently a home festival, celebrated by a ritualistic family meal called the *Seder*.

Jews who have lost all other contact with Judaism as a religion continue in large numbers to celebrate these three major festivals. During the High Holy Days the synagogues in many centers of Jewish population prove inadequate to seat the congregations which assemble, and it is necessary to provide the essentials for worship in temporary quarters.

Some of the other holy days are: *Sukkoth,* a day of thanksgiving for the year's harvests (it usually falls in October); *Hanukkah,* a commemoration of the Maccabean victories (December); *Purim,* a commemoration of the victory over Haman, as told in the book of Esther (March); *Pentecost,* thanksgiving for the "first fruits" of the fields and also for the revelation of the Law (May).

A Jewish boy is circumcised on the eighth day after his birth; "circumcision is still regarded as a religious necessity"[30] by all branches of Judaism. At the age of thirteen, the boy is made *Bar Mitzvah,* that is, son of the commandment, thereby coming into religious adulthood. As part of the ceremonies, he is called to the front of the synagogue on the Sabbath and asked to read from the Torah and to recite a blessing. He may even demonstrate his knowledge of the Law by delivering a discourse. The Bar Mitzvah ceremony is often the occasion for elaborate, birthdaylike celebrations. Will Herberg writes that these celebrations more and more take precedence over the synagogue ceremonies, the emphasis being put on "a lavish and expensive party, with the religious aspect reduced to insignificance, if not altogether ignored."[31] *Bas Mitzvah* (daughter of the commandment) services are used in a few American synagogues. A new type of "confirmation" service for both boys and girls is being widely accepted in all branches of Judaism.

THE ORGANIZATION OF THE SYNAGOGUE

Synagogues are run congregationally. The affairs of each local organization are in the hands of its members, that is, of its men: they control finances, choose prayer books, determine denominational affiliations, hire the rabbi. The women have much influence but little legal authority.

There is no hint of hierarchical control in American Jewry. There are synagogue unions with which local groups may become affiliated. The Reform synagogues support the Union of American Hebrew Congregations; the Conservative synagogues, the United Synagogue of America; the Orthodox, the Union of Orthodox Jewish Congregations. The sharp insistence on ecclesiastical independence is seen in the fact that "after forty-three years of work, the Union of Orthodox Jewish Congregations of America" had not "drawn even one-tenth"[32] of the Orthodox synagogues into its membership; however, the rate of affiliation has been increasing of recent years. The Synagogue Council of America receives support from substantial elements in all Jewish sects and attempts to unite all of them in their public relations.

The Jewish community is outstanding in its support of philanthropy. The Luce magazines have "more than once" said that the United Jewish Appeal is "the No. 1 phenomenon of U.S. philanthropy." The Department of Commerce reported that in a recent year Jewish organizations sent ninety-five million dollars abroad; the corresponding figures in that year were for Protestant organizations seventy-three millions and for Roman Catholic organizations eleven millions.[33] Jewish philanthropy is so firmly established that one astute observer could write that perhaps the American Jewish community "does possess what comes close to being a single hierarchical structure embracing the entire community, and that is the machinery of fund raising and fund allocation."[34]

Traditionally the rabbi was primarily a scholar, a layman whose long hours of study had won him a deep understanding of the Law. He was not a priest. There are no real priests in Judaism. There are cohanim (hence the name Cohen) who presumably are descended from Aaron, the brother of Moses; but today the cohen has no unique functions, except to lead an occasional ceremony in the synagogue. Nor was the traditional rabbi the leader of worship; and he was not, except incidentally, a preacher. His function in the semi-autonomous medieval communities was primarily that of interpreter of the Law. His function in contemporary America continues to be that of student of the Law, especially so in the Orthodox synagogues. But he has assumed many other functions, especially in the Conservative and Reform synagogues. There his role has become much like that of a liberal Protestant minister: he is a leader of worship, a preacher, and the executive of synagogue activities.

The right to act as rabbi could be conferred, traditionally, by any other rabbi. It is now customary for students at the theological schools to receive ordination from their teachers.

The status of women in Judaism is no better than it is in most of the Christian sects. Jews are proud of the fact that in past centuries the status of women in Judaism was somewhat better than in Christendom. Commented one rabbi, "The Christians, who deified Mary, did not allow women to come near the altar." But this same writer said, "Few aspects of Jewish thought and life illustrate so strikingly the need of reconstructing Jewish law as the traditional status of the Jewish woman. In Jewish tradition, her status is unquestionably that of inferiority to the man."[35] In the Morning Service, a man blesses God that He "has not made me a woman"; and a woman blesses God "who hast made me according to thy will." No women have been made rabbis even though the Central Conference of American Rabbis passed a resolution in 1922 declaring that "women cannot justly be denied the privilege of ordination." Women "have been known to be admitted as students in rabbinical schools."[36] In Israel, the Orthodox have fought bitterly against giving women the right to vote. Wives, according to traditional Judaism, may not divorce their husbands; but husbands may divorce their wives "at will."[37] If a man "disappears" without divorcing his wife, she may not remarry unless his death is certain. "It is a well-known fact that twenty-five thousand such unfortunate 'agunahs' [permanent widows] were the result of the last [first] World War."[38] Even among Reform Jews, as in American society generally, few women can be more than the "power behind the throne."

Marriage with gentiles is greatly frowned upon. Some parents whose son or daughter has married outside the fold and embraced Christianity mourn as for the dead. Even the most "emancipated" among the religious Jews retain a strong taboo against intermarriage, unless the gentile spouse becomes a Jew. The number of rabbis who will officiate at a mixed marriage is microscopic.

Judaism is not, today, an active missionary agency; most persons who become Jews do so because of intermarriage. One estimate has it that about two thousand persons are converted in the United States to Judaism each year, that four out of five are women, and that nineteen out of twenty become Jews because of the "involvement of marriage."[39] Converts are asked by the rabbis to "renounce" their former faith.

ISRAEL

Ever since the destruction of Jerusalem by the Romans, Jews have dreamed about the time when Palestine would again be in Jewish hands. "At every public service, morning, afternoon and evening, in each private devotion whether in the grace after meals or on retiring at night, Jews prayed for their return to Zion."[40]

But traditionally Jews did very little about actually getting back to Palestine except pray; Orthodoxy expected God in the fullness of His wisdom to arrange one day for the return. But about sixty years ago, a Viennese journalist by the name of Theodor Herzl became convinced that anti-Semitism would not be conquered until the Jews had a homeland to which they could flee, and a national government which could come to their defense. He was convinced that a homeland could be won only by political means; therefore, he founded the Zionist movement.

Its success was phenomenal. Received at first with almost universal skepticism (wrote Herzl, he "who wants to be right in thirty years must be thought crazy the first two weeks"),[41] Zionism won in an astonishingly short time the support of a majority of the Jews, the sympathy of many Christians, and the official approval of the governments of both England and the United States. Within half a century, under the Nazi spur, "the greatest colonizing enterprise of modern times"[42] was in full swing and the State of Israel had come into being. Immediately the infant state proved the accuracy of Herzl's insight; it became the haven of refuge for the persecuted thousands of European Jewry.

Palestine is a small country, about the size of Vermont, and the Jews have but a part of it. Moreover, its agricultural potential is forbidding to the American eye; the land is stony, the rainfall sparse. But the colonists have irrigated the deserts, terraced the hills, planted trees, drained the swamps, and made their nation again a "land flowing with milk and honey." After a careful survey, Walter Clay Lowdermilk of the United States Department of Agriculture wrote, "This effort is the most remarkable we have seen while studying land use in twenty-four countries." He concluded that by the utilization of all resources, the country could support three million persons in addition to those then living there.[43]

Opposition to Zionism developed from two major sources. One was the Arabs and their sympathizers, chiefly American oil companies and Christian missionaries in Moslem lands. The Arabs had lived in Palestine for some thirteen centuries, it was their home, and they saw no justice in their being asked to take a subordinate place and to turn their government over to Jews. Many Arabs suffered severely, especially because of the war between Israel and Trans-Jordan.

But there is another side to the story. The Government of Israel contends that the suffering during the war was largely the result of policies of the Arab armies, that more Arabs live in the territory of Israel than did before the Jews returned to Palestine, and that the native standard of living is much higher than when the Arabs were in sole possession: they have better working conditions, better wages, better food, better sanitation, better educational facilities, better care during illness, better life expectancy. The Arab case, say the Jews, is not the case of the poor Palestinian farmer, but the case of the feudal landlord, who is fighting to preserve a situation which permits him to exploit the Arab workers.

Opposition to Zionism comes also from Jewish sources. For many years, most Reform Jews in America were set against the movement. But in the thirties, Hitler's savagery altered the opinions of the overwhelming majority. Nevertheless, a few continued their opposition. They contend that Judaism is a religion and should have no entangling alliances. They affirm that the American public must not be given the slightest justification for thinking that Jews give their primary allegiance to a foreign power. An answer to this point of view is stated by one writer as follows:

Methodists, Quakers, Roman Catholics and Christian Scientists maintain relations with sister churches and brother communicants all over the world. British, Irish and Swedish Americans are sentimentally attached to the lands of their origins and keenly interested in their fortunes. Nor is the Quaker, the Catholic, the Swedish or the Irish American the least compromised in his Americanism therefor.[44]

An ardent American Zionist and a great rabbi was Stephen S. Wise. A description of his career will give color to some of the abstract principles of Judaism, make clear some of the tensions that stir the American Jewish community, and demonstrate at how many points the problems of Jewish and Christian leadership are similar.

RABBI STEPHEN WISE

"My sermon this morning will light a million-dollar blaze," said Stephen Wise to his wife one Sunday during the Steel Strike of 1919. In his discourse that day he accused the steel industry of "deluding the nation" and "resorting to every manner of coercion." So many large subscriptions to a projected synagogue building were canceled as a result that the building had to be postponed. Wise wrote in his autobiography:

In opening my address, I reminded my congregation that I knew that some of the members might refuse to lend their help in the building of the synagogue home as a result of what I was about to say but also again made clear that, while it might not be necessary for them to build a synagogue, it was necessary for me to speak the truth as I saw the truth on great issues.[45]

The Board of Trustees meeting shortly afterward refused "in the friendliest terms" to accept his resignation but made clear that the rabbi speaks *to* and not *for* the congregation.

On another occasion Wise delivered a stirring attack on the Associated Waist and Dress Manufacturers, many of whom were members of his congregation. When some of them threatened to resign, "he enthusiastically invited them to do so."[46]

Born in Budapest, the seventh rabbi in direct succession in his family, he was brought at the age of one year to New York City, where his father became rabbi of Temple Rodeph Sholom. Young Wise attended the public schools, the College of the City of New York, and Columbia University from which he received in 1901 the Ph.D. degree.

His choice of the rabbinate as a lifework was made very early; apparently he never gave serious consideration to any other profession. At nineteen he became assistant rabbi at the Madison Avenue Synagogue, and a year later was made its head. From 1900 to 1906, he served a synagogue in Portland, Oregon, "often descending from the temple to the market place to expose civic wrongs."[47] He was the author of the first child-labor law passed in Oregon.

In 1905, the pulpit of Temple Emanu-El of New York, the "Cathedral Synagogue of the country," became empty; he was invited to preach before the congregation.

One who preaches trial sermons lays himself open, as no man with self-respect should, to harassing experiences. . . . I had to listen to such expressions as "Doctor, it was a fine sermon." It was my soul that was tried; I had poured it out in earnest and unafraid appeal to these people to be single-minded and greathearted Jews. They responded to me as if I had been delivering a high-school prize oration.[48]

Like many other congregationally organized synagogues and churches, Emanu-El was controlled by its Board of Trustees. Emanu-El's leading member was the philanthropist, Louis Marshall. So dominant was Marshall in synagogue affairs that one rabbi said, "Temple Emanu-El lives under Marshall law." The Board sent a committee to ask Wise under what conditions he would become the Temple's rabbi. On the committee were leading New Yorkers, including Marshall, M. H. Moses, and Daniel Guggenheim. In his autobiography, Wise tells the story as follows:

I spoke in simple and earnest terms, ". . . You have heard that I have gained for my temple, Beth Israel, my people throughout Oregon, and their rabbi, the respect and for the most part the good will of the entire Northwest community. If I have achieved that, it has been because in my inaugural sermon at Beth Israel, September, 1900, I declared: 'This pulpit must be free.' "

Mr. Marshall, . . . without a moment's hesitation and without even the faintest pretense of consultation with his colleagues, said rather testily, as was his wont, "Dr. Wise, I must say to you at once that such a condition cannot be complied with; the pulpit of Emanu-El has always been and is subject to and under the control of the Board of Trustees." My answer was clear, immediate, unequivocal: "If that be true, gentlemen, there is nothing more to say."

And that would have been the end had not one of Mr. Marshall's colleagues . . . interposed the question, "What do you mean by a free pulpit?"

I replied fully and deliberately, putting my worst foot forward, "I have in Oregon been among the leaders of a civic-reform movement in my community. Mr. Moses, if it be true . . . that your nephew, Mr. Herman, is to be a Tammany Hall candidate for a Supreme Court judgeship, I would . . . oppose his candidacy in and out of my pulpit." I continued, "Mr. Guggenheim, . . . if it ever came to be known that children were being employed in your mines," having reference to his presidency of the famous copper mines, "I would cry out against such wrong."[49]

Later in an *Open Letter,* Wise expounded his position:

The chief office of the minister, I take it, is not to represent the views of the congregation, but to proclaim the truth as he sees it. How can he

serve a congregation as a teacher save as he quickens the minds of his
hearers by the vitality and independence of his utterances? But how can
a man be vital and independent and helpful, if he be tethered and
muzzled? A free pulpit, worthily filled, must command respect and influence; a pulpit that is not free, howsoever filled, is sure to be without
potency and honor. A free pulpit will sometimes stumble into error; a
pulpit that is not free can never powerfully plead for truth and righteousness.[50]

Wise now determined to establish his own free pulpit in New York.
In spite of an offer from his Portland congregation of a "lifetime
contract, at an unprecedented salary,"[51] he went back to New York
and founded the Free Synagogue. For the first year, he received no
salary; and for the second, the sum of three thousand dollars. At first,
Sunday morning services were held in a theater, but in 1910 they
were moved to Carnegie Hall, the leading concert hall of the city. His
dramatic, reforming temper led to much criticism. A Reform journal
denounced him, declaring, "We have from the start protested against
the use of the word 'synagogue' as applied to the movement over
which the Rev. Dr. Stephen S. Wise presides."[52] A prominent New
York rabbi described the Free Synagogue as "a hall, with an orator, an
audience, and a pitcher of ice water." But Wise's vitality, leadership,
and moral earnestness brought large numbers into the movement. Not
only was he dramatizing the need for freedom in the pulpit; he was
also uttering his "deep protest against the lifelessness of what had once
been a great and living Jewish movement, the lifeblood of which had
been pressed out as witnessed by the smugness characterizing New
York temple Judaism. This had ceased to be Reform Judaism without
even ever having become liberal. Its strength, such as it was, lay
merely in its opposition to equally unvital Jewish Orthodoxy."[53]

Wise early leaped into civic and national prominence. His exceptional powers as an orator brought him invitations to speak all over
the nation. But these powers were matched by his abilities as an or-
ganizer. In addition to the Free Synagogue, he founded the Zionist
Organization of America and the Jewish Institute of Religion, a
theological seminary. He also played a prominent role in the creation
of the American Jewish Congress, the World Jewish Congress, and
the Near East Relief. He was the friend of five Presidents of the
United States, and he wielded real influence with two of them: Wood-

row Wilson and Franklin Roosevelt. The day Wilson left office, he mentioned Wise as among "some friends, a very few, [who] have never asked anything for themselves and given me every service."[54]

Wise thrived on controversy. He attacked "evil institutions" and "pernicious ideas," and did not hesitate to name names in order to "serve my people." Some of his clerical colleagues accused him of being a sensation monger and publicity seeker—for a decade and a half *The New York Times* took notice of some activity of his on the average of once a week. It is undoubtedly true that he loved the lime-light and was not at his best when it failed to shine. But his vigorous methods often succeeded where a milder approach would surely have failed. Wise and his causes could not be ignored. He was, for example, so close to the center of the movement which ousted Mayor James J. Walker from New York's City Hall, that Walker's successor, Fiorello La Guardia, could say at the banquet celebrating Wise's sixtieth birthday, "If I am here tonight as Mayor it is because Dr. Wise was in New York many years before." When Dr. Wise "takes an interest in municipal affairs the business of the steamship lines picks up."[55] John Haynes Holmes, New York clergyman, called Wise the first citizen of New York, the greatest Jew of his day, the greatest religious teacher of his time, and added, "No civic leader can be matched with him for sheer eloquence, personal power, moral passion, and idealistic influence among the great masses of the common people."[56]

On the Sunday preceding Christmas, 1925, Wise preached on "Jesus, the Jew." *The New York Times* reported him as saying, "For years I have been led to believe, like thousands of other Jews, that Jesus never existed. 'Jesus was a myth' is the common belief among many Jews. I say this is not so. Jesus was."[57] Then Wise affirmed that "Jesus was man not God; Jesus was a Jew, not a Christian; Jews have not repudiated Jesus, the Jew; Christians have, for the most part, not adopted and followed Jesus, the Jew."[58] For these statements Wise was condemned by many Orthodox groups. One called his position a "grave menace to Judaism," and the Union of Orthodox Rabbis de-clared his statement "threatens to tear down the barrier which has existed between us and the Christian Church for over 1900 years— which may drive our children to conversion."[59]

Wise was excommunicated by the Orthodox rabbis and demands came from several groups that he resign his chairmanship of the United Palestine Appeal, which was then launching a campaign for five million dollars. He presented his resignation, but the Zionist Organization refused to accept it, voting "overwhelmingly" in favor of his retention of the office. The ensuing campaign collected the largest amount raised by the organization up to that time.

Very early in his career, he came to look upon Zionism as the answer to anti-Semitism. "Almost alone among the younger American Jews, Stephen Wise recognized and rallied to his [Herzl's] leadership."[60] Up until the founding of the State of Israel, Wise was easily the American most vocal in furthering the Zionist cause. He organized mass meetings, headed parades, wrote editorials, solicited funds, urged boycotts, denounced compromise, poured out his wrath on politicians who were unsympathetic to the Zionist cause.

Wise claimed he was the first American to call attention to the dangers of Nazism. Six months before Hitler's seizure of power, the American Jewish Congress, under Wise's leadership, sent a message to "about thirty leading Jews" in Germany asking what American Jews could do to help:

[All the Germans,] with one exception, had declared that "Hitler would never come to power." . . . A group of them with utmost sarcasm had sent me the following message: "Say to Rabbi Wise that he need not concern himself with Jewish affairs in Germany. If he insists upon dealing with Jewish affairs in Europe, let him occupy himself with Jewish problems in Poland and Rumania."[61]

Five years later, pleading at a national convention for co-operation from American Jews who feared to jeopardize their security by identifying themselves with Zionism, he said:

In 1933 . . . we offered German-Jewish groups an opportunity to unite. . . . They said they were Germans first, Germans who happened to be Jews. I am a Jew who is an American. I was a Jew before I was an American. I have been an American all my life, but I've been a Jew for 4,000 years.[62]

Against those who would appease the Nazis, Wise said to a World Jewish Conference:

Grievous fate it is to be among the victims of Nazism without help and without redress. Infinitely more tragic it were to come to an understanding with Nazism. To die at the hands of Nazism is cruel; to survive by its grace were ten thousand times worse. We will survive Nazism unless we commit the inexpiable sin of bartering or trafficking with it in order to save some Jewish victims.[63]

Wise lived to see the founding of the State of Israel. But in the last years the fruits of the Zionist labors seemed about to be snatched away by the pressures on the British Foreign Office. Wise led the outcry in America against this danger, voicing the claims and aspirations of his people. Just prior to the Fortieth Annual Convention of the Zionists Organization of America, he wrote an editorial whose basic proposition is a summary of the message of his whole ministry.

The Israel which faltered not in an hour without hope, will not yield one jot or tittle of its rightful aims and claims in this hour. Britain is a mighty empire, but there are forces in the universe mightier and more enduring than empire or dynasty. . . . Empires live in the terms of centuries. Millennia have witnessed Israel's suffering and shall yet crown the triumph of Israel's hope.[64]

JUDAISM AND THE FUTURE

A large portion of American citizens consider the Jews to be a group set apart, a back eddy in the stream of American life. This judgment is surely erroneous. I know of no religious sect whose membership as a whole is more aware of current social trends, more sensitive to current mores, more proud of American achievements, or more anxious to help in maintaining and vitalizing democratic institutions. Most Jews are so well integrated into American culture, that I am emboldened to take sides in the debate on the nature of Judaism: *American* Judaism (Israeli Judaism is another matter) is a religion in the same sense that Catholicism, Lutheranism, Congregationalism, Mormonism are religions; and *American* Judaism is a culture in no other sense than such Christian movements are also cultures. The Jew is every inch an American.

But the bitter facts of anti-Semitism are such that he feels insecure. Statements of opinion could be assembled to the effect that anti-Semitism is decreasing. Even though these opinions may be correct, much discrimination remains. A study of nineteen thousand job orders in the mid-1950's showed that "23 per cent of the orders contained specifications that barred Jews from being considered."[65] A survey in Cincinnati showed that fifteen of the largest public-accounting firms employed only three Jews. A survey in Detroit showed that "most local banks and trust companies hired only white Christians and that 95 per cent of the private employment agencies reported that Jewish applicants face serious barriers in attempting to qualify for jobs."[66] Jews are often excluded "from resorts and the semipublic clubs that have become an important aspect of business and political life."[67] "The power structure of any small city is usually centered in the number-one country club—and rare indeed is the 'prestige' club that admits Jews."[68] A survey of the opinions of 159 Jewish students found that 40 per cent agreed with the statement, "I am convinced that anti-Semitism is likely to interfere with my search for personal success and happiness," and 60 per cent agreed with the statement, "I often worry lest anti-Semitism take on more violent forms in the United States."[69] One Jew wrote:

Any amateur psychologist knows that you cannot build peace of mind on fear. He also knows that the best way to begin the conquest of fear is to examine its source. In our case, the source is anti-Semitism, an anti-Semitism exacerbated by the worldwide Jew-consciousness stimulated by Hitler. . . .

We must recognize that we have developed an inferiority complex from the fact that we are Jewish. Every normal human being has an inferiority complex about something in this neurotic day: he is cross-eyed, color-blind, poor, wealthy, uneducated, a long-haired intellectual, a Polish-American with an unpronounceable name, a hick, a bad athlete, sexually insecure, or bald—and if you touch the sore spot you will get a reaction of pain and embarrassment. In the case of Jews, the sore spot is often the very fact that they are Jews.[70]

This situation undoubtedly affects the contribution which Jews can make to the spiritual revitalization of American culture; it would be surprising if a community which is the object of so much prejudice and discrimination could furnish leadership for creative, nationwide religious experimentation. Moreover, many Jews are spiritually smug. Most of them are spiritually self-isolated, consumed as far as religion goes with things Jewish, eager to keep the relationship among religions in America just what it is. Nevertheless, the Jewish contribution toward quickening our national religious sensitivity can be substantial. Jews uniformly believe in freedom of religion, for others as well as for themselves. They have been made acutely aware by the fate of

European Jewry of the spiritual danger of political fanaticism. They have in Reconstructionism the most mature socioreligious philosophy on the current horizon, even though it is a philosophy couched in terms too exclusively Jewish. And they have a deep sense of internationalism, a sense which can help prevent the glorification of American ideals from deteriorating into a cheap nationalism.

CHAPTER XIII

THE EASTERN ORTHODOX CHURCHES

ALTHOUGH THE majority of Orthodox church members in the United States are by now second- or third-generation Americans, the Orthodox churches give the impression of being strictly old-world institutions, the least Americanized of any large religious group in this country. Orthodox communicants proudly assert their Americanism; but their churches tend to be foreign enclaves in the midst of American society, seldom influencing and not much influenced by the major currents of American thought and mores. The reasons for this situation are many; they run far back into history. An explanation of the distinctive characteristics of the Orthodox churches must be based on an outline of some historical conflicts and on an explanation of relationships with the homeland churches.

THE NATURE OF THE CHURCH

The Christian Church was founded, according to Eastern Orthodox belief, in Jerusalem on the Day of Pentecost by Jesus Christ. He established her as one, holy, catholic, and apostolic. The meaning of the last three of these terms is similar to the meaning given by most other Christian groups. But long ago the Eastern Orthodox and the Roman Catholics began to argue over the meaning of the term *one*. No hint is found in Scripture, say the Orthodox, that Jesus Christ thought of one ecclesiastical organization. One Church means rather a body of people who are one in faith, in discipline, and in type of government.

After Pentecost, local Christian churches were founded by the

Apostles. They had received authority, according to the Orthodox, from Jesus Christ himself and they passed it on to their successors, that is, to the bishops who ruled first over the local churches and then over the dioceses into which the local churches were organized. These bishops were equals in all spiritual matters even though the dignity and prestige of their offices varied considerably, bishops in the large cities usually having more prestige than bishops from the provinces. Final authority in the Church rested with Ecumenical Councils which the bishops attended. The essential matters of doctrine and discipline were debated and settled by these councils in which the bishops all had equal votes. The situation might be likened to an educational conference attended by the presidents of American colleges and universities. In such a group, the president of a little college in a small town would not have the prestige of the president of, say, Columbia University or the University of Chicago; but his vote would be equal to theirs and his influence in debate might be even greater.

Only seven Ecumenical Councils have been held, say the Orthodox; the last one convened in the eighth century. All councils held since that time have failed to meet ecumenical standards; the Council of Trent, for example, which played so prominent a part in the counter reformation and the Vatican Council which enunciated the dogma of papal infallibility were simply synods of a Christian sect. The Orthodox do believe in infallibility, but its seat is not any one person but is rather the body of bishops in the Apostolic Succession sitting as an Ecumenical Council—and none has met for over a thousand years.

The councils did not create dogma; they simply stated the teaching of the Apostolic Church, defended it from error, and prevented anything from being added. Mistakes were avoided through the power of the Holy Spirit. Thus Orthodoxy claims to be the "one saving and infallible church," "the sole infallible holder of the Apostolic tradition," an institution which has "nothing to learn" where *dogma* is concerned.[1]

In the early centuries, the Roman Catholic Church began to break away from the eastern churches by asserting the right of the Roman bishop to rule over all other bishops. Today East and West are completely separate institutions. The breach did not suddenly appear; only after centuries did most Christians recognize it as final. But for the last nine hundred years relations between Eastern Orthodoxy and

Roman Catholicism have usually ranged in the area between frigid formalism and violent hostility. The date of the final break is often placed at 1054; but the Pope's sponsorship of the Fourth Crusade in the thirteenth century was perhaps the injury which most embittered the Orthodox. The crusaders sacked Constantinople, the leading Orthodox city, massacred its Christian population, and carried off its treasure. One Eastern Orthodox wrote, "Soldiers and Latin clergy vied with each other in their attempts to seize some part of these riches for themselves; even the precious Holy Altar of St. Sophia was polluted, broken in pieces and sold."[2]

Today Eastern Orthodoxy is a group of independent churches whose boundaries usually follow the lines of secular states. Thus national Orthodox churches are found in Greece, Russia, Romania, Yugoslavia, Bulgaria, Georgia, Albania, etc. These churches form a loose federation bound together by a common faith, a common tradition, and a common liturgy. No common administrative agency exists. The bishops of the various national churches do meet in councils which exercise great power. For several decades efforts to bring the various churches together "came to nothing mainly because of political obstacles. The absence of the Russian Church . . . during some thirty years of the Communist regime was paralyzing to all inter-Orthodox intercourse."[3] In 1961, however, a Pan-Orthodox "conference" was held and a "full-scale synod" is planned for the near future.

At the head of each national church is an official titled variously: sometimes archbishop, sometimes metropolitan, and sometimes patriarch. The highest office in the churches is that of the patriarch. The leading patriarch in point of dignity has his residence in Istanbul. The number of the faithful in his patriarchate has shrunk to less than one hundred thousand; nevertheless, the memory of the centuries when that city headed the Byzantine Empire is sufficient to give him the honorific title of Ecumenical Patriarch.

Eastern Orthodox churches in the United States retain their national orientations. Thus in this country are found over twenty bodies; the major ones are Russian, Greek, Romanian, Bulgarian, Serbian (Yugoslavian), Syrian Antiochan, and Ukrainian. Culturally all these churches have close affinities for the fatherland; for example, most of them persist in using its language as the chief vehicle for worship. Some of the churches remain under the direct administrative supervision of the fatherland church. Others, however, in view of the Communist domination of most of Eastern Europe doubt the continuing orthodoxy of the church at home and consequently have set up independent organizations. American Russians have split into three groups over the problem of loyalty to the homeland church, and American Romanians into two groups. For many years efforts to form in the United States one Eastern Orthodox Church uniting the many groups received little support either from American Orthodox or from homeland officialdom. However, in 1961, Archbishop Iakovos, Greek Orthodox Primate of North and South America, was reported in the press as forecasting the consolidation of all Eastern Orthodox communions in the Americas—perhaps within two years.

THE FAITH

The basic tenets of Orthodox theology are already familiar to readers of the chapter on Roman Catholicism. Indeed, the Orthodox insist that all sects insofar as they are truly Christian received their dogmas through the undivided Orthodox churches. The source of their theology is believed of course to be Jesus Christ. The teachings of Jesus came down to the Church through two channels: the Bible and tradition. Veneration for the Bible is very high; but veneration for the tradition is equally high. The late Archbishop Michael of the Greek Orthodox Church in North and South America wrote:

The Holy Scriptures and the Apostolic Tradition have the same significance and the same power for the Orthodox. . . . The Apostolic Tradition preceded the Canon of the Holy Scriptures. . . . There exist in Tradition elements which, although not mentioned in the New Testament, as they are in the Church today, are indispensable for the salvation of our souls. . . . Since . . . the books which constitute the Holy Scriptures, in their form and number today, were preserved because of the Apostolic Tradition, it follows that the Apostolic Tradition is the most precious and valuable thing and absolutely useful for the life and existence of the Church. It follows, also, that those who ignore or contempt the Apostolic Tradition as well as those who enlarge Tradition and add to it elements foreign to the fundamental dogmas that have been passed on down to us, suffer a grave loss. To the former belong all the Protestants who have declared an implacable war against Tradition. To the latter belong the Roman Catholics who went as far as embodying in Tradition the primacy of the bishop of Rome.[4]

This reverence for tradition has developed into a suspicion of any move that might result in change. "We do not change the boundaries marked out by our fathers: we keep the traditions we have received,"[5] reads a quotation from an ancient father featured on the cover of an American Orthodox periodical. One standard Orthodox work even asserts that "the first mark of the Church is that all her teachers and pastors agree with each other in everything."[6] As a result of such attitudes there is no "liberal" movement in the Orthodox churches and a frequent observation concerning them is that their theological thinking is uncreative. One Lutheran writer asserted flatly his judgment that Eastern Orthodoxy is in a state of "theological stagnation."[7]

The traditional summary of Orthodox beliefs is the Nicene Creed. (See page 177.) The first two Ecumenical Councils meeting in A.D. 325 and 381 at Nicaea and Constantinople formulated this creed which is the only one the Orthodox recognize. Its basic elements have already been expounded in earlier chapters; here it is necessary only to indicate the major points at which Orthodox beliefs are distinctive. Many readers will consider some of these points to be picayune. But they are not picayune to the Orthodox—nor to the Roman Catholics. The smallest break with revelation is intolerable. Each group has repeatedly made serious efforts to persuade the other of the error of its ways. Differences which seem minor when they are seen at a distance often become great when they are seen right at hand.

The "chief dogmatic difference"[8] between East and West lies in the addition to the creed of the words "and the Son." This phrase is written in Latin *"filioque";* hence the resulting controversy is often called the "Filioque Controversy." The words were added to the creed by the West. The controversy concerns the relations among the three persons of the Trinity. The "procession" of the Holy Spirit is the specific point at issue. Does he proceed from the Father alone, as the Orthodox assert, or from the Father *and the Son,* as the West asserts? The portion of the creed in question should read as follows according to the Orthodox: "I believe in the Holy Spirit, the Lord, the Giver of Life, Who proceeds from the Father." The West makes this sentence read: "Who proceeds from the Father and the Son." (The meaning of "proceeds" in this connection is not known.) The creed says of the Son that he was "begotten" of the Father. The Father is thus the source of the divine nature in both the Son and the Holy Spirit. But the difference between "beget" and "proceed" is not made clear in the revelation. However, say the Orthodox, the revelation makes clear whom the Holy Spirit proceeds from. Jesus Christ himself said, "But when the Comforter is come, whom I will send unto you from the Father, even the Spirit of truth, which proceedeth from the Father, he shall bear witness of me" (John 15:26).

"Rivers of ink have flowed because of this question," wrote a Roman Catholic priest, and he added that the issue "has never yet affected the piety or practical faith of any human being."[9] But an Orthodox theologian wrote, "It [adding the *filioque*] is not only technically illegal and illegitimate, but essentially wrong. . . . Even as a theological opinion it is vicious and inadmissible."[10]

The Virgin Mary, according to Orthodox teachings, became immaculate, free from sin, only at the time of the Annunciation, not at the time of her conception by her mother, St. Anne. A person who accepts the Immaculate Conception believes the Virgin Mary preceded Christ in sinlessness.

Orthodox beliefs about the next life differ somewhat from Roman Catholic beliefs. The condition of man between the particular judgment and the general judgment is said to be that the soul gets a foretaste of the state in which it will finally live, either heaven or hell. The existence of purgatory and limbo are consequently believed to have no foundation in Scripture or in tradition. Indulgences also are thought to have no proper foundation.

In Orthodoxy the sacraments are the familiar seven, and four of them are believed to be necessary for salvation: baptism, confirmation, communion, and penance. So important are these four that the Church uses care not to let sudden death prevent their reception. Infants are baptized forty days after their birth, earlier if necessary, and immediately after baptism are confirmed and receive Holy Communion. Thereafter, infants should be brought to the Church at least four times a year for the reception of Holy Communion, and as soon as they reach the age of accountability and are capable of mortal sin, they go to confession.

The importance of receiving these four sacraments is emphasized by an article written by a Greek priest concerning the conversion of the writer, Elliott H. Paul. Paul had long been a vigorous exponent of agnosticism but finally from his own investigation had become convinced of the correctness of Eastern Orthodoxy. He was in his last illness when he sent for the priest. This official, as soon as he was convinced of the genuineness of Paul's conversion, went into immediate action.

Because of his seriously ill condition . . . I had to make a fast decision. I was told that Mr. Paul could live six hours, six days or six months, as his name remained on the critical list. I, therefore, felt personally responsible. I could not allow him to pass away suddenly without receiving the Sacraments of Crismation (Confirmation), Penance (Confession), and Eucharist (Holy Communion). . . . There was no need for baptism, because Elliott H. Paul, in his early youth was raised in the Congregational Church, a trinitarian faith, which baptizes its communicants. This early baptism was recognized by my Church.

I excused myself for a few moments, . . . conversed with my bishop by telephone, and then requested the head nurse on duty to have the bed clothing changed, and to prepare Mr. Paul for the Sacrament of the Church. In the meantime, I returned to my Church to pick up the necessary ecclesiastical appointments and sacred vessels. Within the hour, I was back at the hospital ready to proceed with the religious ministrations of the Church.[11]

Baptism is performed by triple immersion. This method is vigorously defended: it was the method used by the "undivided church"; sprinkling "was originated by the Roman Church in the fourteenth century"; the abandonment of triple immersion is "an innovation in a sacrament which is basic for the salvation of souls."[12]

A person is confirmed, as has been noted, immediately after his Baptism. The Orthodox consider that Roman Catholics make a major error in withholding confirmation until the adolescent years and in limiting for all practical purposes the right to administer confirmation to bishops. The Roman Catholics err also in withholding Communion until the age of understanding. The spiritual graces granted the soul by these sacraments are needed during our earliest as well as during our later years.

The Orthodox are enjoined to confess to the priest at least twice a year. The impersonal relationships of the Roman Catholic confessional usually do not obtain. The priest "acts like the doctor, who in treating a bodily illness, must know the patient personally."[13] Confession may, however, be made to a priest who is a stranger. The penitent stands facing the East and confesses to God; the priest is witness and counselor rather than judge. The priest does, however, pronounce words of absolution.

Mixed marriages are frowned on, but are nevertheless permitted. "The children born of such a marriage must belong to the Orthodox Church."[14]

Divorce with the privilege of remarriage is also permitted; but a person may marry no more than three times. Parish priests may marry before they are ordained, but not after. The bishops are always chosen from the unmarried clergy.

Holy unction is given to persons who are ill either bodily or spiritually. It is not reserved for persons who are in danger of death. Thus the terms *extreme unction* and *last rites* are not appropriate.

The Orthodox consider that Lent should begin on a Monday.

Theoretically the East sets the date of Easter in the same way as does the West: both celebrate it on the first Sunday after the full moon on or after the vernal equinox. However, one additional regulation by the East (Easter must come after the Jewish Passover) means that most of the time the Orthodox celebrate Easter one or more weeks later than the West does.

ORTHODOX WORSHIP

There is a legend which explains that Russia adopted Eastern Orthodoxy because some Russian envoys attended services in Constantinople and were so impressed they came back saying, "We felt we were in heaven." The aesthetic impact of Orthodox worship on those who know it best is truly substantial. On the other hand, standards of beauty are so subjective that many Americans receive quite another impression; often after their first experience with Orthodox worship, they tend to agree that it is "stiff with gold and gorgeous with ceremonial." Certainly, understanding Orthodox worship is very difficult for the Westerner who comes to it unprepared by a knowledge of any one of the foreign languages in which it is conducted.

The floor plan of an Orthodox church is usually that of a Greek cross; this type of cross has arms of equal length. The crossing of the church is usually covered by a great dome, and often on the interior walls are hung or painted scenes from the Bible and pictures of the saints. The most prominent item of furniture is a large screen called the iconostas which stretches across the whole width of the church and divides the eastern arm of the cross from the rest of the building. The space behind this screen is thought to be so holy that only the clergy and laymen with special responsibilities may enter it. In it are found the altar, the sacred vessels and vestments, and the place for the preparation of the bread and wine which will become the Body and Blood of Our Lord. In the iconostas are three doors. The center one is called the beautiful, or royal gate because it is believed that Jesus Christ passes through it in an invisible manner when the sacraments are conducted. The altar is visible when the door is open. The two other doors are near the ends of the iconostas. Painted on the iconostas are pictures of Christ, of angels, and of saints. These pictures are called icons. They have been especially blessed and are painted after the flat and formal style of the Byzantine school. The Orthodox do not place statues in their churches. This limitation is the result of a long and bitter conflict in the eighth and ninth centuries. At that time, some Christians, called iconoclasts, believed the use of either images or pictures tends toward idolatry. On the other hand, another group favored their use because they were believed to be a valuable aid to worship. The latter group was finally victorious. Thus "representations" are permitted today in Orthodox churches, but "by tacit consent"[15] sculptured figures are no longer used, and icons are pictures in which a studied effort is made to avoid a three-dimensional effect.

Directly in front of the iconostas is a slightly raised platform running the width of the church. On this platform near its center is a pulpit; at the north end is a desk for the lay reader, and at the south end is a throne for the bishop. The choir is located out of sight, usually in a balcony. Near the door as one enters stands a large icon before which worshipers pause as they enter. They kiss it, light a candle, and offer a prayer before giving attention to the ongoing service.

The service, called the Divine Liturgy, is very long, beginning usually about nine o'clock of a Sunday morning and ending about noon. However, attendance for the first two-thirds of it tends to be sparse. In essence it is the same service which other Christians call the Eucharist, the Lord's Supper, the Mass, or Holy Communion. It has much in common with the Roman Catholic Mass, although the outsider finds difficulty at first in seeing the similarities. The term *Mass* is not used.

The service begins with the ritual and ceremonies surrounding the preparation of the bread and wine for consecration. Then comes the Little Entrance: the priest, preceded by altar boys and carrying the Gospel, comes through the north door of the iconostas. They march to the back of the church and then come down the center aisle. A hymn is sung; a passage is read from the Gospel; prayers are said; and the sermon is preached. Then comes the Great Entrance: the priest and the altar boys again come forth from the north door, the priest holding high the bread and wine which are about to be consecrated; again the procession marches to the back of the church and comes down the center aisle; the priest carries the elements through the royal door to the altar; the door is shut; prayers are said; the Nicene Creed is recited; the door is thrown open; the words of consecration are said; the door is again closed; the priest receives Communion; the door is opened, and Communion is offered to the laity. Shortly thereafter comes the dismissal; during this period, the members of the congregation come forward and receive from the priest pieces of bread which have been blessed. The bread is consumed as the worshipers leave the church building.

This brief outline fails to communicate any sense of the elaborate detail which surrounds the Divine Liturgy; special readings from the Scripture, special prayers, special litanies, and special movements abound. The priest is richly vested and the choir seems to be almost constantly active. Its music is strange to western ears. Eastern music often uses modes and intervals to which Americans are quite unaccustomed. Yet a Roman Catholic author could write that the singing in Slavic churches is "probably the most beautiful Church music in the world."[16] Such a level of excellence is difficult to achieve in the far different American environment.

This outline of Orthodox worship also fails to communicate the attitude in which an Orthodox worshiper goes to his church. Nicolas

Zernov, a member of an Eastern Orthodox communion, describes this attitude by contrasting it with the mood in which the average Westerner goes to church. The setting of his description is England; but the mood of Orthodox worship in America is the same, with a minor exception or two which I will note later.[17]

Let us compare, for example, the Sunday service in England with that of an Eastern Orthodox country. In both instances one can usually recognize even in the streets people who intend to take part in Sunday worship, but by quite different marks. An English person going to church is better dressed than others; he often carries a prayer book; and there is something special about his walk also: it is solemn, yet slightly hurried, for he wants to arrive at the church the very moment when the service is due to begin, and not to be too late or too early. On entering the church, an English Christian first kneels for private prayer and then sits down and quietly waits for the service to start. During the service everyone behaves like everybody else; the congregation stand, sit, or kneel at precisely the same moment, and if by chance an individual fails to follow the prescribed movement, he feels most uncomfortable and tries to correct his mistake. The priest or minister occupies the position of a leader who directs the actions of the congregation. As soon as the service is over the people leave the church in a body, with the sense that Sunday duty has been fulfilled. The services are straightforward and well-timed; and there is a sense of obligation attached to them; they foster the spirit of discipline and obedience, and make a strong appeal to the will and moral responsibility of each individual Christian.

The Eastern Christians behave quite differently. They go to church in a slow, leisurely fashion, stopping often to talk to those of their friends who are also going to join in the worship. When they arrive at the church they go in, whatever point the service may have reached; their first act is, as a rule, to light a candle in front of one or the other of the pictures of the saints, and only after this has been done do they join in the worship of the congregation. . . . There is complete freedom and spontaneity of action. But this lack of uniformity, and the fact that all the time fresh people are coming in and others going out, do not destroy the sense of corporateness in an Eastern service. On the contrary, they give the impression of a united family gathering where everyone feels completely at ease and expresses in his own way his share in the common activity. This impression is further enhanced by the large number of children, and even babies in arms. . . .

In the West the Christian acts, feels and thinks as an individual. . . . In the East, a Christian thinks of himself first of all as a member of one big family of all Christian people, both living and departed. . . . Whenever the Church of God is gathered together in an act of worship, it is the Saints and all the faithful departed who form the main body of the congregation, for they are the true worshippers of God, and the Christians who still live here on earth are only joining their company when they come to take part in a service.

American Orthodox churches are tending to deviate from this pattern in the direction of greater unity of action on the part of the congregation. Perhaps the best symbol of this movement is the placing of pews in the churches. Traditionally, few seats were provided, most of the congregation being expected to stand or to kneel throughout the service. But seats are provided in America today—and in some homeland churches. And uniformity of action on the part of the congregation is growing. The worshipers tend to stand and sit and pray and leave (but not to arrive) at the same time. Thus work the syncretistic tendencies of religion. Whatever clergymen may say, religions do borrow from each other. All churches are influenced by the cultures in which they exist.

Leavened bread is used in Holy Communion. The Gospels, according to Orthodox interpretation, make clear that Jesus Christ used leavened bread at the Last Supper since it was held before the beginning of the Passover. Using unleavened bread is, therefore, considered to be an unwarranted innovation. Moreover, laymen as well as priests receive the consecrated wine; "withholding the cup" is clearly contrary to Christ's statement, "Except ye eat of the flesh of the Son of Man and drink his blood, ye have no life in you" (John 6:53). In Communion, the consecrated bread is soaked in the consecrated wine, and then is placed by means of a spoon on the communicant's tongue.

Sermons ordinarily play a relatively small part in Orthodox worship; they lack the authority of tradition. That type of Protestant service which centers in the sermon was attacked by a Greek priest in the following language:

> [This] type of religious service . . . is a comparatively recent development arising at the time of the Reformation in the sixteenth century. . . . If the minister's sermon is warm and inspiring, . . . then the relationship between man and God is heightened, but if the minister's sermon is mediocre, and because there is no liturgy in this type of religious service, the remainder of the hour in church rarely has enough power to produce a definite mood of worship. . . .
> There is another defect which characterizes such a form of worship, and that is the passive role of the congregation. There is little for them to do but to bow their heads and sing the hymns. The rest of the time

they just sit and listen. . . . The main reason for going to church is
to worship God; to worship Him in His presence; to partake of the very
life of Jesus in the form of a sacrament; to worship God together. To
worship God actively the congregation needs definite things to do. . . .
Everyone in the Orthodox Church has a definite role to play and everyone senses that he is needed for the drama that is being enacted. . . .
The central act of the whole Orthodox Liturgy is the sacrament of Holy
Communion in which Christ is actually present; He comes into the lives
of the worshippers who offer their lives to Him and receive His life
in return. It is not sufficient for man to read the Bible and to adhere
to the teachings of Christ, but the very life of Christ Himself is necessary for the development of a healthy personality type.[18]

The Protestant's answer to such charges is of course that from his
point of view Protestant worship demands much more of a worshiper
than does Orthodox worship, that the Orthodox come late and need
not even pay attention, and that the sermon is the most effective means
yet devised for preventing religion from being an escape, for making
it relevant to the personal and social problems of our day or of any
day.

Orthodox theory concerning icons, like Roman Catholic theory
concerning statues, is that no power inheres in the material object. The
Orthodox do believe, however, that there is a supernatural link between the icon and the person it represents, that "an icon is a place of
the Gracious Presence. It is the place of an *appearance* of Christ, of
the Virgin, of the saints, of all those represented by the icon, and
hence it serves as a place for prayer to them."[19] Some of the faithful
believe also that some icons have the power to work wonders—to
cure illness and to bring good fortune. Miracles are sometimes reported in connection with icons. In 1960, for example, the Ecumenical
Patriarchate pronounced three weeping icons—their eyes shed tears—
which were discovered on Long Island, New York, "a sign of Divine
Providence."[20] "Icons not made with hands" are also believed to
exist; for example, the face of Christ has appeared on cloths through
supernatural means, such as when St. Veronica wiped the sweat from
his face as he was carrying the cross to the crucifixion.

Western Christians are often impressed by Orthodox concern for
what seem to be very small matters. Serious controversies have
arisen over such questions as whether the "but" should be in or out of

the phrase "begotten, but not made,"[21] whether the gloria should be preceded by two or three alleluias;[22] whether the gloria should be succeeded by two or three amens;[21] whether the sign of the cross should be made with two or three fingers;[21] whether in the Eucharist the "particle from the rest of the bread in honor of our Lady"[23] should be placed on the altar to the right or to the left of the priest.

INSTITUTIONAL PROBLEMS

The number of Orthodox in the world is sometimes said to be 130 million; a letter from the Office of Information of the Greek Archdiocese claims 300 million. However, even the smaller figure assumes that one hundred million Russians still continue to think of themselves as Christians and that the overwhelming majority of the population in the satellite countries continue to do the same. Such estimates are quite unreliable. Statistics for the United States are often equally unsatisfactory. Writers have sometimes stated the size of American Eastern Orthodoxy by simply estimating the number of Americans who came or whose families came from Eastern Orthodox lands, thus claiming five or six million members. The number of Americans actively participating in Eastern Orthodox church life is much smaller. The total of the reports made by the churches themselves was in 1960 two million seven hundred thousand.

To the outsider, language difficulties seem to reduce considerably the numbers to which the American Orthodox churches can minister effectively. Theoretically, the use of the vernacular in worship is encouraged. But practically, churches under homeland supervision are expected to use the homeland language. The number of Americans who really understand the language of worship is thus small, particularly so since the form of the language used is commonly an old one, one no longer used outside the church even in the homeland. The Greeks seem to be making more serious efforts than the other Orthodox groups to meet American conditions; they maintain, for example, an alert Office of Information. Yet two-thirds of the material published in the Greek denominational magazine, *The Orthodox Observer,* is in Greek. One wonders how many third-generation American Greeks can read these pages. On the other hand, the sermons, especially in the

Greek and Russian churches, are often spoken in English, and in some of the churches on specified occasions English may be used for the entire Liturgy.

The emphasis on foreign languages has meant that Orthodox priests often come from other lands and often have an imperfect knowledge of English. This fact handicaps the American churches and probably results in many misunderstandings or misinterpretations of American customs and institutions. One illustration is the inaccurate statements concerning Protestantism which sometimes appear in Orthodox writings. One author wrote: "To be a Presbyterian . . . one must accept *one man's* interpretation of the Bible, that of John Knox, the founder of Presbyterianism."[24] Another statement reads, "Neither do the Protestants accept the doctrine that the Virgin Mary was virgin."[25]

American Orthodox groups are active members of the ecumenical movement. Participation in the movement, however, seems to be ambivalent. On the one hand, articles are written which point with pride to achievements of the National Council of Churches, to the Archbishop's service as one of the six presidents of the World Council of Churches, to the support the World Council is giving the Ecumenical Patriarch in his "determination to remain in Istanbul irrespective of consequences."[26] On the other hand, the Orthodox have participated in the ecumenical movement not so much to consider with other Christians the basic problems of religion in the twentieth century as to witness to the truth. An editor of the *Christian Century* wrote as follows:

Never a large ecumenical meeting is held that a representative of the Orthodox wing of the World Council does not rise before final adjournment to explain that, while Orthodox Christianity is glad to have been involved in the discussion, it must clearly dissociate itself from the conclusions. Sometimes the disclaimers are calm, sometimes they are lofty, sometimes they are deeply offended. But always they are galling to the Protestant Christian, who can hardly be expected to rejoice in this regular rejection.[27]

However, another attitude may be in the making; the following news item appeared in *Christianity Today* in 1962:

Eastern Orthodoxy came out of the New Delhi Assembly [of the World Council of Churches] playing a new role in the ecumenical movement, according to Archbishop Iakovos. . . . A change in posture toward the ecumenical movement was reflected in the Orthodox hierarchy's decision to discontinue the practice of issuing independent statements on the subject of unity at ecumenical meetings. Archbishop Iakovos characterized the decision as a change of tactics. He said the Orthodox prelates felt they could register their opinions more effectively in helping to shape policy in committee work.[28]

Orthodox churches seem to be only slowly adjusting to America. Orthodox people have become rapidly adjusted in most other areas of American life; persons of Orthodox faith are found among the leaders in American business, politics, science, and art. But the inherent conservatism of Orthodoxy has made for extremely slow adjustments in church life. The continued rejection of the English language as the effective means of communication is a serious self-imposed handicap. Eastern Orthodoxy wishes to become known as one of the "major American faiths"; it has sponsored legislation which declares it to be such. But can it measure up to this standard while it persists in making major use of languages seldom used in America? The use of the vernacular is both dogmatically correct and institutionally wise. The acceptance of English as the major language tool would seem to be an adjustment without which the Orthodox churches can hardly meet fully the needs of their American constituents in the latter decades of the twentieth century.

CHAPTER XIV

THE MORMONS*

THE HISTORIES of both Mormonism and Christian Science developed along the lines of the traditional American success story; after early years of struggle, hardship, and persecution both of these American-founded movements have won respect, influence and a wide following. The Church of Jesus Christ of Latter-day Saints, the official name of the body which early in its history was nicknamed Mormonism, grew from a membership of six at its beginning in 1830 to a membership of one and two-thirds million today. A writer in an official Mormon journal estimated that by the year 2000 the membership will number six million; moreover, he predicted that at that time the city with the largest number of Mormons, instead of being as it is today Salt Lake City, will be Los Angeles.[1]

THE RESTORATION

Joseph Smith, Jr. (1805–1844), the founder of Mormonism, was born at Sharon in central Vermont. When he was ten, his family moved to Palmyra in western New York, a town located about twenty miles southeast of Rochester. At fourteen years of age, Joseph had an experience which Mormons believe changed not only the course of his life but also the course of modern religion. Deeply troubled by the conflicting claims of the rival clergymen in his neigh-

* Most of the material of this chapter deals with doctrine and history from the viewpoint of the Church of Jesus Christ of Latter-day Saints which has its headquarters in Utah; the beliefs of the Reorganized Church of Jesus Christ of Latter Day Saints which has its headquarters in Missouri are dealt with specifically in only a single paragraph.

borhood, Joseph went into the woods to seek the guidance of the Lord. He wrote about his experience:

> . . . I kneeled down and began to offer up the desires of my heart to God. I had scarcely done so, when immediately I was seized upon by some power which entirely overcame me, and had such an astonishing influence over me as to bind my tongue so that I could not speak. Thick darkness gathered around me, and it seemed to me for a time as if I were doomed to sudden destruction.
>
> But, exerting all my powers to call upon God to deliver me out of the power of this enemy which had seized upon me, and at the very moment when I was ready to sink into despair and abandon myself to destruction —not to an imaginary ruin, but to the power of some actual being from the unseen world, who had such marvelous power as I had never before felt in any being—just at this moment of great alarm, I saw a pillar of light exactly over my head, above the brightness of the sun, which descended gradually until it fell upon me.
>
> It no sooner appeared than I found myself delivered from the enemy which held me bound. When the light rested upon me I saw two Personages, whose brightness and glory defy all description, standing above me in the air. One of them spake unto me, calling me by name, and said— pointing to the other—"This is My Beloved Son, Hear Him."
>
> My object in going to inquire of the Lord was to know which of all the sects was right, that I might know which to join. No sooner, therefore, did I get possession of myself, so as to be able to speak, than I asked the Personages who stood above me in the light, which of all the sects was right—and which I should join.
>
> I was answered that I must join none of them, for they were all wrong, and the Personage who addressed me said that all their creeds were an abomination in his sight: that those professors were all corrupt. . . .

Three years later Joseph had another vision. He was praying, he wrote, just after retiring for the night when a light filled the room. Immediately another personage appeared who was "glorious beyond description." This personage said he was Moroni and that he bore a divine message. God had singled Joseph out for a special work.

> . . . [Moroni] said there was a book deposited, written upon gold plates, giving an account of the former inhabitants of this continent, and the sources from whence they sprang. He also said that the fullness of the everlasting Gospel was contained in it, as delivered by the Savior to the ancient inhabitants.
>
> Also that there were two stones in silver bows—and these stones, fastened to a breastplate, constituted what is called the Urim and Thummim—deposited with the plates; and the possession and use of these stones were what constituted "Seers" in ancient or former times; and that God had prepared them for the purpose of translating the book.

Joseph was directed to go to the top of a high hill near Manchester, a town a short distance from Palmyra. He did so and there under a large stone he found the plates deposited in a stone box. But he was not permitted to have possession of them until four years later. When he received the plates, he was given also the Urim and Thummim by means of which he was miraculously enabled to translate the strange language which was written on the plates.

Only eleven persons, aside from Joseph, ever saw the plates. These eleven, however, attested that the plates were real and had "engravings thereon." Moreover, eight of the eleven said, "We did handle [them] with our hands." Six of the eleven later left the Church (two eventually returned) but none ever repudiated his statement that the plates were real. One, however, when he was asked, "Did you see the plates and the engravings upon them with your bodily eyes?" was reported to have replied, "I saw them with the eye of faith."[2]

The translation which Joseph made of the engravings is the *Book of Mormon*. According to this book, which "in no sense supplants the Bible, but supports it,"[3] North and South America were peopled by descendants of ancient Israel who journeyed eastward from Palestine in Old Testament times. With supernatural help they built ships and sailed in them to the Western Hemisphere. Eventually two peoples arose: the Nephites and the Lamanites; the former were a righteous group and the latter a wicked group. Most of the ruined cities of South and Central America probably were Nephite or Lamanite in origin. The *Book of Mormon* discloses how Jesus, shortly after his ministry in Palestine, appeared on earth a second time, in the Western Hemisphere. There he again revealed the Christian gospel, chose twelve disciples, and founded a Church. Many of his teachings were very like those of the New Testament. For example, two of the Beatitudes read as follows:

Yea, blessed are the poor in spirit who come unto me, for theirs is the kingdom of heaven.
And blessed are all they who do hunger and thirst after righteousness, for they shall be filled with the Holy Ghost.

About twenty-five thousand of the three hundred thousand words in the *Book of Mormon* parallel passages in the Old Testament and two thousand parallel passages in the New Testament.[4]

Finally the Lamanites destroyed the Nephites (who had become the more wicked) in a great battle which centered at the hill Cumorah, located near Manchester. Mormon was the general of the defeated armies. His son, Moroni, buried the sacred gold plates at the top of the hill, where Joseph Smith testified he found them.

Non-Mormon writers explain the origin of the *Book of Mormon* in other ways. Some say the book was the work of Joseph Smith's own fervid imagination. One non-Mormon biographer wrote:

Perhaps in the beginning Joseph never intended his stories of the golden plates to be taken so seriously, but once the masquerade had begun, there was no point at which he could call a halt. Since his own family believed him (with the possible exception of his cynical younger brother William), why should not the world?[5]

Mormons hotly deny such assertions. A well-known Mormon theologian wrote: "The fanciful theories of its origin, advanced by prejudiced opponents, are in general too inconsistent, and in most instances too thoroughly puerile, to merit consideration."[6] A Mormon historian asserts that the *Book of Mormon* is criticized only by "poorly trained scholars" who think they "can answer the question by referring to whatever tiny patch of knowledge" they happen to "sit on."[7] However, concerning belief in the historicity of the events narrated, one Mormon wrote, "No competent student maintains that discoveries of archaeologists as yet prove the *Book of Mormon*. The *Book of Mormon,* as every faithful member of the Church knows, is a divine book . . . [which] must be accepted on principles of faith."[8] Whatever one may think about the material printed in the *Book of Mormon,* every fair mind must acknowledge its great influence on American life. Henry A. Wallace, Vice-President of the United States in the New Deal era, wrote:

Of all the American religious books of the nineteenth century, it seems probable that the Book of Mormon was the most powerful. It reached perhaps only one per cent of the people of the United States, but it affected this one per cent so powerfully and lastingly that all the people of the United States have been affected, especially by its contribution to opening up one of our great frontiers.[9]

But the *Book of Mormon* was only one of the pillars on which the Church of Latter-day Saints was founded. Another pillar is the doctrine of the restoration of revelation. Like most Christians, Mormons believe that Jesus Christ established the Christian Church. Mormons contend, however, that soon after its founding the Church fell away from the truth and came under the control of men who were apostates. "The reign of Constantine marks the period when the paganization of Christianity became complete."[10] The so-called Christian priests and ministers have "no more authority to administer Christian ordinances than the Apostate Jews."[11] The Devil became the foundation of the Church.[12] As a result of the apostasy of the churchmen, God withdrew revelation from mankind. Then in these latter days, He chose to re-establish His Church on earth and to begin again giving revelations. Joseph Smith was the first of the latter-day prophets through whom God's revelations came. Beginning on September 21, 1823, when he was eighteen years old, Joseph was the agent God used to instruct His latter-day saints. Some of the revelations which came to Joseph are printed in two books called the *Doctrines and Covenants* and the *Pearl of Great Price*. These two books, and also the Book of Mormon, have the status of scripture and are believed to contain revelation.

Revelation continued, according to Mormon belief, after Joseph Smith's death. Nine men in all have been "modern-day prophets . . . responsible under the inspiration of the Almighty for directing this latter-day marvelous work."[13] Of the present prophet and revelator a Mormon leader wrote, "David O. McKay is the mouthpiece of our Heavenly Father in the earth today who does hold the keys of the kingdom, and the mantle of authority."[14]

The major revelations of the *Doctrine and Covenants* are of course basic Mormon teachings. (These will be outlined shortly.) But the book also contains minor revelations which are a source of considerable difficulty for the non-Mormon reader. Why should God reveal, for example, that Joseph's first wife, Emma, should make a selection of hymns for use by the Church (25:11),* or that Sidney Gilbert should establish a store (57:8), or that William W. Phelps should become the printer for the Church (57:11), or that John Snider and others should build a hotel (124:22–24)? In view of the abuse of liquor on the frontier, understanding why a revelation came which forbade strong drinks is not too difficult; but why did God reveal that

* The numbers refer to sections and verses in *Doctrine and Covenants* (Church of Jesus Christ of Latter-day Saints, 1952).

"hot drinks are not for the body or belly" (89:7, 9.)? On the other hand the correctness of the revelations seems to Mormon thinking to have been fully demonstrated. How else than through revelation could Joseph have prophesied in 1843 that "the commencement of the difficulties which will cause much bloodshed . . . will be in South Carolina" (130:12–13.)? Mormons believe this is an obvious allusion to the Civil War, and to its beginning at Fort Sumter.

PERSECUTION

After the Church's founding in 1830, it grew rapidly; in a year's time one thousand persons had been baptized. Persecution set in almost immediately. Early one morning, for example, while an immersion service was being conducted in a small stream, a mob gathered and threatened the worshipers. Joseph wrote that "it was only by the exercise of great prudence on our part, and reliance in our heavenly Father, that they were kept from laying violent hands upon us . . . we were obliged . . . to bear with insults and threatenings without number."[15] On two occasions Joseph was arrested for disorderly conduct; but he won acquittal both times.

Such experiences led the leaders of the Church to decide to move their headquarters farther west. They chose to go to Kirtland, Ohio, a town near Cleveland where Mormon missionaries had won notable successes. In Kirtland, Joseph received many revelations concerning the Church, its doctrine, organization, and administration. There the saints began at great sacrifice the building of the first Mormon temple, some of the men contributing as much as one-seventh of their time to its construction. There also land fever led some of the saints to indulge in speculation. The economic difficulties which resulted and further persecution by non-Mormons led after a half dozen years to a move from Kirtland and to the establishment of the Church's headquarters in Western Missouri.

Independence, Missouri, in the 1830's was a recently founded boomtown, the take-off place for the old Santa Fe Trail; it was the key town along the whole frontier. Mormons had begun settling in Independence soon after the move to Kirtland. There also they met persecution; their peculiar beliefs, their exclusiveness, their desire to own more and more land, and perhaps above all their opposition to slavery brought down on their heads the wrath of their neighbors. Mobs gathered; they unroofed cabins, whipped men, tarred and feathered leaders. The militia which was called out to restore order was headed by an avowed Mormon enemy. The authorities persuaded the Mormons to disarm as a prelude to a peaceful settlement on the promise that the non-Mormons also would disarm. This promise was not kept. Accordingly, the mob could attack with impunity and worked a systematic havoc. In one night the homes of twelve hundred people were sacked; the men were beaten; and men, women and children were driven out into a November storm with no shelter but the woods which lined the Missouri River. There they shivered for days. Eventually they fled across the river into Clay County. They lived there for a time, but after about three years were asked to leave that locality also. Again they must start anew. This time they acquired largely unoccupied land on the edge of the prairie, organized Caldwell County, Missouri, and established the town of Far West as the county seat.

But persecution continued to dog their efforts to establish a Mormon home. Having tried disarmament, they now began to put their trust in force. Clashes followed which brought them into greater and greater disfavor. Finally the governor of the state declared they "must be treated as enemies, and must be exterminated or driven from the state if necessary for the public peace."

They fled to Illinois and established headquarters a dozen miles north of Quincy on a bend of the Mississippi River. Joseph named the site Nauvoo, a word which he said comes from the Hebrew language and means "beautiful plantation." For a time, security and prosperity seemed finally to have come to the Mormons. Hard work and alertness soon made Nauvoo the largest city in the state and one visitor wrote that it was an "orderly city, magnificently laid out, and teeming with activity and enterprise."[16] Joseph prospered and even ran for President of the United States in the campaign of 1844. At Nauvoo, Joseph also received a revelation sanctioning the practice of polygamy. The heart of it is in Section 132, verses 61 and 62 of the *Doctrine and Covenants:*

And again, as pertaining to the law of the priesthood—if any man espouse a virgin, and desire to espouse another, and the first give her consent, and if he espouse the second, and they are virgins, and have vowed to no other man, then is he justified; he cannot commit adultery for they are given unto him; for he cannot commit adultery with that that belongeth unto him and to no one else.

And if he have ten virgins given unto him by this law, he cannot commit adultery, for they belong to him, and they are given unto him; therefore is he justified.

Although this revelation was not openly acknowledged until 1852, the evidence seems clear that Joseph and other Mormon leaders practiced polygamy in the Nauvoo years; the correctness of the practice continued to be taught in Utah until it was finally abrogated in 1890.

As prosperity grew at Nauvoo, hostility grew in the neighboring sections of Illinois. Exclusiveness, economic competition, rumors about plural marriage, and the unwise destruction of a printing press led to calls on the Governor to summon the militia. This official went to Carthage, the county seat, and demanded of Joseph that he also go there to stand trial. Although Joseph was promised a safe conduct, adequate protection was not provided. At Carthage on June 27, 1844, a mob of one hundred and fifty men, many of them members of the militia, attacked the jail in which Joseph was imprisoned, and he was shot to death.

In the struggle for the succession which followed, several divisions resulted in the Church. Today the Mormons are divided into five groups. Three of them are very small ranging from three thousand down to less than one hundred members. The bulk of the Mormons are divided between two major groups: the Utah church which is much the larger and a church whose official name is Reorganized Church of Jesus Christ of Latter Day Saints. The smaller group won the adherence of Joseph's immediate family, had his son Joseph Smith III as one of its first presidents, and has possession of some of the important properties. This group eventually located its headquarters at Independence, Missouri, and has today one hundred and fifty-five thousand members. The Utah group has an American membership of a million and one half.

After Joseph's death most of the Mormons chose Brigham Young as their leader. About the greatness of this man there can be little dispute. The source of the qualities which produce the natural leader is mysterious: perhaps they are hereditary; perhaps they result from the unconscious learnings of early childhood. Whatever their source, Brigham Young possessed these qualities to a remarkable degree. He had that indefinable something which made men willing to trust his judgment and to put forth extremes of effort. His greatest abilities as a leader were immediately needed. The mobs of western Illinois continued to attack and to persecute. Although the saints had already left three localities in the effort to find a place where their religion could be practiced in safety and in freedom, abandoning Nauvoo seemed the only wise course of action. This time they determined if possible to get out of the range of all possible mob violence and to settle in a land "so unpromising that nobody will covet it." The vast empty spaces of the West seemed providentially provided. Thus they decided to move beyond the Rocky Mountains.

The exodus from Nauvoo began in a manner which presaged the rigors which were to come. Boats began to ferry the refugees across the Mississippi River before dawn on a day in the middle of winter, February 4, 1846. The wind was raw, the temperature twenty below zero, and only a rude shelter was available on the other side. The subsequent trip across Iowa, made in spite of the most forbidding conditions, shows the kind of heroism which had been built into the character of the Mormons; they fought cold, rain, and mud, and for a two weeks' period were even unable to light fires.

After four and one-half months of strenuous effort, the travelers reached the Missouri River and there on a site just north of Omaha established a temporary settlement which they called Winter Quarters. All that summer additional Mormon parties struggled across Iowa until by fall as many as ten thousand people had gathered.

Brigham Young saw clearly that the long trek ahead demanded farsighted planning, firm organization, and careful morale building. At the head of a party of 148 persons, he set out in the spring of 1847 to pioneer the way. On the basis of information he had collected, he hoped that the Salt Lake valley would prove to be their destination. This location had the necessary isolation, and if sufficient water were available, hard work, persistence, and vision could make there the base for a secure and righteous society. Finally on July 24, he got his first view of the valley. He had a vision then of "the future glory of Zion" and said, "This is the place, drive on." July 24 has been celebrated ever since by Mormons as Pioneer Day. On the second day after the first party arrived, potatoes were planted and soon a thriving and prosperous community appeared in the wilderness.

For ten years thereafter the Mormons were relatively free from persecution. A Mormon empire began to develop; skills in irrigation were acquired, a *de facto* theocracy was established, scores of communities were founded, some of them hundreds of miles distant from Salt Lake City. In 1850, Congress established the territory of Utah. Almost immediately conflicts arose between non-Mormon federal officials and Young, who wished to control the policies of the newly organized territory. The conflict finally resulted in the "Mormon War." In 1857, President Buchanan sent a United States army to make federal authority firm. The Mormons organized their own army, and as the federal troops drew near the whole community went onto a war footing. A scorched-earth policy was adopted; outlying settlements were abandoned, grass in the line of federal march was burned, federal wagon trains were set afire, and thirty thousand persons began to abandon Salt Lake City for territory farther south. Finally an agreement was reached between Young and President Buchanan's agents, peace was restored, and the war was over.

In the years which followed, Mormon numbers and prosperity grew, but conflict with non-Mormons was always near the surface. The chief cause was polygamy. The numbers of Mormons involved in this practice was always small; probably at its height only about 10 per cent.[17] But any practice at all was enough to arouse Eastern alarm. Clergymen especially pursued the issue relentlessly and produced definite action. Congress passed laws against polygamy, courts sent polygamous husbands to prison, and Protestant women built a home as an avenue of escape for wives who wished to escape from the "degrading bondage"; (no Mormon woman "ever entered its portals," wrote one Church leader).

Periodically since 1850, Utah had been requesting statehood, a request which had been regularly denied. By the beginning of the last quarter of the nineteenth century, the conflict had reached an impasse. The Mormons refused to deny the correctness of a practice they believed to have been revealed by God, and the federal authorities were embarked on a campaign to enforce the law. The polygamists were forced underground. The national government disfranchised Mormon voters and seized all Church property. Finally the Mormons decided that further resistance was useless and that peace and religious freedom could be won only by abandoning the teaching concerning the correctness of earthly polygamy. Accordingly, the head of the Church officially declared in 1890:

> Inasmuch as laws have been enacted by Congress forbidding plural marriages, which laws have been pronounced constitutional by the court of last resort, I hereby declare my intention to submit to those laws, and to use my influence with the members of the Church over which I preside to have them do likewise.
>
> There is nothing in my teachings to the Church or in those of my associates, during the time specified, which can be reasonably construed to inculcate or encourage polygamy.

Six years later Utah was admitted to the Union. Today the practice of polygamy is definitely opposed by the Church. Evidence does exist that "several thousand" Mormon fundamentalists continue the practice.[18] But "the Mormon Church today is merciless in proceeding against recusants who from time to time may . . . revert to . . . the plurality of wives."[19]

For decades the common saying about the Mormons was, "They are a strange people." This judgment is certainly not accurate today. Insofar as their religious beliefs allow, they try as earnestly now as any group to fit into American mores and to be normal according to American standards.

MORMON THEOLOGY

Although Mormon doctrines clearly fall within the broad general outlines of Christian thought, they contain so large a number of surprising innovations as to make traditional churchmen wonder how they can have sprung from Christian soil. Perhaps the best place to begin an exposition of these innovations is with the doctrines of the eternity of the universe and of the pre-existence of man.

Mormons teach that the universe has always existed, and that God organized rather than created it, a view contrary to the usual Christian teaching, which is that God created the universe out of nothing at the beginning of time. Mormons hold that the matter, energy, and intelligence of the universe go forward eternally in "an on-going process characterized by increasing complexity."[20] They hold that all existence is material, and that spirit in the usual meaning of that word does not have reality. They do use the word "spirit," but mean by it, not a different order of reality, but rather a "more fine or pure" kind of matter. Joseph Smith received a revelation which told how to tell, if a divine messenger comes, whether he is spirit (in the Mormon sense) or has a body of flesh and bones: "Request him to shake hands with you." If he is spirit, he will not shake hands. If he has a body, "you will feel his hand."[21]

All human beings lived before their birth on this earth. Traditional Christians of course believe that each human being had his beginning on earth at or near the time of conception. But Mormons believe in pre-existence. "God, our loving father," writes a Mormon leader, "kept us with Him during our spiritual infancy."[22] In this state men were spirits, they had no bodies. Then God decided that a plan must be adopted by which men might gain greater exaltation. An ecclesiastical council was held where God presented a plan. Lucifer offered to put it into effect, but in such a way as to receive the credit himself, to take away man's freedom, to destroy his hope of eternal progress, and to make him a slave of Satan. Lucifer's offer was rejected and his rebellion came immediately. Jesus Christ then offered to implement God's plan and to give all credit to God. Christ's offer was accepted. God's plan involved creating the earth for men to live on, preserving their freedom, sending to earth those who chose to come, giving them bodies, and thus presenting them the opportunity to work for their own exaltation.

Our pre-existence was our first estate and our experiences on earth are our second estate. Here we come to understand the nature of gross matter, and are able to cultivate our inherent Godlike attributes. Our most important occupation on this earth is to improve and to develop in ourselves those qualities which will lead to the higher degrees of exaltation in the next life. Death apparently is an inherent part of man's destiny "because some sin is inevitable," and death is the result of sin. However, we are freed from the consequences of sin by "a sacrifice of ultimate magnitude, the death of a very God."[23]

Thus we can participate in the resurrection and receive back our bodies because of the atonement by Jesus Christ.

The next life has two phases. The first occurs between death and the resurrection. How long this phase will last for any individual is not known; but Mormons believe that the righteous will be the first to receive back their bodies. In the second phase of the next life, after the resurrection, men will have the benefit both of their experience on earth and of their experience during their pre-existence. The degree of a person's exaltation in this second phase will depend on the quality of his own efforts. There are three degrees of glory in the next life, second phase. The highest will be the celestial; in it one will live literally in the presence of God. The middle degree will be the terrestrial; it will be for those who though they lived honorably failed to keep God's law. They are denied God's presence, but Jesus Christ will come unto them. The lowest degree is the telestial; it is occupied by the largest number of men and women, those who have rejected Christ and have had no interest in his program.

Mormons are universalists; they believe all men (except the very few Sons of Perdition, whose fate is uncertain) will be saved, that is, will finally arrive at the next life. But the degree of one's glory there will depend on oneself, on his works.

Mormon views of God are unusual according to traditional Christian standards. In the godhead are three gods: the Father, the Son, and the Holy Ghost. They are three separate beings, not three in one; they have one purpose but are three beings. The Holy Ghost is "a personage of spirit"; the Father and the Son have bodies of flesh and bone. The corporeality of God is one of the basic Mormon doctrines. To deny it is to "depersonalize" God. Christ himself said, "If ye have seen me, ye have seen the Father," and everyone acknowledges that the Christ had a body when he said it. In the *Book of Mormon* is written the following: "He saw the finger of the Lord; and it was as the finger of a man, like unto flesh and blood."[24]

The principle of eternal progression applies to God as well as to men. Joseph Smith once said, "God himself was once as we are now, and is an exalted man and sits enthroned in yonder heavens."[25] Moreover, He gained his present glory through his own efforts and righteousness. In other words, He is a self-made God.

The logic of these beliefs adds up to polytheism rather than to monotheism. One Mormon apostle wrote:

> During the onward march of the Supreme Being, other intelligent beings were likewise engaged, though less vigorously, in acquiring power over the forces of the universe. . . . Next to God, there may be, therefore, other intelligent beings so nearly approaching his power as to be coequal with him in all things so far as our finite understanding can perceive. These beings may be immeasurably far from God in power, nevertheless immeasurably above us mortal men of the earth.[26]

The possibility that some men can ultimately achieve Godhood is definitely held before the Mormon faithful. A familiar Mormon statement is: "As man is now, God once was; as God is now, man may become." Nevertheless, Mormons think of God's attributes very much as do traditional Christians. He is omniscient, merciful, just, possessor of an infinite authority, deeply concerned for the welfare of all men.

Clearly, Mormonism is an activist religion, one which puts great emphasis on works. What are the works men must accomplish on earth in order to increase the degree of their exaltation in the next life? First, they must have *faith in the Lord Jesus Christ*. Faith is more than a simple, "I believe." It is a living force in a person's life which gives him the power to endure. Second, men must repent. *Repentance* is more than a simple expression of regret. True repentance results in casting out of life the evils which sully it. Faith and repentance must be demonstrated before a person is accepted into church membership, and membership is necessary if a person is to gain any degree of exaltation. The essential step in becoming a church member is *baptism*. This ordinance is performed by immersion, and takes place after a person becomes eight years old, the age when Mormons believe a boy or girl is able to know right from wrong and thus has reached the age of accountability. After baptism a person is confirmed as a church member and receives the *Gift of the Holy Ghost* through the laying on of hands. After receiving this gift, persons of merit have the constant companionship of the Holy Ghost; He increases resistance to temptation, strengthens spiritual and mental abilities, and is a source of deep inspiration. Persons of outstanding merit receive even greater gifts from the Holy

Ghost. These may include prophecy, spiritual healing, speaking in tongues, the power to work miracles, the power to discern spirits. These gifts "are given to bless and not to display to satisfy the curiosity of men."[27]

Another work which males may perform is labor for the Church as a priest. *Ordination* is open to all worthy males twelve years of age or over and is entered into by about half of the eligible men and boys. The priesthood has two orders. The lower order is the Aaronic Priesthood; the higher order is the Melchizedek Priesthood. A young man becomes eligible for the higher order at age nineteen. Both of the priesthoods are divided into three grades or offices; the major positions in the Church are given of course to persons who have the highest office in the Melchizedek order; they are called high priests.

The degree of exaltation which a person can attain in the next life is affected by his family life on earth. President McKay declared, "Parenthood is next to Godhood." Mormon marriage is of two kinds: for time and for eternity. Marriage for time is valid only for this life; at death this kind of marriage ceases. Persons who have been married only for time can have but limited glory in the next world; Joseph Smith wrote that they "are appointed angels in heaven; which angels are ministering servants, to minister for those who are worthy of a far more, and an exceeding, and an eternal weight of glory."[28] But marriage for eternity, also called celestial marriage, is valid in the next life and brings great glory.

> . . . if a man marry a wife by my word . . . and it is sealed unto them by the Holy Spirit of promise . . . [they] shall inherit thrones, kingdoms, principalities, and powers, dominions, all heights and depths . . . they shall pass by the angels, and the gods, which are set there, to their exaltation and glory in all things. . . . Then shall they be gods, because they have no end.[29]

Celestial marriage is performed only for those who are fully worthy and includes the "sealing" of one partner to the other in a service which may be witnessed only by Mormons in good standing.

Mormons are opposed to birth control. The reason for this opposition is that procreation is the only means whereby the yet unborn, pre-existent spirits, whose number is definite, can reach earth, receive bodies, and begin the work of their own self-improvement.

Having children is thus a religious duty. The Mormon birthrate is said to be 50 per cent higher than the birthrate of the rest of the nation.[30]

Vicarious baptism for the dead is a distinctive Mormon practice. Persons who died before they had a chance to be baptized into the Mormon faith can be baptized by proxy, living persons, usually direct descendants, being immersed instead. Since baptism is believed to be essential to salvation and since it can be performed only on the earth, baptism of unbaptized spirits already in the next life is a work of great mercy which brings additional glory. This work also may be witnessed only by Mormons in good standing. The belief in the importance of baptism for the dead, especially for ancestors, leads to a marked interest in genealogy; the Church possesses a large genealogical library and "also a rapidly expanding file of microfilm records of international scope."[31]

Both celestial marriage and baptism for the dead are performed only in temples. The first of the Mormon temples was constructed at Kirtland, Ohio, in the 1830's; it is now in the possession of the Reorganized Church and is open to the public. The temples of the Utah Church, however, are places where only private services take place, and accordingly only worthy Mormons are admitted. About a dozen of these buildings now exist. The majority are in the American West with the most famous located in Salt Lake City. Temples have recently been dedicated in foreign lands. One is now found in Switzerland, one in England, and one in New Zealand. The famous tabernacle in Salt Lake City, a building to which strangers are admitted, is not a temple but a place of public assembly.

At the time of the beginning of the Millennium—it may be near at hand—Christ will come to earth. This period will be an era of peace: the righteous will have full power; Satan will be bound. Children will be born and will "live to the age of a tree" and then be changed from mortality to immortality in the "twinkling of an eye." After the Millennium, Satan will be loosed to tempt men for a little season. Then will come the Final Judgment and men will be assigned to glories according to their works, some to the telestial, some to the terrestrial, and some to the celestial.

Two earthly localities will be designated for the gathering of the saints: one will be in the Eastern Hemisphere at Jerusalem; the other will be in the Western Hemisphere at Zion. Joseph Smith received a revelation which indicated that Zion will be located at Independence, Missouri.

Some of these beliefs are vigorously denied by the smaller of the two main branches of Mormonism. The Reorganized Church of Jesus Christ of Latter Day Saints affirms the following propositions: God is eternal and unchangeable; He is one; it is blasphemy to say He was once a man; baptism for the spirits of the dead is not essential to their salvation; the salvation of all persons depends not on any human action but solely on the saving grace of Jesus Christ; "God promised that Joseph Smith's blessing of prophetic leadership would be placed on the head of his posterity"; "None of the leaders of the Church in Utah have been of the seed of Joseph Smith and thus do not come within the promise of God";[32] temples should be erected only after direct revelation; only the Kirtland Temple has God's sanction; the whole system of temple rituals as practiced by the Utah Church has no necessary relation to salvation; the doctrine of polygamy was vigorously attacked by Joseph Smith; he never received a revelation sanctioning it; he never practiced it; the true restoration movement has always been monogamous.

The rest of the material in this chapter concerns the Utah group.

DAY-BY-DAY RELIGIOUS PRACTICE

Few large religious groups show as much zeal as do the Mormons; concern for their Church and its welfare is second for them only to concern for their families. All worthy Church members give a tithe, that is, they give a tenth of all income to the Church. The percentage of the members which actually meets this high standard is remarkably large. One Mormon author noted that in a recent year the percentage of tithers in his ward, the Mormon name for the local Church, was 68 per cent of the membership.[33]

The wards are run by persons who in most denominations would be called laymen. They receive no salaries from the Church and earn their livings by secular pursuits. At the head of the ward is an official who has the title of Bishop. He holds the rank of high priest, and has two counselors who also are high priests. The wards are organized into stakes. Above the stakes are the General Authorities, the top agency of the whole Church.

The General Authorities are thirty-four men. The ranking body in this group is the First Presidency. It consists of the President of the Church who is also "Prophet, Seer, and Revelator," and two counselors. Next in power and distinction is the Quorum of the Twelve Apostles; when this group acts unanimously, its authority is equal to that of the First Presidency. The Twelve Apostles have eight assistants. Next comes the First Council of Seventy, seven men who preside over the "seventies." Next comes the Presiding Bishopric, three men who administer the temporal affairs of the Church. The thirty-fourth man is the Patriarch; his function is to dispense special, personal blessings to Church members. He is normally a descendant of Joseph Smith, Sr., the father of Joseph Smith, Jr. These General Authorities serve for life, and along with a few assistants and a few heads of missions are the only persons in the whole active leadership of the Church who do not support themselves. Those who do receive support are not paid "salaries"; instead they receive modest "living allowances."

Mormon zeal is nowhere more manifest than in the Church's missionary program. Worthy members are called by the First Presidency to spread the Church's message by accepting full-time missionary service. The usual pattern is for young men between twenty and twenty-five years of age to give two years of service at their own or their family's expense. They must be members of the Melchizedek Priesthood and during their period of service must live under strict discipline. Two by two they go among non-Mormons seeking opportunity to spread the message. Through personal contacts, door-to-door canvassing, speaking, tract distribution and similar activities, they meet potential converts; about 40 per cent of the rapid Mormon growth comes from conversion, mostly the conversion of adults.[34] Today about six thousand persons are serving as full-time missionaries. An additional eight thousand serve as part-time missionaries. The members of this group continue to work at their regular jobs, but give eight or more hours per week, in the evenings and on the weekend, to the work of proselytizing. As a result of this vigorous program, three new chapels are dedicated each week, on the average.

The Mormons do not emphasize the social gospel. A considerable reading of recent Church periodicals published by the Church disclosed, for example, no forthright statement dealing with the problem of war. At the General Conference in 1958 a member of the Presiding Bishopric did say, "Faith is the answer. . . . Because of the righteousness of the people, the Lord has seen fit to protect a nation."[35] Most Christians would not quarrel with this proposition, but it is a long way from a positive program for preserving peace. Concerning the race question, Mormons seem to be silent. In my reading of Mormon periodicals I found no discussion at all of segregation; moreover, while Negroes are admitted to membership in the Church, no Negro may become a member of the priesthood. The justification for such discrimination lies in the belief that Negroes made certain choices before their birth.[36] On economic justice, however, the Church becomes considerably more positive. The emphasis here is on the values of the free-enterprise system. Ezra Taft Benson, Secretary of Agriculture in the Eisenhower Administration and a member of the Council of the Twelve, echoed the general opinion when he said at the General Conference in 1958, "Let us remember that we are a prosperous people today because of a free enterprise system founded on spiritual, not material values."[37] Concern for persons in economic need led during the depression years to the establishment of the famous Welfare Program. No Mormon, it is said, need ever be forced to accept public relief. The Church has an abundance of supplies kept on hand from which help is given to persons who are unable to meet their own needs. Usually goods rather than money are dispensed. The emphasis of the Welfare Program is to help people find the skills and secure the employment which will make them independent.

As a means of increasing the relief resources, faithful Mormons fast for two meals on the first Sunday of each month and give the money which is saved to the Church.

The Church enjoins a strict avoidance of tobacco, liquor, tea, and coffee.

Mormon worship follows frontier Protestant patterns. It is conducted in a chapel where the pulpit is centrally located, contains essentially the same elements as does evangelical worship, is on the informal side, is characterized by vigorous singing, puts a major emphasis on preaching, is openly used as a tool for the building of group morale. The service is conducted by laymen (in the usual meaning of that word) and the various tasks in connection with it are passed freely from one man to another. The sermon, for example, is not preached regularly by the same person; instead many speakers are used. Usually the preacher is an outstanding man from the ward, but sometimes the preacher is a guest from another ward, sometimes a woman, and sometimes a youth. No collection is ever taken. The most solemn part of the service is the Sacrament of the Lord's Supper. In it the elements, which are ordinary bread and water, are usually blessed by the highest order of the Aaronic priesthood, commonly young men sixteen to nineteen years of age. These elements are then passed to the seated congregation by the lowest order of the Aaronic priesthood, boys twelve to fourteen.

Mormon hymns follow the evangelical Protestant pattern; many of them are hymns used by the Protestant Churches. Among the first thirty hymns printed in the Mormon hymnbook are the following: "Abide With Me," "Christ the Lord is Risen Today," "Come, Thou Fount of Every Blessing," "Come, Ye Thankful People, Come," and "All Creatures of our God and King." Of course the Church uses many hymns that are distinctively Mormon. Following is the first verse and chorus of a hymn in praise of Joseph Smith.

> Praise to the man who communed with Jehovah!
> Jesus annointed that Prophet and Seer.
> Blessed to open the last dispensation,
> Kings shall extol him, and nations revere.
>
> Hail to the Prophet, ascended to Heaven!
> Traitors and tyrants now fight him in vain.
> Mingling with Gods, he can plan for his brethren;
> Death cannot conquer the hero again.

Mormons emphasize community recreation. They seem, because of their system of theology, to put major emphasis on the next life; but they do not neglect this life, its goods and its joys. Our life on earth is one essential aspect of our eternal development; the bodily joys are wholesome and profitable; an effort to pursue them righteously is an earthly duty. In harmony with such beliefs, the Church sponsors parties, receptions, outings, dinners, movies, concerts, dramatic productions, athletic contests. In 1956 the all-church basketball tournament attracted twenty-five thousand players and the Church's dramatic program attracted over seventy thousand actors. In that same year over sixty thousand Church members sang in choruses. The world-famous Tabernacle Choir has 375 members. All these recreational activities are undertaken as a part of the religious life. Where else would a college dance be opened and closed with a prayer?[38]

Mormonism is a vital faith. Most observers of the Mormon churches in action are struck by the religious concern, the moral earnestness, and the deep spiritual security of the members; Mormonism is at or very near the top among American religions in the comfort and dynamic which it supplies to its followers.

Mormons insist that their Church is "the sole earthly repository of the eternal Priesthood in the present age" and "the one and only Church possessing a god-given charter of authority."[39] A member of the Council of Twelve wrote:

This gospel has often been spoken of as a way of life. This however is not quite accurate. Consisting as it does of the principles and ordinances necessary to man's exaltation, it is not just *a* way of life, it is *the* one and only way of life by which men may accomplish the full purpose of their mortality.[40]

CHAPTER XV

CHRISTIAN SCIENCE

ALTHOUGH CHRISTIAN SCIENTISTS often insist that spiritual healing is not the most important part of their religion, healing is the best-known part and the most effective means Christian Scientists have of winning converts. Accordingly this chapter will begin with a testimony which relates a startling claim concerning the healing power of Christian Science. Dr. Ernest H. Lyons, Jr., professor of chemistry at Principia College told the following story over television.

I was preparing a compound and I had to start with potassium cyanide. . . . To start the work I melted the potassium cyanide in an iron dish over a powerful burner. I set the iron stirring rod on the edge of the dish and turned away briefly. When I reached for the iron rod again I picked it up by the hot end. I dropped it immediately, but not before my hand was severely burned, and I could see crystals of the poison dislodged from the rod in the wound. But I was not frightened of poisoning for I was conscious of the presence of infinite Life, God, overruling the picture of accident and possible death. I washed my hand in water and wrapped it in a towel, and I was able to go ahead and complete the preparation. Within three days scarcely a scar remained and in less than a week all effects disappeared. Later I talked to an expert on cyanide poisoning and he told me that such an experience would ordinarily prove fatal in a few minutes.

Some viewers of the program doubted Dr. Lyons' competence as a chemist. In reply to one correspondent he wrote that his professional background included a B.S. from the Massachusetts Institute of Technology, a Ph.D. from the University of Illinois, twelve years of service as Chief Chemist for the Meaker Company, and a summer of lecturing in chemistry at the University of Illinois. After learning these facts, a fellow chemist wrote Dr. Lyons a letter which read in part:

As a chemist, you should be loath to give such testimony and make such statements as you did when you have no way of determining the amount of cyanide absorbed into your system at the time of the accident. Many of us in chemistry handle violent poisons, sometimes with our bare hands, and we do not suffer from them if we handle them intelligently.

In reply, Dr. Lyons wrote a letter, a part of which follows:

The accident occurred when I was preparing potassium cyanide in order to carry out the Wohler synthesis of urea. The wound, which extended across my right palm, exposed the tendons. There appeared to be several masses of white crystals in the fluids and burned tissues, but naturally I did not take time for an extended inspection. The wound was so deep that for some days my hand was drawn together.

There appear to be five possible explanations for my survival:

1. The material was not potassium cyanide. An old sample might be completely hydrolyzed to the carbonate. In this event, it would have been impossible to obtain urea. I obtained 10.8 g., whereas Coben's laboratory manual specifies 15 g. There is no reason to expect decomposition of the cyanide on the stirring rod.

2. The shock, pain, and fright might have led me to imagine crystals in the wound, or to mistake pieces of skin for crystals. Since the end of the stirring rod was heavily coated with crystals, it is virtually inconceivable that some were not dislodged and entered the wound, whether I saw them or not. After the first shock, I was surprisingly calm, and I doubt that hysteria influenced my observation.

3. The amount of potassium cyanide absorbed was below the lethal level. This level is so low that such an occurrence is highly improbable. Furthermore, even a much smaller absorption leads to milder symptoms, none of which were observed.

4. The cyanide was washed out of the wound without being absorbed. In the opinion of a safety expert from duPont, sent to check the safety practices of my former employers, who use cyanide in ton lots, this is impossible under the circumstances described. . . .

5. Cyanide was absorbed but rendered harmless by the higher law of life. Although this may sound incredible, such instances are not uncommon to Christian Scientists. There is ample evidence, meeting both legal and scientific standards, that Christian Science healings have occurred which contravene material laws. In many instances, competent medical diagnoses were made before and afterwards.[1]

This illustration shows clearly that from the point of view of ordinary common sense some of the claims of Christian Science are not short of fantastic. Nevertheless, for many thousands of Americans, they are the very essence of a correct view of life. An effort to understand these claims and to see why they are believed by intelligent and alert people should begin with a study of Christian Science's founder and leader.

Mary Baker Eddy

Mrs. Eddy was born in 1821 near the capital of New Hampshire, in the township of Bow. Mark Baker, her farmer father, was a Congregationalist of the stern Calvinist type common in that day. She said in her autobiography that he believed in "a final judgment-day, in the danger of endless punishment, and in a Jehovah merciless toward unbelievers." She added that the minister of the family church "was apparently as eager to have unbelievers in these dogmas lost, as he was to have elect believers converted and rescued from perdition."[2]

Her religious powers developed early, she wrote; at about age eight, she, like the biblical Samuel, heard a voice calling her name, and at age twelve she spoke in the church so earnestly that she confounded her elders and that "even the oldest church-members wept."[3] "Mary Baker Eddy's earthly experiences were and are often likened to those of Jesus."[4]

At age twenty-three, she married George Glover, a building contractor of Charleston, South Carolina. After but six months of marriage, Mr. Glover was suddenly attacked by a severe case of "bilious fever" and in less than two weeks died leaving his wife almost penniless and pregnant; she had to return to the home of her parents where her son was born. After six years her mother also died and within a few months her father married again. Since she did not get along with her stepmother, Mary went to live at the home of her well-to-do sister, Abigail; her son, apparently wanted by neither stepmother nor sister, was sent to live with a neighbor. He grew to maturity in that family.

In an effort to support herself, Mrs. Glover opened a school; but it was unsuccessful. She did a good deal of writing—poetry, essays, and fiction—but had only modest success. Subsequently her health, never robust, went into decline. Her official biographer wrote:

Mary was often confined to her bed for long periods. She was afflicted with a spinal weakness which caused spasmodic seizures, followed by prostration which amounted to a complete nervous collapse. . . . Abigail sought in divers ways to make her sister more comfortable. She had a divan fitted with rockers to give Mary a change from long hours in bed.[5]

Unfriendly biographers have asserted that her illness was hysterical in nature rather than organic, that the "divan fitted with rockers" was in fact a huge cradle, and that Mary was struggling to recapture the infantile comforts of her earliest years.[6]

After nine years of widowhood, she married Dr. Daniel Patterson, a traveling dentist. At the time of the wedding, she was too ill to walk downstairs by herself, and had to be carried down to the ceremony. Now began the most unhappy years of a very rugged and often insecure life. Patterson's practice was not lucrative and he was usually in financial difficulties; they lived at first in a dingy tenement underneath a tailor's shop; and Mary was often prostrate with illness. During the Civil War, Patterson was so unfortunate as to be captured by the enemy and spent months in a Confederate prison. At this time, Mrs. Patterson again went to live with her sister. Since her health continued unimproved, she was sent to Dr. Vail's Water Cure Sanitarium at Hill, New Hampshire, but at the end of three months of treatment was worse off than at the beginning.

But now a new day was about to dawn for her. In Portland, Maine, the fame of Phineas P. Quimby as a mental healer had become firmly established and was spreading rapidly over New England. He was a clockmaker who had had only six months of schooling. But he had witnessed some of the therapeutic values in hypnotism, and later through his own experimentation had stumbled onto the values of simple suggestion. Mrs. Patterson became convinced that he could help her, and in October, 1862, she journeyed to Portland. Under his treatments, she seemed miraculously restored to health, her symptoms disappeared, and her spirits revived. Naturally she was jubilant. In the *Portland Courier* of November 7, 1862, she wrote:

Three weeks since I quitted my nurse and sick room *en route* for Portland. The belief of my recovery had died out of the hearts of those who were most anxious for it. With this mental and physical depression I first visited P. P. Quimby; and in less than one week from that time I ascended by a stairway of one hundred and eighty two steps to the dome of the City Hall, and am improving *ad infinitum*.[7]

After staying in Portland for several weeks, she and her husband, who had escaped from prison, took up residence in Lynn, Massachusetts, where he sought to establish a practice. Her cure unfortunately was not permanent; under pressure of the uncertain Patterson finances, her symptoms began to reappear. Accordingly in order to receive Quimby's treatment she returned to Portland for two other visits.

On the evening of February 1, 1866, in Lynn, occurred the accident which Christian Scientists look upon as the event which led to the discovery of Christian Science. On the way to a meeting, she slipped on the ice and was injured so badly she was carried unconscious into a nearby house. Reports concerning the seriousness of her injury vary sharply. Concerning it she wrote:

Two weeks ago I fell on the sidewalk, and struck my back on the ice, and was taken up for dead. . . . The physician attending me said I had taken the last step I ever should, but in two days I got out of my bed *alone* and *will* walk.

Five years later she wrote:

Dr. Cushing of this city, pronounced my injury incurable and that I could not survive three days because of it.[8]

Thirty-eight years later, Dr. Cushing made an affidavit in which he said in part:

I . . . kept a careful and accurate record, in detail, of my various cases. . . . I found her very nervous, partially unconscious, semi-hysterical. . . . I did not at any time declare, or believe, that there was no hope of Mrs. Patterson's recovery, or that she was in a critical condition.[9]

Whatever were the facts concerning the seriousness of the fall, Mrs. Patterson came to regard her recovery as spiritual rather than physical, and Christian Scientists came to regard it as the birth of their religion. The revelation began at that time. Years later when the movement was well established, she wrote:

On the third day thereafter [after the accident], I called for my Bible, and opened it at Matthew IX. 2.* As I read, the healing Truth dawned upon my sense.[10]

* "And behold, they brought to him a man sick of the palsy, lying on a bed: and Jesus seeing their faith said unto the sick of the palsy, Son, be of good cheer; thy sins are forgiven."

But Mrs. Patterson still had some very hard years ahead. Her marriage had proved so unhappy that finally she and the dentist separated. "For several years he sent her an annual allowance of two hundred dollars;"[11] but eventually even that amount ceased to come, and in 1873 she obtained a divorce. Since she had no home of her own, and since her family refused further help, she lived in rooming houses or with friends. But over and over she was forced to move because of inability to pay her bills or because of conflicts with her hosts. She moved nine times in less than four years and once was put out into the street at night in a heavy rain. These struggles left their mark; how could it have been otherwise? Critics who in later years became sarcastic over her concern for money failed to give adequate consideration to her years of financial desperation.

Her attention now was centered on healing. The new methods fascinated her and she spent a great deal of time studying them. She had profited greatly by her experiences with Quimby and did a great deal of thinking herself about the problem. Everywhere she went in Lynn she talked of healing. Some of her friends also became fascinated and wanted to be taught the new art. Moreover, they were willing to pay for instruction. Thus began the major career of her life. Some healings by her are recorded; but her pre-eminent success was as a teacher of healers. She became outstanding at it. And yet success came slowly and with difficulty. Not until the publication of her first and most important book was her career really launched.

She had always wanted to be an author and had spent much of her energy writing smaller pieces. Now she had a topic worthy of her best efforts; therefore, she determined to write a book. The going was very hard, as it is at the beginning with many an author. She spent four years on the first draft, and then for some years afterward was unable to find a publisher. Finally two of her students put up the money which made publication possible. But sales were extremely slow. So slow in fact that door-to-door peddling was finally resorted to. But in the end persistence won out. Published in her fifty-fourth year, *Science and Health with Key to the Scriptures* eventually went into 382 "editions" and has probably sold close to a million copies, according to one estimate;[12] a Church official, however, wrote of this estimate that it "seems fantastically inadequate."

The book is hard reading for the person who has not had prior introductions to its major teachings. Many a person knowing of its influence and fame has taken it up for study but has been unable or unwilling to read far. Yet Christian Scientists think of it as a revelation from God. One writer in a recent issue of *The Christian Science Journal* said that *Science and Health* "contains . . . the ultimate Word of God."[13] In the same magazine the Directors of the Mother Church wrote, "Hers was indeed the complete and final revelation of Truth."[14]

On the other hand, some non-Christian Scientists have been very caustic about the book. Mark Twain wrote of it: "When you read it you seem to be listening to lively and aggressive and oracular speech delivered in an unknown tongue."[15] An unfriendly biographer wrote, "It was at once a flight from reality and from the self within."[16] A considerable amount of effort has been put into showing that the real source of *Science and Health* was one of the much-discussed movements of that day: Quimbyism, or Transcendentalism, or Hinduism, or Hegelianism. The outright plagiarism of some passages has been asserted and considerable data have been forthcoming to sustain the allegation.[17] However, some scholars who are essentially very critical of her works have either not been willing to accept the assertion or have denied it.[18] One unfriendly biography reads: "There was no question of direct plagiarism."[19] Probably all students other than those with Christian Science leanings would at least agree that the author of *Science and Health* was heavily in debt to the thought of her own times even though she put on her work the stamp of her own creative abilities. Christian Scientists, however, indignantly deny the charge of plagiarism and vigorously assert the complete originality of *Science and Health,* except for its firm footing in the Bible. Its real source is said to be God Himself. She "was under the conviction," writes one painstaking student of Christian Science, "that *Science and Health* had been dictated to her by divine revelation."[20] She once wrote, "What can improve God's work?"[21] And she hinted that the little book in the hand of the angel written about in the Apocalypse, Chapter X, was *Science and Health.*[22]

The first edition of *Science and Health* appeared in 1875. In that same year the holding of public Christian Science meetings began.

Two years later she married Asa Gilbert Eddy, one of her students. He proved to be a person markedly responsive to her moods and deeply concerned for Christian Science. His death after but five years was a serious blow not only to her personal happiness but also to her science of healing. She contended he died not of natural causes but from the malicious mental influences cast on him by her enemies. Apparently she had many enemies at this time for she was engaged in several embarrassing lawsuits. Late in the 1870's she began to teach classes and to hold services in Boston; soon she made that city her headquarters.

At age sixty, Mrs. Eddy founded the Massachusetts Metaphysical College. This institution was remarkable indeed. The campus was one or two rooms in Mrs. Eddy's home, the faculty consisted of no one but herself, the curriculum contained but two courses, and the course met for only twelve sessions. Yet the College was hugely successful. Even though the tuition charged was three hundred dollars, at the end of six and one-half years she had taught over six hundred students. Even counting generous scholarships, a return of well over one hundred thousand dollars must have been realized. Her students became healers and centers for the propagation of Christian Science. When Mrs. Eddy no longer wished to continue teaching, she tried to turn the College over to associates; but they did not have her spark, insight, and classroom presence. Accordingly she closed the College after nine years of operation.

Although Mrs. Eddy had been holding public services for a number of years, first in Lynn and then in Boston, her church was not formally organized until 1879. Growth was slow at the beginning; after three years the membership numbered only about fifty. But after eight more years, the story was different; over a hundred churches and societies had sprung up in various parts of the country. In 1892, the present form of organization was adopted and a charter granted. The name chosen was: The First Church of Christ, Scientist, in Boston, Massachusetts. This organization, called the Mother Church, is the primary Christian Science institution; it is governed by a self-perpetuating Board of Directors. Today the Mother Church has branch churches all over the world.

Mrs. Eddy died in 1910. In the last years of her life, she was one of the most famous women in the world. She grew very wealthy, the newspapers printed a great deal concerning her activities and opinions, and thousands of her followers would have done for her anything she asked.

THE BELIEFS OF CHRISTIAN SCIENCE

A frequent statement about Christian Science is that it is neither Christian nor science. A consideration of each of these accusations is an excellent way to begin an explanation of the essential beliefs of this faith.

The word *Christian* has a wide range of meanings. A "Christian act," for example, is one which has in it much of kindness and concern for others; Christian acts are of course often performed by Moslems, Shintoists, atheists, and other non-Christians. A "Christian nation" is one in which a majority of the citizens call themselves Christians; all Christian nations of course do many things which could not be classed as "Christian acts." In view of such a confusion of meaning, anyone who claims that the word must be restricted to some narrow and partisan meaning is clearly going contrary to current usage. And yet such claims are common. Sectarians of one type or another often say that no person is a Christian unless he has received a certain sacrament, or accepts a certain organization, or holds a certain belief, or can testify to a certain experience. On such grounds, the name *Christian* has been denied to Protestants, to Unitarians, to Quakers, to persons who do not believe in the deity of Jesus, to persons without a Pentecostal experience.

Such efforts to restrict narrowly the meaning of the word are unfortunate. They stem from beliefs that somehow it can be limited to persons whose acts are worthy or whose beliefs are true. A better practice would be to define the word in the same broad and inclusive way that the word *American* is defined, in such a manner as to include many persons whose beliefs and conduct are disapproved of. Such a practice has firm support in present-day usage. I suggest that a Christian can fairly be defined as a person who has an attitude of reverence for the person and teachings of Jesus of Nazareth, and who strives in some degree to follow his example and to be guided by the traditions which have grown up around his name.

On the basis of such a definition, the Christian Scientists are as firmly in the Christian community as is any group. Mrs. Eddy taught that Jesus was born of a virgin, lived a sinless life, was infallible, showed mankind the true way of life, mediated between God and man, and is " 'the resurrection and the life' to all who follow him in deed."[23] She distinguished between Jesus and Christ. "Jesus," she wrote, "is the human man, and Christ is the divine idea."[24] In view of such veneration of Jesus and of Christ, anyone who would deny the term *Christian* to Christian Science no doubt seeks a narrow usage which would deny the term to a large percentage of those whom the world now calls Christian.

The term *science* is also used in several different ways. Popularly it often has a sort of institutional connotation, referring to the men, the organizations, and the equipment which are concentrated on the modern effort to study nature. Somewhat more narrowly, the word is sometimes used to indicate the results of such a study, the facts and the principles which students of nature discover. A still narrower usage emphasizes the methods by which physical nature is studied, the methods of direct observation, experimentation, and induction. If the term *science* is restricted to such usages, then it is misappropriated in the name *Christian Science*. However, there is a meaning of the word, perhaps more in vogue a hundred years ago than today, which applies it to the search for truth, to the effort to discover the nature of reality, to knowledge which is "comprehensive, profound, or philosophical" as *Webster's New International Dictionary* (2nd edition) puts it. This is the meaning which Christian Scientists employ. They deny as emphatically as anyone that their system is scientific in the sense that science centers in the study of material nature. But they are positive that they have discovered the true nature of reality, that they have a knowledge of man and of the universe which is comprehensive and profound, and that this knowledge has been gained empirically, that is, from experience, and is capable of proof through demonstration. Accordingly from their point of view, theirs is the only true science, the only true knowledge.

What is this knowledge?

The basic tenet of Christian Science is the allness of God and the denial that anything truly exists which is not God or His manifestation. Mrs. Eddy wrote:

> God is All-in-all. . . . nothing possesses reality nor existence except the divine Mind and His ideas.
> God is infinite, the only Life, substance, Spirit, or Soul, the only intelligence of the universe, including man.[25]

A corollary to the belief in the allness of God is a belief in the nonexistence of matter. Mrs. Eddy wrote:

> Matter is neither substantial, living, nor intelligent.
> God, Spirit, being all, nothing is matter.
> That matter is substantial . . . is one of the false beliefs of mortals.[26]

Perhaps the best known summary of this point of view is the "Scientific Statement of Being"; it is found on page 468 of *Science and Health* and is recited at each Christian Science Sunday service. This statement has been called "the center of the doctrine of Christian Science."[27]

> There is no life, truth, intelligence, nor substance in matter. All is infinite Mind and its infinite manifestation, for God is All-in-all. Spirit is immortal Truth; matter is mortal error. Spirit is real and eternal; matter is unreal and temporal. Spirit is God, and man is His image and likeness. Therefore man is not material; he is spiritual.

Strange as these ideas may seem to some readers, they have a familiar sound to the student of the philosophy of religion. The "allness of God" is usually called pantheism which is a belief that the universe itself and everything in it is God. Pantheism is a prominent aspect of Hinduism, is basic to some well-known Western philosophical systems (Spinozism, for example), and is close kin to the immanentism of some extreme religious liberals. Mrs. Eddy, however, was not a pantheist even though some of her words may read as though she were. She frequently said she was not and wrote: "Man is not God, and God is not man."[28] "Man is the offspring . . . of the highest qualities of Mind."[29]

The unreality of matter is also an idea which is familiar to the student of the history of thought. A whole school of philosophy, idealism, affirms that the basic reality is mind, and some of the world's greatest philosophers (Berkeley, for example) have held that man can have true knowledge of no other order of existence.

Such a proposition seems to be egregious twaddle to the average common-sense realist. Yet belief in the nonexistence of matter can be based on some very sober thinking. Our knowledge of any material object, a tree for example, is not of the object itself, but of the sensations which are produced in us. We do not experience a tree as such. Rather we see the greenness of its leaves, the brownness of its bark, and the shape of its contour against the sky. We hear the noise of the wind in its branches, smell the fragrance of its blossoms, and taste the fruit it produces. When we have summed up all the sensations produced in us by a tree, we can be certain only of the sensations. We may make a leap of faith and believe that these sensations are produced by a material object. Such a belief is a faith, since the direct experience we have is not of any object but only of sensations. Accordingly, why not stop with the sensations? The simplest belief to adopt, then, is that what we call a tree is not material but rather is a cluster of mental experiences. The logical conclusion follows that all reality is mental and that the existence of matter cannot be proved.

Mrs. Eddy did not emphasize this type of defense of her belief that matter does not exist. Indeed, her lack of knowledge of the history of philosophy is clear. As has been pointed out, Christian Scientists reiterate their contention that her work was completely original, uninfluenced by any other major writing except the Bible. Mrs. Eddy explained that the belief in matter is simply an error of man's mind (of *mortal mind* as she called it), a delusion, a dream. She wrote:

> Mortal existence is a dream; mortal existence has no real entity. . . . A mortal may be weary or pained, enjoy or suffer, according to the dream he entertains in sleep. When that dream vanishes, the mortal finds himself experiencing none of these dream-sensations. . . . Now I ask, Is there any more reality in the waking dream of mortal existence than in the sleeping dream? There cannot be, since whatever appears to be a mortal man is a mortal dream. Take away the mortal mind, and matter has no more sense as a man than it has as a tree. But the spiritual, real man is immortal.[30]

In expanding this theory of Mrs. Eddy's, one of her followers, Neil K. Adam, an English chemist, wrote that matter is

merely a false theory as to the nature of substance. . . . the substance of
the universe is divine Mind, Spirit, which is neither distant nor unsub-
stantial, but is present here and now and is permanent, tangible, and substantial to the spiritually enlightened consciousness. . . . Natural science
assumes things to be outside the observer and the mind that studies them.
. . . In Christian Science, real creations must be observed and studied
from the standpoint of divine Mind, which is continuously creating all
that really is.[31]

Mrs. Eddy dealt with evil in the same way that she dealt with
matter—she denied its reality. Evil is an error of mortal mind. Evil
cannot exist since God is All-in-all. Evil can have reality only in
matter, and since matter is an illusion, evil is also an illusion. "Evil
is nothing, no thing, mind, nor power. As manifested by mankind
it stands for a lie, nothing claiming to be something."[32] This belief
lies at the basis of her system of healing, as will be explained later.

Since all reality is mental, human beings also are mental and not
material. Man is "wholly apart from matter" declares a writer in a
recent Christian Science magazine.[33] In *Science and Health* we read,
"Man is not matter; he is not made up of brain, blood, bones, and
other material elements. . . . Man is idea, the image, of Love; he
is not physique."[34]

Belief in immortality is strong: since man is really in God, he has
always existed and will always exist in "the ever-present now."[35]
Heaven is not thought to be a place nor to begin after "the change
called death"; heaven is rather the attainment of divine Mind, the
gaining of a state which is beyond both time and space. Thus
Christian Scientists deny the reality of death, do not believe in the
existence of hell, and believe that in the end all men will be saved.
In view of Mrs. Eddy's belief in the allness of God, a natural view
for Christian Science to hold would be that in immortal life human
personalities are lost in the allness of God. But Christian Scientists
do not hold this view. They believe strongly in the immortal in-
dividuality of the person, in the survival of each personality.

Man is thought to be inherently good. This belief follows naturally
from the belief that God is all. "God is good, and therefore good is
infinite, is all." "Man . . . in reality has only the substance of good,
the substance of Spirit."[36] One author wrote of the "pre-existent
spiritual perfection of every child of God."[37]

Reflection exposes a serious difficulty with the basic Christian Science dogma. If God is All-in-all, how does it happen that matter, evil, error, could arise? Christian Scientists maintain that the correct answer is a practical demonstration of the fact that evil, like the proposition "two plus two equals five," does not come from anywhere and consequently has no existence. Nevertheless, a person who is not a Christian Scientist is puzzled by the fact that Christian Scientists still struggle against error, which logically could never occur if God is All. Like other Christian denominations, Christian Science has difficulty in reconciling evil with belief in an all-good and an all-powerful God. Theology in all religions often fails to account satisfactorily for important aspects of life. Men have experiences which they are unable to explain adequately. Practice is often better than theory. Christian Science started in experiences and it continues to live and to flourish because men and women continue to have experiences which help them, not because its theology has all the answers.

Mrs. Eddy's ideas about salvation were radical. She attacked traditional ideas of the atonement: "How can God propitiate Himself?" she wrote, ". . . Christ, Truth, could conciliate no nature above his own." "The atonement of Christ reconciles man to God, not God to man."[38] True salvation comes from the realization that all is divine Mind. Sin, sickness, and death are destroyed by the knowledge that matter, the product of mortal mind, is an illusion. Here doubtless is the real radicalism of Christian Science. The belief that matter does not exist is much more defensible than is the belief that ideas alone can effect changes in experience.

Christian Science here expresses its faith in answer to a problem with which all religions deal. The problem is: Can man influence the ultimate issues of his life? Two different answers to this question are prominent in the religions of our day. One is traditional. It says that man can change his circumstances through prayers of petition to God, that God may act in a different way because man asks Him to do so. The other answer is that the only proper prayer is the prayer of communion, the prayer in which man seeks to conform his will to the will of God rather than to bend the will of God to conform to man's will. Prayers of communion are the religious expression of an attitude which is dominant among modern materialistic scientists. They hold that men cannot change ultimates but must instead discover them, use them, conform to them.

This latter view is held by Christian Scientists also; they believe the truth must be discovered and adhered to. The striking difference between the beliefs of Christian Scientists, on the one hand, and the beliefs of materialistic scientists and most of the believers in prayers of communion, on the other, is in the *means* man must use in order to conform to truth and through conforming to succeed in making changes in his experience. Whole-hearted believers in the sufficiency of materialistic science assert that the means are wholly material. Most believers in prayers of communion assert that the means are both material and spiritual. But the Christian Scientist asserts that the means are wholly spiritual, mental. Moreover, he takes an extreme form of this view in asserting that the action of the mind is direct, it needs no mediators either physical or mental between itself and the desired effect. "Man creates his own experience."[39] This, as a leading Christian Scientist says, is truly a "daring" faith.[40] This faith holds that the effect of mental action is to make immediate changes in experience. "As a man thinketh, so is he. Mind is all that feels, acts, or impedes action."[41] Mental facts—realization, knowledge, affirmation, conception, idea—leave at once their mark on human experience. They determine whether a person lives in accordance with divine Mind, Truth, or in accordance with mortal mind, error. The extent to which a person overcomes errors determines the extent to which he avoids evils, sickness, sin, and unhappiness. These things result from ignorance, misconceptions, and denials of the Truth.

The radicalism of the Christian Science faith is so great as to suggest some startling possibilities; the healing of human ills is but the beginning. One careful student of Christian Science wrote, "Any individual who did not share in the erroneous judgment causative to the atomic bomb would have had a possibility of escaping the disaster even though a resident of Hiroshima or Nagasaki."[42] Christian Scientists believe that the change called death can be overcome by any person who like Jesus is *wholly free* from error. Jesus was the way-shower, the most scientific man who ever lived,[43] and what he accomplished we could accomplish if we but had sufficient realization. Mrs. Eddy's discovery "has made available a spiritual power that as greatly exceeds the usual forms of Christian influence as atomic energy exceeds all prior claims of physical power."[44] This power comes from a realization of the allness of the divine Mind, from the effect of a truly scientific understanding.

Mrs. Eddy's conception of God is sometimes said to be impersonal. She did argue against crudely anthropomorphic ideas of God, but many of her statements indicate that she believed in a divine intelligence. For example, she defined God as "The great I AM; the all-knowing, all-seeing, all-acting, all-wise, all-loving, and eternal; Principle; Mind; Soul; Spirit; Life; Truth; Love; all substance; intelligence."[45] She emphasized the feminine as well as the masculine aspects of God, using the term Father-Mother God; she may have learned it from the Shakers, a religious sect who taught it and who had a settlement only five miles from her home during her late adolescence. She believed in the Trinity, but not in its traditional form. "Life, Truth, and Love constitute the triune Person called God," she wrote. She also equated the third person of the Trinity with divine Science.[46]

Christian Scientists do not deify Mrs. Eddy. Some of them may seem to come pretty close, but the emphatic opinion of the Church leaders is simply that she was a highly endowed human being guided by God to the discovery of divine Science. She is held to be a *Leader,* not God. Acknowledgments of her imperfections can appear in the writings of the stanchest Christian Scientists. Robert Peel, a member of the Committee on Publication for the Mother Church, notes that she employed for a time a Unitarian clergyman to help her with her writing, that in the early editions of *Science and Health* the preface began with a dangling participle, that "little Victorian touches . . . sometimes decorated her style," that she could indulge in "a splash of defiant rhetoric," and that she was not the "paragon of academic scholarship" some of her followers thought her to be. But Mr. Peel also wrote, "Years after her death many of those who knew her best . . . could hardly speak of her without tears filling their eyes, even while their faces lighted up with an affection more revealing than anything they could say."[47]

HEALING

The case for spiritual healing is better than most modern Christians and Jews are inclined to think. Persons who believe in its reality point out that it has a prominent place in the Scriptures, and claim that most of the people who reject its possibility have never looked at the evidence, but instead have simply adopted a world view in which by definition anything other than physical healing is impossible. Belief in spiritual healing is probably growing slowly in the Christian denominations. For example: Professor Charles S. Braden sent questionnaires to 982 clergymen in the major denominations asking, "Have you ever as a minister attempted to perform a spiritual healing?" Fourteen per cent replied that they had.[48] The degree of faith in spiritual healing which is reached by some religious leaders is illustrated by the following passage from a chapter by Professor Cyril C. Richardson of Union Theological Seminary in New York.

We are confronted with overwhelming testimony to remarkable cures. However much, moreover, we may discount earlier records, the patient researches of the *Bureau des Constatations medicales,* at Lourdes, make us aware that such things happen in our own day.

People are cured of a wide range of physical sicknesses without the aid of medical or surgical methods. Pulmonary tuberculosis, paralysis, rheumatism, fractures, ulcers, and cancer have been permanently healed. Of that there can now be no doubt. Nor would we be wise to try to explain such cures on the basis of psychosomatic medicine. They are much more far-reaching in their dramatic character. Just as many medical authorities were in error a half a century ago when they denied the existence of such cures, so we now should be in error were we to resolve them solely into psychosomatic terms.[49]

Healings might be classified as taking place through physical, psychological, or spiritual means. Physical healings are accepted as a fact by almost everyone—except of course the Christian Scientists. Psychological healings also receive almost universal acceptance; physicians frequently assert that a majority of the ills they treat are neurotic rather than physical in origin. The problem is with healings by means which are said to be spiritual. A common attitude today is that they are in reality not a different order of healing but rather are simply psychological cures achieved under religious auspices. Some non-Christian Scientist believers in the reality of healing by spiritual means are fully aware of this possibility and accept the probability that what is usually called a spiritual healing is in fact the result of mental and emotional readjustment in the patient's mind. But they also assert that healing does take place in the spiritual dimension, that there do emerge "latent, creative powers within a religious context"[50] which produce physical as well as psychological effects. Such cures are rare.* On the other hand, such spiritual cures as do take place are said to be sudden, to involve no convalescence, and to be permanent. Moreover, it is believed that perhaps they can be produced at a distance through intercessory prayer, although they cannot be produced on order simply by visiting a shrine or a faith healer.[51]

The possibility of spiritual healing receives some support from physicians. For example, Gotthard Booth, M.D., Associate, Columbia University Seminar on Religion and Health, wrote that spiritual healing has not been proved impossible, and that "medicine has known for some time that 'spontaneous cures' of cancer do occur, but so far scientists have not investigated the circumstances under which they took place." However, Dr. Booth stated also the negative side of the picture: "Almost all physicians have encountered sad cases of a curable disease becoming fatal while under some form of spiritual treatment."[52]

Most members of the "major denominations" who defend spiritual healing state emphatically that it should not be relied on to the exclusion of other types of healing. Drugs, antiseptics, anesthetics, surgery, psychiatry, autosuggestion, all have their place—the major place—in healing. However, these persons contend that God can and sometimes does heal through spiritual means alone. Many believers in spiritual healing would also affirm that most of the highly advertised "faith healers" are at best in error and at worst charlatans. They would agree with the famous British clergyman, Leslie Weatherhead, who wrote: "Healing missions produce in many people black

---

* Some students of the cures at Lourdes, the ones which have been most carefully investigated, conclude that only "about one per cent (or less) of the pilgrims" receive physical healing. See Cyril C. Richardson, "Spiritual Healing in the Light of History," in *Healing: Human and Divine,* Simon Doniger, ed. (New York: Association Press, 1957), p. 208.

depression and hopeless despair. Most of those who attend them are not healed, and their last state is often worse than their first."[53]

The beliefs of Christian Scientists about healing vary considerably from the ideas outlined above. In the first place, Christian Scientists of course deny emphatically that healing is physical. But they deny with equal emphasis that healing is psychological; so-called psychosomatic healing assumes the influence of mind over matter, and there is no matter. They believe *all* real healing is spiritual in its nature and they deny that it is "rare," limited to "one percent (or less)." Spiritual forces account for 100 per cent of all genuine healings. The term "mental healing" is not in good favor among Christian Scientists when it is applied to the healing most people call "psychological." Nor do they like the term "faith healing." In the first place, the term tends, they feel, to have been pre-empted by a crude kind of performer who works on his subjects by assembling crowds, choirs, lights, and other paraphernalia in an effort to hypnotize his subjects. And in the second place, the ordinary kind of faith has little to do with healing; "faith must be lifted to spiritual, scientific understanding if one is to experience permanent health and demonstrate God's government of man."[54]

For Christian Scientists, healing is a by-product, a secondary result of rooting out errors produced by mortal mind, of realizing the true nature of reality, of communion with God. Mrs. Eddy herself wrote, "Healing physical sickness is the smallest part of Christian Science. It is only the bugle-call to thought and action." Sin, sickness, and death form the great triad which Mrs. Eddy fought.[55] In the following passage she explains how divine Science enables man to be free of them.

As the mythology of pagan Rome has yielded to a more spiritual idea of Deity, so will our material theories yield to spiritual ideas, until the finite gives place to the infinite, sickness to health, sin to holiness, and God's kingdom comes "in earth, as it is in heaven." The basis of all health, sinlessness, and immortality is the great fact that God is the only Mind; and this Mind must be not merely believed, but it must be understood. To get rid of sin through Science, is to divest sin of any supposed mind or reality, and never to admit that sin can have intelligence or power, pain or pleasure. You conquer error by denying its verity.[56]

Even though getting rid of disease is but one aspect of the Christian Science faith, in actual practice members give more attention to gaining and maintaining health than to any other phase of their faith. A group of professionals called *practitioners* devote full time to healing, make a career of it, and charge fees. Their methods, as has been stated, are strictly mental. Their "practice is based on the observation that a clear knowledge of the correct metaphysical position is a law of destruction to anything unlike it."[57] Practitioners deny that they deal in miracles, in special acts of God which alter the course of nature. They deal rather with nature as it is, with the scientific facts, with mind. They have a deep sense of ministry. One practitioner wrote, "The faithful practitioner . . . can say with the Master (Matt. 11:28), 'Come unto me, all ye that labour and are heavy laden, and I will give you rest!' "[58]

The practitioner tries to bring the realization of Truth, which in turn brings healing. He may work audibly or silently, making denials of error and affirmations of Truth based on a clear sense of the presence of the healing Christ. His effort is not to influence the mind of the patient. His effort is to attain the absolute consciousness of good, and "through the divine energies . . . get out of himself and into God so far that his consciousness is the reflection of the divine."[59] He may work in the presence of the patient, or he may give what is called "absent treatment." The practitioner's own knowledge of the Truth, of the metaphysical realities, has power to heal. Even knowledge of the names, addresses, and supposed diseases of patients is unnecessary and irrelevant "for everyone touching his thought shares in and is blessed by the illumination."[60] Jesus is believed to have healed after this manner. Mrs. Eddy wrote:

Jesus beheld in Science the perfect man, who appeared to him where sinning mortal man appears to mortals. In this perfect man the Saviour saw God's own likeness, and this correct view of man healed the sick.[61]

Denial of illness goes so far that in case of a supposed accident, say a burned finger, the finger should not be looked at or inspected in any way. "Inspection of the finger and the expectation of finding the marks of burning are . . . what produces the familiar effects of burning."[62] Pity and anxiety for illness also aggravate the condition. On the other hand, compassion for the *person* who is under error's control is in order.[63]

Belief in the power of absent treatment is so great as to lead to a general feeling throughout the Church that "addressing the thought" of another person without his consent is improper; his right to privacy should be respected.

Unfortunately, there is a negative aspect to the belief in absent treatment. If good can be effected in this manner, evil can also. Mrs. Eddy was much impressed by this possibility and was greatly troubled by what she thought were the evil influences sent her way. Even though this aspect of Christian Science is much less prominent than formerly, the power of "malicious mental malpractice" is still recognized. Professor Braden wrote, for example, that he has "again and again come across a belief . . . that at certain times Roman Catholic monks deliberately work mentally" for the destruction of Christian Science.[64]

Every realist must admit that however he may explain it, healing does take place under Christian Science auspices. Persons of all classes, degrees of intelligence, and degrees of acquaintance with modern trends have been cured when other methods of treatment have failed. Testimonies to the healing power of divine Science pour by the hundred into the offices of the Mother Church. Of course, Christian Scientists do become ill, suffer, and die. They may upon occasion even call a doctor. Mrs. Eddy herself used spectacles, consulted dentists, probably used morphine.[65] The Church considers that these actions are not an indication of weakness in Christian Science doctrine, but rather of error or sin on the part of human beings who are not yet sufficiently free from the domination of mortal mind. Members of the Church often avoid direct reference to themselves as Christian Scientists, preferring rather to say that they are students of Christian Science.

Contacts with Christian Scientists soon lead one to the conviction that they derive an immense benefit from their religion. On the whole they are a successful, happy brotherhood, looking on the bright side of life. My own observation leads me to the conviction that Christian Scientists receive immense benefit in the area which most persons call preventive medicine. On the other hand, the insistence that no illness is organic in nature is responsible, in the opinion of many observers, for much unnecessary suffering. A former dean of the Yale Divinity

School, Charles R. Brown, is credited with the remark that for every adult who has been benefited by Christian Science methods of healing, a child has been injured by inadequate medical attention. A Christian Science leader answers this charge by saying it is "untrue and unfair; as a group the children of Christian Scientists are notoriously healthy. Furthermore, the statement carries an unfair connotation in its unspoken assumption that other systems may always heal in every case." No available statistics compare the percentage of healing successes by practitioners and by orthodox physicians.

Practitioners now have a legal right to practice in every state in the Union, and the Church claims that hundreds of insurance companies now recognize Christian Science treatment as a substitute for medical care.[66]

Application of the Christian Science belief in the direct power of the mind is made of course to areas other than disease. An editorial writer urges the rejection of the "myth of heredity"[67] and a schoolteacher denies the usefulness of the I.Q.; she wrote, " 'God is All-in-all.' God could not, therefore, manifest Himself in any limited way to His beloved children."[68] Mrs. Eddy wrote in Science and Health: "The daily ablutions of an infant are no more natural nor necessary than would be the process of taking a fish out of water every day and covering it with dirt."[69] On the other hand, one weakness in Christian Science theory and practice seems to be the failure to apply it in some very obvious ways. Members of the Mother Church seem to evidence as much interest in food, clothing, and the ownership of property as do the members of the more earthy sects. Yet logically if the control of disease, the destruction of death, and protection from the action of atomic bombs can be effected by Christian Science, it can also easily dispense with money. Yet I have never seen any writing proposing this theory or heard of any Christian Scientist who tried to put it into practice. Practitioners send bills, the Church seeks monetary contributions, and Mrs. Eddy died a millionaire. A spokesman for the Committee on Publication for the Mother Church counters this line of argument; he wrote in personal correspondence:

To accept the revolutionary logic of Christian Science theoretically does not at once banish the apparent evidence to the contrary. . . . [Full demonstration of Christian Science] is not done at one bound. Our effort is first of all to wipe out what might be considered the abnormal phases of mortal experience, and we would include among these not only sickness but gluttony, excessive love of material possessions, economic anxiety, and so forth.

THE CHURCH

The Board of Directors of the Mother Church was given final authority by Mrs. Eddy. This authority has stood up in spite of a vigorous challenge fought by a dissident group through the courts of Massachusetts. Today the five-member Board rules the Church, even though Mrs. Eddy wrote that for the "divine Principle" there is no "ecclesiastical monopoly"[70] and urged, "Let us serve instead of rule."[71] Professor Braden is of the opinion that this rule is often exercised arbitrarily without sufficient regard for individual freedom, and concludes at the end of a book largely devoted to the history of conflicts within the Church that "there is slowly building up a degree of disagreement with the present leadership and policies of the church which may, one day, and sooner than one might now be inclined to think, eventuate in an explosion that will rock the church."[72]

The branch churches are democratically organized; the local groups elect their own officers, own their own property, and manage their own affairs. Branch officers must, however, be members of the Mother Church and, therefore, are subject to the jurisdiction of the Board. A branch church must have at least sixteen members, four of whom must also be members of the Mother Church, and one must be an active practitioner. Groups which do not meet these requirements are called societies. Members of the Mother Church are expected to reach a higher standard of spirituality and morality than are persons whose membership is limited to the branch churches. Members of the Mother Church and of most of the branch churches are expected to abstain from the use of liquor and tobacco.

Christian Science churches have no pastors, that is, no pastors in the ordinary sense. Mrs. Eddy decreed that no human pastor should minister in the denomination; but she did "ordain" the Bible and *Science and Health* as pastor to the Church. Accordingly there are no talks in the services of worship; instead, passages from these two books are read as the "lesson-sermon." The Sunday service is conducted by two "readers." It consists of hymns, Scripture reading, silent prayer, the Lord's Prayer, a solo, the lesson-sermon, a collection, a benediction. The lesson-sermon is conducted after the following manner: the Second Reader reads passages from the Bible; then the First Reader reads correlative passages from *Science and Health*. These passages are selected by the Mother Church and are on any given Sunday uniform the world over. No prayer is spoken except the Lord's Prayer with its spiritual interpretation by Mary Baker Eddy:

Our Father which are in heaven,
  Our Father-Mother God, all-harmonious,
Hallowed be Thy name.
  Adorable One.
Thy kingdom come.
  Thy kingdom is come; Thou art ever-present.
Thy will be done in earth, as it is in heaven.
  Enable us to know,—as in heaven, so on earth,—God is omnipotent, supreme.
Give us this day our daily bread;
  Give us grace for today; feed the famished affections;
And forgive us our debts, as we forgive our debtors.
  And Love is reflected in love;
And lead us not into temptation, but deliver us from evil;
  And God leadeth us not into temptation, but delivereth us from sin, disease, and death.
For Thine is the kingdom, and the power, and the glory, forever.
  For God is infinite, all-power, all Life, Truth, Love, over all, and All.

A second service of the week is held on Wednesday evenings. This service varies from the Sunday service in that there is no collection, the First Reader alone presides, he selects the readings from the Bible and from *Science and Health,* the Lord's Prayer is spoken without its spiritual interpretation, and a considerable period is given to volunteer testimonies from the local members concerning the value of Christian Science to them. Anyone inclined to doubt the power of Christian Science to bring a deep feeling of spiritual security should attend a series of Wednesday evening services.

The *Hymnal* includes many of the great hymns of the Western world. Some have been slightly altered to bring them into conformity with Christian Science doctrine, and others have been considerably altered. Seven hymns by Mrs. Eddy are in the *Hymnal.* The Church makes no provision for baptismal, funeral, or marriage services. Funeral services are left entirely under the direction of the family.

Some families have no service at all; most of them, however, ask a reader or practitioner to conduct a simple service. When they wish to be married, Christian Scientists ask a Protestant clergyman to perform the ceremony. The church building has a central pulpit which consists of two desks, one for each reader. On the walls in large letters are inscribed sentences from the Bible and from the writings of Mary Baker Eddy.

Christian Science churches are not social institutions. The church supper so familiar in most American denominations has no place. "As a rule there should be no receptions nor festivities after a lecture," wrote Mrs. Eddy. ". . . he who goes to seek truth should have the opportunity to depart in quiet *thought* on that subject."[73]

A Sunday School is conducted for youths up to the age of twenty. This school is the lowest level of the educational system. For adults the Church conducts two kinds of classes: Primary Classes and Normal Classes. They are patterned after Mrs. Eddy's own teaching. Certified teachers may conduct but one Primary Class per year and the number of students who can be accepted is limited to thirty. A fee of one hundred dollars is charged for twelve lessons; the subject matter is limited to the chapter called "Recapitulation" from *Science and Health;* Roman Catholics are not admitted unless they have "the written consent of the authority of their Church"; a student who has attended the class of one teacher is not permitted to attend that of another; and the students of each teacher are organized into an Association whose chief function is to hold an annual meeting. "Class" has sometimes been called the essential institution of Christian Science; no person gets very far into the movement without being "class taught."

A Normal Class is held for the purpose of training teachers for the Primary Classes. Enrollment is strictly limited. Only one Normal Class is held in three years; it meets for only a half dozen sessions, and no more than thirty students are admitted. Admission is controlled by the Board of Directors and is a much coveted privilege. The subject matter is again the chapter called "Recapitulation" to which is added the Christian Science Platform found on pages 330–340 of *Science and Health.* The class is conducted only in English and only "thorough English scholars" are admitted.

In addition to these classes the Church has a Board of Lectureship which annually appoints a number of persons who first write out a lecture designed to be delivered before public audiences, then submit it for approval to the Board of Directors, and finally seek appointments to deliver it, usually for compensation. The number of times such a lecture can be given is not limited. At the close of a public lecture, questions from the floor are not permitted in view of Mrs. Eddy's regulation against "public debating."

Principia College in southern Illinois is operated by Christian Scientists. Its enrollment is limited to five hundred students. It is not in any sense an official institution and is not the center either of theological or of administrative influence. Christian Scientists also operate a half dozen preparatory schools.

The Church has no missionaries in the ordinary meaning of that word. Instead, the primary method of outreach is through the printed page. The several periodicals of the Church are sometimes called "our silent missionaries." One of the periodicals appears in no less than nine foreign languages and is printed in the interesting form of placing the English text on the left-hand page and the translation on the right-hand page. The Church publishes one of the world's greatest newspapers, the *Christian Science Monitor*. Since this paper does not slant the news in a Christian Science direction, its influence extends far beyond the membership of the denomination. Some of the nation's finest commentators write for the *Monitor*. It refuses to emphasize the sordid aspects of life; news of robbery, rape, and murder are absent from its pages. Absent also is news of epidemic and death. An excellent sports page contains no news of horse-racing or of prize-fighting.

In view of an admonition by Mrs. Eddy against "numbering the people," the size of the church membership is not made public except in exceptional circumstances. During World War II when chaplains* were allocated to the armed services on the basis of size of membership, the Church reported 268,915 members in the United States.[74] Perhaps one hundred thousand more Christian Scientists are located

---

* The Church appoints as chaplains persons who have served for three years as a reader, for three years as a practitioner, and have completed sixty hours of graduate study in religion in an accredited institution.

outside the United States. Christian Scientists are found chiefly in cities, and in 1936 their number was largest in California, Illinois, New York, Ohio, Massachusetts, in that order.[75] In 1958, the number of authorized practitioners in the United States was about 8 per cent less than in 1941. The available evidence does not indicate that the present time is a period of rapid growth.[76]

Christian Scientists are not leaders among those who believe the churches should be active in trying to correct social injustices through political or economic action. The emphasis is rather that social injustices like all other kinds of evil will give way before metaphysical work, before the realization that suffering is unreal. Relief funds are maintained for the victims of all sorts of disasters; but I have been unable to locate any other organized effort at social action except a small, recently formed pacifist group. Regular reading of the articles on religion published daily in the *Monitor* during the first six months of 1960 failed to reveal any message which dealt with direct social action even though this was a period during which the student sit-in strikes in the South began and the Summit Conference was wrecked, and even though the *Monitor*'s other contributed articles and editorials on these subjects were full of insight and moral conviction.

A distinctive feature of the Christian Science organization is the regulation that each church shall maintain a public reading room where authorized literature may be read, borrowed, or purchased. These institutions vary from the carefully appointed establishments set up by some large churches in the business sections of cities to such limited efforts as that of the society in the locality where this paragraph is written; this room is about ten feet square, contains perhaps fifty volumes, and is open to the public for only a half hour before and after the Wednesday evening service. But the librarian is a devout and inquiring Christian Scientist, and the influence of the society is undoubtedly enlarged to a considerable degree even by so modest an effort. Christian Scientists sometimes grow lyrical concerning the value of the reading room to them. One wrote that after a crisis which forced him to move to a strange land and city "my first step was to seek out the Christian Science Reading Room."[77] Another wrote, "Often we hear of healings that have come about as the result of an individual's making use of a Christian Science Reading

Room."[78] The following poem by Margaret B. McGee is, to one who
has spent much of his life in libraries, a truly astonishing and moving
testimony:

The Reading Room

Here find the vision of that clear, creative light
of which dark chaos dreamt and still is dreaming,
drawing the mind and heart up and above
illusive cloud of temporal seeming.

Freed from old earthward drag,
here find a new direction;
blessed by this quiet, find
complete divine protection.

Here find solution to each gnawing problem
set by negation,
positive answer to the searching world's
unsatisfied interrogation.

Here is reward for that persistent seeking—
the soul's demanding—
door to the acquisition of that certain peace,
outdistancing all human understanding.[79]

CHAPTER XVI

OTHER RELIGIOUS INNOVATIONS

THE RICHNESS of America's culture springs from the abundance of her freedom. Whenever a people are free to hazard serious and widespread experimentation, culture is enriched; and whenever creative insights are systematically suppressed, culture is impoverished. In the United States, religious life is far more varied and correspondingly far more vital than it would have been without a happy combination of legal guarantees of religious liberty and a social situation, springing out of the frontier, which has tolerated spiritual innovation. The innovators have prodded the traditionalists.

The number of creative groups which have appeared in the past century and a half runs to many scores. In 1936, the Bureau of the Census found well over a hundred, but undoubtedly a considerable number were so small or so ephemeral as not to attract the census taker's attention.

Some of the new sects are often bizarre in their activities. Aimee Semple McPherson roared down the aisle of Angelus Temple one Sunday on a motorcycle. Daddy Grace ("Grace has given God a vacation.") baptized two hundred converts on the streets of Philadelphia with a fire hose. One sensation-seeking clergyman secreted himself on Easter morning in a coffin, had it brought into his church, and in the midst of the service leaped up crying, "Resurrection day has come." An evangelist tells how to make "God's oil" ooze out of the hands. Another evangelist who broadcasts his message over the air instructs his hearers to put their hands on their radios in order to be healed. One church decided that a Biblethon (from marathon) would quicken their spiritual life; seventy-six persons working in fifteen-minute relays, day and night, read the Bible from cover to cover in eighty hours and forty-five minutes. A clergyman in order to advertise his services introduced over the radio the following commercial, sung to the tune of "Little Brown Jug":

> Mother knows what's good for you,
> Go to church like she taught you to.
> Brother we've an empty pew,
> Plainly labeled Y-O-U.

One church gives green trading stamps to people who attend. A radio preacher promised to send inquirers "a genuine, autographed picture of Jesus Christ." Another clergyman sacrificed a lamb on a five-foot cross, a member of the congregation stepping forward to slit the lamb's throat.

On the other hand, no one would deplore such behavior more than some other sectaries; the Oxford Groupers and the followers of Unity, for example, emphasize education and make a systematic use of silence. Clearly the newer sects follow no single pattern. Nor have attempts to classify them into a series of patterns been especially fruitful. Sectarian phenomena are both protean and unsystematic. Therefore, our effort to indicate their essential features must be devoted in the main to brief descriptions of a very few of the groups themselves. But before beginning these descriptions some preliminary observations are in order. These observations should not be taken to be generalizations which apply uniformly to all of the sects, but should rather be viewed merely as introductions to some of the technical terminology in the field and to some factors often, but by no means always, present in sectarian faith and practice.

1. *"Lower" Class Origin.* The formation of new sectarian organizations is usually the work of the poor and is frequently a form of social protest. The majority of our religious innovations have come from persons who find it difficult to make ends meet, who have little influence on public affairs, who tend to be ignored by the established churches, and who often cannot afford the fine clothes most Americans think proper for wearing at church. One discerning writer says that "the divisions of the church have been occasioned more frequently by the direct and indirect operation of economic factors

than by the influence of any other major interest of man."[1] On the other hand, some of our newer groups are of "upper" class origin, the Oxford Group for example. Some sects even make a luxury appeal; one observer wrote concerning the I AM movement, "Buxom middle-aged usherettes, clad in flowing evening gowns, with handsome corsages of orchids and gardenias, bustled around at the morning services in a tabernacle that literally steamed with perfume."[2]

2. *Adventism.* Many of the newer sects put special stress on belief in a Second Coming (Second Advent) of Jesus. They believe that his return to earth is imminent and that the true Christian will live in a state of constant expectation. Adventists have made many attempts to set the date of the Second Coming; more than a hundred persons, for example, ". . . predicted that the advent would occur within the decade following"[3] the Civil War. Elmer T. Clark, author of a standard work on the sects, estimates that there are "probably three or four million" Adventists, counting "Fundamentalists who have not left their denominations."[4]

3. *Holiness.* Some of the sects teach that moral perfection is possible, that through the outpouring of the Holy Spirit some persons can become instantly free from all sin and even free from all desire to sin. This "instantaneous experience" comes after conversion, and is a "second blessing," a sort of advanced degree reserved by the Holy Spirit for some Christians, to whom He grants deep peace and even ecstasy. Holiness sects are frequently puritanical in the extreme. One investigator found groups who proscribed, in addition to the traditional taboos, such things as chewing gum, polishing the nails, attending motion pictures, and participating in athletics.[5] Clark estimates that "nearly ten million persons" are members of groups which believe in holiness. This estimate includes the members of many traditional churches.

4. *Spiritual Gifts.* Ever since the Day of Pentecost, the occasion when the Disciples were so filled with the Holy Spirit that they began to speak in strange languages (see Acts 2), some Christians have sought to exhibit their zeal by inducing physical manifestations, such as shouting, shaking, jerking, dancing, leaping, rolling (however, no sect as far as I know calls itself the Holy Rollers). In present-day America the most prized and frequent spiritual gift is "speaking in

tongues." In recent years immunity from injury after snake bite, after drinking poison or after contact with fire, have also been claimed as gifts of the Holy Spirit. (See pages 6 ff.) These gifts are sometimes called the pentecostal blessing. There are about 5,000 local churches and 400,000 members in sects which emphasize such extreme manifestations.[6]

5. *Psychological Techniques.* Many of the newer sects exploit the discoveries of science concerning the power of the mind over the body. They try to help people out of all sorts of difficulties by the power of suggestion, and claim to be able to present the devotee with success, health, wealth, love and happiness.

6. *Healing.* Many of the sects teach that God, through spiritual rather than physical means, will cure our bodies of their pains and our minds of their anxieties.

7. *Occult Claims.* Some sectaries contend that they have secret knowledge of the supernatural. Through magic, or spiritism, or contact with demigods, they claim to possess keys which will unlock the mysteries of life and death. The leader of one group, with headquarters in Denver, Colorado, said that he is able to transport his spirit to Tibet for conferences with the great souls who dwell there.[7] The leader of another group said that one day when he was hiking on Mount Shasta, he was taken out of his body and shown a number of scenes from his former lives on earth.[8] Still another seer claimed that seventy-five years ago contact was made with a race of supermen with metallic heads who live at the center of the earth and who will help those possessed of the true knowledge to banish war and poverty from the earth.[9]

8. *Fantastic or Fraudulent Claims.* Some sectarian leaders make such extreme statements that they lay themselves open to the charge that their teachings are deliberate attempts to deceive. Several have taught that they will never die, and it has been claimed for others that they are God.[10] It is difficult to determine when ideas such as these are simple delusion and when they are fraud. Fraud does exist in the opinion of well-placed observers, as is evidenced by the sentencing of some I AM leaders to prison,[11] and by the conclusion of a committee for the Society for Psychical Research that Madame Blavatsky, the founder of Theosophy, had produced some of her

startling effects by deception.[12] However, concerning possible fraud, two remarks are in order: (1) A religious leader may be dishonest in some of his teachings and still minister successfully to the needs of many of his followers. The history of churches (and governments) gives ample demonstration of the principle that devotees can receive immense satisfaction and comfort through following the teachings of leaders whom most observers would consider to be charlatans. (2) Fraud is not limited to the newer sects. Elmer Gantrys are to be found in every religious organization.

These eight, frequently-found characteristics of the newer sects also often characterize older religious bodies; moreover—let it be reiterated—some of the newer sects exhibit few and other sects exhibit none of these traits. Every group is a law unto itself.

The newer sects have become prominent enough that the term *third force* has been coined to indicate the belief that they are a new phase of Christendom, parallel with Roman Catholicism and Protestantism. The strength of the third force is uncertain; the estimates vary. One had it that "a generous estimate of the total membership of all the cults put together would be 3,200,000."[13] Another estimate was that six million Americans belong to such sects.[14] Another was that a total of seventeen million "signed up" in the first four decades of this century.[15] However, the ablest writer in this field, Charles S. Braden, says in personal correspondence, "There is simply no way of knowing how many people do belong." One major problem of course is determining just which sects should be listed in the third force.

SEVENTH-DAY ADVENTISTS

In 1818, William Miller, New York farmer, justice of the peace, ex-captain in the United States Army, and painstaking student of the Bible, became convinced that "in about twenty-five years from that time all the affairs of our present state would be wound up."[16] He started his study with the assumption that the Bible must be taken literally at all points except where the language is obviously figurative. Laboring carefully over the prophecies of the end of the world, he finally concluded that a biblical "day" really meant a year; he then computed the date of the Second Advent as falling, "without doubt," somewhere between March 21, 1843 and March 21, 1844.

But Miller did not immediately begin to broadcast his belief. Only after thirteen years of heart searching and continued labor over the prophetic passages did he finally begin preaching. Then his earnestness, his thorough knowledge of the Bible, and the logic of his conclusions (once his assumptions were granted) convinced multitudes. The year came; and though a fiery comet suddenly flamed in the sky late in February, 1843, visible even at midday, the time passed and the affairs of men went on as usual. Miller, saddened and confused, wrote:

> Were I to live my life over again, with the same evidence that I then had, to be honest with God and man I should have to do as I have done. . . . I confess my error and acknowledge my disappointment; yet I still believe that the day of the Lord is near.[17]

One of Miller's followers recalculated the date and set October 22, 1844. Hope was rekindled. As this day approached, excitement ran high: Adventist papers printed their last editions, farmers planted no crops, money was given away. Bitter was the disappointment as the day passed and night came on. Of that night and the next morning, Hiram Edson, the man who found the key which saved the Adventist faith, wrote as follows:

> Such a spirit of weeping came over us as I never experienced before. . . . We wept, and wept, till the day dawn. I mused in my own heart, saying, My advent experience has been the richest and brightest of all my Christian experience. . . . Is there no God, no heaven, no golden home city, no paradise? . . .
> After breakfast I said to one of my brethren, "Let us go and see, and encourage some of our brethren." We started, and while passing through a large field I was stopped about midway of the field. Heaven seemed open to my view, and I saw distinctly and clearly that instead of our High priest coming out of the Most Holy of the heavenly sanctuary to come to this earth . . . He for the first time entered on that day the second apartment of that sanctuary; and that He had a work to perform in the most holy before coming to this earth.[18]

This interpretation of Edson's is still the standard belief among Adventists. They believe that the "records of sins of the people of God through the centuries have been accumulating in the heavenly sanctuary."[19] What happened in 1844 was the beginning of the cleansing of this sanctuary. Christ's return to earth is still expected to take place at any time.

Belief that the seventh day of the week should be celebrated as the Sabbath sprang from the fraternization of some of Miller's followers with some Seventh Day Baptists. The Old Testament directs that the seventh day, not the first day, shall be the Sabbath. A half dozen Christian denominations follow this injunction literally; chief among them is the Seventh-day Adventists, much the largest of these churches.[20]

The Seventh-day Adventists adopt most of conservative Protestant theology; they are "Fundamentalists of the Fundamentalists."[21] They believe in the spiritual gift of prophecy, especially in the prophecies of Ellen G. White, who died in 1915. Seventh-day Adventist speakers " 'buttress' the clear teaching of the Scriptures by various quotations from the writings of Mrs. White."[22] They believe in vegetarianism, in total abstinence from alcoholic beverages and tobacco, in the necessity for tithing, in shunning elaborate dress and jewelry, in avoiding card playing, dancing and attendance at motion pictures.

Men are not inherently immortal, according to Seventh-day Adventist belief. Only the righteous will live in eternity. There is no hell. After death men go to the grave there to sleep until they are resurrected by Christ at his Second Advent, if they are judged to be worthy. This belief is called the doctrine of conditional immortality.

According to the Seventh-day Adventists, the order of events after Christ's return will be as follows: Christ's glory will destroy the wicked who are still living. Then He will call forth from their graves the righteous of the past ages. Then with all the righteous He will ascend into heaven where He will reign for a thousand years. During this period the earth will be desolate, uninhabited by men, the prison house of Satan. At the end of the thousand years, the Holy City will descend to earth, Christ and the saved will take up residence there, Satan will be unchained, the wicked will be resurrected, a great battle will ensue, Satan and his followers will be destroyed, and earth will be restored to the purity and beauty of Eden, whereon the saved will live in bliss, praising God throughout eternity. All these beliefs are based squarely on a literal reading by Seventh-day Adventists of the Bible.[23]

The Church reports over three hundred thousand members in the United States. Its vigorous health is evidenced by such facts as the

following: Seventh-day Adventists give more per member to benevolences than any other group,[24] they claim to have a larger portion of
their members studying in parochial schools than any other group
including the Roman Catholic,[25] they claim to have more active
foreign missionaries than any other group except the Methodists (the
Adventists have over fourteen hundred, the Methodists over fifteen
hundred),[26] they maintain in various parts of the world 107 sani-
tariums and hospitals, their radio programs are broadcast over 860
stations and are heard in 65 languages, their official television program appears on 153 stations in the United States.[27]

The inner satisfaction which comes to the Seventh-day Adventist
is vividly described by a non-Adventist author:

> If . . . the primary desire . . . is security . . . then the Seventh-day
> Adventist must surely be content for his security is assured. They are as
> positive in their own minds as mortal men can be that, if they meet the
> conditions of personal righteousness, their lives not only extend to the
> grave, but far beyond it, forever and ever, in the steady and constant
> unimaginable joy.[28]

At a recent conference, the president of the denomination, standing
under a globe-shaped clock stopped at five minutes to twelve, told
his fellow delegates, "The signs of the times tell us that the prophecies of Scripture have practically run out."[29]

THE PENTECOSTAL HOLINESS CHURCH

There are four levels of Christian experience, according to the
teachings of the Pentecostal Holiness Church: justification, regeneration, sanctification, and baptism by the Holy Spirit.

1. In justification the soul is freed from the condemnation due
for sin.

2. In regeneration the soul is "born again," it is made conscious
of the work of Christ on the cross and begins to live in and by him.

3. In sanctification the soul becomes completely separated from
all evil and "solely and completely devoted to divine service."

4. In Pentecostal baptism a fully cleansed believer, by a definite
act of "appropriating faith," receives from God the greatest of
earthly blessings, a tremendous inner spiritual exaltation, whose first

outward manifestation is speaking in strange tongues, as the Holy Spirit instructs.

The Pentecostal Holiness Church believes in "fervor of spirit" and in joyous manifestations. Worshipers strive to obtain the higher blessings. Elmer T. Clark describes a Pentecostal service as follows:

> The congregation is composed of men and women from the lower ranks of culture. The evangelist preaches on the gift of the Holy Spirit, stresses the possibility and privilege of the pentecostal outpouring for present-day believers, relates experiences thereof, perhaps now and then breaks out in ecstatic jabbering of strange phrases, and points out the barrenness of those who have never been so blessed. . . .
>
> At last conditions become right. Seekers come forward in anticipation of the gift. Confessions are made. Excitement runs high. Various blessings in the form of emotional reactions are secured. Some cry out, others fall in trances or wave their hands or bodies rhythmically in near ecstasy. One feels unduly blessed and rises to testify. He begins speaking, faster and faster, words fail, there is a muttering in the throat, and the subject breaks out in a flood of words that have no meaning to ordinary individuals. The pentecostal power has fallen, the blessing has been received, hallelujahs ring out, persons crowd about the favored saint, a familiar hymn is struck, and a wave of emotion, perceptible even to the unbelieving onlooker, sweeps the company like an electric charge.[30]

The Pentecostal Holiness Church was founded about 1900, is concentrated in the South Atlantic States (and in Oklahoma), and in 1946 reported 26,251 members (justified), 8,043 regenerated, 3,179 sanctified, and 1,724 baptized by the Holy Spirit. In 1960, the number of members reported was fifty-three thousand.

The Church strongly holds to divine healing as the "more excellent way," but does not condemn medical science. The Second Coming is an emphatic teaching; a "thousand signs and events proclaim and signify the immediate end of this present age.[31] The Church's puritanical moral code includes the proscription of attendance at circuses, dance halls and fairs, and the use of tobacco in any form. The government of the Church is strikingly similar to that of the Methodist Church. One important regulation provides that all officers of a local church "shall be in the experience of sanctification at the time of their election." And another regulation insists that all officers above the local level "shall be in the experience of Pentecostal baptism of the Spirit, at the time of their election."[32]

"The Pentecostal people are not without justification in comparing themselves with the early Christians. They also were an underprivileged group."[33] And they also insisted on some spiritual gift as an external evidence of salvation.

THE SALVATION ARMY

The first official invasion of America by the British-founded Salvation Army occurred over seventy years ago in New York City. A landing party of eight officers fought their way through entangling red tape in the immigrant station at Battery Park, emerged from the door of the building, and immediately launched an attack on the Devil and all his ways in an open-air meeting held then and there in the Park.

The enemy in the early years gave no quarter. The Army was ridiculed, its officers were thrown into jail, its meetings were attacked by hoodlums. Worst of all, its campaign was often ignored. But these dauntless soldiers of the Lord fought back, sometimes with eccentricity and sensationalism: "umbrella-swinging men" marched down the street backwards; placarded officers lay "in the snow to attract crowds."[34] The Salvation Army won battle after battle, established firmly its program of social service wed to religion, "soap, soup, and salvation," and became an accepted part of the American scene.

Today the Salvation Army has in America 250,000 soldiers (members), over 5,000 officers (ministers), annually records 75,000 Decisions for Christ (20,000 prayer requests in jails and penitentiaries), holds annually almost a million meetings with a total attendance of 25 million, collects and spends 25 million dollars annually, operates 187 homes, lodges and schools, 33 hospitals, 9 children's homes, 34 residences for unmarried mothers, 54 summer camps, 236 Boys' Clubs and Youth Centers and 122 Social Service Centers. For the Salvation Army, "a man may be down but he is never out." More than any modern religious group it takes seriously the preaching of Jesus: "I was a stranger, and ye took me in: naked, and ye clothed me: . . . in prison, and ye came unto me."

The Salvation Army was organized in 1865 by William Booth, a strong-minded Wesleyan clergyman whose soul had been "well-nigh maddened by the endless miseries of mankind,"[35] and who had left his Church rather than give up the evangelistic methods he found effective with the multitudes who flocked to hear him. The Mid-Victorian churches would have none of Booth or his converts. Thus rather against his will the Salvation Army came into being. It is first of all a church; it ministers to the souls of men. But its leaders are fully aware of the relationship between want and depravity; they rescue sinners from hunger, unemployment and despair. George Bernard Shaw wrote that Salvationists "actually fight the devil instead of merely praying at him."

Military language is used extensively. A local church is a *corps,* a church building is a *citadel,* a Sunday-school class is a *company meeting,* a prayer meeting is a *knee-drill,* a vacation is a *furlough,* a parsonage is called *officer's quarters,* a soldier must sign *Articles of War,* the denominational manual is called *Orders and Regulations for Officers,* ordination is called *commissioning,* death is called *promotion to glory.*

The *line of command* begins with the General whose headquarters are in London. Under him in the United States (the Salvation Army works in over a hundred countries) are Commissioners. Under the Commissioners are thirteen ranks, including: Colonel, Lt. Colonel, Brigadier, Major, Captain, Lieutenant, cadet, soldier. Officers receive and give orders. A writer in *Time,* with some exaggeration, wrote that the Commander for the United States is "vested with more authority than any other U.S. churchman—his command . . . is absolute."[36]

The theology of the Salvation Army is evangelical to fundamentalist. Its doctrine includes holiness, "We believe that it is the privilege of all believers to be wholly sanctified."[37] A major principle of operation is that every soldier must be on *active* duty. Booth early discovered that "the one way in which he could lastingly change men and women was to make them, from the moment of their conversion, seekers and savers of the lost."[38] Many converts, unable or unwilling to meet this vigorous program, join more conventional churches. Women are given a high status in the movement; one of the five generals was Evangeline Booth, daughter of the founder.

The Salvation Army has no program of social change. It seems to be content to do a mopping-up operation, to do all it can to alleviate the sufferings of miserable people. This fact and the undoubted consecration of its leaders may account in large part for its almost universal popularity.

SPIRITUALISM

About three hundred miles above the surface of the earth, according to the beliefs of most Spiritualists, begin the seven spheres of the spirit world. They extend on upward for some eighteen thousand miles, are composed of a kind of matter which is one grade more refined than the stuff of the earth, and emanate from the earth much as fragrance emanates from a rose. The lowest sphere is occupied by the spirits of men who on earth were wicked, barbaric and dominated by material desires; one day these souls will be purified and will progress to higher regions.

Most people on earth go after death to the Third Sphere, the Summerland. There the life is very like the life of earth, except that the evils and sorrows are gone. The spirits in the Summerland wear clothing, live in houses, have animals for pets. Each spirit finds and marries his soul mate (to whom he had been joined before his earthly birth), but the marriage relationship in the heavenly spheres is wholly Platonic.

Beyond the seven earth spheres, say the Spiritualists, are higher and more universal spheres; they surround planets, suns and even solar systems. Farther still are "spheres more awful still and more tremendous." God, Infinite Intelligence, is at the center of the universe.[39]

All these beliefs are founded firmly, Spiritualists believe, on revelations which came straight from the spirit world. Their source is the seance, a meeting around a medium. A medium is "one whose organism is sensitive to vibrations from the spirit world."[40] For a brief description of a seance see pages 4 f.

Spiritualists believe that all the great religious leaders of the past were mediums. The Bible, particularly, is full of accounts of the work of spirits: for example, the annunciation to Mary, the transfiguration, Christ's resurrection appearances, and his appearance to Paul on the road to Damascus.

Fraud dogs the steps of the Spiritualist movement; all its leaders admit that some mediums practice deceit for monetary profit. As a result many unsympathetic observers have concluded that all messages from a "so-called spirit world" result either from fraud or from self-delusion. But the Spiritualists are certain that a multitude of authentic messages come to us from the Great Beyond.

There are about five hundred local churches which follow this faith; they are divided among three national organizations. The polity is congregational. Women overwhelmingly dominate the movement, both in number of mediums and in number of adherents; but men run the national organizations. In the 1936 census of *Religious Bodies,* the membership was listed at about thirty thousand. Today these churches report one hundred seventy-five thousand. George Lawton, author of an able treatment of Spiritualism, wrote thirty years ago, "I should hazard the estimate that for every enrolled member there are at least ten to fifteen non-enrolled ones." Lawton also comments that "the overwhelming majority of them, whether church-goers or not, are drawn from the lower and middle intellectual, social and economic classes."[41]

JEHOVAH'S WITNESSES

"Millions now living will never die," wrote J. F. Rutherford, president of Jehovah's Witnesses in 1920. A dozen years later he insisted that the time of the end is "much less than the length of a generation" away. In 1940, Witnesses were saying, "The Kingdom may be here before Christmas."[42] They are still certain that Armageddon, the last great battle between the hosts of Christ and the hosts of the Devil, is just around the corner. But they no longer set the day.

This sect was founded in 1872 by Charles Taze Russell, a haberdasher of Allegheny, Pennsylvania. He soon became a dynamic preacher, commonly speaking to overflowing crowds as he traveled an average of thirty thousand miles a year. He also became a prolific writer, his followers claiming at the time of his death that no less than thirteen million copies of his books had been sold.

"Pastor" Russell was succeeded in 1917 by "Judge" Rutherford. The Judge soon became a czar. He chose a new name for the movement, Jehovah's Witnesses, substituted his own writings for those of Russell, and launched a strenuous campaign of literature distribution. Two years before his death, which occurred in 1942, he claimed that over three hundred million copies of his writings had been circulated.

This tremendous volume was achieved by the tireless street-corner hawking and house-to-house canvassing of self-effacing yet insistent Witnesses who were "expected" to spend, as their Christian service, no less than sixty hours a month spreading the good news. The actual amount of time spent on the average by over seventy thousand Witnesses, according to the 1948 Yearbook, approaches thirty hours per month. The Witnesses really witness; it is their major religious expression. Persistently they trudge up and down our streets, ringing doorbells, selling books, making "back calls."

Meetings of the group are more like classes than like typical services of worship. The Witnesses believe that they will be saved by knowledge, and that Jehovah has made The Truth available in their literature. They are not trinitarians; for, Jehovah said, "Thou shalt have no other Gods before me." Jesus is thought to be but the *Son* of God. Demons actually exist, according to this sect; their chief is Satan. Hell, however, does not exist, since, Witnesses affirm, it is not taught in the Bible. After the great Battle of Armageddon, when the wicked will be totally destroyed, Christ will set up on earth the divine Theocracy.

The Witnesses are violently opposed to "religion." Religion, they say, comes from the Devil; and the churches are all Devil controlled. The Witnesses condemn the use of Christmas trees (Jeremiah 10:3), the celebration of Mother's Day, the salute to the American flag (Exodus 20:3), and blood transfusions (the Bible forbids "eating" blood). Witnesses refuse to give permission for transfusions even though life may be saved.[43]

Bigotry—obstinate and ill-mannered insistence on the falsity of all beliefs but one's own—characterizes a large percentage of American religious organizations, as the pages of this book show. The Witnesses are as bigoted as any. Generally, they do not even show gratitude to their defenders, for example, to the American Civil Liberties Union, which has energetically championed the legal rights of the sect. On the other hand, "not since the persecution of the Mormons years ago has any religious minority been so bitterly and generally attacked";[44] and probably "no single religious group in the world" displays more zeal.[45] One woman and her husband were able to give full-time service for several years because of a small inheritance. When it was gone, he obtained a part-time job and they continued to give most of their time. When she was asked whether old age did not give them concern, she replied, "The Kingdom will come soon and there is every reason to feel that we shall live to see it."[46] In 1958, no less than one hundred eighty thousand Witnesses attended an international convention in New York City; they filled both the Yankee Stadium and the Polo Grounds.

The Witnesses come chiefly from the "lower" classes; and the position of women in the organization is "one of almost complete subordination to the men."[47] No program of social reform is espoused; Jehovah will one day initiate his own awful program of purifying, healing and destruction. The sect reported in 1960 a quarter of a million members in the United States. One investigator wrote as follows:

While . . . about one half of the membership of the Witnesses is Negro, the leadership is almost completely white. When I visited the national headquarters in Brooklyn in 1952, I was told that there were only two Negroes in the headquarters staff of over four hundred; one Negro worked in the mailing room, the other was a linotypist.[48]

Another investigator wrote:

Nowadays Witnesses of various races intermingle at international conventions. But this is an innovation of recent years. Before that, separate conventions were held for Negroes and whites. Witness publications have repeatedly defended South Africa's official *apartheid*.[49]

UNITY SCHOOL OF CHRISTIANITY

"Whatever man wants he can have by voicing his desire in the right way into the Universal mind,"[50] said the founders of Unity; he can have health, riches, fame, peace of mind. They proved it, according to their followers, in their own lives. In 1887, Charles Fillmore was a bankrupt and a cripple, one leg four inches shorter than the other; his wife, Myrtle, was expected to die of tuberculosis within a few months. They discovered "a mental treatment that is guaranteed to cure every ill that flesh is heir to."[51] Charles discarded the steel extension that lengthened his leg; Myrtle miraculously recovered; and soon they found themselves at the center of a mushrooming religious movement. It grew until today it owns millions of dollars worth of property and touches, according to one objective estimate, five million people.[52]

The secret is thought control, through a proper system of denials and affirmations. "Strong statements of Truth . . . become the lever by which man lifts himself out of the pit. . . . Deny evil; affirm good. Deny weakness; affirm strength. Deny any undesirable condition, and affirm the good you desire."[53] The extreme application of the system is illustrated by Charles Fillmore's rephrasing of the 23rd Psalm:

> The Lord is my banker; my credit is good
> He maketh me to lie down in the consciousness of omnipresent abundance;
> He giveth me the key to His strong-box
> He restoreth my faith in His riches
> He guideth me in the paths of prosperity for His name's sake. . . .
>
> Surely goodness and plenty will follow me all the days of my life;
> And I shall do business in the name of the Lord forever.[54]

Unity thinks of itself as a religious school. Its chief tool is the United States mails. Every month over four million booklets, leaflets and magazines—every item blessed—are sent out from the Kansas City (or Lee's Summit), Missouri, headquarters. But the unique method of the organization is called Silent Unity. Workers are always on call, ready to help through prayer and affirmation, anyone who appeals to them. And appeals come by mail, telephone and telegraph from persons who are ill, in debt, having family troubles. The workers pray with these persons for a month, or more. Many times a long correspondence is entered into. But there is no charge. However, a "love offering" is gratefully received. Silent Unity received in one year six hundred thousand requests for help, a large percentage coming from the members of orthodox churches. One writer declared that Unity is "undoubtedly the greatest 'supplementary' faith in the world."[55] Another even asserted that "some evangelical ministers, Roman Catholic priests, and rabbis leave Unity tracts in hospitals."[56]

The movement ministers primarily to the middle classes, and it has no program of social reform. It has no creed. Its leaders think of God as omnipotent mind; He is everywhere about us, always accessible. He is a Trinity. The leaders also believe in reincarnation, and hold that death is but a temporary abandonment of a physical body. All mankind will ultimately be saved. We keep being reborn on earth until our spirituality develops to the point that our physical body is replaced by its true spiritual body. Thus we achieve eternal life. Charles and Myrtle Fillmore hoped that they had achieved eternal life and would not die. But they both have. Some followers of Unity declare they unselfishly renounced their already attained eternal bodies in order to continue their earthly labor for mankind.

THE OXFORD GROUP (MORAL RE-ARMAMENT)

"Religion in a tuxedo" is a phrase sometimes used to describe the movement founded in the early 1920's by Frank Buchman. Just as the Salvation Army ministers chiefly to the down-and-outers, the Oxford Group ministers chiefly to the "up-and-outers." It uses methods calculated to appeal first to society's leaders, to "key men," to big industrialists, to leading politicians, to champion athletes. Instead of store fronts, the movement uses swank hotels; instead of street meetings, houseparties; instead of day coaches, jet airliners.

A seminary classmate of Frank Buchman (1878–1961) told him he was ambitious. Stung by this accusation, he asked after graduation to be assigned to a church in a very poor section of Philadelphia, a church whose building *was* located in a "corner shop." After serving there for six years, he became Secretary of the Y.M.C.A. at Pennsylvania State College. In this position he was notably successful. During his tenure, scores of students went to regional conferences, hundreds enrolled in voluntary Bible study groups, and the Y membership increased from 35 per cent of the student body to 75 per cent.

Beginning in 1916, Buchman spent a few not very successful years at Hartford Seminary lecturing to theological students on personal evangelism. During this period he received several leaves of absence which he spent in evangelistic work, traveling as far as England and

China. Then he determined to cut all formal ties with established institutions, to found a movement of his own, and to "live on faith."

The first major focus of his attack was the prestige colleges on the eastern seaboard. At Yale, Harvard, Princeton, and Williams, Buchmanism, as his movement came to be known, made deep impressions. Students attended houseparties, sought God's guidance, confessed their sins, and were "changed." The numbers reached were not large but they always included the "key student leaders" if Buchman could get them, and he often did. The methods used at the houseparties were so sensational that widespread discussion ensued. Public confession of sin, including sexual sins, was urged. The resulting controversy over the wisdom of such practices sometimes reached a "violent pitch"; and at Princeton the President of the University asked Buchman to discontinue activities on the campus. "It was not until a college generation had passed [at Princeton] that the marks of the controversy ceased to trouble college religious activities."[57] Thus was success followed by failure; after a few years Buchmanism was almost unknown on American campuses.

But in the late 1920's the movement had begun to gain headway in England, especially at Oxford where meetings were held in which public witness was encouraged. After a time, Buchman made friends with Queen Marie of Romania and through her friendship gained contacts with many socially prominent persons. In 1928, a team of Buchmanites from Oxford went to South Africa where someone referred to "that group from Oxford." Buchman seized on the phrase and, much to the chagrin of many persons holding positions at Oxford University, officialized the name Oxford Group. (It should not be confused with the Oxford Movement, a nineteenth-century group which included Edward Bouverie Pusey, John Keble, and John Henry Newman.) Later Buchman also used for his movement the name Moral Re-Armament; this name is often shortened to M.R.A.

The decade of the 1930's was a time of great success. Houseparties (conferences) grew larger and larger, the coverage given by the press grew better and better, the territory traveled by the leaders grew broader and broader. The movement extended to Germany, France, Switzerland, China, Japan, Denmark, Holland, and Norway. In Norway a bishop declared the effect to be "the greatest spiritual movement since the Reformation."[58] Many famous persons lent their influence: noblemen, members of parliament, delegates to the League of Nations, the foreign ministers of Holland and France, the President of the Norwegian Parliament, and even a couple of professors at Oxford.

In the middle of the 1930's, efforts in the United States were renewed. At Stockbridge, Massachusetts, a nationally publicized houseparty was held; it was followed by a big meeting in the Metropolitan Opera House in New York City, which in turn was followed by a nationwide tour. But even bigger things were in store. A couple of years later, following a great meeting at Madison Square Garden, one thousand of Buchman's followers climbed onto a twenty-two car railroad special, made a two-month tour of the nation, and wound up at the Hollywood Bowl where thirty thousand people heard the message, "New Men, New Nations, a New World," and ten thousand were said to have been turned away.[59]

Buchman thought that the message of Moral Re-Armament brought to the world's leaders would dispel the war clouds which were gathering—but the war came in spite of his best efforts. During the war the movement lost ground badly. Some of the groupers tried to get draft deferment and received much unfavorable publicity as draft dodgers; Buchman suffered a serious illness; and worst of all, he was accused of Nazi sympathy. The chief item in this accusation was the report of a press interview held in 1936 in which he was quoted as follows:

I thank heaven for a man like Adolf Hitler who built a first line of defense against the Anti-Christ of Communism. . . . Think what it would mean to the world if Hitler surrendered to God. . . . Through such a man God could control a nation overnight and solve every last bewildering problem.[60]

The Group eventually rode out these storms and after the war was over its fortunes revived. Today it is active in many countries. Emphasizing the name Moral Re-Armament, it holds annual meetings at its centers in Caux, Switzerland, and Mackinac Island, Michigan. Also it sends deputations and dramatic productions all over the world.

The major aim of the Oxford Group is to *change* the individual, that is, to convert him, to persuade him to abandon his sinful life and to adopt a truly Christian way of living. The Group teaches that true Christianity demands Four Absolutes: absolute honesty, purity, unselfishness, and love. Many critics have pointed out what they consider to be elementary failures to meet the first absolute, absolute honesty. They say that even the name adopted by the movement gives a false impression of connection with Oxford University and that the Group's publicity is persistently exaggerated and inaccurate. In 1961, in full-page advertisements, it printed statements in which persons who had come under its influence claimed that M.R.A. had prevented Japan from going Communist; that if Nationalist China had had M.R.A., mainland China would never have been lost; and that if M.R.A. had not been in the Congo, an even "more terrible catastrophe" would have resulted.

At the time of Buchman's death, an M.R.A. team brought his body back to eastern Pennsylvania for burial, and used the occasion for the start of a campaign. Concerning it a clergyman in the area wrote: "One has to be subjected to an M.R.A. campaign to appreciate the deft touch and highly developed skill in public relations the teams bring to a 'new' area. . . . one executive said that never before had he met with such dishonest practices as were employed by the teams."[61]

A major technique of the Group is the *quiet time,* a period of individual or group silence during which one prays and *listens* for the guidance of God. Groupers believe firmly that God will direct the lives of persons who will listen for His guidance with the intention of acting on God's direction. During a quiet time a worshiper sits with pencil and pad in hand jotting down the ideas which come to him. Guidance comes on all kinds of topics, large and small; groupers are directed to take a trip, to call on an industrial giant, to arrange for a series of articles, to wear a certain necktie, to dine at a certain restaurant. Just as Quakers have difficulty in applying the doctrine of the inner light, so groupers have difficulty in applying the theory of guidance. Confusion sometimes arises, for example, when people think they are guided of God to do conflicting things. A practical test to determine whether guidance is truly from God is to subject it to the scrutiny of the Four Absolutes.

The Group is run by a small circle of insiders who "live by faith," that is, they accept no salaries and make no open appeals for funds, but instead have faith that God will guide wealthy followers to provide for living expenses. Occasionally the living provided has been Spartan, but usually funds flow generously. "Good food and good Christianity go together," say the groupers. Buchman managed to live in the manner to which he had become accustomed.

In the early days, houseparties consisted of a few dozen students and adults gathered in comfortable surroundings for a weekend. Later much larger groups convened in big hotels and in exclusive summer resorts. Even though the average size of the meetings has increased greatly, the Group still strives for the intimate and personal. Buchman gave himself without stint to helping individuals, counseling them, praying with them, seeking with them God's guidance—and so do his followers. The object of the big meeting is to change individuals. "It's no use throwing eye medicine out of a second-story window," they say.

Public confession, called *sharing,* is urged. This term refers to what some other religious groups call *witnessing*. At a houseparty or an assembly, some noteworthy person, a baroness or a senator, or an admiral, or a Davis Cup winner, or a columnist, or of later years even a labor leader will get on his feet and "share" with those present. These talks usually have a lot of hilarity about them as well as a lot of earnestness and moral concern. Humor of the self-depreciating type is a standard technique. *Confession* is urged on everyone. It need not be public, but it must be to some person. Honest facing of one's moral condition is considered to be the first step on the road to spiritual attainment. *Restitution* for any injuries one has done other people is also considered essential for spiritual growth.

The Oxford Group is not creative theologically; it has adopted the theology of a rather conservative type of Protestantism. Buchman's background was Lutheran. He believed in a fairly literal interpretation of the Bible, and one of his colleagues on the faculty of Hartford Seminary complained of his premillennial views.

Buchman and the leaders who gathered around him were master advertisers. They have used radio, television, full-page ads, billboards, parades, searchlights, airplanes towing banners. Following are some slogans which have been given prominence by the movement:

Woo, Win, Warn
Hate, Confess, Forsake, Restore
Confidence, Correction, Confession, Conversion, Continuance
Sin blinds, sin binds.

P owerful	*J* ust
R adiograms	*E* xactly
A lways	*S* uits
Y ours	*U* s
	S inners

THE BLACK MUSLIMS

The Original Man was the black man. He was the primogenitor of all other races. The white man's history goes back only six thousand years; but the black man's history goes back to the creation. Moreover, "the Black Man by nature is divine."

The era of white domination is about over. The Black Muslims know this to be true because Allah (God) declared it in Detroit during the Great Depression of the 1930's. He appeared then as a human being. Men took Him to be an Arab, a man who went by the name of W. D. Fard (also Mr. Farrad Mohammad, Mr. F. Mohammad Ali, and Professor Ford) and who earned his living as a peddler of silks and artifacts. But in reality He was the Most High God, the "Supreme Ruler of the Universe," the being orthodox Muslims call *Allah*.

After four years Fard disappeared (the police never could discover what happened to him) and leadership of the Black Muslims was won by Elijah Muhammad—born Elijah Poole in 1897 in Sandersville, Georgia, one of thirteen children. His father was a Baptist minister. In 1923, Poole moved to Detroit where he came into contact with Fard and was converted. Two years later Poole, now rechristened Elijah Muhammad, established a temple in Chicago; after Fard's disappearance it became the headquarters of the movement. Today Muhammad is known as the Prophet, the Messenger of Allah to the Lost-Found Nation of Islam in the Wilderness of North America.

The dynamic behind the Black Muslims comes from the indignities, the brutalities, the hatreds suffered by Negroes in the United States. These realities in American life can be illustrated by allusion to the experiences of Malcolm X, the number two leader and the heir apparent to Muhammad's office. Malcolm X (originally named Malcolm Little) was born in Omaha, Nebraska, one of eleven children. His father also was a Baptist minister. When Malcolm X was very young, the family moved to Lansing, Michigan. His father's outspoken ideas on race got him into trouble and the family home was "burned by the Ku-Klux Klan" when the son was six years of age. "The firemen," says Malcolm X, "came and just sat there without making any effort to put a drop of water on the fire." Later the father, trying to become economically independent, began to build his own store. Also, the family moved into a block where they were the only black family. "My father was found with his head bashed and his body mangled under a streetcar."

When Malcolm was in the eighth grade, he wanted to become a lawyer, but was told the law was not a suitable profession for a Negro; carpentry was suggested instead. He left school, moved East, and became a Harlem hoodlum. As "Big Red" he peddled dope and bootleg whisky, and engaged in "diverse forms of hustling." The law caught up with him and he went to prison seven times. Finally in the maximum-security prison at Concord, Massachusetts, he came into contact with the Black Muslims and was converted. Today he is the minister of Temple No. 7 in Harlem in New York City.

In 1960, the Black Muslims numbered about one hundred thousand persons; there were sixty-nine temples in twenty-seven states. Temples are located in the South as well as the North and West.[62] C. Eric Lincoln, author of *The Black Muslims in America,* says the sect is "growing daily in size and power." The membership is predominantly young, male, lower class, and ex-Christian; it is wholly "Black," predominantly American Negro but includes some Arabs, Japanese, and others.

The Black Muslims (they do not use the spelling Moslem and they prefer the pronunciation moose'lem) have little connection with the institutions of traditional Islam. Consequently they are rejected by the traditional Islamic Mosques in this country. But they have been accepted by Muslims in Egypt and Arabia, and Elijah Muhammad and his two sons were permitted to make the pilgrimage to Mecca and were aided in their efforts to do so.

Theologically the movement is poorly developed. Allah is believed to be the Supreme Black Man, and apparently all Blacks have a kind of divinity. Allah's recent incarnation was for the purpose of exposing the injustices which the black man suffers and of putting into operation forces which will put him in his rightful position. Sometime in a rather dim future the great Battle of Armageddon will be fought. Fully observant Black Muslims pray five times a day facing toward the holy city Mecca, as do traditional Muslims; these prayers are preceded by proper ablutions. The Quran (Koran), the holy book of Islam, is revered. The Bible is held in suspicion, but has a little worth if properly interpreted. The Christian religion is fought. It was "organized and backed by the devils for the purpose of making slaves of black mankind."[63] "Christianity is the white man's religion."[64] Expectation of a future life is not a part of the teaching. Christian names are changed to Muslim names. The word *Negro* is disliked; it is "a label the white man placed on us to make his discrimination more convenient."[65] The term *so-called Negro* is sometimes used.

The most important dogmas of the Black Muslims have to do with a social program. They look for the day when the "Lost Nation of Islam in North America" will dominate the nation. They seek to receive, at some time in the indefinite future, a part of the United States, say one-fifth of the nation's territory, as their own. They are dead set against integration. "Why integrate with a dying world?" They have contempt for the sit-in movement, and, writes Lincoln, "reject the ultimate integration—racial intermarriage—as sternly as any Southern white, and for much the same reasons."[66] They plan to set up their own economy and urge all black men to "buy black." They have established parochial schools in which the achievements of black men are given major consideration.

The American press has often declared the Black Muslims to be a hate group. They do teach that "there is no white man a Muslim can trust." But they do not openly teach hate or aggression. Rather they teach retaliation. They revere the *lex talionis*, "an eye for an eye." If the Black Muslim is attacked by the white man, he must return all that is sent, laying down his life if necessary. "They are obsessed with the humiliation of white supremacy."[67]

This determination to retaliate does not today lead to overt aggressive behavior. The leaders deny, for example, any part in the riot staged by American Negroes in the balcony of the United Nations when Adlai Stevenson made his maiden speech there.[68] Instead of aggression the leaders emphasize law observance and acceptance of proper authority. The record of the Muslims in the field of personal morality is outstanding. Many of their converts are former criminals, pimps, prostitutes, alcoholics, and drug addicts. (Some of their temples are located inside prison walls.) Their success with such persons is remarkable, especially so since their code of personal morals is both puritanical and stern. Muslims are forbidden to philander, gamble, smoke, drink alcoholic beverages, buy on credit, overeat, eat pork, eat cornbread or other items of a "slave diet." The movement encourages thrift, cleanliness, honesty, steady employment, and a strong family life. "Delinquency, juvenile or adult, is almost unheard of."[69]

Muslims are expected to attend meetings twice a week. Before being admitted to these meetings all visitors are carefully searched for concealed weapons—guns or knives—and two guards are stationed at each door of the hall. The chief feature of the meeting is a two- or three-hour talk in which a leader tries to present all the major teachings. The Muslims are falsely accused of "liking to strut in gaudy costumes"; the approved dress is a dark suit and a red tie. Men and women sit separately. Each Muslim is expected to give a definite percentage of his earnings to the movement; at one time this percentage was as high as one-third.

One Negro columnist wrote concerning the movement:

Mr. Muhammad may be a rogue and a charlatan, but when anybody can get tens of thousands of Negroes to practice economic solidarity, respect their women, alter their atrocious diet, give up liquor, stop crime, juvenile delinquency and adultery, he is doing more for the Negro's welfare than any current Negro leader I know.[70]

A hundred stories could be told of the same type as these eight: Seventh-day Adventists, Pentecostal Holiness, Salvation Army, Spiritualism, Jehovah's Witnesses, Unity, the Oxford Group, and the Black Muslims. A better indication of the essential religiousness of Americans would be hard to find. If men have freedom, they seek to satisfy their fundamental urges in a multitude of ways.

Yet the impact of the newer sects on our larger national life is small. With but few exceptions, they present no program for the reconstruction by human beings of the social order. These sects suffer so much from public ridicule that they develop a defense of indifference to problems whose solution depends on public opinion. The emphasis is on personal concerns.

Moreover, the newer sects are often autocratic in their type of government. The amount of will power necessary to pioneer a new religion is so great that ordinarily the founder develops a pattern of almost regal dominance. He is also under constant attack, as he sees it, from dissenters who would alter past all recognition the spiritual essence out of which the movement sprang. Thus he is forced, so he thinks, into ruthlessness, which usually takes the form of claiming to protect a divine revelation; the end result is old-fashioned autocracy.

These two characteristics—the usual lack of active concern for the larger social welfare and the frequent lack of democratic procedures—limit the contributions which these sects can make to our national spiritual problem. But they have made a signal contribution in preparing men's minds for change. The idea that all the good in religion originated centuries ago has been dispelled for many Americans. They are open to suggestion, looking for new light. The avidity with which they take up with the newer cults demonstrates the intensity of their spiritual hunger. Many are ready to follow a dynamic spiritual leader. Unfortunately, they are ready to follow him in almost any direction.

CHAPTER XVII

SOME NONECCLESIASTICAL
SPIRITUAL MOVEMENTS

THE PREVIOUS chapters complete our outline of movements Americans ordinarily consider to be religious. Yet there are very important aspects of the American spiritual situation which have not been considered. A religion of some kind is the possession of practically every sane adult; that is to say, practically every person's life is integrated—in some degree—around a set of values which he considers to be part of the very nature of things and essential to the good life for him. Some people give their lives to escaping hell and gaining heaven; others, to succeeding in business, or to self-expression through poetry, or to chasing the goddess of luck at the racetracks. Religion, as I use the term, is the pursuit of whatever a man considers to be the demands of the Ultimate on him, however he may conceive the Ultimate. The completely religious man—if there be such—is the man who devotes himself completely to meeting these demands. The completely irreligious man—if there is one—would be the person who thinks there are no ultimate demands or who is completely cynical about the possibility of meeting them.

This conception of religion is broad; it includes what might be called nonecclesiastical as well as ecclesiastical patterns of behavior. These two types of religious pattern often differ in the view of the universe around which they center. But they are similar in the way they function, in the things they do for people. Take Catholicism and Communism, for example. To their adherents, they both are a way of life, they supply a code to live by. They both furnish peace of

453

mind, they give a sense of personal worth and spiritual security. They both supply a frame of reference, they provide a vantage point from which to judge men and events. They both are certain that their teachings reflect ultimate reality, they teach with unmitigated dogmatism that they have *the* answer to the mystery of life.

Some students object to including a movement such as Communism, which is atheistic, among the religions, saying that it should be called a political ideology, or a philosophy of life; a religion, they contend, must center in God, teach belief in the hereafter, and expound supernaturalism. The answer to this point of view is that many of the world's religions have not been centered in supernaturalism. Early Buddhism was agnostic. Confucianism was a type of humanism. Jainism is atheistic. An orthodox Hindu can be an atheist. In contemporary American Judaism and Protestantism, there is a very small but perhaps increasing number of active clergymen and rabbis who are naturalistic or humanist. If the term *religion* is not to be used in a narrow and provincial sense, if it is to be defined in such a way as to include something more than Christianity and Judaism (to include, for example, the polytheism of primitive Africans or the sun worship of the ancient Egyptians), then it ought also to be defined so as to include the beliefs and behavior of those members of more complex civilizations who attempt to explain life in nontheistic terms.

Our interest in this volume is in a broad rather than a narrow view of religion because our concern is focused on the basic motivations of the American people. A man's major driving power comes from his ultimate faith. Religious convictions—basic convictions about the kind of universe we live in and the kind of life we must strive for in order to be in harmony with the universal forces—give the power to endure. The extremes of behavior—complete consecration to a task, voluntary poverty, permanent continence, indomitable resistance to oppression, martyrdom—spring from religious faith, from faith that the value one seeks is part of the very nature of things.

Since our major interest is to explore the full range of spiritual activity in America in order to indicate the role that religion plays in shaping national destiny, this chapter will be devoted to a brief outline of some major nonecclesiastical spiritual movements. Unfortunately, this aspect of America's religious life has been the object of but little serious study. Most of these movements do not gain their chief impetus from an organization, their major tenets are almost never set down in formal creeds; and statistics on the number of adherents are either meager or nonexistent. Consequently, the material of this chapter lacks precision. Yet such knowledge as we have forces us to the conclusion that these nonecclesiastical movements furnish the vital spiritual orientation of multitudes of the American people, including many who hold formal membership in a church or a synagogue, and must be taken into account in any realistic appraisal of the religious situation in the United States.

ASTROLOGY

One such movement is astrology, a system which holds that the progress of any person (or animal, plant, business, career, or even building) is determined by the positions of the heavenly bodies at the time of birth (or beginning). Most of the people who buy horoscopes (diagrams showing the position of the stars at a given time) permit astrology to play but a minor role in their hierarchy of belief. Most of them are simply confused by the complexity of their lives, fearful of their future, and unable to control the forces which push them around. They wish merely to play it safe. Waiting for the "auspicious time," they reason, can do no harm. If they invest a little money and patience, perhaps large returns will follow.

But there are some Americans who do make of astrology a full-blown religion. One writer asserted that "astrology is filling a need of millions of people who have passed the point where they can receive much help from the clergy or the churches." One star-reader from Kansas City declared that "being an astrologer is next to being God." Another reminds us that " a star led the Magi to our divine savior." Full believers contend that they are dealing with fact, scientific fact. They place the same kind of faith in their work that a scientist puts in his research.

But the scientists are violently opposed to astrology. They claim that no demonstration of stellar influence on human life has ever been made. They point, for example, to a study of several thousand musicians, painters and scientists which showed that the birth dates of these men were distributed throughout the calendar year in the same pattern as the birth dates of the general population.[1]

But in spite of the opposition of the scientists, the number of followers of astrology is startlingly large. *Life* magazine estimated that there are in this country three million buyers of horoscopes and twenty-five thousand horoscope casters.[2] Another estimate had it that the American public spends over two hundred million dollars a year on astrology.[3] It is said that one practitioner had as clients four hundred business firms, some of which paid him an annual fee of two thousand dollars.[4]

Devout followers of astrology really order their lives by the star patterns, believing that they thus become in tune with the infinite. A former president of the American Association of Astrologers is reported to note carefully the time whenever his telephone rings. Some astrologers claim to be able to discover what pets are most suitable for given persons. In some cities astrological furniture, neckties, jewelry and face cream are for sale. There are also specialties in astrology, including such esoteric skills as vocational, cosmic, glandular and sexual astrology.

NATURALISTIC HUMANISM

A nonecclesiastical religion of a completely different type is naturalistic humanism. This faith is frequently called *atheism*. But that term is inadequate, for three reasons. First, atheism, as the word is commonly used, has an odious connotation. Secondly, many naturalistic humanists are not atheist; they do not positively assert that there is no personal God. Rather they are agnostic, they declare that no man can know whether there is a personal God or not. Thirdly, naturalistic humanism is far more complex than simple atheism or agnosticism.

Naturalistic humanism holds that nothing exists but Nature and that, as far as we know, human beings are its highest form. Correspondingly, the pursuit of human welfare is the highest obligation any person has. Naturalistic humanists have the highest "respect for personality." They deny the existence of a soul and the possibility of a future life. They affirm that the universe is eternally self-existent, that experience, coupled to reason, is our only source of knowledge, and that co-operative effort can achieve a social order in which all men share the earth's goods.

The religious nature of this faith can be seen by observing how the word *God* can be substituted for the word *Nature* in the following statement, a statement to which most naturalistic humanists would subscribe; it was written by a colleague of mine. When the word *God* is used, the statement becomes acceptable to most theists.

Nature is eternally self-existent, uncreated, the ground of all life. All existence is in Nature, and outside of Nature there is nothing. Man must put all his trust in Nature; he will experience life at its best and fullest when he comes to understand and live in accordance with the laws of Nature.

Among humanists the percentage of belief in the infallibility of their faith is high, just as it is high in most of the religious groups. Many humanists are certain that any type of supernaturalism is naïve, amusing, obscurantist or superstitious and that we will have universal faith in humanism just as soon as we have honest leaders and unprejudiced education. "We are convinced," wrote the promulgators of the *Humanist Manifesto,* "that the time has passed for theism, deism, modernism, and the several varieties of 'new thought.'"[5]

How many persons hold this faith? No one knows. Doubtless the number is relatively small; but it includes many alert minds and spirits. An English Bishop has been quoted as saying that naturalistic humanism is the religion of half the intellectuals of the modern world.[6] That this estimate is not too wide of the mark is shown by the researches of James H. Leuba. In the thirties, he polled the opinions of several thousand American scientists, writers and "students in the colleges which are 'intellectually superior.'" He found that substantially less than a majority affirmed belief in "personal immortality" and in a "God to whom one may pray in the expectation of receiving an answer." In addition, he found that the percentage who deny these beliefs had increased substantially in the twenty years since a previous study, that scientists had a greater percentage of denial than businessmen, and that the "more distinguished" scientists had a greater percentage of denial than the ones "less distinguished."[7]

Surely no one can look honestly at our contemporary scene without acknowledging that naturalistic humanism (perhaps under another name) is the faith of a substantial portion of our nation's intellectual and artistic leaders. The men and women who write our leading editorials, produce our best sellers, man our universities, conduct our scientific investigations, provide our entertainment, and set the pace in painting, sculpture, music, poetry and the drama are as apt to be humanist as theist, especially in the higher brackets of achievement.

For some men this faith is a gloomy affair. Bertrand Russell wrote in a famous passage: "Brief and powerless is Man's life; on him and all his race the slow, sure doom falls pitiless and dark." "Omnipotent matter rolls on its relentless way." But for others, confidence in nature and the need to work for human welfare furnish full hope in the future and rich joy in living. The last lines of Shelley's "Prometheus Unbound" express the mood:

> Gentleness, Virtue, Wisdom, and Endurance,
> These are the seals of that most firm assurance . . .
> To love, and bear; to hope till Hope creates
> From its own wreck the thing it contemplates;
>   Neither to change, nor falter, nor repent;
> This, like thy glory, Titan! is to be
> Good, great and joyous, beautiful and free;
> This is alone Life, Joy, Empire, and Victory!

An immense potential for good is present in naturalistic humanism. It powers many a pioneering movement and many an indomitable worker. It awakens latent sympathies and challenges smug indulgence. Its critics claim that it is a fair-weather philosophy. They say that its adherents lack staying power and that the will to endure comes only from faith in a personal God. "Humanism is not self-reproducing," wrote an ardent supernaturalist.[8] This claim is surely false in view of the history of the groups and individuals who have lived, and died, in the humanist faith. In past centuries, multitudes of Chinese, for example, have lived by humanist principles and have given full evidence of the power to see life through. In our own time, Communists in the Russian armies, and elsewhere, have shown a staying power which surpasses that of most Christians and Jews.

In the United States some of the most devoted social workers, labor leaders and economic reformers are avowed humanists. If theists are honest in their reading of history, they must admit that it is not alone faith in God which produces courage in the face of adversity. The same result can come from faith in impersonal Nature.

But the high spiritual and ethical potential of naturalistic humanism is not realized in American society. The primary reason for this failure is a lack of organization among humanists for spiritual labor. Humanists seem to fancy that their spiritual needs can be adequately cared for through haphazard contacts. They have discarded the Church's discipline as well as its theism. America does have a few humanist societies; but they are ignored by most of the humanists. Among these societies is the American Ethical Union, often called the Society for Ethical Culture. Almost a century ago it was founded in New York City by the brilliant and sensitive Felix Adler. It established a realistic program of education and worship (by *worship* in this connection I mean the effort to glorify and revitalize spiritual values). The nobility of the Union's program has been widely acknowledged. But the movement, as a movement, has all but stood still. In 1960, it reported twenty-eight local units and less than seven thousand members. For every humanist who has supported the Ethical Union scores have not even been aware of its existence. If they had been, most of them would not have joined; membership would have smacked too much of churchgoing and of every-member canvasses.

The consequences of this rejection of spiritual organization have not been happy, especially with the second generation. The first generation has retained from its church and synagogue training a substantial amount of ethical stamina. But the children, denied their birthright of systematic spiritual instruction, have frequently concluded that there is nothing in our ethical heritage that is really worth struggling to preserve. They shrug their shoulders and go in for "self-expression," which too often is a weak rationalization of downright selfishness. Humanists need to learn that spiritual values, like all other values, demand the fostering climate of a sheltering institution, if they are to flourish.

HEDONISM

Another religious orientation which is widespread among Americans is the pursuit of pleasure. Many believe that the way to get the most out of life, the way to be truly in harmony with the patterns of the universe, is to seek the maximum of enjoyment. There are two kinds of happiness, suggests Bertrand Russell, "plain and fancy, or animal and spiritual."[9] The kind of happiness we are dealing with here is the plain animal satisfaction. Most persons, of course, wish to eat zestfully, to sleep warmly, to dress attractively. But some people make these satisfactions the *end* of living, not a *means* to living.

Lin Yutang, an author widely read in America, held at one stage of his career the following view:

All human happiness is biological happiness. This is strictly scientific. At the risk of being misunderstood, I must make it clearer: all human happiness is sensuous happiness. . . . Happiness for me is largely a matter of digestion. . . . To me, for instance, the truly happy moments are: when I get up in the morning after a night of perfect sleep and sniff the morning air and there is an expansiveness in the lungs, when I feel inclined to inhale deeply and there is a fine sensation of movement around the skin and muscles of the chest, and when, therefore, I am fit for work; or when I hold a pipe in my hand and rest my legs on a chair, and the tobacco burns slowly and evenly.[10]

A newspaper's report of a gourmets' dinner describes a type of search for sensuous pleasure. A hundred members of the *Les Amis d'Escoffier,* with napkins tucked in at the neck, ate the finest food they could procure. Each dish, with its appropriate wine, was received in absolute silence, since "one cannot appreciate good food in the midst of uproarious noise." Members were expelled for breaking the silence, and no smoking was permitted during the meal in order that the senses might not be dulled.[11]

But for every man who would go to such lengths in pursuing pleasure, a thousand would seek it in more usual ways: the hedonists' interests may lead to sports, to sex, to the social round, to nature study, to travel, to theatergoing, to music, to art, and to many combinations of these interests. Of course, not everyone who pursues them does so religiously. These interests become religious when they are pursued as an ultimate goal. And they are hedonist when they are pursued primarily for *the pleasure they give.*

A passion for possessions is probably the commonest form which the religion of hedonism takes in America. One writer estimates that a century ago "there were not more than 200 different items urged on the average man by the seller; today there are something like 32,000."[12] Another estimate has it that there are enough motor vehicles in the United States for every inhabitant to be on the road at the same time. A President of the United States dangled before American eyes the possibility of a trillion dollar national income.

The passion for things is so pervasive that it influences the values of everyone. Who has not been tempted to see the Kingdom of Heaven, at least temporarily, in terms of a shiny, new, two-tone, super-de-luxe hardtop, with extra-large luggage compartment; or in terms of a new, fifty watt, stereophonic hi-fi capable of making sounds as though a train were going right through the living room; or in terms of a new, aluminum, bagless, chromium-trimmed, streamlined vacuum cleaner, with special attachments for reaching into difficult spaces. Hedonism adulterates the Christianity and the Judaism of most persons who attend church and synagogue regularly. According to a distinguished foreign observer:

No church [in America] which urged the desirability of asceticism had any hope of influence or much hope of survival. . . . The schools . . . are almost wholly devoted to the exposition of a faith which makes "getting on in the world" practically an article of religious creed.[13]

The average American habitually judges pursuits, occupations, movements, men and even churches by their material value; and often his deepest allegiance, whatever the religion to which he gives formal adherence, goes, not to the Judeo-Christian God, but to the Dollar.

ALCOHOLICS ANONYMOUS

"I am my brother's keeper and he is mine and that's the heart of it," explained one member of Alcoholics Anonymous. "The central idea," writes a physician, "is that of a fellowship of ex-alcoholic men and women banded together for mutual help. Each member feels duty bound to assist alcoholic newcomers to get up on their feet."[14]

A recent estimate—"and it is probably a serious underestimate"— counts that "one man in every fifteen over the age of 20"[15] in the

United States is an alcoholic. The total is probably close to five millions, and an even larger number of nonalcoholic wives, children, and other relatives live in the same home with an alcoholic and are the victims of his recurring depressions, tempers, and unemployment. "The alcoholic is like a tornado roaring his way through the lives of others."[16] Excessive use of liquor is surely one of America's major social problems.

Many kinds of attack on this problem have been made, none with complete success. They range all the way from prohibition to the cures-for-a-fee advertised by private sanatoria; universities conduct research and analysts delve into the unconscious. A.A. is an effort by alcoholics themselves, an unusually successful effort. "They almost inadvertently," writes a physician, "made a major medical and social discovery."[17]

In the spring of 1935, Bill (whose last name is withheld), a compulsive drinker who was making heroic efforts to stay sober, went to Akron, Ohio, where he tried to win control of a small machine-tool company. This effort was unsuccessful and Bill felt the urge to escape from his defeat through liquor. "Then I panicked. That was really a gift! I had never panicked before at the threat of alcohol. . . . I thought, 'you need another alcoholic to talk to. You need another alcoholic just as much as he needs you.' "[18] A friendly clergyman gave Bill ten names. He called them by telephone. Nine were not available or not interested. But the tenth opened a door which brought Bill into contact with Dr. Bob. Together they found they could stay sober, and together they founded A.A. After three months, the fellowship had increased by one member; after two years it had ten members; after four years, a hundred members. Today it has two hundred thousand members and seven thousand groups in seventy countries and United States possessions. Most American communities of reasonable size have a group which can be contacted through a telephone listed under Alcoholics Anonymous.

The early struggles of the founders, like the early struggles of Joseph Smith and Mary Baker Eddy, would have defeated men of less character and ability. Public understanding and financial resources were hard to win. But eventually articles in the *Cleveland Plain Dealer* and the *Saturday Evening Post* brought such a deluge of calls and letters from alcoholics and their families that the principle of giving personal attention (sometimes by mail) to each case was threatened. But this principle was persistently clung to and it is the cornerstone of A.A. today. The real secret of the group's tremendous success has been the willingness of men who have once been in the grip of alcohol or who are still struggling to free themselves from it to give their time, their money, and their brotherly concern to others like themselves. The person struggling against the temptation to drink knows he is only a telephone call away from help. Moreover, willingness to be on call for this "dime therapy" is a major factor in keeping the older members of A.A. sober. Like the Salvation Army, A.A. has discovered the power of "active duty." "Once a recovered drinker slows up in this work he is likely to go back to drinking, him-self."[19] Success is by no means one hundred per cent. Members of A.A. believe they succeed with 75 per cent of "those who really try." Studies by outside observers "report figures in the 30-to-40 per cent range."[20]

In order to stay sober, the alcoholic is not asked to take a pledge. Instead he is urged to take Twelve Steps. These steps constitute one of the most notable sets of spiritual exercises ever formulated. They are not theory, concocted in the minister's or physician's study. They emerged from the hot war where happiness and even life were at stake. Bill wrote: "Unless each A.A. member follows to the best of his ability our suggested Twelve Steps of recovery, he almost certainly signs his own death warrant. Drunkenness and disintegration are not penalties inflicted by people in authority; they are results of personal disobedience to spiritual principles. We *must* obey certain principles, or we die."[21] The Twelve Steps apply to the concerns of men and women who never drink, and will repay serious study by any person who is confronted by anxiety or who is harassed by trials which seem greater than he can bear.

Step One: We admitted that we were powerless over alcohol—that our lives had become unmanageable.
Step Two: Came to believe that a Power greater than ourselves could restore us to sanity.
Step Three: Made a decision to turn our will and our lives over to the care of God *as we understood Him.*
Step Four: Made a searching and fearless moral inventory of ourselves.

Step Five: Admitted to God, to ourselves, and to another human being the exact nature of our wrongs.

Step Six: Were entirely ready to have God remove all these defects of character.

Step Seven: Humbly asked Him to remove our shortcomings.

Step Eight: Made a list of all persons we had harmed, and became willing to make amends to them all.

Step Nine: Made direct amends to such people whenever possible, except when to do so would injure them or others.

Step Ten: Continued to take personal inventory and when we were wrong promptly admitted it.

Step Eleven: Sought through prayer and meditation to improve our conscious contact with God *as we understood Him,* praying only for knowledge of His will for us and the power to carry that out.

Step Twelve: Having had a spiritual awakening as the result of these steps, we tried to carry this message to alcoholics, and to practice these principles in all our affairs.

These steps make clear the religious character of A.A. God *"as we understood Him"* or at least a "Power greater than ourselves" is leaned on. Men who call themselves agnostics do belong, but they are not of the I-am-the-captain-of-my-fate variety. A.A. had in fact a religious beginning. In the months just before Bill met Dr. Bob, he was attending meetings of the Oxford Group, was much helped by them, and later used many of their methods. Only after three or four years was a definite line drawn between the two movements.

All the theological propositions in the movement are contained in the Twelve Steps. Moreover, it has no ritualistic forms except the repetition of the Lord's Prayer at meetings. In it Roman Catholics, Protestants, and Jews stand shoulder to shoulder, and it has the support of priests, rabbis, and ministers. The only requirement for membership is a sincere desire to stop drinking. The Rev. Harry Emerson Fosdick has written: "The meetings of Alcoholics Anonymous are the only place, so far as I know, where Roman Catholics, Jews, all kinds of Protestants and even agnostics get together harmoniously on a religious basis. . . ."[22]

The first of the Twelve Steps—recognition of complete personal defeat—is essential, according to A.A. "Deflation at depth," Bill calls it and he quotes from William James, *Varieties of Religious Experience:* "Self surrender has been and always must be regarded as the vital turning point of the religious life." Analogous is the revivalist preacher's practice of beginning his protracted meetings with a series of sermons in which he strives to bring his listeners to a "conviction of sin." Frank Buchman, the founder of the Oxford Group, made a similar emphasis.

There are no professionals in A.A. "Special workers" are employed to write letters, answer telephones, cook hamburgers, and sweep out meeting halls; but no one is ever paid for twelfth-step work, that is, for face-to-face care of a fellow alcoholic. The movement refuses all outside gifts; it will not accept the gifts of philanthropists. Each local group is autonomous. Elected regional or national committees have only an advisory relationship to local groups.

A.A. is a one-purpose movement; it exists "to carry its message to the alcoholic who still suffers." Therefore, it does not endorse any other type of social movement, including prohibition and all other efforts to deal with alcoholism.

Unique, as far as I know, for a large movement is the insistence on anonymity. Bill observes that alcoholics often are overcome by power drives and that safety for the movement can come only through a deliberate effort to avoid personal recognition. Consequently one of the main planks of the A.A. platform reads: "Anonymity is the spiritual foundation of our traditions, ever reminding us to place principles before personalities."

BEAT ZEN

The problem is how to "let go" and "go with"; how to "get high and let it spill"; how to "swing" but not to "flip too far out." The beatniks try many twists in the effort to "get in the groove": jazz, poetry, alcohol, sex, marihuana. Some of them dabble in Zen, the religion of a small Japanese sect.

Zen Buddhism, out of which grew Beat Zen, is said to have had its beginning in the sixth century A.D. An Indian Buddhist scholar traveled to China where he interviewed the Chinese Emperor, who inquired how much merit flowed from his many benefactions. He received the gruff reply, "No merit at all." This incident illustrates a basic Zen tenet that efforts such as gifts to charity, support of the order, performing rites, believing dogmas, studying scriptures are worthless for furthering one's "enlightenment." Dr. Daisetz Suzuki, a

Japanese Zenist who lives in America, says that a major obstacle to gaining enlightenment is the "taint of intellection" and speaks of the books he has written as "my sins." The Zen path to enlightenment is meditation. One ancient story has it that the founder of Zen demonstrated the path by spending nine years sitting with his face to a wall.

According to Zen, proper methods of meditation require great discipline. It is said of an American woman who became the head of a Zen temple in Japan that when she first began meditation "she rose at 5 A.M. to meditate for two hours before breakfast, then went to the temple to meditate from 9:30 A.M. to 5:30 P.M., with a few minutes off for a meager lunch. After supper at home she would return to the temple for meditation with the monks until 9:30 at night, then return home, take a bath, and meditate until bedtime, around midnight."[23] During these long, long hours many precautions are taken against "drifting" and "looseness" of mind. Proper posture and proper breathing are carefully studied. "At intervals, the sitting posture is interrupted, and the monks fall into ranks for a swift march around the floor between the platforms to keep themselves from sluggishness."[24] Thus Zen is not something to believe, not a church to join, not a ritual to follow. But it is discipline. Years of disciplined meditation are required for even the first flash of enlightenment.

From time to time the monk interrupts his meditation to consult his master. He asks a question or perhaps is asked a question. These questions together with their answers are called *koans*. A *koan* is by nature irrational. For example: the question, "What is the true nature of the Buddha?" may be answered by a slap in the face, or by a smile, or by the word "rubbish." Some of the questions are themselves irrational, such as, "You know the sound of two hands struck together; what is the sound of one hand?" Such *koans* are the subject of long meditation.

Followers of Zen hold that the distinction between right and wrong, good and evil, is pointless; in fact

Fair is foul, and foul is fair.

The oldest Zen poem contains the following lines:

> If you want to get the plain truth,
> Be not concerned with right and wrong.
> The conflict between right and wrong
> Is the sickness of the mind.[25]

These paragraphs give but a hint of what Zen is all about. Zenists claim it is unknowable. "To know Zen is not to know it," is a well-worn proverb. One disciple describes Zen as "something round and rolling, slippery and slick, ungraspable and indescribable."

Zen missionaries began working in America in the early decades of the twentieth century and have made a considerable impact on *avant garde* artistic and intellectual circles. Authentic Zen, however, is seldom practiced. Most American Zenists are either "square" or "beat," according to Alan W. Watts, a leading exponent of Zen and a former professor in the American Academy of Asian Studies. By Square Zen he means "a new form of stuffiness and respectability"; by Beat Zen he means "a revolt from the culture and social order."[26]

Perhaps *beat* means "done in," and perhaps it means "beatific." In any case the beatniks are in revolt. Persons who are not "hip" suppose that the beatnik revolt is the same as the revolt of any other decade, that going for "kicks" in the fifties was barely distinguishable from "making whoopee" in the twenties. But the beatniks are sure they are different. Theirs is a "total rejection of American lifeways and values," and an adoption of "poverty as a voluntary act of dedication."[27] They believe they have a new and true set of ultimate values. Their revolt is standardized in a beat orthodoxy, an orthodoxy which is all mixed up with beards, joblessness, blue jeans, illicit sex, aversion to soap, and undressing in public. The Zen of these people is a fad. It bears little resemblance to the Zen of the Japanese monastery. Disciplined contemplation is not a part of beat orthodoxy. John Ciardi says, "There is in all of them an innate fidget. . . . the Beats talk endlessly about serenity, detachment, and mangled Zen, but the last thing they know how to do is sit still."[28] The attraction of Zen for the beatniks is chiefly twofold. First it seems to buttress sexual license through its doctrine that right and wrong are indistinguishable. "They have . . . raided from Zen whatever offered them an easy rationale for what they wanted to do in the first place"; they believe in the "holiness of the personal impulse."[29] Second, Zen helps them gain enlightenment. But the kind of enlightenment they are after is not the Zen Buddhist kind. The monks seek a loss of self in the oneness of the all, an immediate insight into the heart of reality, a true experience of being without perception or conception. The beatniks, on the other hand, are after a personal sensory experience. They want serenity and bliss: not the bliss of the opium eater but that of mild intoxication. The "cool cats" avoid getting too "high." They prefer not to really "flip." They are after a "low-high," a smooth "marihuana-float" that leaves them in possession of their elementary faculties, and heightens their awareness of sounds, smells, and colors. "The chicks stare off glassy-eyed into the Ultimate-All and keep saying "Yeh! . . . hyeh! . . . hyeh!"—long drawn out, ecstatic, and aspirate. I mean like real cosmic, man."[30] In less ecstatic times, their characteristic pastime is sitting in small groups in a dingy "pad" (home) listening to sounds: blaring jazz, throbbing drums, animated "jive" (talk). Their talk of Zen in these sessions is a set of words which veneer their real religion, a religion which is a mixture of revolt, gregariousness, exhibitionism, and sensationism.

NATIONALISM

Another important item in the American system of faiths is love of country; it is a sentiment which influences the values of us all. Perhaps few would agree with Eddie Rickenbacker, "I say to you frankly, gentlemen, there is nothing in the world, nothing that I love more— not even life—that compares with my love for America."[31] But most Americans would place service to the nation as one of their highest obligations, one which can upon occasion take precedence over all others. In conflict situations, loyalty to the nation generally supersedes loyalty to the local community, to the church, to the family, and to such ideals as honesty, honor and fair play.

Some readers will question the religious character of nationalism. It may be clear to them that certain movements in foreign countries have been motivated by religious fanaticism: Fascism, Nazism, Communism, Shinto. But it is not so clear to them that in the United States patriotism sometimes takes on a religious cast. Carlton Hayes, eminent historian and a Catholic layman, defends this thesis:

Nationalism, viewed as a religion, has much in common with other great religious systems of the past. It has, for example, a god, who is either the patron or the personification of one's *patrie,* one's fatherland, and one's national state. . . .

On his own national god the modern religious nationalist is conscious of dependence. Of His powerful help he feels the need. In Him he recognises the source of his own perfection and happiness. To him, in a strictly religious sense, he subjects himself. . . .

To the modern national state, as to the mediaeval church, is attributable an *ideal,* a *mission.* . . . The nation is conceived of as eternal, and the deaths of her loyal sons do but add to her undying fame and glory. . . .

The ritual of modern nationalism is simpler than that of certain other great historic religions, probably because sufficient time has not yet elapsed for its elaboration, but, considering its youthfulness, it is already fairly well developed. Nationalism's chief symbol of faith and central object of worship is the flag, and curious liturgical forms have been devised for "saluting" the flag, for "dipping" the flag, for "lowering" the flag, and for "hoisting" the flag. Men bare their heads when the flag passes by; and in praise of the flag poets write odes and children sing hymns. . . .

"My country, right or wrong, my country!" Thus responds the faithful nationalist to the magisterial call of his religion, and thereby he intends nothing dubious or immoral. He is merely making a subtle distinction between governmental officials who may go wrong and a nation which, from the inherent nature of things, must ever be right. . . . The most impressive fact about the present age is the universality of the religious aspects of nationalism. . . . Nationalism has a large number of particularly quarrelsome sects, but as a whole it is the latest and nearest approach to a world-religion.[32]

The belief that service to country is an ultimate demand is perhaps best seen in such organizations as the American Legion and the Daughters of the American Revolution. Probably no man can be a successful politician in America today unless he does obeisance to nationalistic gods, certainly not if he attacks them. The high priests of this religion can probably be found in the Pentagon. These men evidence a genuinely religious zeal in their devotion to the national interest, as they see it; they exhibit the same narrowness of purpose, the same ruthlessness in dealing with obstacles, the same dogmatism in thinking, the same persistence in the effort to expand the frontiers of their power as has characterized the successful hierarchies of every age. They have sold "deterrence" to the American people; it has become an avowed national faith. So committed is the United States to this policy that opposition to it is ignored, or scorned, or irrationally attacked. The logic of the present world situation points inexorably to the conclusion that in the long run the policy of deterrence cannot preserve the safety and probably not even the existence of the country. Disarmament and world government are essential goals for responsible national leadership. But the religious nationalists have a fixation. They put their trust solely in preparedness and do their best to prevent serious consideration of disarmament. Some of them attack the United Nations and all participation in its work by the United States. They would echo Walter Lippmann who wrote that "we must consider first and last the American national interest. . . . We shall succeed in so far as we can become fully enlightened American nationalists."[33]

These six faiths—astrology, naturalistic humanism, hedonism, Alcoholics Anonymous, Beat Zen, and nationalism—give an indication of the nature of nonecclesiastic religions in America. A complete survey would include many more such interests: scientism, nature study, scholarship, aestheticism, organized fraternalism, white supremacy, humanitarianism, socialism, Communism, native fascism, democracy. Limitations of space will not permit the description here of further movements; but a large enough sampling has been made to make the point clear: the *de facto* religion of a large percentage of Americans is "nonecclesiastical" in type; it has no necessary connection with churches or synagogues; and, moreover, the *de facto* religion of a large percentage of church and synagogue members contains important elements of a "nonecclesiastical" character.

The real religion of a person is usually a blend of many components. Sectarian leaders almost universally contend that their particular ism furnishes all necessary spiritual sustenance and deserves an exclusive devotion. Most people find, however, that our twentieth-century religions are not as completely satisfying nor as mutually exclusive as their most ardent devotees imagine. Honest introspection would convince most people that the set of values to which they give devotion in their heart of hearts, and from which they gain their chief motivation, is a complex matter, derived from many sources. Many a hedonist is also a nationalist, believes with religious passion in white supremacy, and is a member of a church. Many an active clergyman holds fervently to a religion in which are united aspects of such faiths as Neo-orthodoxy, socialism, democracy, hedonism, and aestheticism.

If it is true that religion furnishes the major motivation for both individuals and groups, then a correct assessment of religious conviction in America is a matter of extraordinary importance for the person who attempts to judge the trend of the future. In the effort to make such an assessment we must keep constantly in mind the fact that vital religious movements are to be found outside the churches and synagogues. Indeed, from the point of view of human destiny on this earth the most important religions in the United States—and in the world—are not Christianity and Judaism. The most important religions are political movements which receive from their followers the same type of impassioned devotion which true Christians and Jews in every age have lavished on their faiths.

CHAPTER XVIII

THE ROLE OF RELIGION IN SHAPING AMERICAN DESTINY

SURPRISING AS it may seem, Catholicism, Protestantism and Judaism are all stronger in the United States than anywhere else in the world—in number of *active* adherents, in financial resources, in support of missionary enterprises. The opinion polls have consistently shown that an overwhelming majority of the American people believe in God; for example, a Gallup poll released in 1954 indicated that 96 per cent of the Americans who were asked the question, "Do you, yourself, believe in God?" answered "Yes."*

The percentage of the population attending worship services is probably greater in the United States than in any other large nation. The National Opinion Research Center reported in 1945 that 81 per cent of Catholics, 62 per cent of Protestants, and 24 per cent of Jews said they attended church or synagogue at least once a month, and only 18 per cent of the population said they never or hardly ever attended. The figures for regular *weekly* attendance were Catholics 69 per cent, Protestants 36 per cent, Jews 9 per cent.** (The terms

---

* The report was released December 18, 1954. When the answers were broken down according to the age of the respondents, the percentages of "Yes" answers were: 21-29 years, 93%; 30-49 years, 95%; 50 years and over, 98%. When the answers were broken down according to "education" the results were: college, 92%; high school, 96%; grade school, 97%. When the answers were broken down according to "city-size" the results were: over 100 thousand, 93%; 25,000-100,000, 98%; rural, 98%.

** Another poll, conducted by Ben Gaffin and Associates for *Catholic Digest*, reported in 1952 the following percentages of persons who said they attended church or synagogue weekly: Catholics, 62%; Protestants, 25%; Jews, 12%. See *Information Service*, Feb. 14, 1953.

Catholic, Protestant and Jew in this report include persons of these backgrounds as well as *bona fide* members.) Releases by the Gallup organization[1] indicate that the percentages of the population attending church during a typical week have been: 1940, 37 per cent; 1950, 39 per cent; 1954, 46 per cent; 1955, 49 per cent; 1956, 46 per cent; 1957, 51 per cent; 1958, 49 per cent; 1960, 47 per cent. Gallup reported in 1957 as follows:

Women were found to be better churchgoers than men. College people are better churchgoers than those who did not go beyond grammar school. People in the middle years—30 to 49—attended in greater proportion than younger adults—21 to 29—or older people—50 years and over.

Churchgoing was about the same in all sections of the country except in the Far West where substantially fewer persons had attended.

The figures:

	% Attended [on a typical Sunday]
Men	43
Women	57
21–29 years	51
30–49 years	53
50 years and over	49
College	53
High school	52
Grade school	48
East	52
Midwest	51
South	53
West	42
Protestants	44
Catholics	76

The attendance figures in European countries are lower. In England, according to Mass-Observation, "15 or 20 per cent . . . go to church regularly, another 40 per cent do so occasionally."[2] Poll results of regular church attendance for other European nations were reported as follows: Netherlands, 50 per cent; Czechoslovakia, 20 per cent; Italy, 12 per cent.[3] These data were gathered in the 1940's. In France, according to one estimate, less than six million of the forty million inhabitants are practicing Catholics or Protestants.[4] South American nations are often said to be 100 per cent Catholic; actually the percentage of practicing Catholics in South American nations is much smaller.[5]

There are about three hundred thousand local churches in the United States; they are served by about two hundred and fifty thousand pastors.[6] A few years ago, some students of religious statistics held that "the trend of adult church membership in relation to adult population has been slightly but definitely downward since the 1920s."[7] Other students, however, affirm the proposition that the proportion of church and synagogue members in the population has been steadily rising. In 1800, the proportion was about 5 per cent; and by 1835, it had risen to about 14 per cent.[8] Following are the percentages reported by the *Yearbook of American Churches* for the past century.[9]

1850	16%	1910	43%
1860	23%	1920	43%
1870	18%	1930	47%
1880	20%	1940	49%
1890	22%	1950	57%
1900	36%	1959	63%

The past decade, according to the common opinion, has been a time of religious revival in the United States. In addition to the evidence of increased church and synagogue attendance and membership are such items as heavier enrollments in college classes in religion, increased numbers of articles on religion appearing in popular journals, larger sales of religious books, the openly expressed piety of some very highly placed politicians, and the rising prestige of the clergy. The prestige of the clergy was measured in various polls by Elmo Roper. He found in 1942 that 17 per cent of his sample thought that religious leaders were the group doing the most good for the country. In 1947, the percentage was thirty-three. By 1957, it had climbed to forty-six. "No other group—whether government, congressional, business, or labor—came anywhere near matching the prestige and pulling power of the men who are the ministers of God."[10] In fact, clergymen received more votes than all other groups combined.[11]

The reality of this revival has been challenged. For example, the statistics on church membership are called into question. One student claims that if errors are eliminated which are due to sometimes including and sometimes not including persons under thirteen years of age, then church membership since 1900 has been relatively constant and in fact declined until 1940.[12] Another student shows how some churches play their own brand of "numbers game"; in one year, 1952, different denominations reported the following improbable gains: from 400,000 to 750,000, from 209,615 to 1,500,000, from 682,172 to 1,112,123.[13] The gains in church attendance are also called into question as "due to unreliability of the . . . poll[s], or to sampling at different times in the year."[14]

Some students, even though they may acknowledge the probability of statistical increases in religious activities since 1940, question the depth of these activities. One investigator concludes that we have had merely an "upsurge of *interest* in religion."[15] One poll showed that 53 per cent of the Americans queried could not name even one of the four Gospels. Moreover, organized religion receives a relatively small and apparently declining share of the national income. From 1909 to 1941, the percentage of consumer expenditures given to religious organizations decreased by nearly one-half.[16] Church and synagogue members give slightly less than 2 per cent of their disposable personal incomes to religious institutions.[17] In 1960, Americans gave only four and two-thirds billion dollars to religion *and welfare,* but spent nearly seven and one-half billion on tobacco and nearly ten billion on alcoholic beverages.[18] The sales of religious books increased from 1951 to 1961 only half as much as the sales of books in the general trade.

Whatever position students take on the reality of the revival, most of them agree that it is about over. *The Christian Century* published an editorial in which 1958 was designated as "the year the revival passed crest." A cartoonist in *Esquire* designated 1955 as the year the intellectuals were most fascinated with religion. Enrollments in Protestant theological schools dropped over 5 per cent in one year, 1959 to 1960.[19] In 1960, the membership gains reported by the churches shrank for the first time in fifteen years to about the percentage of the estimated population gain. Cries that the churches are "in retreat" are beginning to appear.

The conclusion can hardly be avoided that as institutions the churches and synagogues play less determining roles in American culture than in former decades. Corporate life is steadily growing more and more complex; religious institutions struggle to hold their own against a multitude of competing agencies. A large number of Americans find their religious satisfactions outside the Judeo-Christian tradition; they make naturalistic, or nationalist, or hedonist assumptions. Indeed, a sizable percentage of formal church members (including some who are church leaders) find their major spiritual drive—their goals, their values, their view of life—outside the teachings and dogmas of the churches and synagogues. The number of Americans who hold such points of view is probably increasing.

The situation might be summarized as follows: At one end of the scale are many millions of Americans who support the churches and synagogues. At the other end of the scale are a few millions who are humanists and naturalists. Both of these groups appear to be growing at the expense of a large group in the middle which is spiritually unaware or indifferent. Some of the spiritually unaware hold formal membership in a church or synagogue.

No doubt there is much in the religious situation which could lead to pessimistic conclusions. The level of knowledge about religion is woefully low. The barriers separating faith from faith are often high. The gulf between religious ideal and religious practice is sometimes tragically wide. Creeds often inadequately express present convictions. Leaders of worship often think that professional skill is an adequate substitute for lay participation. Lay conservatism often inhibits clerical expression of the deepest moral and spiritual insights. Churches and synagogues are frequently class institutions. The list might go on and on.

Yet I find many reasons for optimism in our current religious situation, as it applies to *individuals*. The American citizen is presented with a large variety of religious opportunities; he can select a faith which fits his need and temper. Religious liberty is a reality; individuals can change their faith without civil disability. The sects can engage without hindrance in missionary work; and individuals are free to ignore it. Church and synagogue morality is seldom sullied by corruption. Religious competition puts churches and synagogues on their mettle; clergymen are intelligent, alert, sincere. The churches and synagogues are loyal to democracy; many of them teach and practice popular control. The churches and synagogues teach a high standard of ethics; they work persistently to increase the level of honesty, loyalty, candor and moral courage.

It is often said that Americans are an unspiritual people. If by *spiritual* is meant *mystical,* no doubt the accusation is correct. But if by *spiritual* is meant "persistent awareness of one's life ideal," I judge individual Americans to be no less spiritual than individuals living in other countries. Men everywhere fail by a large margin to fulfill their religious aspirations. Americans probably succeed as well as most others. Unfortunately, the aspirations of many are low. The average citizen is commonly unaware of, or uninterested in, the deepest religious possibilities (theist or humanist) just as he is unaware of, and uninterested in, high standards in most other fields: literature, music, scholarship, vocational achievement. The many sectarians who are dissatisfied with this situation have full opportunity to do missionary work.

These generalizations characterize the religious scene as it applies to the individual American and his private spiritual needs. But the American religious problem has another equally important aspect. It concerns the nature and vitality of the values which lie at the center of American culture. At numerous points in the previous pages it has been asserted that a culture is above everything else a faith, a set of shared convictions, a spiritual entity. This conception can perhaps be clarified by analyzing religious functioning into the following three categories: *private, denominational,* and *societal.* Private religion would be the religion a person shared with very few persons, only with his intimates. Denominational religion would be shared with the members of a whole denomination, a much larger group. Societal religion would be shared with the members of a whole society. The reality of such levels as these of religious functioning (whatever names be given them) can hardly be doubted by the careful observer. Religious values (values believed to derive from the Ultimate) pervade human conduct and thought. Every type of association and every size of group can be guided or influenced by religious values. The careless assumption of the past has too often been that religion is what goes on in and around the churches. Theorizing about religion has too consistently been left to clergymen and theologians. As a result the role of denominational religion has been overstressed; indeed a frequent assumption has been that this is the only kind of religion which exists. This view is parochial. Another great agency through which individual persons share religious values is society, that is, the whole group of which they are a part. "Every functioning society," writes the sociologist, Robin M. Williams, "has to an important degree a *common* religion." "A society's common-value system—its 'moral solidarity'—is always correlated with and to a degree dependent upon a shared religious orientation."[20] A society has unity to the extent that the individual persons who compose it share common values. A society has vitality and staying power to the extent that some of these values are believed to be religious, to be ultimate.

Previous pages have indicated how for centuries societies in the West assumed the principle of religious uniformity. The safety of the nation was believed to depend on uniform denominational faith and membership. In these societies denominational and societal religion merged. As a result of long and bitter conflicts, the belief in religious uniformity was abandoned, and the principle of religious pluralism was adopted.

The change, however, was only of *denominational* religion. Uniformity of societal religion continued. It is a false view to read the record as providing for complete pluralism of religion. The separation of *church* and state does not mean the complete separation of *religion* and the state. To affirm this meaning is to identify religion with the churches. Every viable state in our intense and uncertain world is in fact based on a societal religion, on a set of values believed by the dominant group to be ultimate.

The lines dividing private, denominational, and societal religion are of course blurred; the phenomena of religious experience, like the phenomena of most human experiences, are not sharply differentiated and are often difficult to classify, a situation which ought not to blind us to the value of the attempt to classify. One confusing fact is that all three categories of religious value are usually supported by the churches. This point is made by the church historian, Sidney E. Mead, in an article which describes American Protestantism since the Civil War.

During the second half of the nineteenth century there occurred a virtual identification of the outlook of this denominational Protestantism with "Americanism" or "the American way of life." . . .

The United States, in effect, had two religions, or at least two different forms of the same religion, and . . . the prevailing Protestant ideology represented a syncretistic mingling of the two. The first was the religion of the denominations. . . .

The second was the religion of the democratic society and nation. . . .

The free churches eventually found themselves entangled in a more subtle form of identification of Christianity, nationalism, and economic system than Christendom had ever known before.[21]

The process here described was not peculiar to America in the second half of the nineteenth century. All denominational groups in whatever society at least tolerate most societal religious values, and most denominational groups in fact actually support these values. No denominational group in any going society could long survive if the denomination genuinely threatened the societal religion. Thus most American denominations in practice give vigorous support to most of the American societal religious values.

And yet the student of religious life has not achieved a true picture of the total situation unless he realizes that in a pluralistic situation most denominational religion is a great deal more complex than is societal religion. The denominations cover many spiritual and moral needs to which the larger group is quite indifferent. The denominations set higher moral and spiritual standards and ask faith in metaphysical propositions which often are very different from those of the average citizen.

Like most institutions churches strive to expand the area of their control. Thus they seek dominion over the whole of religion. They tend to religious imperialism, claiming that denominational religion is the whole of religion. This contention cannot be true in societies which are in fact denominationally pluralistic—if "every functioning society" has a *common* religion."

What are the values which constitute the "societal religion" of this country? That is, what are the values which the dominant group in America accept to the level of believing and acting as though they are ultimate? No definitive answer is possible. This is not the kind of topic to which research techniques have been applied. It is true that a number of investigators have studied the values of the American people and have constructed lists. But none of them, as far as I know, deal with the religious dimension, that is, none distinguish religious values from values which are believed to be merely convenient temporal arrangements, human choices with no necessary grounding in beliefs about the ultimate nature of things. On the other hand, the results of investigations by sociologists and anthropologists of American values are the best indicators we have of the nature of societal religion in this country. Robin Williams submits the following as "major value-orientations"; the phrases listed here are sharp abridgments of Professor Williams' statements.

Stress on personal achievement
Stress on activity and work
The world viewed in moral terms
Humanitarianism, concern for the needs of others
Efficiency and practicality
Progress, optimism, an emphasis on the future
Emphasis on material comfort
Avowal of equality, and often its practice
Freedom
External conformity
Science and secular rationality
Nationalism and patriotism
Democracy
The worth of the individual personality
Racism and the granting of privilege to individuals on the basis of race or particularistic group membership[22]

Williams cautions that these value-orientations are, even with his careful definitions and qualifications, "subject to numerous exceptions" and "represent *tendencies* only."[23] He does not assert that these values are religious. Yet surely some of them are, that is, some of them do reflect values believed to be ultimate, values which would be clung to when others were discarded.

Probably the best single word to describe the societal faith of Americans, the *ideal* toward which they strive, would be democracy. A. Powell Davies wrote thus about it:

The democratic faith is a belief that man, if he resolves upon it, can raise the level of his life indefinitely, making the world increasingly more happy, more just, and more good; no fate has made him prisoner of his circumstances, no natural weakness has condemned him to be ruled by tyranny. He is meant to be free. Through the power of reason he can form intelligent opinions, and by discussion and debate can test them.

Knowing that truth is precious above all things and the only safe guide to purposes and aims, the right to seek it must be held inviolate.

And the democratic faith declares that human rights are by their nature universal: that liberty is such a right, and that without liberty there cannot be justice; that, to ensure justice, the people should make the laws under which they live; that besides justice there should be benevolence and sympathy; that those doctrines of religion which beseech mankind to practice brotherhood are right; that love must expel hate, and good will take the place of malice; that as well as zeal there must be patience and forbearance, and that persuasion is better than coercion; that none should hold the people in contempt, or profane the sacredness of conscience, or deny the worth of human life; and finally, that God and history are on the side of freedom and justice, love and righteousness; and man will therefore, be it soon or late, achieve a world society of peace and happiness where all are free and none shall be afraid.[24]

What is the health of societal religion in America? There are many opinions on this subject, but few data. For my part I do not know where data are to be had. The problem is so large, the present century so different, and the immediate situation so fluid that the collection of definitive data is next to impossible. Yet we are dealing here with the factor most important in determining national destiny.

The opinion is widespread that there is in America, and in the West generally, a dedication to the democratic way of life which is insufficient to carry through the crises of the coming decades. Whether the present dedication is more or less than that of previous centuries is irrelevant. Whether American culture is more or less integrated than formerly is also irrelevant. The question is: Does American culture in the second half of the twentieth century have enough integration, and do Americans have enough faith, courage and stamina to preserve what democracy they possess, to gain more, and to play a democratic role on the world stage where they now find themselves?

A pessimistic answer has much justification. The United States is large, the backgrounds of its citizens are various, the national development has been haphazard. The values most generally shared in American culture reflect a low rather than a high understanding of and faith in democracy. Americans are united in the conviction that the continuous bettering of the standard of living is an important goal. Most citizens agree that the capitalist system is superior to any alternative. And the overwhelming majority would unite in the national defense. How much further does American ideological integration go?

The will to survive is strong. But the will to survive and the will to preserve and increase the democratic way of life are not the same thing. Preparations for the national defense seem often to be made in defiance of democratic principles. In the attempt to overcome dictators, Americans often propose and adopt dictatorial ways. Americans do not even have a clear common conception of what the democratic ideal is. Frequently it is equated with such things as private enterprise, a system of intense competition, opposition to Communism, white supremacy, nationalism, the detection of subversion, resistance to change in political forms, and the way the boys do things down at the city hall.

A low level of public morals threatens the stability of democratic institutions. Political corruption is widespread in American cities. Racketeers exert a large influence. Politicians often disregard the ideals they are sworn to defend. The government is honeycombed by officials whose primary concern is service, not to the general public, but to some pressure group. Businessmen often expect to pay bribes before being awarded contracts. Tax evasion is widespread. Tax assessment is often dishonest. Some universities hire their football players. What wonder is it that many college students cheat on examinations and some college athletes accept bribes?

Corruption, cynicism, and a low level of ideological integration are menaces at any time; they are doubly so in a time of world revolution. The peril of our position is increased by the fact that the devotees of Communism evidence all the earmarks, not merely of religious dedication, but of religious fanaticism. Communist infiltration of neutralistic countries cannot be successfully countered by career-minded Americans who go abroad bearing ideals no stronger than the profit motive, and who won't go at all unless they can take along an electric refrigerator. The real battle is in these neutralist nations. Communist missionaries win many Asians and Africans to such devotion that they risk and sometimes give their lives in efforts to bring Communism to their country. Apart from the Peace Corps and missionaries sent by churches, the men we send are primarily persons with a business point of view. They appeal chiefly to native owners and managers, and in a heavy percentage of the cases have entrenched undemocratic regimes and greatly increased graft and corruption. A President of the United

States has said, "Our greatest adversary is not the Russians. It is in our own unwillingness to do what must be done."

Walter Lippmann has expressed the view that the integration which America and Western society once had around the values of democracy has been largely dissipated. He calls this core of values *the public philosophy,* and nowhere calls it a religion. And yet he often uses language in such a fashion as to indicate a public faith about *ultimate* values.

The democracies are ceasing to receive the traditions of civility in which the good society, the liberal, democratic way of life at its best, originated and developed. They are cut off from the public philosophy and the political arts which are needed to govern the liberal democratic society. They have not been initiated into its secrets, and they do not greatly care for as much of it as they are prepared to understand. . . .

The men of the seventeenth and eighteenth centuries who established these great salutary rules would certainly have denied that a community could do without a general public philosophy. They were themselves the adherents of a public philosophy—of the doctrine of natural law, which held that there was law 'above the ruler and the sovereign people . . . above the whole community of mortals.'* . . .

The public philosophy . . . is the premise of the institutions of the Western society, and they are, I believe, unworkable in communities that do not adhere to it. Except on the premises of this philosophy, it is impossible to reach intelligible and workable conceptions of popular election, majority rule, representative assemblies, free speech, loyalty, property, corporations and voluntary associations. . . . Increasingly, the people are alienated from the inner principles of their institutions. The question is whether and how this alienation can be overcome, and the rupture of the traditions of civility repaired. . . . For several generations it has been exceptional and indeed eccentric to use this philosophy in the practical discussion of public policies. . . .

The ancient world, we may remind ourselves, was not destroyed because the traditions were false. They were submerged, neglected, lost. For the men adhering to them had become a dwindling minority who were overthrown and displaced by men who were alien to the traditions, having never been initiated and adopted into them. May it not be that while the historical circumstances are obviously so different, something like that is happening again? . . .

The freedom which modern men are turned away from, not seldom with relief and often with enthusiasm, is the hollow shell of freedom. The current theory of freedom holds that what men believe may be important to them but that it has almost no public significance. The outer defenses of the free way of life stand upon the legal guarantees against

* Otto von Gierke, *Political Theories of the Middle Age.*

484 WHAT AMERICANS BELIEVE AND HOW THEY WORSHIP

the coercion of belief. But the citadel is vacant because the public philosophy is gone, and all that the defenders of freedom have to defend in common is a public neutrality and a public agnosticism. . . . To come to grips with the unbelief which underlies the condition of anomy, we must find a way to re-establish confidence in the validity of public standards. We must renew the convictions from which our political morality springs. . . . Given the practical need which is acute, and the higher generalities, which are self-evident, can we develop a positive working doctrine of the good society under modern conditions? The answer which I am making to this question is that it can be done if the ideas of the public philosophy are recovered and are re-established in the minds of men of light and leading. . . . Political ideas acquire operative force in human affairs when, as we have seen, they acquire legitimacy, when they have the title of being right which binds men's consciences. Then they possess, as the Confucian doctrine has it, "the mandate of heaven."[25]

I see no escape from the conclusion that, in the present world situation, America runs a grave danger from lack of attention to the spiritual core which is the heart of her national existence. If we are to avoid this danger, democracy must become an object of religious dedication. Americans must come to look on the democratic ideal (not necessarily American practice of it) as the Will of God or, if they prefer, the Law of Nature. They must strive for a common understanding of this ideal and for a devotion which taps the deepest motivations. The times demand a dedication which is comparable to Christian and Jewish dedication at their best, and superior to Communist or Fascist dedication; superior because it attempts mastery of self rather than of others. Such a dedication would cause Americans to give their lives if need be to preserve and enhance the democratic ideal and democratic practice wherever they are found, but would not use violence to coerce other peoples into the acceptance of democratic patterns.

Ignoring the lack of spiritual integration invites disaster. Relying on the haphazard methods of the past will not meet the need. Insisting that the problem has nothing in common with religion will mean that America may fall prey to the designs of totalitarian leaders who know how to quicken the religious passions through a mass appeal. The Nazi victory over the German mind and heart was deliberately engineered. Shinto was deliberately raised to a national Japanese cult. Our current Communist hysteria is probably the result of deliberate control of the means of communication.

If the nation is to achieve a vitalization of her spiritual center, it will be necessary to mobilize many agencies. The churches and synagogues are obviously first on the list. They already teach religion. They have experience and technical skill. They urge a high level of personal integrity. They have the loyalty of multitudes. Many church and synagogue leaders already make the teaching of democracy a major objective. But other religious leaders raise anguished objections to such a proposal.

One objection is that their religion teaches a universal ideal and democracy is a national ideal; their religion (so runs the argument) teaches a way of life for men all over the earth, under all types of government. Against this point of view it can be said that it is a mistake to suppose that democracy is merely a national ideal; democracy is a way of life for all men, the truest vision of social ethics which mankind has dreamed. Human beings can live happily under many kinds of social order, even under slavery, so great is our capacity for adaptation. But it is surely true that given something like equal opportunities to experiment with social orders, human beings everywhere will live more happily under personality-regarding social systems than under systems which treat humans as means. Democracy *is* superior to Communism, Fascism, monarchy, feudalism, slavery, industrial plutocracy, colonialism. The contention of some sectarian leaders that their faith must not be "encumbered by political ideology" is basically an aspect of the position that the function of religion is individual and not social salvation.

Another objection made by some sectarian leaders to the churches and synagogues teaching democracy as one item of their religious dogma ("It is the Will of God") is that religion is pulled down to the level of "mere ethics." This objection may take the form of contending that man's eternal salvation is so important that his salvation in this life shrinks into insignificance; or it may take the form of declaring that the most precious aspects of religion are such things as the mystical experience, reception of the sacraments, membership in a blessed fellowship, knowledge of God's revelation, study of the Scriptures. I have deep sympathy for this view. A societal religion focused on public morals and morale promotes merely a highest common denominator kind of religion, one that spiritually minded persons can soon transcend in their private and denominational experiences. The right and wisdom of sectarian efforts to help us all achieve levels higher than a religious experience of societal ethics should be unquestioned. But that is not the same thing as saying that focusing attention on these ethics is unnecessary or unimportant. The foundation of a house may not receive as much attention as the living room; nevertheless, the foundation must be kept in order. Teaching the basic values on which society is founded may appear to most churchmen to be elementary; but as every teacher knows, men must first master the elementary aspect of a subject before they can go on to higher achievement.

Objections from the churches and synagogues to teaching democracy as religion often take the form of an especially rigid exclusivism. It is the conviction of a multitude of orthodox groups that they alone have received religious truth. Accordingly they are religious imperialists. They claim dominion over all forms of the religious life and would of course deny the validity of the analysis of religion into such categories as private, denominational, and societal. They fear the societal religion and call it colorful names: "a fourth faith," "a form of nationalism," a "conspiracy" against Catholicism, Protestantism and Judaism, "a vague and somewhat sentimental religious syncretism," a "superreligion," "a particularly insidious kind of idolatry."[26] Such epithets stem from hasty reading or from an intense exclusivism which so blinds the reader he is unable to see whatever of truth may lie in propositions which challenge his cherished dogmas.

I see no escape from two basic facts about the nature of society: (1) a society is founded on a set of shared values; (2) in our kind of world some of these values must be religious if the society is to endure.

These, I say, are facts concerning the nature of our world. They cannot be disposed of either by ridicule or by indifference. They should not be deplored and fought against. They should be understood and used. Churchmen who persist in claiming that Christ must always be above culture are in fact trying to return to the bygone day when one church dominated society, the day when denominational and societal religion were merged; yet in a society religiously pluralistic, denominational exclusivism can apply only to denominational religion; the churches can properly take an exclusivist stance only before their own members and only in the denominational arena. This truth is the meaning of freedom of religion. Freedom of religion does not mean an ignoring of the societal religion, nor support of the claim that the religious values central to a society's life are only a private or a denominational concern.

Nor does freedom of religion mean that the denominations can have the whole field of religion to themselves. The exclusivists would like it that way. This hope is the genesis of the name calling which dubs the societal religion "a fourth faith." But societal religion is not another denomination among the many denominations; it is a common element of faith among persons of all the denominations—and of *no* denomination.

The term *syncretism* is also misapplied. Societal religion does not mean choosing a little from this denomination and a little from that and creating a new and competing denomination. Historically societal religion antedated denominational religion. Today societal religion creates the framework in which denominations work.

The term *superfaith* as applied to the societal religion results from a careless reading. In a pluralistic society, the societal religion is simply a common faith, not a superfaith; the denominations have the protection of the societal religion in going their distinctive ways, and are not supervised by the societal religion.

The accusation that societal religion is a form of nationalism has some substance. The most vital large-scale societies in the modern world are nations. Until such time as a more closely knit international society shall emerge, nations will continue to be the dominant agencies in the world. Thus national societal religion is inescapable. And yet a major evil in the modern world is the kind of nationalism which has developed in most countries. This fact is but a challenge to develop for modern nations a higher type of societal religion than the faith which is now called nationalism. The evil in nationalism is not an excuse for denying the reality of the societal religion, nor for advancing exclusivist claims that one's own denomination can fulfill all the functions of societal religion.

The preceding paragraphs in addition to answering some of the attacks made by exclusivists on the conception of societal religion have, I hope, clarified the conception of societal religion and have shown how imperceptive is much of the attack which exclusivist writers make on the view that the health and vigor of the societal religion is basic to the welfare of the American people. These writers make the common error of the misplaced accusation. They are disturbed by the failure of the churches and synagogues to present a distinctive, sectarian message. This failure is frequent, even usual. Many churches have sunk to the level of the going societal religion. They have adopted the attitude of the public relations fraternity. They want to offend no one. Their values are not much different from the values of the rest of the respectable community. Professor Peter L. Berger has written: "The religious institution does not (perhaps one should say 'not any longer') generate its own values; instead, it ratifies and sanctifies the values prevalent in the general community."[27] This situation is unfortunate. But it will not be corrected by colorful attacks on the thesis that a societal religion exists or by trying to persuade the churches to ignore this religion or to fight against it.

The thesis I am here defending is that the churches and synagogues should not only promulgate their own denominational values, but in addition should support those broader values which are essential to the continuance and betterment of society as a whole. This support should take two forms. One is constructive criticism. The churches and synagogues should evaluate the societal religion, should bring it to judgment. They should be the conscience of society. They should be the prophets who challenge society to higher and higher values. The second form of support the churches and synagogues should give the societal religion is active promulgation of its essential values. In America the churches and synagogues should teach faith in democracy as one item of their creed.

But they could not carry the whole load of religious education for democracy. The state must be brought into the picture. Since an enduring state (as opposed to social anarchy) in the present world requires widespread faith in a set of values believed to be ultimate, the state itself cannot wisely permit inculcation of these values to be left to the haphazard ministrations of agencies whose primary purpose is something else. Governmental agencies must teach the democratic ideal *as religion*. The churches deal effectively with but half the population; the government deals with all the population. The churches receive but voluntary attention; the government may require attention. The resources of the churches are limited; the resources of governments are ample.

Vigorous objections arise to so untraditional a suggestion as conscious assumption by governments of religious functions even though they be limited to societal religion and do not deal with denominational religion. One objection would be that state interference in religious teaching would threaten our liberties, since the state necessarily would deal in indoctrination. This objection fails to recognize that absolute liberty in religion is as impossible as is absolute liberty in economics, or in education, or in any other area of basic human need. At those points where religion is wholly or chiefly a private or denominational affair, the historic interpretation of freedom of religion is a most essential aspect of American tradition (though in practice we tend to have freedom, not for children, but for parents and sects). But at those points where religion is a public matter, those areas which contain the ethical propositions essential to corporate welfare, society will only at its peril allow individuals and sects to indulge their dogmatic whims. Systematic and universal indoctrination is essential in the values on which a society is based, if that society is to have any permanence or stability. The only way we can preserve our liberties in private and denominational religion is to forgo some liberties in societal religion.

Another objection is that indoctrination is contrary to the democratic ideal. This statement is surely in error. Indoctrination, as such, is contrary to anarchy, but not to democracy. When a more perfect conception of the democratic ideal shall have been achieved, it will surely include the systematic inculcation of whole populations with the ideal itself, since the absence of such indoctrination would mean the destruction of the ideal. Democratic living is not following out a preconceived principle to a logical conclusion, *reductio ad absurdum*, but rather meeting the practical problems of living.

A further objection is that giving the power of wholesale religious indoctrination into the hands of politicians would be a very dangerous business. The power of governmental authorities is already too great; making them priests as well as kings could prove disastrous. The cogency of this contention cannot be denied. I see no definitive answer. Yet stressing the undoubted danger neglects the greater danger of the alternative. If America fails to quicken the democratic faith, dedication and behavior of her people, she will probably lose most of the democracy she already has. Unless the basic values at the center of American culture attain the level of religious convictions, they lack staying power, and society is menaced by every demagogue who manifests the power to awaken religious fervor.

Moreover, there are reassuring considerations. (1) The ideal of democracy is itself antithetical to the usurpation of governmental power by an individual or by a group. The more successful indoctrination in a high type of democratic ideal proved to be, the greater the likelihood that America would possess a multitude of dedicated individuals who would recognize threats to the democratic order and rise to meet them. (2) The deeper the understanding and dedication to democracy the greater the likelihood that society would root out the kind of social injustice and inequality which furnishes the opportunity for the demagogue. (3) Ways could be found through which to protect society from the assumption of religious functions by the executive branches of governments; primary responsibility for teaching democracy as religion might very well be given to other agencies of the government—the public school, for example, a thesis I have defended in my *The New Education and Religion*.[28] No agency in this country is in as strategic a position as is the public school when it comes to teaching democracy. Of course, no agency is in as strategic a position to arouse religious devotion to any other social philosophy—Fascism for example—as is the public school, and doubtless that obligation will be forced upon it if the majority of our people come to possess something less than a religious devotion to the democratic way of life.

Public officials, especially the schoolmen, almost without exception applaud the proposition that one of their chief functions is to teach democracy. But their universal reaction is "What's all the shouting about? We are doing that already." There is much truth in this contention. They talk about democracy in season and out of season. But let there be no mistake about it: democracy is not now treated as an item of *religious* faith, except accidentally and unsystematically. Public school efforts lack at least two elements which must be present before *religious* faith will be awakened: metaphysical sanctions and ceremonial reinforcement.

By the term *metaphysical sanctions* I mean an open indoctrination of the faith that the democratic ideal accords with ultimate reality, whether that reality be conceived in naturalistic or supernaturalistic terms. Americans must be brought to the conviction that democracy is the very Law of Life, and that conduct in accord with that Law will in the end prove more satisfying than conduct which runs counter to democracy. The thesis is that public agencies must find ways to present democracy as a metaphysical reality, without teaching a complete metaphysical system.

That it is possible to behave in such a fashion will come as a new thought to most readers. Yet surely the members of a community can agree on the absolute necessity of their living according to certain ethical principles, without agreeing on a philosophy which supports the principles. Already in the United States the many sects agree on a few basic moral ideals—the ethical portions of the Ten Commandments, for example—although the sects could not agree on any statement of the reasons why these ideals are ultimate in their demands. Governmental agencies can teach that democracy is a Law of Life, without evoking the specters of the naturalist-supernaturalist debate.

The second element which must be present before regard for democracy results in religious dedication is ceremonial reinforcement, the periodic revitalization of the democratic ideal. No set of values can long remain virile unless it is nourished by experiences of deep emotional power. An effective ceremonial of this type would include: (1) the recall and glorification of the set of values believed to have metaphysical sanction, (2) self-appraisal in the light of these values, (3) rededication to living according to the standards sanctioned by these values.

The churches and synagogues have no monopoly on this pattern of behavior. It is used by every effective religious agency. The great Nazi mass meetings were ceremonials of tremendous power. Many of our own public meetings, especially in time of war, are effective ceremonies of reinforcement, particularly if they appeal for patriotic service as an *ultimate* obligation. We can discover, although it would not be easy, how to conduct on public occasions nonsectarian meetings which would produce dedication and joy, and bring to the service of our democratic institutions steadfastness and the spirit of self-sacrifice.

The power of religious devotion has been abundantly illustrated in these pages; it is the most cogent agency at the service of modern leaders—whether they are Communist, Fascist or democratic. If we are to keep as much democracy as we have and gain more, if we are to preserve our freedom of private religious belief and worship, we must find ways to awaken in the hearts of multitudes of Americans a devotion to democratic ideals like the devotion given by ardent believers in every age to the traditional religions.

NOTES

CHAPTER I. *Preview*

1 *A Study of History* (Oxford, 1939), IV, 119 ff.
2 *Ibid.*, V, 339.
3 Quoted by Paul Van Dyke, *Ignatius Loyola* (Scribner's, 1927), p. 168.

CHAPTER II. *The Roman Catholic Church*

1 Michael Williams, *The Catholic Church in Action* (Macmillan, 1935), pp. 44–45.
2 May 12, 1946, p. 3.
3 *The New York Times*, June 3, 1957, p. 32.
4 Quoted by E. Boyd Barrett in *Journal of Religion*, IX, 22.
5 *Essays of a Catholic* (Macmillan, 1931), p. 78.
6 *The Pilot* (Boston), Mar. 10, 1951, p. 4.
7 Claris Edwin Silox and Galen M. Fisher, *Catholics, Jews, and Protestants* (Harper, 1934), p. 23.
8 Camille M. Cianfarra, *The Vatican and the War* (Dutton, 1944), p. 60.
9 II, 581.
10 *Apologetics* (Herder, 1938), pp. 73–74.
11 Bertrand L. Conway, *The Question Box* (Paulist Press, 1929), pp. 39–40.
12 *Ibid.*, p. 47.
13 *The New York Times*, Oct. 4, 1958, p. 18.
14 P. J. Toner in *Catholic Encyclopedia*, V, 711.
15 *New York Herald Tribune*, Aug. 21, 1949.
16 Francis L. Filas in *Our Sunday Visitor*, Mar. 17, 1946, p. 12.
17 *Ibid.*, editorial, May 26, 1946, p. 11.
18 See Matt. 12:46, 13:55–56; Gal. 1:19.
19 Filas, *loc. cit.*
20 J. A. McHugh in *Catholic Encyclopedia*, VIII, 551.
21 Sidney A. Raemers, trans., J. Berthier, *A Compendium of Theology* (Herder, 1932), II, 74.
22 Quoted by C. Anderson Scott, *Romanism and the Gospel* (Westminster, 1946), p. 116.
23 *Ibid.*, p. 117.
24 *The Treasures of the Mass* (Benedictine Convent of Perpetual Adoration, 1944), p. 65.
25 Leo R. Ward, *Catholic Life, U.S.A. Contemporary Lay Movements* (Herder, 1959), p. 26.
26 George W. Casey, "The Vernacular Movement," *The Pilot*, Oct. 15, 1960, p. 4.

27   Conway, *op. cit.*, p. 220.
28   Berthier, *op. cit.*, II, 72.
29   J. Elliot Ross in *Religions of Democracy* by Louis Finkelstein, J. Elliot Ross, William Adams Brown (Devin-Adair, 1943), p. 127.
30   *Ibid.*, p. 130.
31   Donald Attwater, ed., *The Catholic Encyclopædic Dictionary* (Macmillan, 1931), p. 408.
32   *Op. cit.*, pp. 295–97.
33   LXI, 823–24.
34   *The New York Times*, Sept. 18, 1957, p. 1.
35   Howard M. Bell, *Youth Tell Their Story* (American Council on Education, 1938), p. 21.
36   Judson T. Landis, "Marriages of Mixed and Non-mixed Religious Faith," *American Sociological Review*, XIV (1949), 401 ff. See also H. Ashley Weeks, "Differential Divorce Rates by Occupations," *Social Forces*, XXI (1943), p. 336.
37   Paul Blanshard, *American Freedom and Catholic Power* (Beacon Press, 1949), p. 166.
38   Purnell Handy Benson, *Religion in Contemporary Culture* (Harper, 1960), p. 738. Benson cites J. L. Thomas.
39   *Our Sunday Visitor*, July 1, 1945, Religious Section, p. 5.
40   Leland Foster Wood, *If I Marry a Roman Catholic* (Federal Council of Churches, 1945), pp. 12, 14, 21.
41   *Time*, Nov. 4, 1946, p. 71.
42   Conway, *op. cit.*, pp. 335 f.
43   *Ibid.*, p. 340.
44   Gerald C. Treacy, *Sex—Sacred and Sinful* (Paulist Press, 1941), p. 14.
45   CXII, 265.
46   *The New York Times*, April 11, 1960, p. 1.
47   *The Pilot*, May 21, 1960, p. 18.
48   Quoted by John von Rohr from a statement by the Roman Rota (1944), in "Christianity and Birth Control," *The Christian Century*, Sept. 28, 1960, pp. 1115 f.
49   Personal correspondence with the Federation. See also Eugene J. Kanin, "Value Conflicts in Catholic Device-Contraceptive Usage," *Social Forces*, XXXV, Mar., 1957, 238 ff.
50   Berthier, *op. cit.*, II, 482.
51   Conway, *op. cit.*, p. 340.
52   Kanin, *op. cit.*, p. 238.
53   *Fortune*, Aug., 1943, p. 24.
54   Ronald Freedman, Pascal K. Whelpton, Arthur A. Campbell, *Family Planning, Sterility, and Population Growth* (McGraw-Hill, 1959), p. 10.
55   *Ibid.*, p. 155.
56   *Ibid.*, pp. 183, 180.
57   *Ibid.*, p. 185.
58   May 10, 1946, p. 85. See also Albert J. Mayer and Sue Marx, "So-

cial Change, Religion, and Birth Rates," *American Journal of Sociology*, Jan., 1957, pp. 383 ff.

59 Editorial, CII, Jan. 16, 1960, 444.
60 Article by a Passionist, *Our Sunday Visitor*, Jan. 13, 1946, p. 6.
61 *Christianity Today*, July 4, 1960, p. 840.
62 Conway, *op. cit.*, pp. 315–316.
63 *Catholic Encyclopedia*, XII, 750, 752.
64 Allen Sinclair Will, *Life of Cardinal Gibbons* (Dutton, 1922), p. 661.
65 *Ibid.*, p. 48.
66 *Ibid.*, p. 53.
67 Pp. 119 and xvi f.
68 Quoted by Will, *op. cit.*, pp. 245, 309.
69 Quoted by *ibid.*, p. 314.
70 Quoted by *ibid.*, p. 712.
71 Quoted by *ibid.*, p. 359.
72 *Ibid.*, p. 569.
73 *Ibid.*, p. 926.
74 P. xvii.
75 *Catholic Encyclopedia*, X, 342.
76 John J. Foley, *The Pilot*, July 30, 1960, p. 1.
77 Carmelite Fathers, *Our Lady's Promise* (1942).
78 IX, 390.
79 *Our Sunday Visitor*, Jan. 27, 1946, p. 6.
80 *Ibid.*, Feb. 3, 1946, p. 6.
81 Conway, *op. cit.*, pp. 67–68.
82 Quoted editorially by *The Christian Century*, Feb. 23, 1949, p. 228, from Orestes Brownson, *Growth and Development of the Catholic School System in the United States.*
83 *Our Sunday Visitor*, Aug. 17, 1947, p. 12.
84 *The Pilot*, March 4, 1961, p. 7.
85 *Ibid.*, July 2, 1960.
86 *Ibid.*, July 16, 1960, p. 6.
87 James Deakin, "Very, Very Educational," *New Republic*, Aug. 7, 1961, p. 13.
88 News letter from Commission on Law and Social Action, American Jewish Congress, Dec. 29, 1961.
89 *The Christian Century*, Oct. 27, 1948, p. 1154.
90 *Time*, June 20, 1960, p. 40.
91 Emmett McLoughlin, *American Culture and Catholic Schools* (Lyle Stuart, 1960), p. 184.
92 Anne Roe, *The Making of a Scientist* (Dodd, Mead, 1952). For the quoted phrases, see pp. 23, 231.
93 John Tracy Ellis, "The American Catholic and the Intellectual Life," *Thought*, Autumn, 1955; cited by Thomas F. O'Dea, *American Catholic Dilemma: An Inquiry into the Intellectual Life* (Sheed and Ward, 1958), pp. 5 f.

94   *Yearbook of American Churches, Edition for 1962* (National Council of Churches of Christ in U.S.A., 1961), pp. 247 ff.
95   Bureau of the Census, *Current Population Reports*, Series P-20, No. 79, Feb. 2, 1958.
96   *The World Almanac, 1961* (*New York World-Telegram*, 1961), pp. 712, 466.
97   Henry P. Van Dusen in "Letters to the Times," *The New York Times*, May 25, 1956, p. 22.
98   *Our Sunday Visitor*, Aug. 22, 1948, p. 2.
99   *Yearbook of American Churches, Edition for 1962*, National Council of the Churches of Christ in the U.S.A., pp. 248, 277 f.
100 *The Christian Century*, July 30, 1947, p. 932.
101 *Methodism's 1960 Fact Book*, Albert C. Hoover, ed. (Statistical Office of The Methodist Church, 1960), p. 139.
102 *The Protestant Church as a Social Institution* (Harper, 1935), p. 10.
103 John C. Bennett, *Christianity and Crisis*, Sept. 19, 1960, p. 125.
104 *Social Action*, Jan. 15, 1948, p. 16.
105 George P. Howard, *Religious Liberty in Latin America* (Westminster, 1944), chap. i.
106 Renwick C. Kennedy in *The Christian Century*, June 5, 1946, p. 717.
107 C. C. Cawley, "The Outlaws," *The Christian Century*, Apr. 3, 1957, pp. 420 ff.
108 *Op. cit.*, p. 86.
109 *The Christian Century*, Mar. 9, 1949, p. 308.
110 *Ibid.*, June 8, 1949, p. 701.
111 Harold E. Fey, *Can Catholicism Win America?* reprint from *The Christian Century*, p. 7.
112 *Peace Notes*, Fellowship of Reconciliation, Nov. 2, 1949, p. 1.
113 Paul Blanshard, *My Catholic Critics* (Beacon Press, 1952), p. 50.
114 "An Advertisement of the Beacon Press," *The Christian Century*, Feb. 1, 1956, p. 151.
115 Thomas Sugrue, "What Happens When a Catholic Speaks His Mind," *Advance*, May 26, 1952, p. 6.
116 Blanshard, *op. cit.*, p. 199.
117 S. E. W., "Elmer Gantry," *Christianity Today*, Aug. 1, 1960, p. 918.
118 John A. Ryan and Francis J. Boland, *Catholic Principles of Politics* (Macmillan, 1940), p. 316 f.
119 Quoted by George H. Dunne, *Religion and American Democracy* (America Press, 1949), p. 42.
120 See, for example, *Protestant Voice*, Jan. 10, 1949.
121 *The Christian Century*, June 24, 1959, p. 741.
122 William Attwood, *The Nation*, Aug. 28, 1948, p. 223.
123 *Christianity Today*, Sept. 26, 1960, p. 1058.
124 *Op. cit.*, p. 41.
125 Dunne, *op. cit.*, p. 27.

126   Charles R. Andrews, "A Catholic President: Pro," *The Christian Century*, Oct. 26, 1960, p. 1241.
127   Quoted by *Christianity Today*, June 20, 1960, p. 799.
128   Daniel J. Callahan, "Freedom and Authority in Roman Catholicism," *Christianity and Crisis*, Oct. 3, 1960, p. 136.
129   *America*, June 25, 1960, p. 390.
130   *The Commonweal*, June 3, 1960, p. 245.
131   Quoted by Robert C. Hartnett, *Federal Aid to Education* (America Press, 1950), p. 36.
132   Quoted by *ibid.*, p. 37.
133   Ryan and Boland, *op. cit.*, p. 320.
134   *The Christian Century*, Aug. 24, 1960, p. 982.
135   A. F. Carrillo de Albornoz, *Roman Catholicism and Religious Liberty* (World Council of Churches, 1959), p. 8.
136   Quoted by Kenneth W. Underwood, *Protestant and Catholic, Religious and Social Interaction in an Industrial Community* (Beacon, 1957), pp. 154 f.
137   Gregory Baum, *The Commonweal*, Dec. 27, 1963, p. 396.
138   *Christianity and Crisis*, Dec. 9, 1963, p. 231.
139   Francis J. Connell in *The American Ecclesiastical Review*, July 1944, p. 108.

CHAPTER III. *Protestantism*

1   Alfred Fawkes, *Encyclopedia of Religion and Ethics*, IX, 621.
2   *The Rise of Modern Religious Ideas* (Macmillan, 1915), p. 2.
3   Roland H. Bainton, *The Church of Our Fathers* (Scribner's, 1941), p. 167.
4   Theodore Maynard, *The Story of American Catholicism* (Macmillan, 1942), p. 91.
5   Winthrop S. Hudson in personal correspondence.
6   *What is Faith?* (Macmillan, 1925), p. 216.
7   *Ibid.*, p. 248.
8   *Ibid.*, pp. 249, 250.
9   Quoted by Gaius Glenn Atkins, *Religion in Our Times* (Round Table Press, 1932), p. 251.
10   *The New York Times*, Mar. 13, 1960, p. 62.
11   Edward John Carnell, "Post-Fundamentalist Faith," *The Christian Century*, August 26, 1959, p. 971.
12   *Christianity Today*, Oct. 10, 1960, p. 27.
13   David E. Adams, *Man of God* (Harper, 1941), p. 120.
14   *My Idea of God* (Little, Brown, 1927), pp. 237–40.
15   "Our Christ." Used by permission of Mrs. Harry Webb Farrington.
16   Bruno Lasker, *Race Attitudes in Children* (Holt, 1929).
17   Margaret Mead, *Cooperation and Competition Among Primitive Peoples* (McGraw-Hill, 1937).

498 NOTES

18 Margaret Mead, *Sex and Temperament in Three Primitive Societies* (Morrow, 1935).

19 L. Harold DeWolf, *The Case for Theology in Liberal Perspective* (Westminster Press, 1959), p. 11. Used by permission.

20 Reinhold Niebuhr, *The Nature and Destiny of Man*, 1941, I, 165.

21 W. Burnet Easton, Jr., *The Faith of a Protestant* (Macmillan, 1946), p. 20.

22 *Loc. cit.*

23 John C. Bennett, "Professor's Column," *Alumni Bulletin of Union Theological Seminary*, Jan. 1956.

24 *The New York Times*, Apr 6, 1959, p. 8.

25 *Christianity Today*, Mar. 28, 1960, p. 27.

26 Bainton, *op. cit.*, p. 191.

27 *The Works of the Right Honourable Edmund Burke* (Bell & Daldy, 1869), I, 466.

28 Ernest Sutherland Bates, *American Faith* (Norton, 1940), p. 9.

29 Herbert Parrish, *The Atlantic Monthly*, Mar., 1927, p. 303.

30 Alfred Kramer, "Racial Integration in Three Protestant Denominations," *The Journal of Educational Sociology*, Oct., 1954, pp. 59 ff.

31 Nancy Lawrence, "Racially Inclusive Churches—a Progress Report," *The Crisis*, Mar., 1955, p. 140.

32 John Wicklein, *The New York Times*, July 8, 1959, p. 20.

33 Vance Packard, *The Status Seekers* (David McKay, 1959), p. 196.

34 *Information Service*, Federal Council of Churches, May 15, 1948, p. 3.

35 Appraisals of the Ecumenical Era," *Information Service*, Feb. 6, 1960, p. 1.

36 Maynard L. Cassady, *Crozer Quarterly*, Jan., 1949, p. 18.

37 Sept. 24, 1947, p. 8.

38 *Religious Education*, Nov., 1947, p. 324. Concerning Roman Catholic eligibility for the National Council see: Peter Day, "The National Council of Churches: An Evaluation," *Christianity and Crisis*, May 16, 1960, p. 69.

39 Lloyd J. Averill, "In Defense of Christian Pluralism," *The Christian Century*, June 1, 1960, pp. 666, 665.

40 *Judaism as a Civilization* (Macmillan, 1935), p. 123.

41 R. E. Wolseley, "The Plight of Religious Journalism," *Crozer Quarterly*, XXIII, p. 217.

42 Hornell Hart in *Recent Social Trends* (McGraw-Hill, 1933), pp. 382–442.

43 Alfred McClung Lee in *Annals of the American Academy of Political and Social Science*, Mar., 1948, p. 121.

44 Wolseley, *op. cit.*, pp. 218, 220.

45 Kenneth Scott Latourette, *Toward a World Christian Fellowship* (Association Press, 1938), p. 42.

46 Kenneth Scott Latourette, *A History of the Expansion of Christianity* (Harper, 1945), VII, chap. xvi.

47 Quoted by Charles Henry Robinson in *An Outline of Christianity*, Shailer Mathews, ed. (Dodd, Mead, 1926), III, 464.
48 *Information Service*, Feb. 4, 1961.
49 M. Searle Bates, "What Are Missionaries Really Doing?" *Union Seminary Quarterly Review*, May 1960, pp. 320 ff.
50 Kenneth Grubb, *World Christian Handbook, 1957 Edition* (World Dominion Press, 1957), p. XXI.
51 Alan Walker, "For a World Christian Mission," *The Christian Century*, Oct. 11, 1961, p. 1203.
52 News Release for June 4, 1949, by Missions Public Relations. Data based on *World Christian Handbook*, Kenneth G. Grubb, ed. (World Dominion Press, 1949).
53 *Essays of a Catholic* (Macmillan, 1931), p. 128.
54 *Ibid.*, p. 318.
55 Herbert Parrish in *Atlantic*, Mar., 1927, p. 302.
56 *Protestantism*, William K. Anderson, ed. (The Methodist Church, 1944), p. 270.
57 *Ibid.*, p. 218.
58 H. Paul Douglass and Edmund deS. Brunner, *The Protestant Church as a Social Institution* (Harper, 1935), chap. ii.
59 Compare *The Canada Year Book 1946* (Dominion Bureau of Statistics), p. 107, with the *Year Book of American Churches, 1943 Edition* (Federal Council of Churches), p. 94.
60 Peter L. Berger, "The Second Children's Crusade," *The Christian Century*, Dec. 2, 1959, p. 1399.
61 "Protestantism in a Post-Protestant America," *Christianity and Crisis*, Feb. 5, 1962, 3.

CHAPTER IV. *Lutheran Churches*

1 William H. T. Dau, "What is Lutheranism?" in Vergilius Ferm, ed., *What is Lutheranism?* (Macmillan, 1930), p. 207.
2 Theodore G. Tappert and John W. Doberstein, trans., *The Journals of Henry Melchoir Muhlenberg* (Muhlenberg Press, 1942), I, 28.
3 *Ibid.*, p. 54.
4 Quoted by M. L. Stoever, *A Memoir of the Life and Times of Henry Melchoir Muhlenberg* (G. W. Frederick, 1883), p. 55.
5 Tappert and Doberstein, *op. cit.*, I, 183.
6 Stoever, *op. cit.*, p. 65.
7 J. L. Neve, *History of the Lutheran Church in America* (Lutheran Literary Board, 1934), p. 56.
8 Tappert and Doberstein, *op. cit.*, pp. 721–22.
9 Dau, *op. cit.*, p. 211.
10 George Park Fisher, *History of the Christian Church* (Scribner's, 1936), pp. 421–22.
11 Abdel Ross Wentz, "What is Lutheranism?" in Vergilius Ferm, ed., *op. cit.*, p. 92.
12 Quoted by *The Christian Century*, editorial, Oct. 19, 1938, p. 1252.

13 Quoted by T. Benton Perry in *The Christian Century*, May 8, 1940, p. 606.

14 "20,870 Clergymen on War and Economic Injustice," *The World Tomorrow*, May 10, 1934, XVII, No. 10.

15 R. L. Moellering, "Rauschenbusch in Retrospect," *Concordia Theological Monthly*, Aug., 1956, p. 614.

16 Sept., 1951, pp. 2 ff.

17 Robert K. Menzel, "Lutheran Church—Missouri Synod," *The Christian Century*, July 15, 1959, p. 834.

18 William A. Kramer, "Are Our Schools Worth the Cost?" *Advance*, Sept., 1960, p. 6.

19 Willmar L. Thorkelson in *The Christian Century*, Aug. 10, 1960, p. 933.

20 Theodore A. Gill, "Missouri in Motion," *The Christian Century*, July 18, 1956, p. 846.

21 George Dugan, "Problems Raised by Heresy Trials," *The New York Times*, Feb. 12, 1956, p. 85.

22 Theodore A. Gill, "Lutheran Answers Leave Questions," *The Christian Century*, Feb. 8, 1956, pp. 167 f.

CHAPTER V.   *The Protestant Episcopal Church*

1 William Alva Gifford, *The Story of the Faith* (Macmillan, 1946), p. 371.

2 *Ibid.*, p. 375.

3 Roland F. Palmer, *The Living Church*, Feb. 10, 1946, p. 11.

4 John Wicklein, "Clerics Polled on Virgin Birth," *The New York Times*, July 10, 1960, p. 76; *Christianity Today*, Oct. 10, 1960, p. 28.

5 James A. Pike, "Three-pronged Synthesis," *The Christian Century*, Dec. 21, 1960, pp. 1496 f.

6 *The New York Times*, Jan. 29, 1961, p. 31.

7 Marshall M. Day, *The Living Church*, Feb. 16, 1947, p. 4.

8 Bradford Young in *The Living Church*, Jan. 5, 1947, p. 13.

9 Alec R. Vidler in *The Witness*, Apr. 29, 1948, p. 9.

10 Young, *loc. cit.*

11 Charles D. Kean in *The Witness*, June 12, 1947, p. 3.

12 Angus Dun, *Prospecting for a United Church* (Harper, 1948), p. xi.

13 Vidler, *op. cit.*, p. 8.

14 Palmer, *op. cit.*, p. 10.

15 Peter Day, "The National Council of Churches: An Evaluation," *Christianity and Crisis*, May 16, 1960, p. 70.

16 Nov. 6, 1946.

17 Quoted by Alexander C. Zabriskie, *Bishop Brent, Crusader for Christian Unity* (Westminster Press, 1948), p. 27.

18 *Ibid.*, p. 122.

19 *Ibid.*, pp. 169–170.

20 Quoted by Dun, *op. cit.*, p. 96.

21 Quoted by Zabriskie, *op. cit.,* p. 169.
22 Quoted by *ibid.,* p. 163.
23 John Howard Melish in *The Churchman,* Nov. 1, 1948, p. 10.
24 *The Lutheran,* Apr. 3, 1946, p. 5.
25 From an editorial in *The Presbyterian Tribune,* quoted in a news article by *The Witness,* Jan. 9, 1947, p. 3.
26 *The Witness,* Oct. 10, 1946, p. 9.
27 Quoted in a news article in *The Living Church,* Feb. 15, 1948, p. 9.
28 *The Living Church,* Dec. 6, 1959, p. 14.
29 "No Reordination Please," *The Living Church,* Mar. 6, 1960, p. 7.
30 *The Christian Century,* Dec. 6, 1961, p. 1454.
31 James A. Pike, "That They May Be One," *The Christian Century,* Jan. 13, 1960, p. 47.

CHAPTER VI. *Presbyterian Churches*

1 Unidentified author quoted by William Thomson Hanzsche, *The Presbyterians* (Westminster, 1934), p. 37.
2 William Thomson Hanzsche, *Our Presbyterian Church* (Board of Christian Education of the Presbyterian Church, U.S.A., 1939), p. 15.
3 Varnum Lansing Collins, *President Witherspoon* (Princeton University Press, 1925), I, 11.
4 Quoted by *ibid.,* II, 168.
5 Quoted by *ibid.,* II, 184.
6 Edmund Cody Burnett, *The Continental Congress* (Macmillan, 1941), p. 425.
7 Collins, *op. cit.,* II, 121.
8 *Ibid.,* I, 4.
9 *Ibid.,* II, 223.
10 Quoted by Lewis Bevens Schenck, *The Presbyterian Doctrine of Children in the Covenant* (Yale University Press, 1940), p. 119, from Henry B. Smith, *System of Christian Theology,* 1884.
11 John Paul Jones, "Does a Little Dishonesty Matter?" unpublished typescript.
12 *The Christian Century,* Apr. 2, 1958, p. 396.
13 Harold E. Fey, "171st GAUPCUSA Meets," *The Christian Century,* June 10, 1959, p. 694.
14 John H. McComb as reported in *The New York Times,* Feb. 17, 1947, p. 14.
15 D. P. McGeachy in *The Christian Century,* June 24, 1942, pp. 817–18.
16 See "20,870 Clergymen on War and Economic Injustice," *World Tomorrow,* May 10, 1934.
17 William H. Hudnut, Jr., in *The Christian Century,* Apr. 26, 1944, pp. 522–23.
18 Theodore A. Gill, "Grown-Up Christian Education," *Presbyterian Life,* Oct. 15, 1960, p. 44.

CHAPTER VII.   *The United Church of Christ*

1   William E. Barton quoted by Gaius Glenn Atkins and Frederick L. Fagley, *History of American Congregationalism* (Pilgrim Press, 1942), p. 299.
2   R. W. Dale, *History of English Congregationalism* (Hodder and Stoughton, 1907), p. 5. See also Douglas Horton, *Congregationalism, A Study in Church Polity* (Independent Press, 1952), p. 9; and Albert Peel and Douglas Horton, *International Congregationalism* (Independent Press, 1949), p. 7.
3   Bradford's *History of the Plymouth Settlement*, rendered into modern English by Harold Paget (Dutton, 1920), p. 8.
4   *Ibid.*, p. 20.
5   Charles M. Andrews, *The Colonial Period of American History* (Yale, 1934), I, 254.
6   Bradford, *op. cit.*, p. 58.
7   *Ibid.*, p. 77.
8   *Ibid.*, pp. 90, 93.
9   "Winslow's Relation," *Chronicles of the Pilgrim Fathers* (Everyman's Library, Dutton, 1910), p. 336.
10 *Op. cit.*, p. 320.
11 Ernest Sutherland Bates, *American Faith* (Norton, 1940), p. 126.
12 See Alice M. Baldwin, *The New England Clergy and the American Revolution* (Duke University Press, 1928).
13 William Warren Sweet, *The Story of Religion in America* (Harper, 1950), p. 211.
14 Atkins and Fagley, *op. cit.*, p. 147.
15 June 16, 1937, p. 782.
16 Russel Henry Stafford in *The Christian Century*, LV, 841.
17 Malcolm K. Burton, *What Really Happened at Oberlin*, leaflet, Aug. 20, 1948, pp. 9, 4.
18 Edith Helen Wolfe, "Petticoats, Preachers and Prejudice," *Advance*, May 16, 1956, p. 11.
19 Barbara Graymont, in "Letters," *United Church Herald*, Mar. 31, 1960, p. 32.
20 Thomas Alfred Tripp in *Advance*, Jan., 1949, p. 16.
21 Atkins and Fagley, *op. cit.*, p. 342.
22 James E. Wagner, *The Evangelical and Reformed Church*, reprint from the *National Council Outlook*, Jan., 1956.
23 *Ibid.*
24 *Advance*, Jan. 31, 1958, p. 7.
25 *Christianity Today*, July 20, 1959, p. 24.
26 Henry David Gray, *Christianity Today*, July 20, 1959, p. 25.
27 Howard Conn, "Our Historic Fear of Ecclesiasticism," *The Congregationalist*, June, 1959, p. 4.
28 Henry David Gray, "The Sin in Ecumenicity," *The Congregationalist*, July, 1960, p. 8.

29 Douglas Horton, "Now the United Church of Christ," *The Christian Century*, June 12, 1957, p. 734.

CHAPTER VIII.   *Baptist Churches and Christian (Disciples) Churches*

1   W. E. Garrison and A. T. DeGroot, *The Disciples of Christ, a History* (Christian Board of Publication, 1948), p. 38.

2   Henry C. Vedder, *A Short History of the Baptists* (American Baptist Publication Society, 1941), p. 183.

3   Samuel Hugh Brockunier, *The Irrepressible Democrat—Roger Williams* (Ronald, 1940), pp. 47–48, 53–55, 78–79.

4   Charles M. Andrews, *The Colonial Period of American History* (Yale, 1934), I, 358n.

5   Quoted by Frederick L. Schuman, *New Republic*, Sept. 26, 1949, p. 4.

6   James Ernst, *Roger Williams* (Macmillan, 1932), pp. 283–84.

7   Ernest Sutherland Bates, *American Faith* (Norton, 1940), p. 131.

8   *Publications of the Narragansett Club*, III, 249–50.

9   Quoted by Ernst, *op. cit.*, pp. 345, 436, 244, 435, 245.

10 Vedder, *op. cit.*, p. 306.

11 Quoted by W. W. Sweet, *The Story of Religion in America* (Harper, 1950), p. 183.

12 Garrison and DeGroot, *op. cit.*, p. 60.

13 Conrad H. Moehlmann, "Baptists and Their Allied Groups," in *An Outline of Christianity*, III, 366.

14 Hillyer H. Straton in *The Christian Century*, Mar. 29, 1944, p. 394.

15 *Loc. cit.*

16 Garrison and DeGroot, *op. cit.*, p. 441.

17 William Roy McNutt, *Polity and Practice in Baptist Churches* (Judson Press, 1935), p. 5.

18 Straton, *op. cit.*, p. 394.

19 Editorial, *Missions*, June, 1944, p. 323.

20 Phillip J. McLean in *The Christian Century*, Sept. 6, 1944, p. 1018.

21 *Christianity Today*, June 6, 1960, p. 758.

22 S. W. Badgett, "Correspondence," *The Christian Century*, Jan. 24, 1951, p. 115.

23 Harold E. Fey, reporting the Convention in *The Christian Century*, May 31, 1944, p. 679.

24 William Coolidge Hart, "Southern Baptist Power and Portent," *The Christian Century*, June 3, 1959, p. 663.

25 *The Christian Century*, Dec. 5, 1956, p. 1432.

26 *Ibid.*, Nov. 23, 1960, p. 1384.

27 *Missions*, Dec., 1960, p. 9.

28 *The Christian Century*, Dec. 26, 1956, p. 1519.

29 Louis A. McCord, "What Do Southern Baptists Fear?" *The Christian Century*, Nov. 21, 1956, p. 1353.

30 *Ibid.*, June 11, 1958, p. 700.

504

NOTES

31 Harold E. Fey in *The Christian Century*, June 9, 1948, p. 577.
32 *Ibid.*, May 31, 1944, pp. 672, 678.
33 *Time*, May 18, 1953; quoted in *The Christian Century*, Nov. 21, 1956, p. 1353.
34 June 23, 1948, pp. 623–24.
35 McLean, *op. cit.*
36 William Coolidge Hart, *op. cit.*, p. 663.
37 Virgil E. Lowder, "Southern Baptists," *The Christian Century*, June 11, 1958, p. 702.
38 See Leslie G. Thomas, "History, Doctrine, and Organization of Churches of Christ," *Religious Bodies, 1936*, Bureau of the Census, II, Pt. 1, 469–70.
39 Quoted by Garrison and DeGroot, *op. cit.*, pp. 344–45.
40 *Christianity Today*, Nov. 21, 1960, p. 164.
41 Garrison and DeGroot, *op. cit.*, p. 440.
42 Harold E. Fey, "Disciples Now 'Christians,'" *The Christian Century*, Oct. 30, 1957, p. 1279.
43 Paul M. Harrison, *Authority and Power in the Free Church Tradition* (Princeton University, 1959), p. 60.
44 *Ibid.*, p. 92.
45 *Ibid.*, pp. 62, 79, 86 f., 117, 183.
46 *Ibid.*, p. 164.
47 *Ibid.*, p. 183.
48 *Ibid.*, pp. 157, 116.
49 *Ibid.*, p. 151.
50 F. F., "United Church Adopts Ambiguous Confession," *Christianity Today*, Aug. 3, 1959, pp. 32 f.
51 *Op. cit.*, p. 175.
52 Graham R. Hodges, "Letters," *United Church Herald*, June 9, 1960, pp. 31 f.
53 Douglas Horton, *Congregationalism, A Study in Church Polity* (Independent Press, 1952), p. 17.
54 *Advance*, Jan. 31, 1958, p. 7.
55 Editorial, *The Christian Century*, Nov. 21, 1951, p. 1333.
56 William Coolidge Hart, *loc. cit.*

CHAPTER IX. *The Quakers*

1 *The Journal of George Fox* (Dutton, 1924), pp. 3, 8, 9.
2 Elbert Russell, *The History of Quakerism* (Macmillan, 1942), p. 124.
3 William Warren Sweet, *The American Churches* (Epworth, 1947), p. 15.
4 Kenneth Ives, "Our Diminishing Society of Friends, Part 1," *Friends Journal*, Oct. 5, 1957, p. 645.
5 Robert O. Blood, Jr., "Seekers in Buddhist Clothing," *Friends Journal*, Jan. 1, 1961, p. 4.
6 Robert J. Leach referring to a statement by Gerald Heard, "Letter from Geneva," *Friends Journal*, Jan. 16, 1960, p. 37.
7 *Ibid.*, June 29, 1957, p. 428.

8 *Ibid.*, Jan. 1, 1961, p. 13.
9 Joseph Holmes Summers in *The Christian Century*, Sept. 22, 1943, p. 1068.
10 Quoted by Russell, *op. cit.*, p. 54.
11 Howard H. Brinton, *The Nature of Quakerism* [reprint of chap. ii of *Quaker Education in Theory and Practice*] (Pendle Hill, n.d.), p. 3.
12 See Russell, *op. cit.*, chap. vi and p. 351.
13 Douglas V. Steere in *Friends Intelligencer*, Third Month 20, 1937, p. 190.
14 *Handbook of the Religious Society of Friends* (Friends World Committee for Consultation, 1941).
15 Howard H. Brinton in *Friends Intelligencer*, First Month 24, 1948, p. 51.
16 Rufus M. Jones, *The Flowering of Mysticism* (Macmillan, 1939), p. 251.
17 *Op. cit.*, p. 17.
18 See James H. Leuba, *The Psychology of Religious Mysticism* (Harcourt, Brace, 1925), chap. ii.
19 James B. Pratt, *The Religious Consciousness* (Macmillan, 1921), pp. 404 f.
20 Steere, *op. cit.*, pp. 189 f.
21 Russell, *op. cit.*, p. 286.
22 Brenda Kuhn, First Month 17, 1948, p. 42.
23 Henry J. Cadbury in *The Christian Century*, Aug. 16, 1944, p. 954.
24 Harry Elmer Barnes, *The Repression of Crime* (Doran, 1926), p. 28. See also chap. iv.
25 Edith M. Lerrigo in personal correspondence.
26 Quoted by Merle Curti, *Peace or War* (Norton, 1936), p. 61.
27 Russell, *op. cit.*, pp. 516n., 517.
28 *Time*, Oct. 11, 1943, pp. 46–47.
29 Sidney Lucas in *Friends Intelligencer*, Fifth Month 29, 1948, p. 308.
30 *The Christian Century*, Jan. 2, 1957, p. 4.
31 *Ibid.*, Nov. 14, 1956, p. 1338.
32 Friends Peace Committee, Philadelphia, "If I Were Eighteen," *Friends Journal*, Mar. 30, 1957, p. 203.
33 Charles C. Walker, "On Vigil at Fort Detrick," *The American Friend*, June 13, 1960, pp. 180 f.
34 *Newsweek*, Nov. 10, 1947, p. 68.
35 *The Christian Century*, Feb. 19, 1958, p. 211.
36 Summers, *op. cit.*, pp. 1069–70.
37 See Kirby Page, "20,870 Clergymen on War and Economic Injustice," *The World Tomorrow*, May 10, 1934.

CHAPTER X. *The Methodist Church*

1 Quoted by Charles A. Beard and Mary R. Beard, *The Rise of American Civilization* (Macmillan, 1930), II, 399 f.
2 Editorial, *Life*, Nov. 10, 1947, p. 38.

3 John M. Moore, *The Long Road to Methodist Union* (Abingdon, 1943), p. 24.
4 See William Warren Sweet, *The American Churches* (Epworth, 1947), pp. 15–17. See also Halford E. Luccock and Paul Hutchinson, *The Story of Methodism* (Abingdon, 1949), p. 219.
5 Quoted by George Park Fisher, *History of the Christian Church* (Scribner's, 1936), p. 515.
6 Luccock and Hutchinson, *op. cit.*, p. 173.
7 *Ibid.*, p. 236.
8 Sweet, *op. cit.*, p. 14.
9 Sweet, *The Rise of Methodism in the West* (Methodist Book Concern, 1920), pp. 41 f.
10 Quoted by Luccock and Hutchinson, *op. cit.*, p. 230.
11 Quoted by Sweet, *The Rise of Methodism in the West*, p. 47.
12 Luccock and Hutchinson, *op. cit.*, p. 229.
13 *Doctrines and Discipline of The Methodist Church, 1948* (Methodist Publishing House), p. 32.
14 Quoted in *Methodism*, William K. Anderson, ed. (Methodist Publishing House, 1947), p. 128.
15 See George Herbert Betts, *The Beliefs of 700 Ministers* (Abingdon, 1926).
16 "The American Clergy and the Basic Truths," *Christianity Today*, Oct. 10, 1961, pp. 27 f.
17 S. Paul Schilling, *Methodism and Society in Theological Perspective* (Abingdon, 1960), p. 280.
18 J. Paul Williams, *Social Adjustment in Methodism* (Bureau of Publications, Teachers College, Columbia University, 1938), p. 14.
19 For a discussion of this tendency, see *ibid.*, chap. i–iii.
20 See *ibid.*, pp. 30 ff.
21 "The Constitution," *Doctrines and Discipline of The Methodist Church, 1948* (Methodist Publishing House), pp. 10, 16.
22 Gerald Kennedy, *The Christian Century*, March 30, 1960, p. 387.
23 *Discipline, 1948*, p. 10.
24 Quoted by J. Paul Williams, *op. cit.*, p. 29.
25 R. P. Marshall, "Trends in Modern Methodism," *Christianity Today*, Jan. 4, 1960, p. 266.
26 Quoted by J. Paul Williams, *op. cit.*, p. 29.
27 *The New York Times*, May 30, 1928, p. 38.
28 A. C. Hoover, ed., *Methodism's 1960 Fact Book* (Statistical Office of the Methodist Church), pp. 27, 36.
29 Raymond E. Balcomb, "Another Methodist on the 'Central,'" *The Christian Century*, Mar. 9, 1960, p. 287.
30 Gerald Kennedy, *The Christian Century*, Mar. 30, 1960, p. 387.
31 Charles C. Parlin, *Daily Christian Advocate*, Apr. 29, 1960, p. 80.
32 Walter G. Muelder, "Methodism and Segregation: A Case Study," *Christianity and Crisis*, Apr. 4, 1960, p. 42.
33 Ralph Lord Roy, "Church Race Policies Compared," *The Christian Century*, May 30, 1956, p. 665.

34 Michael Daves, "Methodism's 'Central' Problem," *The Christian Century*, March 2, 1960, p. 255.

35 *The Christian Century*, July 31, 1957, p. 909.

36 R. P. Marshall, "Trends in Modern Methodism," *Christianity Today*, Jan. 4, 1960, p. 266.

37 Quoted by Elmore Brown in *Zions Herald*, Feb. 15, 1950; p. 154.

38 *The Christian Advocate*, June 9, 1960, p. 3.

39 Arthur J. Moore, Jr., "The Methodist General Conference, 1960," *Christianity and Crisis*, May 30, 1960, p. 78.

40 *The Christian Advocate*, Sept. 18, 1947, p. 1192.

41 Albert C. Hoover in *The Christian Advocate*, Aug. 18, 1949, p. 1069.

42 *Together*, July, 1960, p. 72.

43 Peter H. Odegard, *Pressure Politics* (Columbia, 1928), p. 18.

44 *Ibid.*, pp. 21, 22.

45 Francis W. McPeek in *Social Action*, Apr. 15, 1950, p. 7.

46 Thomas Keehn and Kenneth Underwood in *Social Action*, June 15, 1950, p. 7.

47 Paul F. Douglass in *Zions Herald*, Feb. 1, 1950, p. 108.

48 *The Christian Century*, Dec. 10, 1958, pp. 1419 f.

49 Roger Burgess, "One in 12 Now Alcoholic!" *Together*, May, 1960, p. 6.

50 *Missions*, Feb., 1959, p. 11.

51 John Wicklein, "Methodists Ease Rigidity of Code," *The New York Times*, June 3, 1961, p. 10.

52 News item, *ibid.*, May 26, 1961, p. 22.

53 Kirby Page, "20,870 Clergymen on War and Economic Injustice," *The World Tomorrow*, May 10, 1934, pp. 230 f. "The questionnaire was sent to approximately 100,000 Protestant ministers and Jewish rabbis."

54 *12,854 Clergymen on War and Peace* (pamphlet, Methodist Bishop James C. Baker, chairman sponsoring committee, no publisher or date indicated). "Replies came in response to a questionnaire sent to about 100,000 Protestant and Jewish clergymen."

55 "The American Clergy and the Basic Truths," *Christianity Today*, Oct. 10, 1960, p. 28.

56 D. E. Lindstrom and W. W. Riffe, *The Effectiveness of Methodist Teachings* (Department of Agricultural Economics, University of Illinois, n.d., 1959 or later), pp. 17 f.

57 Jasper E. Crane in *Faith and Freedom*, May, 1950, pp. 7 f.

58 Calvin W. Franz in *Christian Economics*, Mar. 13, 1951, p. 1.

59 Stanley High in *The Christian Century*, Jan. 19, 1949, p. 74.

60 Norman L. Trott and Ross W. Sanderson, *What Church People Think about Social and Economic Issues* (Association Press, 1938), p. 42.

61 Liston Pope, *Annals of the American Academy of Political and Social Science*, Mar., 1948, p. 88.

62 J. Howard Pew, *Presbyterian Life*, Aug. 1, 1960, p. 13.

63 Editorial, *The Christian Century*, Apr. 3, 1957, p. 412.

64 "American Conservatives and Religion," *Information Service*, Jan. 14, 1956, p. 1.

65 James A. Pike, "Should the Pulpit Be a Rostrum?" *The New York Times Magazine*, Aug. 14, 1960, p. 15.

66 In a sermon delivered at Mount Holyoke College, 1961.

67 Quoted in *Friends Journal*, Aug. 20, 1960, p. 1.

68 Jack R. McMichael in *Social Questions Bulletin*, Apr., 1950, p. 19.

69 Harris Franklin Rall, ed., *Religion and Public Affairs; In Honor of Bishop Francis John McConnell* (Macmillan, 1937), p. 116.

70 Commission of Inquiry, *Report on the Steel Strike of 1919* (Harcourt, Brace, 1920), pp. 11 f., 85.

71 *Literary Digest*, Nov. 19, 1921, p. 31.

72 See Commission of Inquiry, *Public Opinion and the Steel Strike* (Harcourt, Brace, 1921), chap. ii.

73 *Report on the Steel Strike*, pp. 248 f.

74 Commission of Inquiry, *op. cit.*, p. vi.

75 Jeremiah W. Jenks in Foreword to Marshall Olds, *Analysis of the Interchurch World Movement Report on the Steel Strike* (Putnam, 1922), pp. vi f.

76 Quoted by Rall, *op. cit.*, p. 19.

77 *The New York Times*, Dec. 31, 1927, p. 3.

78 *Ibid.*, Mar. 30, 1928, p. 8.

79 Editorial, *The Christian Century*, May 24, 1928, p. 655.

80 Heber Blankenhorn in Rall, *op. cit.*, p. 36n.

81 *Democratic Christianity* (Macmillan, 1919), p. 11.

82 *Forum*, XCIV (Nov., 1935), 270.

83 *Ibid.*, p. 271.

84 *The New York Times*, Feb. 12, 1934, p. 13.

CHAPTER XI. *The Unitarian Universalists*

1 Earl Morse Wilbur, *Our Unitarian Heritage* (Beacon Press, 1925), p. 8.

2 *Ibid.*, p. 9.

3 *Ibid.*, p. 294.

4 *The Christian Register*, July, 1946, pp. 306, 308; also *Newsweek*, July 22, 1946, pp. 78–79.

5 Correspondence in *The Christian Century*, Nov. 19, 1947, p. 1423.

6 *Ibid.*

7 December, 1948, p. 5.

8 John Nicholls Booth in *The Christian Register*, May, 1949, p. 16.

9 *Ibid.*, p. 18.

10 Richard M. Steiner, *Unitarian Christianity*, p. 3.

11 Peter H. Samson, "A Faith for the Few?" *The Unitarian Register*, Jan., 1959, pp. 10, 11.

12 Jack Mendelsohn, *Why I Am a Unitarian* (Nelson, 1960), pp. 141 f.

13 Irving R. Murray, "A Case for Merger," *The Unitarian Register*, Mar., 1959, p. 30.

14 Robert B. Tapp, "Universalist-Unitarian Merger," *The Christian Century,* June 15, 1960, p. 730.

15 *Christian Science Monitor,* Feb. 8, 1960, p. 7.

16 Walter Donald Kring, "A Case against Merger," *The Unitarian Register,* Mar., 1959, p. 31.

17 Donald Harrington, quoted in *The Unitarian Register,* Dec., 1959, p. 20.

18 John Nicholls Booth, "A Greater Light Beyond Christianity," *The Unitarian Register,* Apr., 1960, p. 5.

19 Robert B. Tapp, *The Christian Century,* June 15, 1960, p. 732.

20 Walter Donald Kring in *The Christian Register,* Dec., 1948, p. 21.

21 *The Christian Century,* Jan. 3, 1945, p. 21.

22 News item in *The Christian Century,* May 13, 1959, p. 588.

23 Peter L. Berger, "Religious Liberalism and the Totalitarian Situation," *Bulletin, Hartford Seminary Foundation,* Mar., 1960, p. 8. Berger's terminology is confusing. However, his conclusion supports my judgment which concerns *theological* liberalism.

24 George Herbert Betts, *The Beliefs of 700 Ministers* (Abingdon, 1926). The 500 clergymen were distributed as follows: Baptist, 50; Congregational, 50; Episcopalian, 30; Evangelical, 49; Lutheran, 104; Methodist, 111; Presbyterian, 63; all others (13 denominations), 43.

25 J. Paul Williams, *Social Adjustment in Methodism* (Bureau of Publications, Teachers College, Columbia, 1938), p. 131.

CHAPTER XII. *Judaism*

1 Louis Finkelstein, *The Pharisees* (Jewish Publication Society, 1940), I, ix.

2 Mordecai M. Kaplan, *Judaism as a Civilization* (Macmillan, 1935), p. 270.

3 Arieh Tartakower, "The Problem of European Jewry," *The Jews, Their History, Culture, and Religion,* Louis Finkelstein, ed. (Harper, 1949), p. 288.

4 Morris S. Lazaron, "Palestine and the Jew," *The Christian Century,* Nov. 19, 1947, p. 1406.

5 Will Herberg, *Protestant-Catholic-Jew* (Doubleday, 1955), p. 210.

6 *American Jewish Yearbook* (Jewish Publication Society of America, 1961), p. 12.

7 Leo Jung in *The Orthodox Union,* Oct., 1944, p. 4, and in *Modern Trends in American Judaism* (reprinted from the Mizrachi Jubilee Publication, 1936), p. 11.

8 David de Sola Pool, in *The American Jew,* Oscar I. Janowsky, ed. (Harper, 1942), p. 34.

9 *Ibid.,* pp. 37 f.

10 Albert I. Gordon, *Jews in Transition* (University of Minnesota, 1949), p. 84.

11   Lee J. Levinger, *A History of the Jews in the United States* (Union of American Hebrew Congregations, 1935), p. 403.

12   *A Program for Jewish Life Today*, reprinted from *The Reconstructionist*, Feb. 24, 1950, p. 5.

13   *Ibid.*, p. 8.

14   Will Herberg, "The Postwar Revival of the Synagogue," *Commentary*, Apr., 1950, p. 318.

15   Erich Unger, "Modern Judaism's Need for Philosophy," *Commentary*, May, 1957, p. 423.

16   Samuel Price, *Outlines of Judaism* (Bloch, 1946), p. 34.

17   Quoted in *ibid.*

18   Milton R. Konvitz in *The Jews, Their History, Culture, and Religion*, Finkelstein, ed., p. 1094.

19   Mordecai M. Kaplan, *The Future of the American Jew* (Macmillan, 1948), pp. 211 ff.

20   For a discussion of these propositions see, for example, Ralph Linton, *The Study of Man* (Appleton-Century, 1936), or Ruth Benedict and Gene Weltfish, *The Races of Mankind* (Public Affairs Committee, 1943).

21   Milton Steinberg, *Basic Judaism* (Harcourt, Brace, 1947), p. 124.

22   Beryl D. Cohon, *Judaism in Theory and Practice* (Bloch, 1948), p. 133.

23   Joseph H. Hertz, *The Authorised Daily Prayer Book* (Bloch, 1948), p. xxi.

24   Israel Abrahams, quoted by Cohon, *op. cit.*, p. 132.

25   G. E. Biddle, quoted by Hertz, *op. cit.*, p. xxiii.

26   Finkelstein, ed., *The Jews, Their History, Culture, and Religion*, p. 1357.

27   Herman Wouk, *This Is My God* (Doubleday, 1959), p. 149; see also p. 250.

28   Finkelstein, ed., *op. cit.*, p. 1331.

29   Gordon, *op. cit.*, p. 104.

30   *Ibid.*, p. 122.

31   Will Herberg, *op. cit.*, p. 211.

32   Pool, *op. cit.*, p. 39.

33   Milton Himmelfarb, "In the Community," *Commentary*, Aug., 1960, p. 159.

34   Will Herberg, *op. cit.*, p. 212.

35   Kaplan, *The Future of the American Jew*, pp. 404, 402.

36   Finkelstein, ed., *op. cit.*, p. 1384.

37   Cohon, *op. cit.*, p. 166.

38   Samuel Gerstenfeld in *The Orthodox Union*, Oct., 1943, p. 4.

39   *Information Service*, Jan. 22, 1955, p. 3.

40   Milton Steinberg, *A Partisan Guide to the Jewish Problem* (Bobbs-Merrill, 1945), pp. 222 f.

41   Quoted by Abram Leon Sachar, *A History of the Jews* (Knopf, 1932), p. 354.

42   Permanent Mandates Commission of the League of Nations quoted by Milton Steinberg, *A Partisan Guide*, p. 240.

43 *Palestine: Land of Promise* (Harper, 1944), pp. 5, 227.

44 Steinberg, *A Partisan Guide,* p. 238.

45 Stephen S. Wise, *Challenging Years* (Putnam, 1949), p. 68.

46 Louis Minsky in *The Christian Century,* Apr. 18, 1934, p. 527.

47 *Current Biography, 1941* (H. W. Wilson), p. 930.

48 Wise, *op. cit.,* pp. 82 f.

49 *Ibid.,* pp. 83 f.

50 *Ibid.,* pp. 91 f.

51 Justine Wise Polier and James Waterman Wise in Stephen S. Wise, *op. cit.,* p. xv.

52 Quoted by Minsky from the *American Hebrew,* in *The Christian Century,* Apr. 18, 1934, p. 526.

53 Wise, *op. cit.,* pp. 96 f.

54 *Ibid.,* p. 168.

55 *The New York Times,* Mar. 20, 1934, p. 11.

56 *Ibid.,* July 23, 1934, p. 9.

57 *Ibid.,* Dec. 21, 1925, p. 24.

58 Wise, *op. cit.,* p. 283.

59 *The New York Times,* Dec. 29, 1925, p. 9.

60 Polier and Wise, *op. cit.,* p. xiii.

61 *Ibid.,* pp. 234 f.

62 *Current Biography, 1941,* p. 931.

63 *Op. cit.,* p. 260.

64 *As I See It* (Jewish Opinion Publishing Corp., 1944), p. 162.

65 Oscar J. Toye, *Report on the Conference on Job Equality and the Jewish Worker* (Jewish War Veterans of the U.S.A., 1954), p. 6.

66 Mrs. Abraham Schnee, *Congress Opinion Check List with Answers* (National Committee on Membership, Women's Division, American Jewish Congress, n.d., but 1953 or after), p. 8.

67 Allyn P. Robinson, "Roots of Anti-Semitism in American Life," *Social Action,* Nov., 1960, p. 5.

68 William Attwood, "The Position of the Jews in America Today," *Look,* Nov. 29, 1955, p. 33.

69 Gordon, *op. cit.,* p. 60. The data were collected in 1942.

70 David Bernstein in *Commentary,* Feb., 1948, pp. 126 f.

CHAPTER XIII. *The Eastern Orthodox Churches*

1 Quoted by an editorial writer in *The Christian Century,* June 8, 1949, p. 700.

2 Nicholas Zernov, *The Church of the Eastern Christians* (London: S.P.C.K., 1942), p. 12.

3 George Petrovich Fedotov, "The Eastern Orthodox Church," Vergilius Ferm, ed., *Living Schools of Religion* (Littlefield, Adams), 1956, p. 178.

4 "The Orthodox Theology," *The Orthodox Observer,* Mar., 1957, pp. 55 f.

5 *Ibid.,* Jan., 1957.

6 Quoted by R. M. French, *The Eastern Orthodox Church* (Hutchinson's University Library, 1951), p. 165 from Metrophanes Kritopoulos, *Confession.*

7 J. L. Neve, *Churches and Sects of Christendom* (Lutheran Literary Board, 1940), p. 75.

8 Frank Gavin, *Some Aspects of Contemporary Greek Orthodox Thought* (Morehouse, 1923), p. 126.

9 Adrian Fortescue, *The Orthodox Eastern Church* (Catholic Truth Society, 1929), p. 372.

10 Quoted by Gavin, *op. cit.,* pp. 126 f., from Rhôsse, *Dogmatics,* pp. 254 f.

11 John A. Limberakis, "A First Hand Account of the Conversion of Elliott Paul to Orthodoxy," *The Orthodox Observer,* Sept., 1958, p. 260.

12 Athenagoras Kokkinakis, *Christian Orthodoxy and Roman Catholicism* (Greek Archdiocese of North and South America, 1952), p. 20.

13 Archbishop Michael, "The Guide for the Christian of the Greek Orthodox Church of America—IV," *The Orthodox Observer,* July, 1958, p. 201.

14 Archbishop Michael, "The Guide for the Christian of the Greek Orthodox Church of America—V," *The Orthodox Observer,* Sept., 1958, p. 261.

15 Kenneth Scott Latourette, *A History of Christianity* (Harper, 1953), p. 297.

16 Adrian Fortescue, *op. cit.,* p. 410.

17 Zernov, *op. cit.,* pp. 26–28. By permission of The Macmillan Co.

18 C. N. Dombalis, "The Makings of a Permanent Decision," *The Orthodox Observer,* July, 1957, pp. 161 f.

19 Quoted by R. M. French, *op. cit.,* p. 133, from S. Bulgakov, *The Orthodox Church,* pp. 162 f., 1935.

20 News items in *The Living Church,* May 1, 1960, p. 17; July 3, 1960, p. 8.

21 Neve, *op. cit.,* p. 77.

22 Latourette, *op. cit.,* p. 906.

23 Fortescue, *op. cit.,* p. 413.

24 Anthony Coniaris, "What Makes Us Greek Orthodox Christians," *The Orthodox Observer,* July, 1957, p. 159.

25 Archbishop Michael, "The Guide for the Christian of the Greek Orthodox Church in America—VII," *The Orthodox Observer,* Nov., 1958, p. 329.

26 Peter T. Kourides, "A Report to the 14th Biennial Congress," *The Orthodox Observer,* Sept., 1958, p. 257.

27 Nov. 20, 1957, p. 1374.

28 Jan. 5, 1962, p. 339.

CHAPTER XIV. *The Mormons*

1 Albert L. Zobell, Jr., *The Improvement Era,* Sept., 1959, pp. 664 f.

2 Fawn M. Brodie, *No Man Knows My Story* (Knopf, 1945), p. 78.

3 Quoted by Charles Samuel Braden, *These Also Believe* (Macmillan, 1949), p. 438, from Paul Hanson, *Jesus Christ among Ancient Americans* (Independence, Mo., 1945), p. 143.
4 Brodie, *op. cit.*, p. 58.
5 *Ibid.*, p. 41.
6 James E. Talmage, *Articles of Faith* (Church of Jesus Christ of Latter-day Saints, 1925), p. 269.
7 Hugh Nibley, *The Improvement Era*, Nov., 1959, pp. 856, 854.
8 Harold Lundstrom, *The Instructor*, Aug., 1956, p. 238.
9 Quoted by Brodie, *op. cit.*, pp. 67 f.
10 B. H. Roberts, in "Introduction" to Joseph Smith, *History of the Church of Jesus Christ of Latter-day Saints* (Deseret, 1951), I, lxviii.
11 Orson Pratt, *Divine Authority of the Book of Mormon*, a series of pamphlets published in 1850–51, quoted by Thomas F. O'Dea, *The Mormons* (University of Chicago Press, 1957), p. 133.
12 *The Book of Mormon*, 1 Nephi 13:6.
13 John Longden, *The Improvement Era*, June, 1959, p. 444.
14 Henry D. Taylor, *The Improvement Era*, June, 1959, p. 447.
15 *History of the Church*, I, 88.
16 Quoted by Brodie, *op. cit.*, pp. 293 f.
17 O'Dea, *op. cit.*, p. 244; see also Wallace F. Bennett, *Why I Am a Mormon* (Nelson, 1958), p. 70.
18 O'Dea, *op. cit.*, p. 248.
19 *Ibid.*, pp. 139 f.
20 *Ibid.*, p. 122.
21 *Doctrine and Covenants:* 131:7; 129:4, 5.
22 Bennett, *op. cit.*, p. 172.
23 *Ibid.*, pp. 175 f.
24 *The Book of Mormon*, "The Book of Esther," 3:6.
25 Quoted by O'Dea, *op. cit.*, pp. 55 and 122; from Joseph Smith, *The King Follett Discourse*, ed. B. H. Roberts (Salt Lake City, 193?).
26 Quoted by O'Dea, *op. cit.*, p. 126, from John A. Widtsoe, *Rational Theology* (Salt Lake City, 1915), pp. 23–25.
27 Bennett, *op. cit.*, p. 210.
28 *Doctrine and Covenants*, 132:16.
29 *Ibid.*, 132:19–20.
30 Hartzell Spence, "The Mormons," *Look*, Jan. 21, 1958.
31 Bennett, *op. cit.*, p. 123.
32 Russell F. Ralston, *Fundamental Differences between the Reorganized Church and the Church in Utah* (Herald House, 1960), p. 15.
33 Bennett, *op. cit.*, p. 106.
34 *Ibid.*, p. 146.
35 Thorpe B. Isaacson, *The Improvement Era*, Dec., 1958, p. 948.
36 Ralston, *op. cit.*, pp. 230 f.
37 *The Improvement Era*, Dec., 1958, p. 957.
38 Hartzell Spence, *op. cit.*
39 James E. Talmage, *op. cit.* (1939 edition), p. 403.
40 Marion G. Romney, *The Improvement Era*, Dec., 1958, p. 952.

CHAPTER XV. *Christian Science*

1 Quoted by Robert Peel, *Christian Science, Its Encounter with American Culture* (Holt, 1958), pp. 218 ff.
2 Mary Baker Eddy, *Retrospection and Introspection* (Trustees under the Will of Mary Baker G. Eddy, 1891), pp. 13 f.
3 *Loc. cit.*
4 Charles S. Braden, *Christian Science Today* (Southern Methodist University Press, 1958), p. 360.
5 Sibyl Wilbur, *The Life of Mary Baker Eddy* (Christian Publishing Society, 1907), p. 54.
6 See Ernest Sutherland Bates and John V. Dittemore, *Mary Baker Eddy, The Truth and The Tradition* (Knopf, 1932), pp. 40 f. and Edwin Franden Dakin, *Mrs. Eddy, The Biography of a Virginal Mind* (Scribner's, 1930), p. 19.
7 Quoted by Bates and Dittemore, *op. cit.*, p. 109.
8 Quoted by *ibid.*, p. 109.
9 Quoted by *ibid.*, pp. 111 f.
10 Mary Baker Eddy, *Miscellaneous Writings* (Trustees under the Will of Mary Baker G. Eddy, 1896), p. 24.
11 Bates and Dittemore, *op. cit.*, p. 116.
12 Braden, *op. cit.*, p. 279.
13 Vernon H. Blair, Dec., 1959, p. 632.
14 July, 1959, p. 375.
15 Mark Twain, *Christian Science* (Harper, 1907), p. 30.
16 Dakin, *op. cit.*, p. 81.
17 See Braden, *op. cit.*, pp. 31 ff; see also Dakin, *op. cit.*, pp. 50 ff; see also Bates and Dittemore, *op. cit.*, pp. 248 ff., 385 ff.
18 Braden, *op. cit.*, pp. 32 ff.
19 Bates and Dittemore, *op. cit.*, p. 145.
20 Henry W. Steiger, *Christian Science and Philosophy* (Philosophical Library, 1948), p. 50.
21 Braden, *op. cit.*, p. 279; reference not given.
22 *Science and Health*, pp. 558 f.
23 *Ibid.*, p. 31.
24 *Ibid.*, p. 473.
25 *Ibid.*, pp. 331, 330.
26 *Ibid.*, pp. 113, 275, 278.
27 Steiger, *op. cit.*, p. 83.
28 *Science and Health*, p. 480.
29 *Ibid.*, p. 265.
30 *Ibid.*, p. 250.
31 *The Herald of Christian Science* (German edition), Apr., 1960, pp. 81 f.
32 *Science and Health*, p. 330.
33 Marian A. Sorenson, *Christian Science Sentinel*, Nov. 14, 1959, p. 1997.

34 *Science and Health,* p. 475.
35 Elizabeth G. Baker, *Christian Science Sentinel,* May 2, 1959, p. 761.
36 *Science and Health,* pp. 399, 301.
37 Robert Peel, *op. cit.,* p. 124.
38 *Science and Health,* p. 18.
39 Steiger, *op. cit.,* p. 190.
40 Peel, *op. cit.,* p. 202.
41 *Science and Health,* p. 166.
42 Steiger, *op. cit.,* p. 54.
43 *Science and Health,* p. 313.
44 Peel, *op. cit.,* p. 182.
45 *Science and Health,* p. 587.
46 *Ibid.,* p. 331.
47 Peel, *op. cit.,* pp. 49, 75, 142, 54, 74.
48 Charles S. Braden, "Study of Spiritual Healing in the Churches," *Pastoral Psychology,* May 1954, pp. 9–15.
49 Cyril C. Richardson, "Spiritual Healing in the Light of History," in *Healing: Human and Divine,* Simon Doniger, ed. (Association Press, 1957), pp. 206 f.
50 *Ibid.,* p. 215.
51 *Ibid.,* p. 218.
52 Gotthard Booth, "Science and Spiritual Healing," in *Healing: Human and Divine,* Simon Doniger, ed., pp. 217 ff.
53 Quoted by Wade H. Boggs, Jr., *Faith Healing and the Christian Church* (John Knox Press, 1956), p. 29.
54 Helen Wood Bauman, "Science and the Body," *Christian Science Sentinel,* May 16, 1959, p. 856.
55 Mary Baker Eddy, *Rudimental Divine Science* (Trustees under the Will of Mary Baker G. Eddy, 1891), p. 2.
56 *Science and Health,* p. 339.
57 Steiger, *op. cit.,* p. 220.
58 Loy Elizabeth Anderson, "The Christian Science Practitioner," *The Christian Science Journal,* Dec., 1959, p. 624.
59 Mary Baker Eddy, *Miscellaneous Writings* (Trustees under the Will of Mary Baker G. Eddy, 1896), p. 352.
60 John R. Dunn, "Some Thoughts on Obtrusive Mental Healing," *Christian Science Sentinel,* XLIX, 1075 ff., quoted by Braden, *op. cit.,* p. 347.
61 *Science and Health,* pp. 476 f.
62 Steiger, *op. cit.,* p. 157.
63 Braden, *op. cit.,* p. 341.
64 *Ibid.,* p. 344.
65 Bates and Dittemore, *op. cit.,* pp. 206, 373, 389.
66 Peel, *op. cit.,* pp. 151 f.
67 *Christian Science Sentinel,* June 20, 1959, p. 1075.
68 *Ibid.,* Nov., 28, p. 2078.
69 *Science and Health,* p. 413.
70 *Ibid.,* p. 141.

71 Quoted by Braden, *op. cit.*, p. 43.
72 *Ibid.*, p. 401. For other summary statements by Professor Braden, see p. 43 and pp. 98 f.
73 Mary Baker Eddy, *Manual of The Mother Church* (Trustees under the Will of Mary Baker G. Eddy, 1895), p. 94.
74 Braden, *op. cit.*, p. 272.
75 *Religious Bodies, 1936* (Bureau of the Census), II, Pt. 1, 392.
76 Braden, *op. cit.*, pp. 270, 275.
77 Leishman, *Why I Am a Christian Scientist* (Thomas Nelson, 1958), p. 149.
78 Kathryn E. Davis, "Our Reading Room," *Christian Science Sentinel,* Apr. 11, 1959, p. 624.
79 *The Christian Science Journal,* Nov., 1959, p. 584.

CHAPTER XVI. *Other Religious Innovations*

1   H. Richard Niebuhr, *The Social Sources of Denominationalism* (Holt, 1929), p. 26.
2   Carey McWilliams in *The Atlantic Monthly,* Mar., 1946, p. 109.
3   Elmer T. Clark, *The Small Sects in America* (Abingdon, 1949), p. 33.
4   *Ibid.*, p. 34.
5   Arthur Huff Fauset, *Black Gods of the Metropolis* (University of Pennsylvania, 1944), pp. 59, 74.
6   *Yearbook of American Churches, Edition for 1962* (National Council of Churches, 1961), p. 269.
7   *Ibid.*, p. 234.
8   Charles Samuel Braden, *These Also Believe* (Macmillan, 1949), pp. 257 ff.
9   McWilliams, *loc. cit.*
10 Braden, *op. cit.*, chap. i, p. 179.
11 *Ibid.*, p. 286.
12 J. N. Farquhar, *Modern Religious Movements in India* (Macmillan, 1915), p. 255.
13 Harold Lindsell, "Are Cults Outpacing Our Churches?" *Christianity Today,* Dec. 19, 1960, p. 223.
14 *Life,* June 9, 1958, p. 113.
15 Frank S. Mead, "The Lunatic Fringe of Religion," *American Mercury,* Feb., 1941, p. 167.
16 Quoted by Francis D. Nichol, *The Midnight Cry* (Review and Herald, 1944), p. 33.
17 Quoted by *ibid.*, p. 171.
18 Quoted by *ibid.*, pp. 247, 458.
19 Carlyle B. Haynes, *Seventh-day Adventists* (Review and Herald, 1940), p. 48.
20 See *Religious Bodies, 1936* (Bureau of the Census), II, Pt. 2, 1691.
21 Haynes, *op. cit.*, p. 9.
22 Walter R. Martin, "Seventh-day Adventism," *Christianity Today,* Dec. 19, 1960, p. 234.

23 See *Religious Bodies, 1936,* II, Pt. 1, 28 ff.
24 *Yearbook of American Churches, Edition for 1961,* p. 278.
25 Richard H. Utt, in "Letters to the Editor," *The Christian Century,* Mar. 1, 1961, p. 276.
26 Walter R. Martin, *loc. cit.*
27 *Ibid.,* p. 233.
28 Booton Herndon, *The Seventh-day: The Story of the Seventh-day Adventists* (McGraw-Hill, 1960), quoted by Walter R. Martin, in a book review in *Christianity Today,* Mar. 27, 1961, p. 574.
29 *Time,* July 31, 1950, p. 43.
30 *Op. cit.,* pp. 94 ff.
31 *Discipline of the Pentecostal Holiness Church,* 1949 (Board of Publication), p. 25.
32 *Ibid.,* pp. 36, 38.
33 A. T. Boisen in *Social Action,* Mar. 15, 1939, p. 27.
34 Don Pitt, *Pilgrim's Progress: 20th Century; The Story of Salvation Army Officership* (The Salvation Army, 1950), p. 21.
35 Harold Begbie, *The Life of General William Booth* (Macmillan, 1920), p. 278.
36 Dec. 26, 1949, LIV, 41.
37 *Handbook of Salvation Army Doctrine* (The Salvation Army, 1923), p. 122.
38 Begbie, *op. cit.,* p. 311.
39 G. V. Owen, *The Life beyond the Veil,* quoted by George Lawton, *The Drama of Life after Death* (Holt, 1932), p. 37.
40 *Spiritualist Manual* (National Spiritualist Association, 1948), p. 172.
41 *Op. cit.,* pp. 156, 176.
42 Quoted by Herbert Hewitt Stroup, *The Jehovah's Witnesses* (Columbia, 1945), pp. 55, 142.
43 *The New York Times,* Oct. 23, 1959, p. 19.
44 Quoted by Stroup, *op. cit.,* p. 64, from American Civil Liberties Union, *The Persecution of the Jehovah's Witnesses.*
45 Braden, *op. cit.,* p. 370.
46 G. Norman Eddy, "The Jehovah's Witnesses," *Journal of Bible and Religion,* Apr., 1958, p. 118.
47 Stroup, *op. cit.,* p. 32.
48 Werner Cohn, "Jehovah's Witnesses and Racial Prejudice," *The Crisis,* Jan., 1956, p. 6.
49 Werner Cohn, *The Christian Century,* Sept. 17, 1958, p. 1055.
50 Quoted by Marcus Bach, *They Have Found a Faith* (Bobbs-Merrill, 1946), p. 229.
51 Charles Fillmore, *A Sure Remedy* (leaflet, Unity School of Christianity, n.d.), p. 1.
52 Braden, *op. cit.,* p. 175.
53 *The Unity School of Christianity* (pamphlet, Unity School of Christianity, n.d.), p. 10.
54 Quoted by Braden, *op. cit.,* pp. 171 f., from *Prosperity,* p. 69.

55 Marcus Bach, "Pioneer in 'Positive Thinking,'" *The Christian Century*, Mar. 20, 1957, p. 358.

56 Jan Karel Van Baalen, "Unity," *Christianity Today*, Dec. 19, 1960, p. 231.

57 Walter Houston Clark, *The Oxford Group; Its History and Significance* (Bookman Associates, 1951), pp. 57, 71.

58 Geoffrey Williamson, *Inside Buchmanism* (Watts, 1954), p. 147.

59 Marcus Bach, *They Have Found a Faith* (Bobbs-Merrill, 1946), pp. 127 f.

60 Quoted by Charles Samuel Braden, *These Also Believe* (Macmillan, 1949), p. 410.

61 Russell Barbour, "'Absolute Honesty' in Easton," *The Christian Century*, Jan. 17, 1962, pp. 81 f.

62 C. Eric Lincoln, *The Black Muslims in America* (Beacon, 1961), p. 249.

63 *Ibid.*, p. 78.

64 *Ibid.*, p. 29.

65 *Ibid.*, p. 68.

66 *Ibid.*, p. 89.

67 *Ibid.*, p. 247.

68 *Newsweek*, Mar., 13, 1961, p. 59.

69 Lincoln, *op. cit.*, p. 25.

70 George S. Schuyler quoted by Lincoln, *op. cit.*, p. 142.

CHAPTER XVII.   *Some Nonecclesiastical Spiritual Movements*

1 Gene O'Mere, *American Mercury*, LX, 765.

2 Dec. 30, 1946, p. 45.

3 Adolph E. Meyer in *American Mercury*, Jan., 1945, p. 81.

4 *Fortune*, Nov., 1945, p. 249.

5 *The Christian Century*, June 7, 1933, p. 744.

6 Willard L. Sperry, *Religion in America* (Macmillan, 1946), p. 256.

7 *The Reformation of the Churches* (Beacon Press, 1950).

8 David de Sola Pool in *The American Jew*, Oscar I. Janowsky, ed. (Harper, 1942), p. 52.

9 *The Conquest of Happiness* (Allen & Unwin, 1932), p. 143.

10 Lin Yutang, *The Importance of Living* (John Day, 1940), pp. 126 ff.

11 *The New York Times*, May 5, 1936, p. 8.

12 Halford E. Luccock, *Jesus and the American Mind* (Abingdon, 1930), p. 102.

13 Harold J. Laski, *The American Democracy* (Viking Press, 1948), pp. 13, 22.

14 W. D. Silkworth, "A New Approach to Psychotherapy in Chronic Alcoholism," in A Co-founder, *Alcoholics Anonymous Comes of Age* (Harper, 1957), p. 302.

15 H. Jack Geiger, "Anonymous Struggle for 25 Years," *The New York Times Magazine*, June 5, 1960, p. 26.

16 *Alcoholics Anonymous* (rev. ed.; Alcoholics Anonymous Publishing, Inc., 1955), p. 82.

17 H. Jack Geiger, *loc. cit.*

18 A Co-founder, *op. cit.*, pp. 65 f.

19 Jack Alexander, *Alcoholics Anonymous* (pamphlet reprint from *Saturday Evening Post,* Mar. 1, 1941, distributed by Alcoholics Anonymous), p. 9.

20 H. Jack Geiger, *op. cit.*, p. 74.

21 A Co-founder, *op. cit.*, p. 119.

22 Harry Emerson Fosdick, *The Living of These Days* (Harper, 1956), p. 287.

23 *Time,* May 26, 1958, LXX, 65.

24 Alan W. Watts, *The Way of Zen* (Pantheon, 1957), p. 157.

25 Quoted by Watts, *ibid.,* pp. 117, 115.

26 *Time,* July 21, 1958, LXXII, 49.

27 Lawrence Lipton, *The Holy Barbarians* (Messner, 1959), p. 264.

28 John Ciardi, *Saturday Review,* Feb. 6, 1960, p. 11.

29 *Ibid.,* p. 12.

30 *Ibid.*

31 *Vital Speeches,* Jan. 1, 1950, p. 167.

32 Carlton J. H. Hayes, *Essays on Nationalism* (Macmillan, 1926), pp. 104 ff.

33 Walter Lippmann, *U.S. Foreign Policy* (Little, Brown, 1943), p. 137.

CHAPTER XVIII. *The Role of Religion in Shaping American Destiny*

1 American Institute of Public Opinion, Apr. 14, 1957 and Dec. 30, 1960.

2 *Information Service* (Federal Council of Churches), Apr. 30, 1949.

3 *The Lutheran,* Mar. 21, 1947.

4 Paul Hutchinson in *The Christian Century,* Sept. 18, 1946, p. 1112.

5 George P. Howard, *Religious Liberty in Latin America* (Westminster, 1944), Chap. iii.

6 *Yearbook of American Churches* (edition for 1962; National Council of Churches, 1961), p. 248.

7 J. Frederic Dewhurst and Associates, *America's Needs and Resources* (Twentieth Century Fund, 1947), p. 332.

8 H. Paul Douglass and Edmund deS. Brunner, *The Protestant Church as a Social Institution* (Harper, 1935), pp. 22, 26.

9 Edition for 1961, p. 279.

10 "A New Trust in Religious Leaders," *Information Service,* Apr. 3, 1954, p. 3.

11 Will Herberg, "There Is a Religious Revival," *Review of Religious Research,* Fall, 1959, p. 48.

12 Michael Argyle, *Religious Behavior* (Routledge and Kegan Paul, 1958), p. 28.

13 Winthrop S. Hudson, "Are Churches Really Booming?" *The Christian Century*, Dec. 21, 1955, p. 1494.

14 Michael Argyle, *op. cit.*, p. 29.

15 A. Roy Eckhardt, *The Surge of Piety in America* (Association Press, 1958), p. 17.

16 J. Frederic Dewhurst and Associates, *America's Needs and Resources* (Twentieth Century Fund, 1947), pp. 326, 81.

17 Calculated for 1955 from statistics published by The American Association of Fund-Raising Counsel in *Giving USA*, 1961 edition, p. 27, and by the U.S. Dept. of Commerce in *Statistical Abstracts of the United States*, 1956, p. 292. This calculation was based on the assumption that the disposable personal incomes of church members average the same as the incomes of nonchurch members. The percentage of church members in the population was taken at 61 per cent.

18 U.S. Department of Commerce, *Survey of Current Business*, July, 1961, p. 14.

19 John Wicklein, "3 Faiths Find Lag in Clerical Students," *The New York Times*, Apr. 16, 1961, pp. 1, 82.

20 Robin M. Williams, Jr., *American Society* (Knopf, 1951), pp. 312, 342.

21 Sidney E. Mead, "American Protestantism Since the Civil War. I. From Denominationalism to Americanism," *The Journal of Religion*, Jan., 1956, pp. 1 f. "American Protestantism Since the Civil War. II. From Americanism to Christianity," *The Journal of Religion*, Apr., 1956, p. 67.

22 Robin M. Williams, Jr., *op. cit.*, Chapter xi.

23 *Ibid.*, p. 388.

24 A. Powell Davies, *Man's Vast Future* (Farrar, Straus and Cudahy, 1951), pp. 27 f.

25 Walter Lippmann, *Essays in the Public Philosophy* (Little, Brown and Company, 1955), pp. 96, 97, 101, 102, 103, 104, 113, 114, 115, 181. By permission of Little, Brown and Company.

26 Martin E. Marty, *The New Shape of American Religion* (Harper, 1958), pp. 85 f.; see also Will Herberg, *Protestant-Catholic-Jew* (Doubleday, 1955), p. 102.

27 Peter L. Berger, *The Noise of Solemn Assemblies* (Doubleday, 1961), pp. 40 f.

28 Association Press, 1946.

INDEX

Aaronic Priesthood, 392, 397
abortion, 44
absolution, 46 f.
Act of Toleration, 198, 200
Adams, John, 204; quoted, 252
Adler, Felix, 459
Adventism, 106, 429, 431 ff., 435, 439
agnosticism, 456
Alcoholics Anonymous, 461 ff.
altar, 38, 124
America, 59, 86
American Baptist Convention, 253 ff., 263, 265, 302
American Board of Commissioners for Foreign Missions, 229
American Civil Liberties Union, 440 f.
American Council of Christian Churches, 139
American Ecclesiastical Review, 56
American Ethical Union, 459
American Freedom and Catholic Power, 83
American Friends Service Committee, 283
American Home Missionary Society, 234
American Jewish Congress, 75, 356, 358
American Lutheran Church, 136, 153 f.
American Revolution, *see* Revolutionary War
American Unitarian Association, 296, 302, 313, 318
Ames, Edward S., quoted, 113
Amida, 346
Amish Mennonites, 241
Anabaptists, 129, 240 f.
angels, 28 f.
Anglican Communion, 173
Anglo-Catholic, 174, 178 f., 187
Annual Conference, Methodist, 294 ff.
annulment, 52 f., 172
antenuptial agreement, 54 ff.
anti-Catholicism, 63, 81 ff.

antimergerite, 237
Anti-Saloon League, 299 f.
anti-Semitism, 333 f., 352, 358 ff.
Apostles' Creed, 160 f., 176 f., 208, 232, 317
Apostolic Delegate, 26, 52
Apostolic Succession, 3, 17, 174, 178 f., 363
Arab, 352 f.
archbishop, 24 f., 364
Archbishop Iakovos, 365
Archbishop Michael, quoted, 365 f.
Archbishop of Baltimore, 63 ff.
Archbishop of Canterbury, 172, 184
Archbishop of St. Louis, 75
Archbishop of York, 55
Archbishop Temple, quoted, 190
archdioceses, 24
areas, Methodist, 294
Ark, 344, 346 f.
Arnold, Thomas, quoted, 163
Asbury, Francis, 289 f.
Assumption of Virgin, 30 f.
Astrain, P. Antonio, quoted, 13
astrology, 455 f.
Athanasian Creed, 160
atheism, 454, 456
Augsburg Confession, 157, 161 ff., 234
Augustana Evangelical Lutheran Church, 154, 159
Authority and Power in the Free Church Tradition, 263

Babson, Roger, 229
Bach, Johann Sebastian, 165
baptism, 29, 43 ff., 128 f., 240 ff., 245, 367 f., 391, 393, 434 f.
Baptists, chap. VIII, 128 f.
Bar Mitzvah, 349
Bas Mitzvah, 349
Beat Zen, 465 ff.
Belloc, Hilaire, quoted, 20 f., 146
Benedictine Fathers, 29
Benson, Ezra Taft, quoted, 396
Berger, Peter L., quoted, 488

Bernadette, 70
Bible, 71 f., 101 ff., 119 f., 123, 160, 173, 196, 199, 330 f., 335 f., 346, 365, 431
bigotry, 82, 100, 440 f.
birth control, 56 ff., 392 f.
birth rate, 256, 393
bishop, 24 f., 45, 59, 81, 182 ff., 294 f., 363 f.
Black Muslims, 448 ff.
Black Muslims in America, The, 449
Blanshard, Paul, 83; quoted, 82
Blavatsky, Madam, 430 f.
Bloudy Tenent of Persecution, 250 f.
Boland, Francis J., quoted, 85, 87
Boleyn, Anne, 172
Board of Home Missions, 229
Book of Common Prayer, 173, 175 ff., 218
Book of Mormon, 380 ff.
Booth, Evangeline, 437
Booth, Gotthard, quoted, 416
Booth, William, 436 f.
Bowne, Borden P., quoted, 292, 298
Braden, Charles S., 415; quoted, 419, 431
Bradford, Robert F., 318
Bradford, William, quoted, 218, 220, 223, 244
Brady, William O., 74
Brent, Charles Henry, 185 ff.
Brewster, Elder, 222
Briggs, Charles A., 210
Brooks, Phillips, 189
brothers of Jesus, 30
Brown, Charles R., 419 f.
Bryan, William Jennings, 101, 108
Buchman, Frank, 443 ff., 465
bureaucracy, 229 f., 258 f., 278 ff., 318 f.
Burke, Edmund, quoted, 129

Caesarean operation, 44
calendar, Jewish, 331 n., 348 f.
Calvin, John, 98, 163, 196 ff., 233, 291
Campbell, Alexander, 260 f.
cantor, 345
cardinal, 23 f.
Cardinal Cushing, 21
Cardinal Gibbons, *see* Gibbons, James
Cardinal Spellman, 82
Carey, William, 143

Carmelite Fathers, 69
Catherine of Aragon, 172
catholic, definition of, 15–16 n., 96
Catholic Digest, 472 n.
Catholic Press Association, 76
Catholic Speaks His Mind, A, 83
celestial marriage, 392 f.
celibacy, 51, 60 f.
Central Churchmen, Protestant Episcopal, 175 ff.
Central Conference of American Rabbis, 301
Central Jurisdiction, Methodist, 294 ff.
Chaplains Corps, 82, 189, 424 f.
Charles I, 199 f.
Charles II, 271 f.
charms, 69
chosen people, 341 f.
Christian, definition of, 407 f.
Christian Century, The, 50, 89, 140, 169 f., 227 ff., 258 f., 300, 324, 475
Christian (Disciples) Churches, chap. VIII, 135, 226
Christian Freedom Foundation, 304
Christian Register, The, 318 f.
Christian Science, chap. XV
Christian Science Journal, The, 405
Christian Science Monitor, 424 f.
Christianity Today, 109 f., 168, 178, 286–87 n., 302 f., 328
Christos, 333
church and state, 65, 86 f., 256, 267 f., 488 ff.
church attendance, 78, 146 f., 472 ff.
Church of England, 171 ff., 201, 217 f., 287, 289
Church of South India, 192
Church, theory of, 16 ff., 95, 99, 159 f., 171 ff., 175 ff., 179 f., 196, 199, 239 f., 250, 255, 280 f., 362 ff., 382 f.
church unity, 95, 134 ff., 190 ff., 212 f., 262 f., 291, 323; *see also* Ecumenical Movement
Churches of Christ, 242 f., 262
circuit rider, 289 ff.
circumcision, 349
Civilian Public Service, 282
Clark, Elmer T., quoted, 429 f., 435
Clement VII, 172
Cleveland Plain Dealer, 462
close communion, 178, 181, 254, 259
Coe, George A., quoted, 138

Coke, Sir Edward, 243
College of Cardinals, 23 f.
Colonial period, 78, 100, 139, 141, 154, 173 f., 219 ff., 271 f., 334
Commonweal, The, 59, 86 f.
Communion, 36 ff., 41 f., 52, 127 f., 176, 181 f., 367 f.; *see also* Mass
Communism, 453 f., 482 ff.
confession, 46 ff., 367 ff., 447
Confessions, Lutheran, 159
confirmation, 45, 176, 349, 367 f., 391
Congregational Christian Churches, 135, 216, 226; *see also* chap. VII
Congregationalists, 216 ff.
Congregations, Roman Catholic, 23 f.
Connell, Francis J., quoted, 56
conscientious objectors, 281 f.
Conservative Judaism, 339 f.
Continental Congress, 158 f., 204 ff.
contraception, 56 ff.
conversion, 78 f., 122, 395
Conway, Bertrand L., quoted, 49 f., 56
Council of Trent, 65, 98, 363
councils of churches, 136 ff., 256 f.
Cowley Fathers, 186 f.
creeds, 39, 159 ff., 176 f., 208 ff., 232 f., 253 ff., 316 f., 366 ff.
Cromwell, Oliver, 200
Crucifixion, The, 37
Cumberland Presbyterian Church, 212 f.

Dahlberg, Edwin T., 306
Darrow, Clarence, 108; quoted, 309
Davies, A. Powell, quoted, 480 f.
Day of Atonement, 348
definition of *catholic*, 15–16 n., 96; of *Christian*, 407 f., of *protestant*, 96; of *religion*, 454; of *science*, 408
Deism, 314
democracy, 10 f., 84 f., 129 ff., 150 f., 170, 182 f., 224, 245 ff., 255 ff., 263 ff., 278 ff., 297 f., 480 ff.
demons, 29
Devil, 51, 106, 115, 440
Dewey, Thomas E., 305
dietary laws, 335 f., 338
dioceses, 24 f., 184
Disciples, 113, 128 f., 136, 260 ff.; *see also* chap. VIII, Christian Churches

District Conference, Methodist, 293 f.
divisions in Protestantism, 95
divorce, 51, 171, 228, 351, 369
Doctrine and Covenants, 382
Doctrines and Discipline of the Methodist Church, 293 f.
Doheney, E. L., 307
Douay Version, 31
Douglas, H. Paul, quoted, 80
Duck River Baptists, 253
Dun, Angus, quoted, 180
Dunne, George H., quoted, 86

Easter, 369
Eastern Orthodox Churches, chap. XIII, 77, 136
Ecumenical Council, 363
ecumenical creeds, 160 ff., 232
Ecumenical Movement, 134 ff., 168 f., 180 ff., 185, 190 ff., 212 f., 216, 235 ff., 262 f., 291, 321 ff., 376
Ecumenical Patriarch, 364, 376
Eddy, Mary Baker, 401 ff.
Edson, Hiram, quoted, 432
education, 72 ff., 79, 82, 85, 89, 103, 107 ff., 139 ff., 160, 213, 226, 348, 423, 488 ff.
Edward VI, 173
Eliot, Frederick M., quoted, 325
Elizabeth, Queen, 173, 198 f., 243
Elmer Gantry, 84
Emerson, R. W., quoted, 14
Esquire, 475
established church, 173 f., 224, 287
Ethical Culture, 459
Eucharist, *see* Communion
Evangelical and Reformed Church, 135, 216, 233 ff.; *see also* chap. VII, United Church of Christ
Evangelical Synod of North America, 233 f.
Evangelical United Brethren Church, 135
evangelicalism, 121 ff.
Evangelicals, Protestant Episcopal, 174, 179 ff.
evolution, 107 ff.
exclusivism, 89 f., 168 f., 181, 256 f., 262 f., 398, 440, 486 ff.
exorcism, 29
extreme unction, 45, 50 f., 369

Faith of Our Fathers, 64 f.
Farrington, Harry Webb, quoted, 113

fasting, 41, 396
Federal aid to schools, 73 ff.
Federal Council of Churches, 55, 137, 169, 193, 306, 325
Fey, Harold E. quoted, 169 f., 258
Filioque Controversy, 366 f.
Fillmore, Charles and Myrtle, 441 ff.
financial support of churches, 23, 475
First Church of Christ, Scientist, 405
first communion, 41, 45
Five Years Meeting, 278
foreign missions, 142 ff., 434
forgiveness of sins, 3; see also Penance
Fortune, 58 f.
Fosdick, Harry Emerson, 211; quoted, 121, 464
Fox, George, 269 ff.; quoted, 276 f.
fraud, 430 f., 439
"free" churches, 263
Free Synagogues, 356
freedom of religion, 20 ff., 84 ff., 108, 177, 226, 248 f., 252, 267 f., 292, 318 ff., 489
French Revolution, 333, 337
Friday evening service, 347
Friends, chap. IX, 1 f., 326
Friends Committee for National Legislation, 284
Friends Intelligencer, 279
Friends Journal, 272
frontier, 207, 252, 260, 289 ff.
Fundamentalism, 105 ff., 138 f., 292
future life, see immortality

Gallup Poll, 472 ff.
General Assembly, Presbyterian, 202, 210 ff., 301
General Authorities, 395
General Conference, Methodist, 295 ff., 301, 310
General Convention, Protestant Episcopal, 184 f., 190 ff.
General Rules, Methodist, 293
General Six Principle Baptists, 253
General Synod, 228, 236, 264 f., 302
genuflection, 38, 179
ghetto, 333, 337 f.
Gibbons, James, 18, 62 ff.; quoted, 20
Gill, Theodore, A., 210
Glenn, Paul J., quoted, 27
Government of churches and synagogues: Baptist, 227 ff., 255 f., 263 ff.; Christian Science, 421 ff.;

Congregational, 226 ff.; Disciples (Christian Churches), 227 ff., 255 f., 263 ff.; Eastern Orthodox, 364 f.; Friends, 227, 278 ff.; Jewish, 349 ff.; Lutheran, 167 f.; Methodist, 293 ff.; Pentecostal Holiness, 435; Presbyterian, 211 ff.; Protestant, 129 ff.; Protestant Episcopal, 182 ff.; Roman Catholic, 17 ff.
Grace, Daddy, 427
Grace, God's, 41, 43 ff., 48, 106, 162, 240
Greek Orthodox Church, 365
Greeley, Horace, quoted, 13

Hall, A. C. A., 186 f.
Hanukkah, 348
Harrison, Paul M., quoted, 263 ff.
Hayes, Carlton, quoted, 468 f.
healing, 111 f., 415 ff., 430, 435, 442
heaven, 32, 44, 106, 115, 163, 209, 367
Hedley, George, 193
hedonism, 460 f.
Heidelberg Catechism, 233 f.
hell, 32 f., 44, 46, 106, 115, 163, 209, 367
Henry VIII, 171 ff.
Herberg, Will, 150, 349
Herzl, Theodor, 352, 358
Hicksite Friends, 273, 279
hierarchy, Roman Catholic, 22 ff.
High Church, 174, 178 f.
higher education, 73, 75 f., 141, 202 f., 207
High Holy Days, 348
Hillel, Rabbi, quoted, 340
holiness, 6 ff., 429, 434 ff.
Holmes, John Haynes, quoted, 357
Holy Club, 288
Holy Communion, 127 f., 178 f., 367, 373; see also Communion, Mass
Holy Ghost, see Trinity
holy oil, 50
holy water, 29
Host, 36 ff., 123, 173
House of Bishops, Protestant Episcopal, 184
House of Deputies, Protestant Episcopal, 184 f.
humanism, 317 ff., 322, 324, 456 ff.
Humanist Manifesto, 457
Hutchins, Robert M., quoted, 248
Hutchinson, Paul, quoted, 298 f.

Hymnal, 422
hymns, 164 ff., 242, 347, 397 f., 422

I Am Movement, 429 f.
iconostas, 370 f.
icons, 370 f., 374
Immaculate Conception, 43, 367
immanence of God, 112
immortality, 4 f., 32 ff., 115, 322, 341, 367, 390, 411, 433, 438, 443, 456 f.; see also heaven, hell
Independent, 141
indulgences, 48 ff., 367
infallibility, papal, 17 ff., 363
Inman, Samuel Guy, quoted, 307
Inner Light, 274 f.
Institutes of the Christian Religion, The, 197
Interchurch World Movement, 308 f.
interfaith relations, 89 f., 100 f; see also bigotry, exclusivism
intermarriage, see mixed marriage
International Convention of Christian Churches, 261, 265
intolerance, 19 ff., 100 f.; see also bigotry, exclusivism
Introit, 164 f.
Israel, 334, 351 ff.

James I, 218 f., 241
James, William, 328 f.; quoted, 464
Jehovah's Witnesses, 439 ff.
Jesus Christ, 16, 26 ff., 96 f., 113 f., 330, 341, 357, 413 f.
Jewish festivals, 348 f.
John XXIII, 24
Judaism, chap. XII
judgment day, 29, 33 f., 106
Jurisdictional Conference, Methodist, 294 f.

Kansas City Statement, 232 f.
Kaplan, Mordecai M., quoted, 139
King James Version, 31, 313
Know-Nothing movement, 63, 80
Knox, John, 196 ff.
kosher, 336 ff.
Ku-Klux Klan, 80, 259, 449

La Guardia, Fiorello, quoted, 357
Lamanites, 380 f.
Lamb, Charles, quoted, 118
Lambeth Conference, 184 f., 190, 192

language of worship, 234, 365, 375 ff.
last judgment, 33 f.
last rites, 50 f., 369
Latourette, Kenneth S., quoted, 146
Latter-day Saints, 378, 381
Lawton, George, quoted, 439
Legion of Decency, 83 f.
Leo XIII, quoted, 66, 85
letters of dismissal, 180 f., 262
Leuba, James H., quoted, 457
levels of certainty, 19
Liberal Evangelicals, 174
Liberalism, 102, 110, 116, 292, 313 ff., 326 ff.
liberty, see freedom
Life, 286, 456
limbo, 32 f., 44, 367
Lincoln, Abraham, quoted, 61 f.
Lincoln, C. Eric, 449; quoted, 450
Lippmann, Walter, quoted, 470, 483
Lord's Prayer, 31 f., 346, 422
L'Osservatore Romano, 86
Lourdes, 70, 416 n.
Low Church, 174, 178 f.
Lowdermilk, Walter Clay, quoted, 352
Luther, 49 f., 166; quoted, 98, 101, 152 f., 166
Lutheran, The, 137 f., 185
Lutheran Church in America, 136
Lutheran Churches, chap. IV
Lutheran, Missouri Synod, 153, 168 f.
Lutheran, United, 135, 154, 159, 167, 169
Lutheran, Wisconsin Synod, 169
Lyons, Ernest H., Jr., quoted, 399 f.

Machen, J. Gresham, 210; quoted, 106
magic, 68
Maimonides, thirteen principles of, 340 f.
Malcolm X, 449
man, nature of, 114, 118 f., 341; see also original sin
maps, 90 ff.
marriage, 51 ff.; see also mixed marriage
Marshall, Louis, 355
Mary, Queen of England, 173
Mary, Queen of Scots, 198
Mass, 3, 34 ff., 127 f., 164, 176, 178, 371
Massachusetts Bay Colony, 223 f.

Massachusetts Metaphysical College, 406
Matthias, Virginia P., quoted, 6 ff.
Mayflower Compact, 220 f.
McConnell, Francis J., 306 ff.
McGee, Margaret B., quoted, 426
McGiffert, Arthur C., quoted, 100
McKay, David O., 382; quoted, 392
McLoughlin, Emmett, 83
McNicholas, John T., quoted, 87
McPherson, Aimee Semple, 427
Mead, Sidney E., quoted, 478 f.
medals, 69
meditation, 466
Melchizedek Priesthood, 392, 395
Melish case, 184
membership statistics, see statistics
Mencken, H. L., quoted, 8
Mennonites, 240 f.
merit, 49
Messiah, 333, 340
Methodist Board of Temperance, 300
Methodist Church, chap. X, 135, 327 f.
Methodist Episcopal Church, 289, 291
Methodist Episcopal Church, South, 291
Methodist Protestant Church, 291
metropolitan, 364
mezuzah, 344
Midrash, 332
Mill, John Stuart, quoted, 10
Millennium, 393 f.
Miller, William, quoted, 431 f.
miracles, 15 ff., 68 ff., 105, 111 f., 119, 374
Mishnah, 332
missal, 38
missionary societies, 261
missions, 142 ff., 319 f., 351, 395, 434
Missouri Synod, see Lutheran, Missouri Synod
mixed marriage, 53 ff., 351, 369
monastic orders, 61 f., 172 f., 179, 186 f.
monotheism, 26, 341, 391
monsignor, 26
monthly meeting, 278
Moral Re-Armament, see Oxford Group
morals, 147 f., 335, 482
Mormons, chap. XIV
Moroni, 379

Morrison, Charles C., quoted, 140
mortal sin, 46, 367
Mother Church, 406, 419 ff.
motion pictures, 83 f.
Muhammad, 448 f.
Muhlenberg, Henry Melchior, 154 ff.
Muhlenberg, Peter, 158
mysticism, 122, 276 ff.
myth, 119 f.

Nation, The, 83
National Baptist Convention, 253
National Catholic Welfare Conference, 26, 75 f., 87, 301
National Conference of Christians and Jews, 81
National Council, Congregational, 228
National Council, Lutheran, 169
National Council of Churches, 136 f., 168, 301, 325
National Council of Presbyterian Men, 304 f.
National Opinion Research Center, 472
nationalism, 468 ff.
natural law, 15, 111
naturalistic humanism, 457 ff.; see also humanism
nature of man, 114 f., 117 ff., 341, 411 f.; see also original sin
Negro, 114, 132 ff., 253, 259, 294, 296 f., 343, 396, 441, 448 ff.
Neo-liberalism, 121
Neo-orthodoxy, 116 ff.
Nephites, 380 f.
Neumann, Theresa, 71
New Education and Religion, The, 490
New Testament, 330 n., 380
New Year, Jewish, 348
New York Herald Tribune, 82
New York Times, The, 83, 357
Nicene Creed, 39, 160, 176 f., 366
Niebuhr, Reinhold, quoted, 83, 116
nonexistence of matter, 409 ff.
Northern Baptists, see American Baptist Convention
Northern Presbyterians, see Presbyterians, U.S.A.
nun, 61 f.

oath, 245, 270
O'Brien, John A., 74

occult, 430
Old Roman Catholic Church, 94
Old Testament, 330 n., 332
open communion, 180, 259, 263
open membership, 254 f., 259, 263
Oral Law, 298
ordination, 2 ff., 59 f., 175, 186, 201 f.,
    249 f., 277, 316, 372
organs, 261 f., 344, 347
original sin, 43 f., 100, 114 f.
Orthodox Friends, 263, 269
Orthodox Judaism, 334 ff.
Orthodox Observer, The, 375
Our Sunday Visitor, 19, 22
Outlook, 141
Oxford Group, 428 f., 443 ff., 464 f.
Oxnam, G. Bromley, quoted, 19, 193,
    305 f.

pacifism, 281 f., 288
Packard, Vance, quoted, 133
Palestine, 330 f., 352 ff.
Pan-Orthodox conference, 364
pantheism, 409
Park, Charles Edward, 317
parochial schools, 73 ff., 139, 161,
    434, 485 ff.
particular judgment, 33, 367
Passover, 336, 348
pastor, 25 f., 59 f., 131 f., 275, 397,
    421 f.
patriarch, 364
Paul, Elliott H., 368
Peace Corps, 482
peace movement, 281 ff.
Peel, Robert, quoted, 414
Penance, 3, 45 ff., 367
Penn, William, 271 f.
Pentateuch, 332, 344, 346
Pentecost, 45, 349, 362 f., 429
Pentecostal Holiness Church, 434 ff.
People's Padre, 83
Pew, J. Howard, quoted, 304 f.
pews, 373
Pfeffer, Leo, quoted, 75
Pharisees, 331
Philadelphia Association, 252
Pike, James A., 193; quoted, 57, 194
Pilgrims, 219 ff.
Pius IX, quoted, 84 f.
Pius XII, 19, 23 f., 28 f., 31, 51, 57
Plan of Union, 225
Planned Parenthood Federation, 57 f.

Plymouth Colony, 218 ff.
Polish National Catholic Church, 94
political office held by members of
    denominations, 286 f.
polls of opinion, 77, 292, 302 f., 328,
    360, 472 ff.
polygamy, 384 f., 387 f., 394
Poole, Elijah, 448
Pope, 17 ff., 22 ff., 97, 101, 171 ff.,
    364
Portland Courier, 402
Power, Tyrone, 52
practitioners, 418 ff., 425
prayer, 30, 114, 122, 125 f., 412 f.
prayer book, 38, 60, 125, 164 f.,
    175 ff., 345
prayers, written, 125 f., 175 f., 345
preaching, *see* sermons
predestination, 163, 208 ff., 291 f.
Presbyterian Church, U.S., 207
Presbyterian Church, U.S.A., 135,
    191 ff., 207, 213
Presbyterian Churches, chap. VI
Presbyterian Life, 305
presbytery, 201, 211 ff.
press, religious, 76, 82 f., 141 f.
priesthood of all believers, 97, 164,
    180
Primitive Baptists, 253
Princeton University, 202 ff.
Principia College, 424
prohibition, 300
Protestant, definition of, 95 f.
Protestant Episcopal Church, chap. V,
    136, 320
Protestantism, chap. III, 198
public schools, 72 ff., 83, 85, 103,
    107 ff., 139 ff., 490 f.
purgatory, 30, 32 f., 46, 48 f., 69, 152,
    367
Purim, 348 f.
Puritanism, 199 f., 223 f., 287

Quakers, *see* Friends
Quarterly Conference, Methodist,
    293 f.
Quimby, Phineas P., 402, 404
Quorum of the Twelve Apostles, 395

rabbi, 344, 350 f., 353
race differences, 342 f.
reading room, Christian Science,
    425 f.

Reconstructionism, 339 f.
rector, 183
Reform Judaism, 336 ff.
Reformation, the, 49, 72, 97 ff., 103, 123, 142, 163, 166, 171 ff., 196 ff., 239 ff., 314
Reformed churches, 200 f., 233
Register, The, 42
religion, definition of, 454, 477 ff.
religion, role of, 475 ff.
Religious Bodies, 439
religious education, 72, 79, 89, 139 ff., 168, 213, 226, 423 ff.; *see also* public schools, education
religious journals, *see* press
religious statistics, *see* statistics
Reorganized Church of Jesus Christ of Latter Day Saints, 385, 395
resurrection, *see* immortality
revival meetings, 107
Revolutionary War, 158 f., 174, 201, 203 ff., 224, 252, 280, 289, 321
rhythm system, 58 f.
Richardson, Cyril C., quoted, 415
Rickenbacker, Eddie, quoted, 468
Robinson, John, quoted, 220
Roman Catholic Church, chap. II, 363 ff.
Roman Catholicism and Religious Liberty, 87
Roosevelt, Franklin D., 357
Roosevelt, Theodore, 188; quoted, 286
Roper, Elmo, 474
Rosary, 31, 38, 72
Rush, Benjamin, 202
Russell, Bertrand, quoted, 458, 460
Russell, Charles Taze, 439
Russian Orthodox Church, 137
Rutherford, J. F., 439 f.
Ryan, John A., quoted, 85, 87

Sabbath observance, 148, 199, 335, 337 f., 433
sacrament, 44 f., 255, 344 f., 367, 397
sacrifice of Mass, 37, 40
saints, 29 ff.
St. Anne de Beaupré, 70
St. Blaise, 69
St. Joseph, 31
St. Peter, 16 f.
St. Peter's Church in Rome, 49
salvation, 106 f., 128, 162 ff., 208 f., 291 f., 322, 390 ff., 412, 436 f., 443

Salvation Army, 436 ff.
Sanger, Margaret, 88
Saturday Evening Post, 462
scapular, 69
science, definition of, 408
Science and Health with Key to the Scriptures, 404 ff.
science and religion, 108 ff., 117, 150 f.
Scopes, John, 107 f.
second coming, 34, 105 f., 241, 393 f., 429, 431 f., 435, 439; *see also* Adventism
secret societies, 89
Seder, 348
Seekers, 250, 269, 271
segregation, 132 f., 259, 294 ff., 396, 441
separation, marital, 53
Separatism, 199, 217 ff., 241 ff., 244 ff., 250
sermons, 39, 100, 126, 160 f., 180, 347, 373 f., 397
session, Presbyterian, 211
Seventh-day Adventists, 431 ff.
Seventh Day Baptists, 253, 433
Shakers, 414
Shaw, George Bernard, quoted, 342, 437
Shelley, P. B., quoted, 458
Shema, 346
sin, 46 ff., 105 f., 432
singing, 166, 242, 347, 371, 396 f.
sinlessness, 18 f., 367
Slovak Lutheran Church, 168
Smith, Joseph, 378 ff.
Smith, Joseph, III, 385
Smyth, John, 241 f.
social action, 66 f., 115, 120, 166 f., 188 f., 203 ff., 280 ff., 299 ff., 396, 425, 437 f.
social classes and religion, 132 ff., 428 f., 441
social conservatism, 284 f., 304 ff., 310
Social Gospel, 115, 167, 299 ff.
social pronouncements, 301 f.
social position, 133 f., 226, 280, 428 f.
social service, 280 ff., 342 ff.
societal religion, 477 ff.
Society for Ethical Culture, 459
Society for Psychical Research, 430 f.
soul competence, 254
Southern Baptist Convention, 253, 255 ff., 265

Southern Presbyterian, *see* Presbyterian Church, U.S.
spiritual gifts, 6 ff., 392 f., 429 f., 433, 436
Spiritual Mobilization, 304
Spiritualism, 4 f., 438 f.
State Conference, Congregational, 228 f.
Statement of Faith, United Church, 238
statistics of religious bodies: discussion of problem, 76 ff.; Baptist, 242 f., 253; Canadian, 147; Christian Churches (Disciples), 262 f.; Christian Scientist, 424 f.,; Congregational (United Church of Christ) 225; Eastern Orthodox, 77, 375; Evangelical and Reformed (United Church of Christ), 235; Friends, 271 ff.; growth in membership, 472 ff.; Jehovah's Witnesses, 441; Jewish, 77, 334, 338 f.; Lutheran, 153 f.; Methodist, 286 f.; missionary, 143 ff.; Mormon, 385; Pentecostal Holiness, 435; Presbyterian, 207; Protestant, 76 f., 94 f., 146 f.; Protestant Episcopal, 174, 185; Roman Catholic, 76 ff.; Salvation Army, 436; Seventh-day Adventist, 433; Southern Baptist, 256 f.; Spiritualism, 439; Unitarian Universalist, 315 f., 323; Unity, 442; *see also* Maps, 91 ff.
Status Seekers, The, 133
Steel Strike of 1919, 307 ff., 354
Stevenson, Adlai, 282, 451
stigmatic, 70 f.
Student Christian Movement, 5 f., 181 f.
Sugrue, Thomas, quoted, 83
Sukkoth, 348
Sunday, Billy, 107
Sunday schools, 140, 423; *see also* religious education
symbolism, 123 f., 179
synagogue, 340 ff., 349 ff.
synod, 211

tabernacle, 38, 41, 393
Taft, William Howard, 188
Talmud, 332
temperance, 300 f.
Temple Emanu-El, 354 f.

Tennyson, Alfred, quoted, 116
Tetzel, 49 f., 152
theology: Baptist, 239 ff., 253 ff.; Black Muslims, 449 ff.; Christian Science, 408 ff.; Congregational, 232, 238; Eastern Othodox, 365 ff.; Friends, 273 ff.; Jewish, 334 ff., 337 f., 340 ff.; Lutheran, 159 ff.; Methodist, 291 ff., 327 f.; Mormon, 388 ff., 396 ff.; Pentecostal Holiness, 434 f.; Presbyterian, 207 ff.; Protestant, 104 ff.; Protestant Episcopal, 185; Roman Catholic, 26 ff.; Seventh-day Adventists, 431 ff.; Unitarian Universalist, 316 ff., 327 f.
Theosophy, 430
third force, 431
Thirty-nine Articles of Religion, 178, 182, 292 f.
Thorndike, Edward L., 299
Time, 258, 437
tithes, 256, 394, 433
tolerance, 21, 100 f., 328
Torah, 332, 335, 346 f.
Toynbee, Arnold J., quoted, 10
transcendence of God, 112
transubstantiation, 3, 35 f., 39
Trinity, 27 f., 43, 113, 122, 225, 313, 341, 366, 414, 440
Trueblood, Elton, 282
Truman, Harry S., 259
Twelve Steps, 463 ff.
Two-Seed-in-the-Spirit Predestinarian Baptists, 253

Union of American Hebrew Congregations, 349
Union of Orthodox Jewish Congregations, 349
Union of Orthodox Rabbis, 357
Unitarian Christian Fellowship, 318
Unitarian Universalist Association, chap. XI, 135 ff.
United Church of Canada, 193
United Church of Christ, chap. VII, 135, 264 f.
United Jewish Appeal, 350
United Lutheran Church, 135, 154, 159, 169
United Presbyterian Church, 207, 213
United Presbyterian Church in the U.S.A., 135, 207, 213

United Synagogue of America, 349
Unity School of Christianity, 428, 441 ff.
universalism, 320 f.

Van Dusen, Henry P., quoted, 146, 192
Vatican City, 22 f.
Vatican Council, 363
venial sin, 46
vestments, 38
vestry, 183
vicar, 183
Virgin Birth, 30 f., 43, 113 f., 177 f., 210
Virgin Mary, 30 ff., 43, 70, 178, 364, 367
Voltaire, quoted, 272

Walker, James J., 357
Wall Street Journal, 336
Wallace, Henry A., quoted, 381
wards, 394 f.
Wedel, Theodore O., quoted, 192
weekday religious education, 140 f.
Welfare Program, Mormon, 396
Wesley, John, 287 ff.
Westminster Confession, 200, 207 ff., 232
White, Ellen G., 433
Williams, Robin M., quoted, 478, 480
Williams, Roger, 243 ff.
Wilson, Woodrow, 22, 307, 356 f.
Winchester Profession, 321

Wise, Stephen S., 353 ff.
witchcraft, 69
Witherspoon, John, 201 ff.
women, status of, in religious organizations: 61 f., 168, 185, 212, 230 ff., 274, 295 f., 350 f., 437, 439, 441
Wood, Leland Foster, quoted, 55
Wordsworth, William, quoted, 277
World Christian Handbook, 144
World Conference on Faith and Order, 190
World Council of Churches, 87, 136 f., 168, 182, 191, 325
World Jewish Congress, 356
World Missionary Conference, 189 f.
worship: Christian Science, 421 ff.; Eastern Orthodox, 369 ff.; Friends, 1 f., 273 ff.; Holiness, 6 ff.; Jewish, 343 ff.; Lutheran, 164 ff.; Mormon, 396 ff.; Pentecostal Holiness, 435; Protestant, 123 ff.; Protestant Episcopal, 175 ff.; Roman Catholic, 34 ff.
Wouk, Herman, quoted, 347
written prayers, 125 f., 175 f., 344 f.

Yearbook of American Churches, 78, 227 f., 278
yearly meeting, 278 f.
Young, Brigham, 385 ff.
Yutang, Lin, quoted, 460

Zen Buddhism, 465
Zionism, 351 ff., 356, 358 f.

Format by Barbara Luttringhaus
Set in Linotype Times Roman
Composed, printed and bound by American Book–Stratford Press
HARPER & ROW, PUBLISHERS, INCORPORATED